W9-AVJ-805

Publisher George Provol
Acquisitions Editor Gary Nelson
Product Manager Kathleen Sharp
Developmental Editors Cinda Cheney/Anita Fallon
Project Editor D. W. Salisbury
Production Manager Darryl King
Art Director Jeanette Barber

Address for Orders
The Dryden Press
6277 Sea Harbor Drive
Orlando, FL 32887-6777
1-800-782-4479

Address for Editorial Correspondence
The Dryden Press
301 Commerce Street, Suite 3700
Fort Worth, TX 76102

ISBN: 0-03-024511-7

Library of Congress Catalog Card Number: 97-66400

Printed in the United States of America

7 8 9 0 1 2 3 4 5 6 039 9 8 7 6 5 4 3 2 1

The Dryden Press
Harcourt Brace College Publishers

To Jerry, Steve, and Bari, who share my love of learning.

THE DRYDEN PRESS SERIES IN ECONOMICS

Baldani, Bradfield, and Turner
Mathematical Economics

Baumol and Blinder
Economics: Principles and Policy
Seventh Edition
(also available in Micro and Macro paperbacks)

Baumol, Panzar, and Willig
Contestable Markets and the Theory of Industry Structure
Revised Edition

Breit and Elzinga
The Antitrust Casebook: Milestones in Economic Regulation
Third Edition

Brue
The Evolution of Economic Thought
Fifth Edition

Edgmand, Moomaw, and Olson
Economics and Contemporary Issues
Fourth Edition

Gardner
Comparative Economic Systems
Second Edition

Gwartney and Stroup
Economics: Private and Public Choice
Eighth Edition
(also available in Micro and Macro paperbacks)

Gwartney and Stroup
Introduction to Economics: The Wealth and Poverty of Nations

Heilbroner and Singer
The Economic Transformation of America: 1600 to the Present
Third Edition

Hess and Ross
Economic Development: Theories, Evidence, and Policies

Hirschey and Pappas
Fundamentals of Managerial Economics: Theories, Evidence, and Policies
Sixth Edition

Hirschey and Pappas
Managerial Economics
Eighth Edition

Hyman
Public Finance: A Contemporary Application of Theory to Policy
Fifth Edition

Kahn
The Economic Approach to Environmental and Natural Resources
Second Edition

Kaserman and Mayo
Government and Business: The Economics of Antitrust and Regulation

Kaufman
The Economics of Labor Markets
Fourth Edition

Kennett and Lieberman
The Road to Capitalism: The Economic Transformation of Eastern Europe and the Former Soviet Union

Kreinin
International Economics: A Policy Approach
Eighth Edition

Lott and Ray
Applied Econometrics with Data Sets

Mankiw, *Principles of Economics*
(also available in Micro and Macro paperbacks)

Marlow
Public Finance: Theory and Practice

Nicholson
Intermediate Microeconomics and Its Application
Seventh Edition

Nicholson
Microeconomic Theory: Basic Principles and Extensions
Seventh Edition

Puth
American Economic History
Third Edition

Ragan and Thomas
Principles of Economics
Second Edition
(also available in Micro and Macro paperbacks)

Ramanathan
Introductory Econometrics with Applications
Fourth Edition

Rukstad
Corporate Decision Making in the World Economy: Company Case Studies

Rukstad
Macroeconomic Decision Making in the World Economy: Text and Cases
Third Edition

Samuelson and Marks
Managerial Economics
Second Edition

Scarth
Macroeconomics: An Introduction to Advanced Methods
Third Edition

Stockman
Introduction to Economics
(also available in Micro and Macro paperbacks)

Walton and Rockoff
History of the American Economy
Eighth Edition

Welch and Welch
Economics: Theory and Practice
Sixth Edition

Yarbrough and Yarbrough
The World Economy: Trade and Finance
Fourth Edition

PREFACE

If it's wild to your own heart, protect it. And fight for it, and dedicate yourself to it, whether it's a mountain range, your wife, your husband, or even (heaven forbid) your job. It doesn't matter if it's wild to anyone else: if it's what makes your heart sing, if it's what makes your days soar like a hawk in summertime, then focus on it. Because for sure, it's wild, and if it's wild, it'll mean you're still free. No matter where you are.

Rick Bass, Wild to the Heart

The question of why study environmental and resource economics is posed in Chapter 1. A more basic question is, "Why write a book about environmental economics?" The answer to the second question is much more complex than the first, but it revolves around the sentiments expressed in the above quotation by Rick Bass. Quite simply, the environment matters and the field of environmental economics offers a unique way of thinking about the environment. A former colleague of mine, a biologist, used to kid me that "environmental economics" is an oxymoron. That always provoked a reaction from me. I believe very strongly that it is possible to be passionate about both the environment and economics, and this book tries to reflect these two passions and show that they are not mutually exclusive.

Organization of the Book

This book does several things differently from other textbooks in this area. The first is the division into theory (Chapters 1–5) and application (Chapters 6–18). The second organizational difference that distinguishes this textbook from others is the integration of environmental and resource economics.

The theory of environmental and resource economics is initially presented in the dedicated theory chapters but is constantly revisited in the application chapters to reinforce the theoretical concepts and to demonstrate their practical importance. For example, the theory of marketable pollution permits is extensively developed in Chapter 3, but applied to specific environmental problems in many different chapters, such as Chapter 6 (Global Change); Chapter 7 (Acid Rain); Chapter 8 (Energy Use and the Environment); Chapter 14 (Water Resources); and Chapter 17 (The Environment and Development in Third World Countries).

The second organizational difference between this book and most others is the integration of environmental and natural resource economics. This integration is important because the pressing environmental and resource issues of the 1990s and beyond all have to do with the environmental implications of resource production and use. Although traditional resource economic issues (such as the optimal rotation of a forest, the optimal amount of effort in a fishery, and how the market allocates an exhaustible resource over time) are all covered, more emphasis is placed on the interaction of resources with the environment. For example, Chapter 9 discusses exhaustible resources, but it places the discussion in the context of materials policy. The chapter looks at the problem from the perspective of extraction of minerals to the disposal of the materials that are made with minerals. Similarly, the forestry chapters look not only at the question of harvesting, but also focus on the questions of the causes and consequences of deforestation, and whether old growth forests should be harvested or preserved. Chapters on global warming, acid rain, biodiversity, agriculture and the environment, the environment and the macroeconomy, tropical forests, and issues relating to the environment in developing countries also distinguish this from other existing textbooks.

LEVEL

The theory chapters present the economic theory of environmental and resource economics in a format useful to students with varied backgrounds in economics. Environmental and resource economics is a required course for students from many different majors, yet intermediate microeconomic theory is usually taken only by economics students. Consequently, this book does not require an intermediate microeconomics prerequisite. New economics concepts such as consumers' and producers' surplus, demand and supply analysis, opportunity cost, and marginal analysis are thoroughly explained as they are encountered, providing a review for economics students and a foundation for students with less experience in economics. The presentation is designed to give economics majors an opportunity to apply their economic learning to an interesting and topical area, while noneconomics students will find the new perspective an interesting and different approach from other perspectives, such as ecology. The key vehicle for maintaining student interest and providing intellectual challenge are the application chapters. The material from the theory chapters is pulled into the application chapters in a manner specifically tailored to the applications. The student thus learns how economic theory can be used to understand both the causes of specific environmental problems and the process for developing specific solutions for each environmental problem.

One of the important features of the book is the boxed application, which contributes to the understanding of how to use environmental and resource economics to solve real-world problems. The main goal of the boxed application is to show how the principles of environmental and resource economics are applied in very specific situations. For example, boxed applications include gold mining in the Amazon, the protection of elephants in Africa, the pollution of fisheries in Europe, and water allocation in New York City. The book contains a total of 33 boxed applications.

ORIENTATION

Another feature of this textbook that distinguishes it from other textbooks in the market is its orientation. Many textbooks take the perspective that environmental and resource economics are a subset and straightforward application of microeconomics. While microeconomics is extremely important in studying environmental and resource issues, other perspectives must also be utilized. The macroeconomic perspective also generates important insights into these issues. In addition, one must understand the basic principles of the natural and physical sciences that govern the natural world, in order to properly apply economic principles that also govern the human-environmental interaction.

Part of developing an appropriate economic perspective on an environmental problem is understanding the underlying natural and physical science. In an effort to place economics in an interdisciplinary context, the applications chapters of this book include basic information on the natural and physical science dimensions of each application topic. The emphasis, however, is on the economics and its role in the development of environmental policy. Students who desire more in-depth understanding of the physical and natural sciences should take a course or read a book that directly emphasizes the physical and natural sciences component of environmental studies.

WHAT'S NEW IN THE SECOND EDITION

The most important revisions made in the second edition were to incorporate the many changes in the environment and in environmental policy that have occurred since the first edition was written. Of particular importance in this regard are the changes in Chapters 6 (Global Environmental Change), 7 (Acid Deposition), 16 (Agriculture and the Environment), and 17 (The Environment and Economic Growth in Third World Countries). Chapter 5 (The Macroeconomics of the Environment) has been completely rewritten to incorporate new literary contributions, including discussions of the double dividend of environmental taxation and the economic theory of sustainable development.

Another important feature of the second edition has been a response to the valuable suggestions made by students and professors who used the first edition. Sections of the theory chapters, which in the last edition unintentionally required intermediate microeconomics as a prerequisite, have been rewritten to make them more accessible. More in-depth discussion of many issues have been developed, and references to Internet resources have been added to the text in every application chapter. New boxed examples and applications have also been added to every chapter.

WEB SITE

An important feature of the textbook is a Web page (Enviro-Economics Gateway), which is available at http://funnelweb.utcc.utk.edu/~jrkahn/book.html. The Web page serves primarily as a link to resources throughout the Internet. One section of this gateway lists general Internet resources, such as the primary Web pages of the U.S. Environmental Protection Agency and the World Bank. Another section of the Web page lists Internet resources related to the topic of each application chapter of the textbook. In addition, the Web page lists information about the book, such as new features of the second edition and has example outlines of the course for environmental economics classes, natural resource economics classes, and combined environmental and resource economics classes. The Web page is supported by a short Internet Appendix at the end of the textbook, which provides basic information for students (or professors) who are inexperienced in the use of the Internet.

LEARNING TOOLS

Each chapter contains many tools to help students better understand and master environmental economics. Throughout both the theory and application chapters, boxed applications pose environmental questions and issues with analysis relevant to the topics being discussed. Each chapter ends with a *Summary* recapping the main topics, concepts, and current debates. The *Review Questions* help to test student's comprehension of the material presented in the chapter, while the *Questions for Further Study* provide a more general framework for reviewing material, tying the material in the current chapter to broader themes and to material presented in earlier chapters. The *Suggested Paper Topics* are offered to help the student with the difficult task of getting started on a research paper. In addition to suggesting topics revolving around different approaches to issues, policies, problems, effects, and consequence; both electronic and print resources are listed to help the students begin their research. *Works Cited and Selected Readings* provide a sample of writings on current issues, as well as a listing of established sources.

INSTRUCTOR'S MANUAL

The book has an Instructor's Manual/Test Bank (IM/TB) that provides instructors with a valuable teaching tool. Written by Joy L. Clark at Auburn University at Montgomery, the IM/TB gives lecture outlines for every chapter, along with additional bibliographic sources and answers to all of the review questions in the text. The IM/TB includes 500 test questions with 10 short-answer questions for every chapter.

ACKNOWLEDGMENTS

This book could not have been written without the help of many people to whom I owe immeasurable thanks. The many students I taught at the University of Tennessee, the Universidade do Amazonas, and SUNY-Binghamton helped shape my approach to teaching environmental and resource economics and have substantially influenced the book. My approach to environmental economics has also been influenced by my colleagues at the University of Tennessee, Oak Ridge National Laboratory, and Universidade do Amazonas, as well as my colleagues throughout the profession. I would especially like to thank John Cumberland, Wallace Oates, and Kerry Smith who have been my mentors throughout my career and who continue to encourage and guide me even as I enter middle age. I owe special thanks to John Cumberland for his constant and continual support. When I was a graduate student, it was John who taught me that it is possible to be passionate about the environment and still be a careful economist.

Many people reviewed the book and used drafts in their classes. They all made substantial contributions to the development of the book. Without the help of these colleagues, the book would not have progressed as quickly or as well as it did. The reviewers include: William W. Hall, University of North Carolina-Wilmington; Bruce Rettig, Oregon State University; Keith Willett, Oklahoma State University; Morteza Rahmatian, California State University-Fullerton; Bill Provencher, University of Wisconsin-Madison; Gary Anderson, California State University-Northridge; James Shatava, University of Wisconsin-River Falls; Frederick Bell, Florida State University; Kevin Rask, Colgate University; Richard Rosenberg, Pennsylvania State University; Amin U. Sarkar, SUNY-Fredonia; Nancy Williams, Loyola College-Baltimore; Robert Kling, Colorado State University; John Griffin, Worcester Polytechnic Institute; Jonathan Rubin, University of Tennessee; and Joy L. Clark, who wrote the Instructor's Manual/Test Bank. In addition to the contributions of the reviewers, many professors and students who used the first edition made valuable suggestions to improve the second edition. In particular, I would like to thank Lynne Bennett, Andrew Brod, Dina Franceschi, Rebecca Grantz, and John Whitehead for their help.

I would like to thank my colleagues at The Dryden Press who guided my efforts and graciously tolerated many missed deadlines. First, I would like to thank Anita Fallon, my developmental editor who was always positive and who improved the quality of the second edition in numerous ways. Her guidance was always on the mark and much appreciated. I would like to thank Emily Barrosse, acquisitions editor, whose enthusiasm for the project was appreciated. I would also like to thank Cinda Cheney, developmental editor; Dee Salisbury, project editor; Darryl King, production manager; Jeanette Barber, art director; Susan Van Buren, permissions editor; and Kathleen Sharp, product manager, for their efforts.

People who have not written a book may not be aware of the tremendous costs that the author's family bears. I thank my sons, Steve and Jerry, for tolerating my preoccupation and for encouraging me rather than complaining. Last and most importantly, I thank my wife Bari, whose patience, support, and encouragement was much appreciated. Bari not only helped create the time for me to complete this book, but she went above and beyond the call of spousal duty by reading every single word I wrote. She provided the first edit of the book, which substantially speeded the project and improved the quality of exposition.

JAMES R. KAHN

ABOUT THE AUTHOR

Professor James Kahn received his Ph.D. in 1981 from the University of Maryland, concentrating on environmental economics. From 1980 to 1991, Professor Kahn was on the faculty of the economics department at the State University of New York at Binghamton (now Binghamton University), where he taught both graduate and undergraduate environmental and resource economics. Since 1991, Professor Kahn has had a joint appointment as a professor in the economics department at the University of Tennessee and as a collaborating scientist at the Oak Ridge National Laboratory. At the University of Tennessee, Professor Kahn continues to teach graduate and undergraduate environmental and resource economics. He recently organized an interdisciplinary graduate program in environmental policy. Professor Kahn also has an appointment as an international collaborator at the University of Amazonas in Manuas, Brazil. Professor Kahn has published over 40 journal articles, books and book chapters. His research interests include nonmarket valuation, marine resources, modeling the interaction between ecosystems and economic systems, the interaction between the environment and the macroeconomy, sustainable development, and the role of the environment in the economic development of tropical countries.

Professor Kahn has had an interest in the environment since he was a young child. He enjoys hiking, camping, fishing, landscape gardening, and golf. He is a former soccer player and is very active in the American Youth Soccer Organization. Professor Kahn has been married to Bari Kahn for 21 years. The couple have two children, Steven, 17 and Jerry, 12.

BRIEF CONTENTS

Preface vii

PART I THEORY AND TOOLS OF ENVIRONMENTAL AND RESOURCE ECONOMICS 1

Chapter 1 *Introduction* 3

Chapter 2 *Economic Efficiency and Markets: How the Invisible Hand Works* 11

Chapter 3 *Government Intervention in Market Failure* 39

Chapter 4 *Valuing the Environment for Environmental Decision Making* 87

Chapter 5 *The Macroeconomics of the Environment* 132

PART II EXHAUSTIBLE RESOURCES, POLLUTION, AND THE ENVIRONMENT 157

Chapter 6 *Global Environmental Changes: Ozone Depletion and Global Warming* 159

Chapter 7 *Acid Deposition* 193

Chapter 8 *Energy and the Environment* 225

Chapter 9 *Material Policy: Minerals, Materials, and Solid Waste* 257

PART III RENEWABLE RESOURCES AND THE ENVIRONMENT 285

Chapter 10 *Fisheries* 287

Chapter 11 *Temperate Forests* 326

Chapter 12 *Tropical Forests* 352

Chapter 13 *Biodiversity and Habitat Preservation* 375

Chapter 14 *Water Resources* 393

PART IV FURTHER TOPICS 413

Chapter 15 *Toxins in the Ecosystem* 415

Chapter 16 *Agriculture and the Environment* 435

Chapter 17 *The Environment and Economic Growth in Third World Countries* 456

Chapter 18 *Prospects for the Future* 480

APPENDIX THE INTERNET: WHAT IS IT, HOW TO USE IT, WHAT'S ON IT 493

CONTENTS

Preface vii

PART I THEORY AND TOOLS OF ENVIRONMENTAL AND RESOURCE ECONOMICS 1

CHAPTER 1 INTRODUCTION 3

Introduction 3
A Taxonomy of Resources 4

CHAPTER 2 ECONOMIC EFFICIENCY AND MARKETS: HOW THE INVISIBLE HAND WORKS 11

Introduction: The Working of the Invisible Hand 11
Market Failure When the Invisible Hand Doesn't Work 14
The Invisible Hand and Equity 27
The Invisible Hand and Dynamic Efficiency 28
Summary 32
Appendix 2.A Discounting and Present Value 34
Appendix 2.B Dynamic Efficiency 37

CHAPTER 3 GOVERNMENT INTERVENTION IN MARKET FAILURE 39

Introduction 39
Should the Government Intervene to Correct Environmental Externalities? 40
Resolution of the Issue of Government Intervention 48
Types of Government Intervention 49
Choosing the Correct Level of Environmental Quality 51
Pursuing Environmental Quality with Command and Control Policies 62
Pursuing Environmental Quality with Economic Incentives 70
Summary 84

CHAPTER 4 VALUING THE ENVIRONMENT FOR ENVIRONMENTAL DECISION MAKING 87

Introduction 87
What Is Value? 88
Techniques for Measuring the Value of Nonmarket Goods 92
The Use of Value Measures in Determining Environmental Policy 107
Other Approaches to Value 122
Summary 127

CHAPTER 5 THE MACROECONOMICS OF THE ENVIRONMENT 132

Introduction 132
Conceptual Model of the Environment and the Economy 133
Environmental Taxation and Macroeconomic Benefits 146
The Environment and International Economic Issues 149
Summary 152

PART II EXHAUSTIBLE RESOURCES, POLLUTION, AND THE ENVIRONMENT 157

CHAPTER 6 GLOBAL ENVIRONMENTAL CHANGES: OZONE DEPLETION AND GLOBAL WARMING 159

Introduction 159
The Depletion of the Ozone Layer 161
Greenhouse Gases and Global Climate 167
The Economic Consequences of Global Warming 175
Global Warming Policy 179
Summary 189

CHAPTER 7 ACID DEPOSITION 193

Introduction 193
What Causes Acid Deposition? 194
The Impacts of Acid Deposition 198
Acid Deposition Policy 214
Summary 222

CHAPTER 8 ENERGY AND THE ENVIRONMENT 225

Introduction 225
The Historical Development of U.S. Energy Policy in the Post-World War II Era 226
Energy and the Environment 237
Energy Policy and the Environment 241
Nuclear Power Issues 249
Transition Fuels and Future Fuels 251
Energy and the Third World 253
Summary 254

CHAPTER 9 MATERIAL POLICY: MINERALS, MATERIALS, AND SOLID WASTE 257

Introduction 257
The Economics of Mineral Extraction 259
Solid Waste and Waste Disposal 267
Waste and Recycling 273

A Comprehensive Materials Policy 276
Summary 282

PART III RENEWABLE RESOURCES AND THE ENVIRONMENT 285

CHAPTER 10 FISHERIES 287

Introduction 287
Fisheries Biology 289
The Optimal Harvest 291
The Gordon Model 293
Incorporating Consumers' and Producers' Surplus into Fishery Models 301
Current Fishery Policy 304
Other Issues in Fishery Management 314
Summary 321
Appendix 10.A 324

CHAPTER 11 TEMPERATE FORESTS 326

Introduction 326
Forest Ecology 327
The Privately Optimal and Socially Optimal Management of Forests 330
Maximizing the Physical Quantities of Harvested Wood 330
The Optimal Rotation 335
Multiple Use Management 340
Below-Cost Timber Sales 342
Ancient Growth Forests 345
Summary 349

CHAPTER 12 TROPICAL FORESTS 352

Introduction 352
Deforestation 358
Rain Forests as a Global Public Good and International Policy 369
Summary 372

CHAPTER 13 BIODIVERSITY AND HABITAT PRESERVATION 375

Introduction 375
Anthropogenic Causes of Species Extinction 376
The Costs of Losses of Biodiversity 381
Costs of Losses of Habitat 382
Policies for Maintaining Biodiversity 382
Summary 390

CHAPTER 14 WATER RESOURCES 393

Introduction 393
Water Consumption 395
Water and Property Rights 400
Degrading Uses of Water 402
U.S. Policy toward Water Pollution 403
International Water Issues 408
Summary 411

PART IV FURTHER TOPICS 413

CHAPTER 15 TOXINS IN THE ECOSYSTEM 415

Introduction 415
The Nature of the Market Failure 417
The Government and Improper Waste Disposal 425
How Much Should We Pay for Cleanup? 428
Summary 432

CHAPTER 16 AGRICULTURE AND THE ENVIRONMENT 435

Introduction 435
The Effect of Environmental Quality on Agriculture 436
The Effect of Agriculture on the Environment 439
Agriculture and Public Policy 442
A Comprehensive Set of Environmental Policies for Agriculture 451
Summary 454

CHAPTER 17 THE ENVIRONMENT AND ECONOMIC GROWTH IN THIRD WORLD COUNTRIES 456

Introduction 456
The Role of Population Growth in Environmental Degradation and Economic Development 460
Traditional Models of Development and Their Flaws 468
The Impact of the Exclusion of Subsistence Agriculture from Measures of GDP 469
Sustainable Development 471
Summary 476

CHAPTER 18 PROSPECTS FOR THE FUTURE 480

Introduction 480
Absolute Scarcity 481
Price, Scarcity, and Neo-Classical Economics 484
System-Wide Change and Sustainability 488
Summary 491

APPENDIX THE INTERNET: WHAT IS IT, HOW TO USE IT, WHAT'S ON IT 493

Credits 501
Name Index 503
Subject Index 507

PART I

Theory and Tools of Environmental and Resource Economics

This textbook develops an economic approach to environmental and natural resources in an interdisciplinary context. Part I is designed to develop the economic theory and analytical tools that are necessary to understand this approach.

Introduction

. . . Are we so unique and powerful as to be essentially separate from the earth?

Al Gore, Earth in the Balance, *p. 1.*

INTRODUCTION

The last decade of the twentieth century represents a fascinating time to be studying environmental and natural resource economics. At this juncture in time, environmental degradation is regarded by many observers (Brown, 1990) as at its historic worst. On the positive side, however, public attention to the problem has also reached a zenith (Brown). What will be the actual outcome as we enter and proceed into the third millennium? Will environmental degradation rather than nuclear war be the source of the apocalypse, with the entire world coming to resemble the horrendously polluted industrial cities of eastern Europe—where the air is a dirty yellow brown, where all surfaces are covered with black soot, and where pollution is a leading cause of illness and death? Will rain forests disappear, and temperate forests succumb to acid rain? Will the greenhouse effect and loss of the ozone layer lead to massive changes in climate?

Although many may regard these predictions as extreme, they are not out of the range of possibility. However, even if the planet is currently moving in this direction, these outcomes are not etched in stone and can be avoided.

The policies implemented on the national and international level during this and the next decade will, to a large extent, determine both the nature of the planet's physical and natural environment and the standard of living of the planet's inhabitants. How those policies should be made is the primary focus of this book.

A TAXONOMY OF RESOURCES

The title of this book indicates that both environmental and resource issues will be examined. Most economics courses in this area are entitled "Environmental and Natural Resource Economics." Does the presence of the words "environmental and natural resources" imply a dichotomization, or are these terms synonyms, and the two terms merely represent a redundancy?

It is difficult to answer this question in a universal sense, as definitions are in themselves quite arbitrary. However, to provide an internal consistency, three concepts of resources will be defined by this text. These are:

1. natural resources
2. resource flows
3. environmental resources

NATURAL RESOURCES

Natural resources are those resources provided by nature that can be divided into increasingly small units and allocated at the margin. Examples of such resources would include barrels of oil, cubic meters of wood, kilograms of fish, liters of drinking water, and so on. It is important to note that while these resources are provided by nature in the sense that they are found in or on the earth, they cannot be utilized without the provision of other inputs, such as capital and labor.

The stocks of natural resources may be fixed or may have regenerative capability. This characteristic is used to divide natural resources into renewable resources and exhaustible resources. Resources such as oil, minerals, and coal are generally included among exhaustible resources, while living resources such as plants and animals are generally included among renewable resources. Although new deposits of oil and coal are currently being formed, these do not meet the definition of renewable for two reasons. First, the rate of regeneration of oil is so small relative to the consumption of oil that, for practical purposes, it can be regarded as being equal to zero. Second, and more importantly, the regeneration is not due to the current stock but due to new deposits of organic materials subject to pressure.[1] It is the property of the stock being responsible for the growth or regeneration that characterizes a renewable resource, and this property also makes for extremely interesting management decisions. (How many fish should we catch today, and how many should we leave to produce new fish for tomorrow?)

[1]It is interesting to note that the oil of the distant future may be from today's landfills.

Resource Flows

Solar energy, wind power, and similar resources are often called renewable resources, but these do not fit the above definition of a renewable resource, for even though they can't be exhausted, they do not have regenerative capabilities. Solar energy may be naturally stored by trees, fossil fuels, or algae, and solar energy may be artificially stored by batteries or hot water tanks, but the stock of solar energy is the sun itself. Our consumption of solar energy has no effect on the stock or our ability to consume solar energy in the future. These resources, which do not exist as a stock, but have never-ending flows, will be termed resource flows to differentiate them from renewable resources.

Environmental Resources

Environmental resources are those resources provided by nature that are indivisible. For example, an ecosystem, an estuary, the ozone layer, and the lower atmosphere can't be allocated unit by unit in the same fashion that you could allocate barrels of oil or tons of copper. These environmental resources can be examined at the margin in terms of quality but not in terms of quantity. For example, if one is formulating environmental policy for an estuary such as the Chesapeake Bay, the marginal units are not quantity units such as volume of water, but quality units such as dissolved oxygen levels or concentrations of herbicides. Another distinguishing property of environmental resources is that the resources are not consumed directly, but people consume the services that these resources provide. These services can be very broad in scope, from basic biological life support services to aesthetic benefits.

It is possible that some resources may be classified into more than one category. For example, trees can be viewed as a renewable resource that can be harvested, but the forest ecosystem should be viewed as an indivisible environmental resource. Similarly, salmon swimming up a river can be viewed as a harvestable renewable resource, but the existence of salmon in the river can be viewed as an environmental resource.

Why Study Environmental and Resource Economics? The question of why one should study environmental and resource economics hinges on whether existing academic disciplines are adequate for examining the environmental problem. One can look toward conventional economics and one can look toward the natural sciences, but neither are independently capable of analyzing and developing solutions to environmental and resource problems.

Study of natural sciences, such as ecology, is not sufficient to completely analyze the problem, because these sciences do not include analysis of human behavior. Although understanding the natural sciences is essential to the understanding of human activity, natural science studies do not include how human activity responds to changes in the economic and natural environment.

On the other hand, economics is often defined as the study of the allocation of scarce resources. If so, why do we need to study environmental and natural resources separately? Aren't the guiding principles developed in microeconomics sufficient to correctly allocate our environmental and natural resources?

The answer is that there are important differences between environmental resources and conventional goods, which need to be examined differentially. For example, the rules that define optimality in the allocation of private goods are essentially static in nature. Today's decision of how many VCRs to produce in the present period does not substantially affect the ability to produce VCRs in some future period. However, the decision of how much oil to produce and consume in the present period has important implications for the future. First, the amount of oil taken out of the ground today affects our ability to take oil out of the ground in the future. Also, the amount of oil consumed today affects the level of carbon dioxide in the atmosphere, which will lead to a future warming of the earth's climate.

The fact that many decisions regarding environmental resources are irreversible further complicates analysis, particularly when viewed in the context of the dynamic issues that are discussed above. For example, if the market produced insufficient VCRs in one period, this would not interfere with the ability to produce VCRs or enjoy using VCRs in the future. Any loss in social welfare that was created by that production decision need not be carried forward to future periods. However, let's assume that the present demand for preserving giant redwood forests (each tree is several hundred, to over 1,000 years old) is low, so we decide to cut them for export to Japan. This is an irreversible action. Once these forests are cut, it will take many hundreds of years for them to become reestablished, and there is a significant probability that they will never become reestablished. No matter how high the future demand is for intact giant redwood forests, it is impossible to provide the forests. Other examples of irreversible events or actions include the generation of nuclear wastes (which retain their radioactivity for hundreds of thousands of years), the destruction of tropical rain forests, global warming, the extinction of species, and the release into the environment of toxic substances such as dioxins (toxic chemicals) and DDT (a persistent pesticide).

In addition, there is another critical factor that differentiates environmental and natural resources from typical goods. Market failure, which is assumed not to exist in the basic market models, is obviously important, as pollution is a form of market failure, and environmental and natural resources may be public goods or open access resources. The importance and pervasiveness of market failure will be discussed in depth in Chapter 3.

Finally, and most importantly, environmental and natural resources require separate examination, because optimal allocation requires an understanding of more than just economic behavior. It also requires an understanding of the whole ecological system and how the ecological system responds

FIGURE 1.1

Economic and Social System

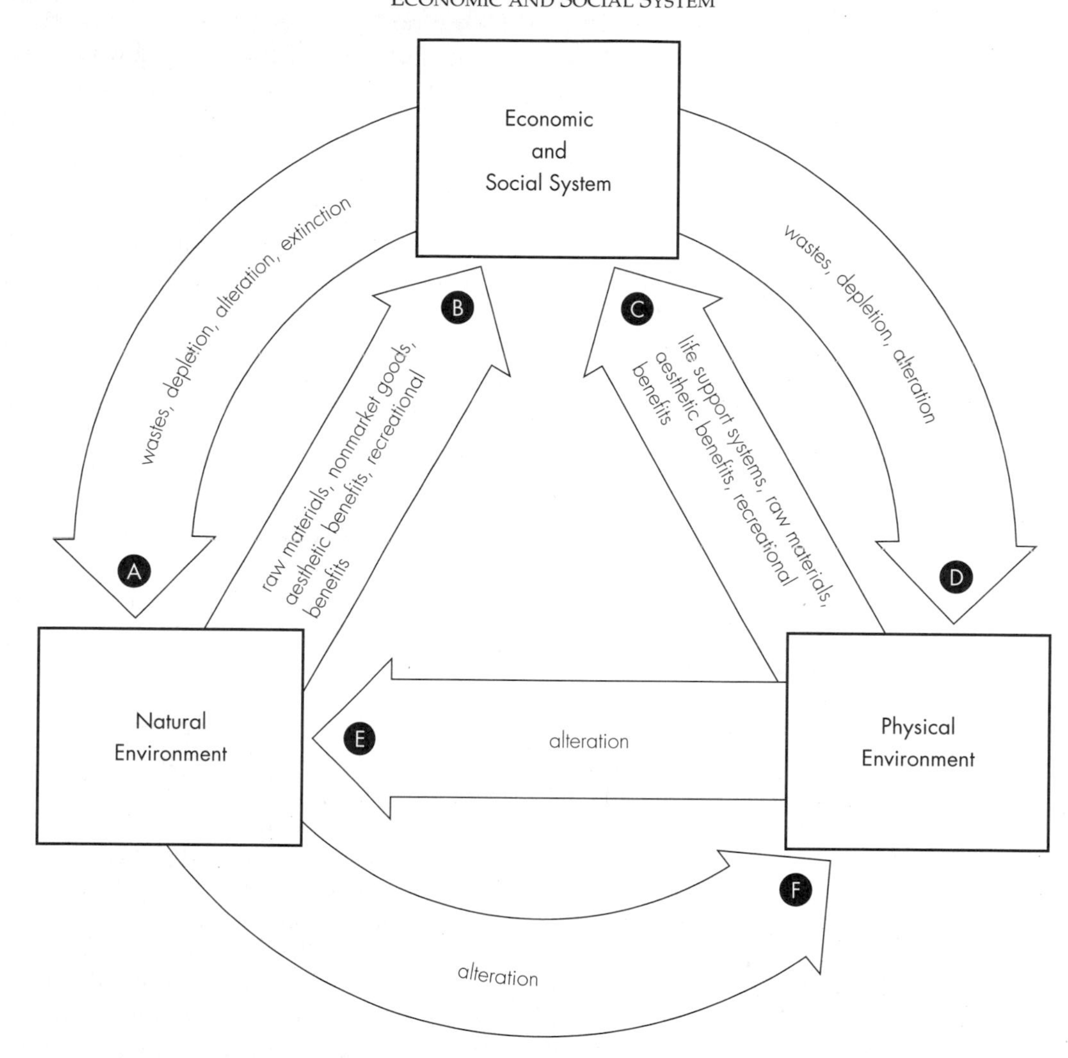

to changes in both the economic system and the ecological system. The interrelationships between these two systems are some of the keys to understanding environmental and resource economics.

Figure 1.1 represents a schematic diagram of all relationships between the economic system, the physical environment, and the ecosystem. It may seem surprising to see separate demarcations for the physical environment and the natural environment, as these two systems are obviously very closely intertwined. They have been separated for the purposes of the discussion to allow more precise focus on these intertwining relationships. The physical environment would include all the nonliving aspects of the environment, such as the climate, the chemical composition of air, soil, and water and the mechanical systems (wind, evaporation, tides, earthquakes, and so on), that influence these nonliving components of the environment. The natural environment, as defined for the purposes of this book, includes all the living components of the environment. As stated earlier, the two are closely intertwined, as climate, soil, and water have an effect on the living components of the ecosystem, and, in turn, the living components alter the physical systems. Of course, the living components have effects on each other, as do the physical components.

Now, let us assume that we want to figure out the effects on social welfare of an environmental problem such as acid rain, which is caused by the emission of sulfur dioxide and nitrogen oxides during the combustion of fuels such as oil, gasoline, and coal. Figure 1.2 shows just some of the relationships that must be determined in order to make this assessment. Traditional economics has only been concerned with relationships within the economic systems; environmental and resource economics directly focus on relationships B, C, D, and E (relationships between the economic system and the natural and physical environment). The purpose of this illustration is to show that one must also understand the importance of relationships within the natural environment, within the physical environment, and between these two closely related systems.

In order to develop a good understanding of environmental and resource economics, one does not need to become an expert in the natural and physical sciences, but one must develop an understanding of the contributions of other disciplines. For example, in order to understand how pollution affects an individual's utility function, some knowledge of the role of nutrient cycling must be obtained. An environmental and resource economist, particularly if anticipating a career in the public policy area, must be somewhat interdisciplinary and understand enough of how the physical and natural environment works to be able to talk to scientists in these fields. The same thing can be said for people who wish to make a contribution in other areas of environmental study. In order to be a good ecologist, political scientist, resource manager, and so on, one must develop an interdisciplinary perspective, including an understanding of environmental economics.

Economics has some very important contributions to make to the solution of our environmental and resource problems, contributions that become even more significant when integrated into an interdisciplinary overview. It is hoped that this book will not only help the reader to develop an understanding and appreciation for the theory and application of environmental resource economics, but also help to develop an insight into all aspects of our most pressing environmental and resource problems.

ECONOMIC AND SOCIAL SYSTEM

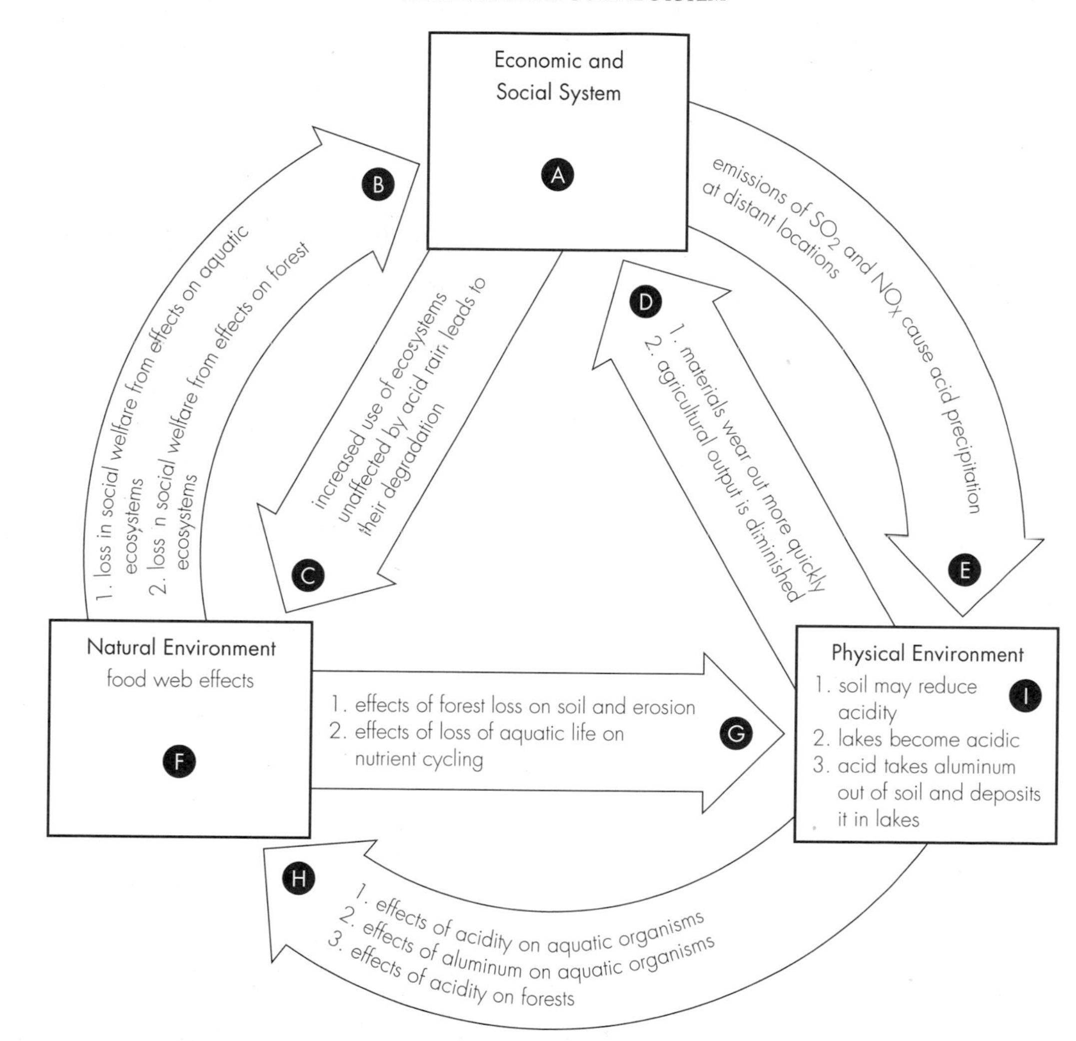

The first section of the book (Chapters 2–5) outlines some basic principles of economic theory and applies them to environmental and natural resources, to develop a foundation for the study of environmental and resource economics. The second section (Chapters 6–9) uses the theory to look at the problems associated with exhaustible resources and the environment. The third section of the book (Chapters 10–14) looks at the relationship between renewable resources and the environment, while the fourth section (Chapters 15–18) looks at additional topics such as the relationship between agriculture and the environment, Third World environmental problems, toxins in the ecosystem, and prospects for the future.

WORKS CITED AND SELECTED READINGS

1. Brown, Lester. *State of the World.* New York: Norton, annual editions are available beginning in 1984.
2. Gore, Albert. *Earth in the Balance.* New York: Houghton–Mifflin, 1991.
3. Kahn, Herman, and Julian Simon. *Global 2000 Revised.* Heritage Foundation, 1982.
4. Krutilla, John. "Conservation Reconsidered." *American Economic Review* 57 (1967): 777–787.
5. Myers, Norman. *Population, Resources and the Environment: The Critical Challenge.* United Nations Population Fund, 1991.
6. Pearce, David, et al. *Sustainable Development: Economics and Environment in the Third World.* London: Edward Elgar Publishing, 1990.

Economic Efficiency and Markets: How the Invisible Hand Works

. . . and by directing industry in such a manner as its produce may be of the greatest value, he intends only his own gain and he is in this, as in many other cases, led by an invisible hand to promote an end which was no part of his intention. Nor is it always the worse for society that it was no part of it. By pursuing his own interest he frequently promotes that of society more effectually than when he really intends to promote it.

Adam Smith, The Wealth of Nations, 1976

INTRODUCTION: THE WORKING OF THE INVISIBLE HAND

The above passage is the famous excerpt from Adam Smith's *The Wealth of Nations* in which he describes why markets lead to a socially efficient allocation of resources. His argument is very simple: People acting in their own best interests tend to promote the social interest. By allocating the resources under their control in a fashion that maximizes their well-being, they maximize society's well-being.

The functioning of the invisible hand as an efficient allocator of resources will seem obvious to some students and an impossibility to other students. The divergence of the two views will become more clear as the invisible hand is further examined. The key to understanding the problem is understanding what kinds of costs and benefits are generated by the good or activity in question. Even though our ultimate interest is the study of the environment, we will begin our examination of the way the market works with an every day private good. After examining how the market works in this simple case, we can add the complexity associated with the impact of market activity on the environment, and the impact of the environment on market activity and society in general.

Let us assume that the good we are talking about is blue jeans. The marginal costs of blue jeans can be represented by the upward sloping function in Figure 2.1. These consist of the costs of production such as labor, energy, capital, and materials. The marginal cost function is upward sloping

FIGURE 2.1

MARKET EQUILIBRIUM

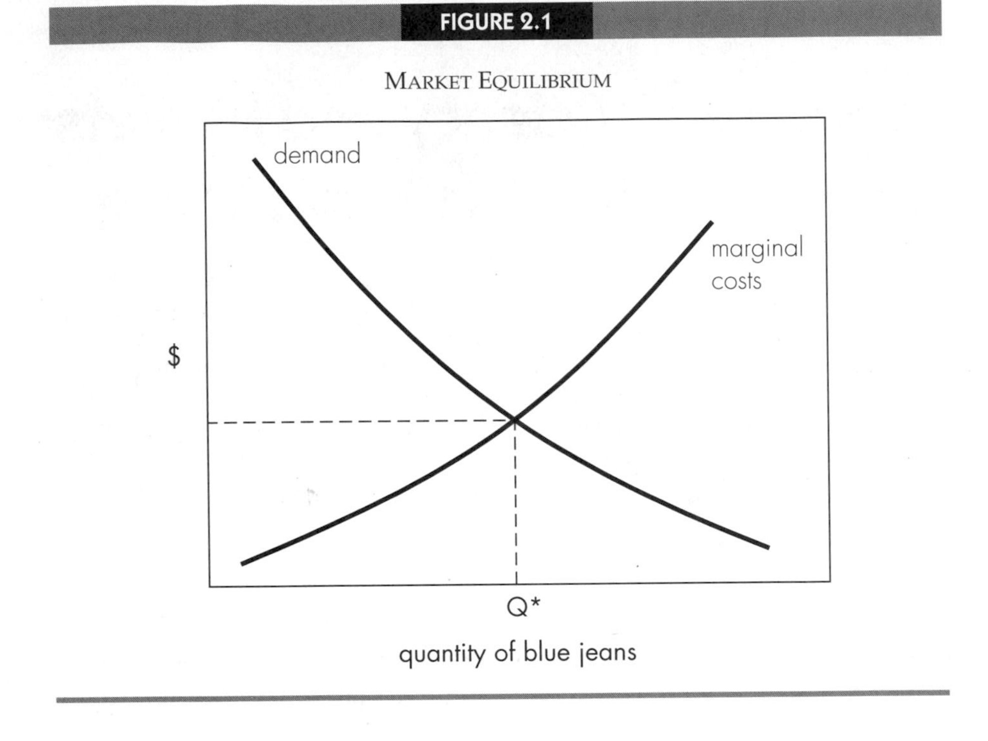

to reflect the increasing costs of production. The demand curve represents how much people are willing to pay for an additional pair of blue jeans, given a level of consumption. The demand or willingness to pay function is downward sloping to reflect the fact that the greater the level of consumption of the good, the less people are willing to pay for an additional unit of the good. We also know that the market will arrive at equilibrium, with quantity demanded equal to quantity supplied at Q*.

At the same time that the market forces set quantity demanded equal to quantity supplied, they also set marginal costs equal to marginal willingness to pay. The allocative properties of the invisible hand can be better understood if one focuses on this equating of marginal costs and marginal willingness to pay. The demand (marginal willingness to pay function) can also be thought of as a marginal value or benefit function, since individual willingness to pay will be based on the benefits to the individual.

We will start by making two simplifying assumptions that will later be relaxed. These assumptions are that all the costs associated with blue jeans are incorporated into the supply function and that all the benefits associated with blue jeans are incorporated into the demand function. This, then, implies that the market forces that equate quantity demanded and quantity supplied also equate marginal benefits and marginal costs.

A basic microeconomic principle is that the equating of marginal benefits and marginal costs will maximize total net benefits.[1] If marginal costs are greater than marginal benefits (quantity is greater than Q* in Figure 2.1), then reducing quantity reduces costs by more than it reduces benefits, so total net benefits would increase. On the other hand, if quantity is less than Q*, then increasing quantity increases benefits by more than it increases costs (marginal benefit is greater than marginal cost), so total net benefits would increase. It is only at Q*, where marginal costs and marginal benefits are equal, that it is impossible to increase net benefits by changing quantity. If it is impossible to increase net benefits, then net benefits must be at their greatest.[2] Although this argument has been developed for the market for an output, similar arguments could be constructed for the market for an input, such as the cotton that is used to make the blue jeans.

The above discussion shows how the market can maximize net benefits by equating marginal benefits with marginal costs. What has not been discussed is the nature of the costs and benefits that comprise these net benefits that have been maximized. To understand this, one must examine the behavioral forces that give rise to the demand and supply curves. As stated earlier, the demand curve can be viewed as a marginal value function that is based on willingness to pay for an additional unit of consumption. Since the market demand curve is the sum of every individual demand curve, marginal value is ultimately based on individual willingness to pay, which is based on how much an individual thinks the particular good or service will contribute to his or her utility in comparison with other goods and services. Thus, the demand curve reflects private benefits.

The supply curve, as stated above, reflects the costs of producing the good or service. These costs are those incurred in production, such as labor, capital, energy, and materials, and they can also be viewed as private, in the sense that all these costs of production are borne by the suppliers. Since the demand curve embodies private benefits and the supply curve embodies private costs, it follows that the net benefits that are maximized by market forces are private net benefits. In order for social net benefits to be maximized, it is necessary that private marginal benefits be identical to social marginal benefits and

---

[1]Net benefits are equal to total benefits minus total costs. When used without a qualifier, the term benefits will be used to signify benefits before costs have been subtracted. If benefits minus costs are being discussed, the adjective "net" will always appear in this text.

[2]Readers who are familiar with calculus can see this quite readily. Let NB represent net benefits, TC represent total costs, and TB represent total benefits, all of which are functions of Q. Then:

$$NB(Q) = TB(Q) - TC(Q).$$

To maximize NB, take the derivative (dNB/dQ) of net benefits and set equal to zero.

$$dNB/dQ = dTB/dQ - dTC/dQ = 0$$

Since $dTB/dQ - dTC/dQ = 0$, then $dTB/dQ = dTC/dQ$.
Since dTB/dQ = marginal benefits, and dTC/dQ = marginal costs, then net benefits are maximized when marginal benefits and marginal costs are equal.

private marginal costs be equal to social marginal costs. Then, market forces that equate marginal private costs with marginal private benefits will also equate marginal social benefits and marginal social costs.

Returning to the quote by Adam Smith at the beginning of this chapter, it appears obvious that Smith believed that private costs were identical to social costs, and private benefits were identical to social benefits. In a somewhat tautological fashion, this is how economists define a perfectly competitive market, and perfectly competitive markets automatically allocate resources efficiently. The critical questions are to what extent does this type of market exist in the real world and to what extent do deviations from this type of market impact the environment and lead to losses in social welfare?

MARKET FAILURE: WHEN THE INVISIBLE HAND DOESN'T WORK

An inability of the market to allocate resources efficiently is called a market failure. It is important to note that "failure" does not imply a barrier to market clearing (quantity demanded equaling quantity supplied); rather, it means that the market clearing forces do not maximize social net benefits by equating marginal social benefits with marginal social costs. A market failure may create a divergence between private costs and social costs. For example, the production of steel generates labor, land, capital, and material costs for the producers, which are costs to both the steel producer and society as a whole. Hence, these land, labor, capital and material costs are components of both private (producer's) costs and social costs. In addition, there is a set of costs attributable to the pollution generated by the steel production that is borne by society as a whole and not by the individual steel producers. This creates the disparity between private costs and social costs in Figure 2.2. Social costs are greater than private costs because they both include the private costs of production (land, labor, capital, etc.), and social costs include additional costs consisting of the damage generated by pollution. It should also be noted that in this case, nothing has happened to create a gap between private benefits and social benefits.

The steel producers respond to private costs and price; the steel consumers respond to private benefits (which in this example are equal to social benefits and price). Thus, the market forces will generate an equilibrium where marginal private cost is equal to marginal private benefit at a level of output equal to Q_1. This level is greater than the socially optimal level of output that is equal to Q^*, where marginal social cost is equal to marginal social benefit. Note that for the output between Q^* and Q_1, the benefits of the good are less than the costs associated with the good. This excess cost is the shaded area in Figure 2.2 and represents the costs to society of having this higher than optimal level of output. Pollution is not the only phenomenon

FIGURE 2.2

AN EXAMPLE OF MARKET FAILURE

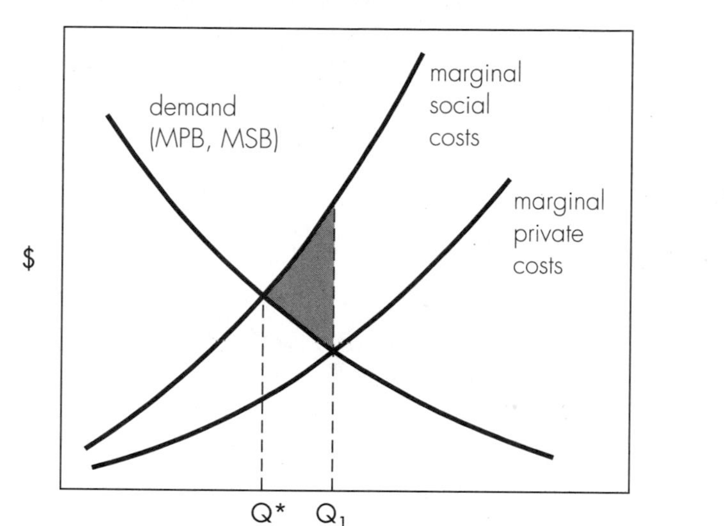

which creates a disparity between social costs and private costs. There are five categories of market failure, all of which have some importance for environmental and natural resources. These categories are:

1. imperfect competition
2. imperfect information
3. public goods
4. inappropriate government intervention
5. externalities

IMPERFECT COMPETITION

Imperfect competition is the term used for markets where the individual actions of particular buyers or sellers have an effect on market price. The importance of imperfect competition is that in such markets the marginal revenue of the firm becomes different from market price, and this tends to generate an equilibrium where marginal social cost is not equal to marginal social benefit. Figure 2.3 contains a market equilibrium for a market characterized by an extreme form of imperfect competition, monopoly (only one seller). In this case, the monopolist chooses the level of output which will maximize profit. In contrast, output in a competitive market is determined by market forces. The profit-maximizing level of output for the monopolist

FIGURE 2.3

MARKET FAILURE DUE TO MONOPOLY

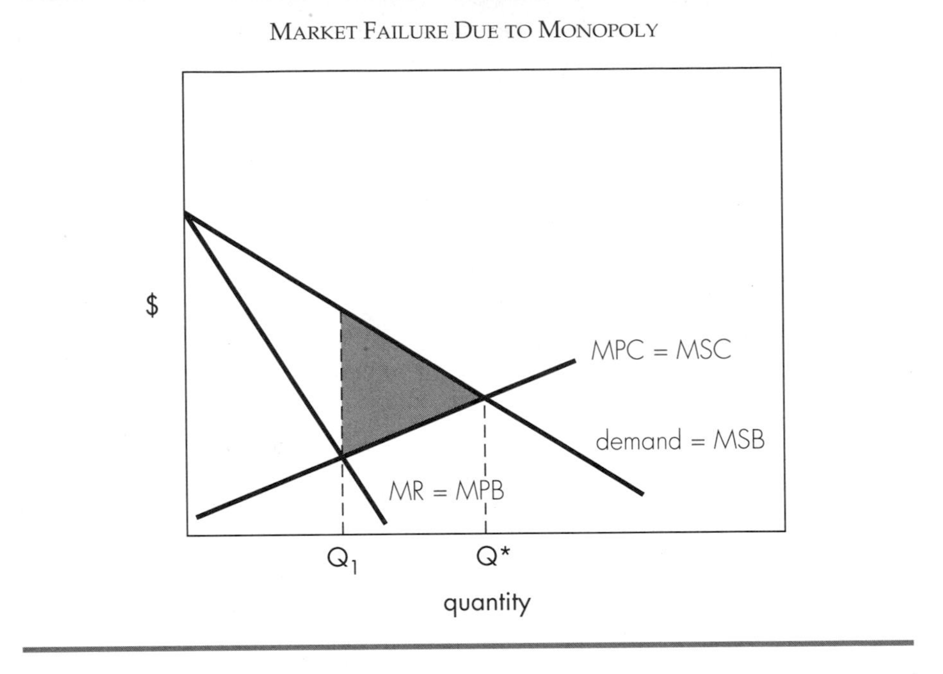

is lower than the competitive market level of output, and it occurs at the point where marginal costs are equal to marginal revenues. Marginal revenue measures how much more money the monopolist receives from an additional unit of output. Market failure occurs because, in the case of monopoly, marginal revenue is different from marginal social benefit. The marginal revenue function is below the average revenue function (the average revenue function is the same as the demand function and the marginal social benefit function). The monopolist's marginal revenue function is below the demand function because when the monopolist increases output, the price it receives for output becomes smaller. In contrast, in perfect competition, an individual firm's output level has no effect on the price the firm receives.[3] In the case of monopoly, the market output (Q_1) is less than the socially optimal level of output (Q^*), and the cost of the market failure is equal to the shaded area. The market failure of imperfect competition is important in the study of environmental and natural resources, as many extractive industries may be characterized by imperfect competition. Some industries, such as electric power and natural gas distribution, are regulated monopolies. Other indus-

[3]A more detailed explanation of this point can be found in any "Principles of Economics" textbook.

tries, such as oil and coal, are regarded by the general public as oligopolistic (only a few sellers who have price-setting ability).

IMPERFECT INFORMATION

Imperfect information means that some segment of the market—consumers or producers or both—does not know the true costs or benefits associated with the good or activity. If this is the case, then one would not expect the forces of supply and demand to equate marginal social benefits with marginal social costs. The importance of imperfect information to the study of natural and environmental resources can't be understated. For example, labor markets may efficiently allocate the on-the-job exposure to toxic substances if wage differentials between high exposure jobs and low exposure jobs adequately reflect the cost to workers of increasing their probability of contracting cancer or other health risks. The prospect of paying these compensating differentials will lead employers to pursue risk reduction measures that are less costly than risking compensating payments for the higher risk, and an optimal level of on-the-job safety should be achieved. However, this market for risk will not give the optimal amount of risk if workers do not adequately understand the health consequences of exposure to toxic substances. If they do not understand the true nature of the risk, they will require too much or too little compensating payment and the market mechanism will generate too much or too little risk.

One can think of many other instances in environmental and natural resource economics where imperfect information may be an important factor, including global warming, acid rain, the effect of exposure to radon in the home, and the hazards of using chemicals in the home (pesticides, solvents, and so on). One must be careful, however, to distinguish between imperfect information involving a public good or an externality and imperfect information involving a private good. The first two examples listed above (global warming and acid rain) involve public goods and externalities. There is already a market failure in the global warming problem in that carbon dioxide is emitted into the atmosphere based on comparisons of private costs and benefits, not social costs and benefits. There is also imperfect information in that the exact relationship between atmospheric build-up of carbon dioxide and global warming is not known. It is important to note that this imperfect information does not cause the market failure but makes it more difficult to develop public policy dealing with the market failure. On the other hand, radon (radioactive gas caused by the decay of naturally occurring uranium in the soil) leaks into homes, but this is not an externality; it is a natural feature in certain geographic regions of the country. The market failure occurs if people do not understand the true health consequences associated with radon leaks and do not take proper mitigative measures (sealing cracks in foundations, ventilating basements, moving to another location). If people understood the health consequences of radon leaking into their homes, in the long run there would be an adjustment of housing prices and institution of mitigative measures to generate the optimal level of exposure to radon.

PUBLIC GOODS

The third class of market failure deals with public goods. Although one tends to think of goods and services provided by the government when thinking of public goods, government provision merely implies that goods are classified as collectively provided. These goods are not necessarily public goods, as public goods may be collectively or privately provided, while private goods may be privately or collectively provided. The market failure exists because the market fails to provide the socially optimal level of public goods.

Public goods are distinguished from private goods by two primary characteristics: nonrivalry and nonexcludability in consumption.

Nonrivalry means that one individual's consumption of the public good does not diminish the amount of the public good available for others to consume. Nonexcludability means that if one person has the ability to consume the public good, then others can't be excluded from consuming it. These properties can be more easily understood by looking at national defense, one of the most frequently cited examples of a public good. The property of nonexcludability holds for national defense as, in protecting one citizen in a region from a missile attack, every citizen is simultaneously protected. The property of nonrivalry or nonexhaustibility also holds, as one citizen's consumption of protection does not reduce the production available to other citizens in the same geographic region.

National defense also is an example of a pure public good. A pure public good enjoys these properties completely. In contrast, a pure *private* good is completely exhaustible and completely excludable. An environmental resource that has pure public good characteristics is climate. All people in a geographic location experience the same climate, and none can be excluded from experiencing it.

Most of what we think of as public goods are not pure public goods, as they have some degree of exhaustibility and excludability. For example, the Grand Canyon is often thought of as one of our environmental resources that is a public good. While it is true that it has a certain degree of nonrivalry and nonexcludability, these properties are not present to the same extent as with national defense or climate. For example, it would be technically possible to exclude a subset of the population by building a fence (albeit a long and expensive one) around the canyon. Also, as the use of the canyon increases, the quality of the experience in the canyon declines (because of congestion and environmental degradation, such as erosion and littering), so the Grand Canyon is not completely nonrival in consumption. Rather than trying to categorize a particular good as strictly a pure public good or strictly a pure private good, one should look at where it lies on a spectrum with pure public

FIGURE 2.4

THE SPECTRUM OF PUBLIC AND PRIVATE GOODS

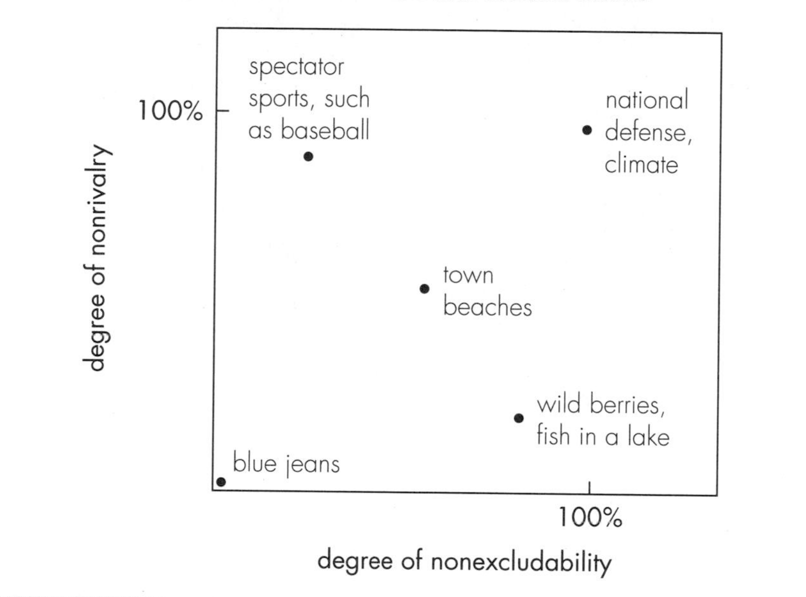

goods and pure private goods at the extremes. Figure 2.4 illustrates a two-dimensional spectrum, with the degree of nonrivalry on the vertical axis and the degree of nonexcludability on the horizontal axis. Goods like national defense and climate, which are pure public goods, are in the upper right-hand corner. This location signifies complete nonrivalry and nonexcludability. Goods like blue jeans are pure private goods and are at the origin, signifying no public good characteristics. Spectator sports, such as major league baseball, are mixed goods in that they have some public good properties. These spectator sports are nonrival in the sense that one spectator's consumption of the game does not diminish the quantity of the good available to other spectators. It should be noted that this good is not located at a position of complete nonrivalry because congestion can diminish the quality of the consumption available to spectators. Note that spectator sports are located in a position of a low degree of nonexcludability. Fans can be easily excluded from stadium attendance (they need one of a limited number of tickets for admission). If one was looking at spectator sports in a broader perspective including television consumption of spectator sports, then these sports would have a higher nonexcludability score, as spectators cannot be excluded from network (nonpay TV) broadcasts of the game. Other goods are mapped on this spectrum to show their public good properties.

FIGURE 2.5

INAPPROPRIATE GOVERNMENT INTERVENTION

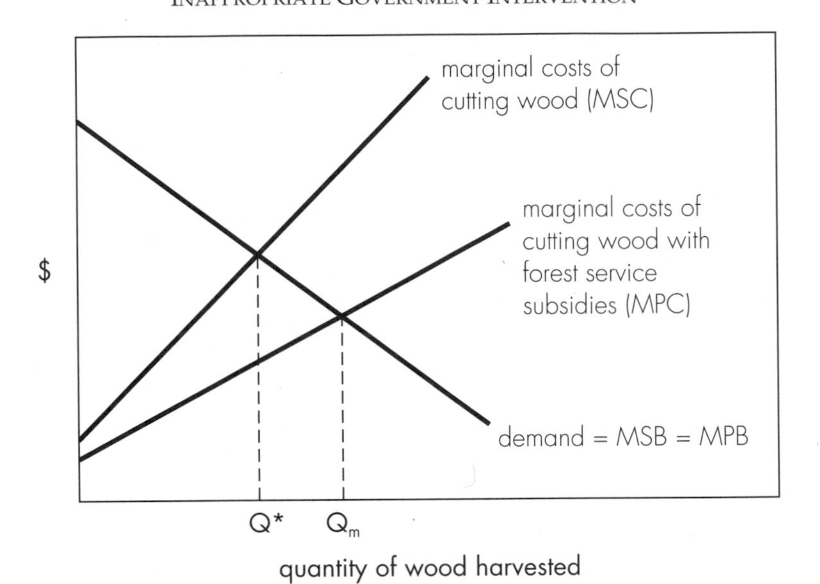

INAPPROPRIATE GOVERNMENT INTERVENTION

Another example of market failure is inappropriate government intervention. (*Note:* In many textbooks, inappropriate government intervention is not included as a market failure, since it represents constraints imposed on buyers and sellers, rather than direct actions of the buyers and sellers themselves. However, in this book, they will be included as market failures as the government intervention is a source of disparity between private and social values, which is the particular focus on market failure.) This means that the government intervenes in the economy not to correct a divergence between private costs and social costs, but for some other purpose, which can cause a divergence between private and social costs. A prime example of this, which we shall examine in greater detail in Chapter 11, is the U.S. Forest Service policy concerning the leasing of timbering (wood harvesting) rights in national forests. In this case, the Forest Service is treating the forests as private goods (ignoring their public good characteristics, such as recreation sites, wildlife habitat, wilderness, and so on). For the time being, let's assume that it is wise to treat at least a portion of national forests as private goods for use in wood production. Then the appropriate policy for the Forest Service would be to lease the rights to the highest bidder, allowing timbering if the bid is positive (this assumes that there is no opportunity cost to cutting down the forest). However, what the Forest Service does is to make roads through the forest at no cost to the companies who are leasing the cutting rights. Figure 2.5 shows how this distorts the market. The higher marginal cost function (MSC) contains all the costs of cutting wood, including road building. However, since the timber companies do not pay the cost of road building, their marginal private costs (MPC) are lower. While the socially optimal level of cutting is Q^*, the market solution based on the lower costs (MPC) to the harvesters generates an inefficiently high level of harvests (Q_m). Of course, if there are additional social costs associated with cutting trees (such as loss of ecological services), then the socially optimal level of harvest will be even lower.

EXTERNALITIES

Externalities are perhaps the most important class of market failures for the field of environmental and resource economics. In fact, pollution is probably the most often cited example of an externality in principles of microeconomics textbooks. In these textbooks, externalities are probably described as spillover costs or benefits, unintended consequences, or unintended side effects (either beneficial or detrimental) associated with market transactions. If there is an unintended detrimental consequence (such as pollution) associated with a good or an activity, then its marginal private cost function will be below its marginal social cost function, generating a market failure of the type analyzed in Figure 2.2. As explained when this graph was introduced, the market failure (in this case a detrimental externality) creates a disparity between marginal social costs and marginal private costs, so that when the invisible hand equates marginal private benefits and marginal private costs, it generates an excessive level of the activity of Q_1, rather than a socially optimal level of Q^*.

Most people think of externalities as detrimental, but it is also possible for externalities to be beneficial. For example, when parents have their child vaccinated against measles, they also protect other children, since their vaccinated child can't spread the disease. In other words, the private benefits of vaccination are less than the social benefits. Similarly, when suburban landowners generate private benefits by planting trees, they also generate social benefits by reducing erosion, increasing air quality, reducing global warming, and improving neighborhood aesthetics. Since landowners make the tree-planting decision by equating marginal private cost and marginal private benefits, the market level of suburban trees will be Q_1 in Figure 2.6, whereas the optimal level is Q^*.

While the above discussion does give some insight into what is meant by an externality, a more complete definition would be helpful to specify exactly what is meant by the term externality. The definition that will be employed in this textbook is that of Baumol and Oates (1988, p. 17).

FIGURE 2.6

OPTIMAL QUANTITY OF A GOOD WITH PARTIAL PUBLIC GOOD CHARACTERISTICS

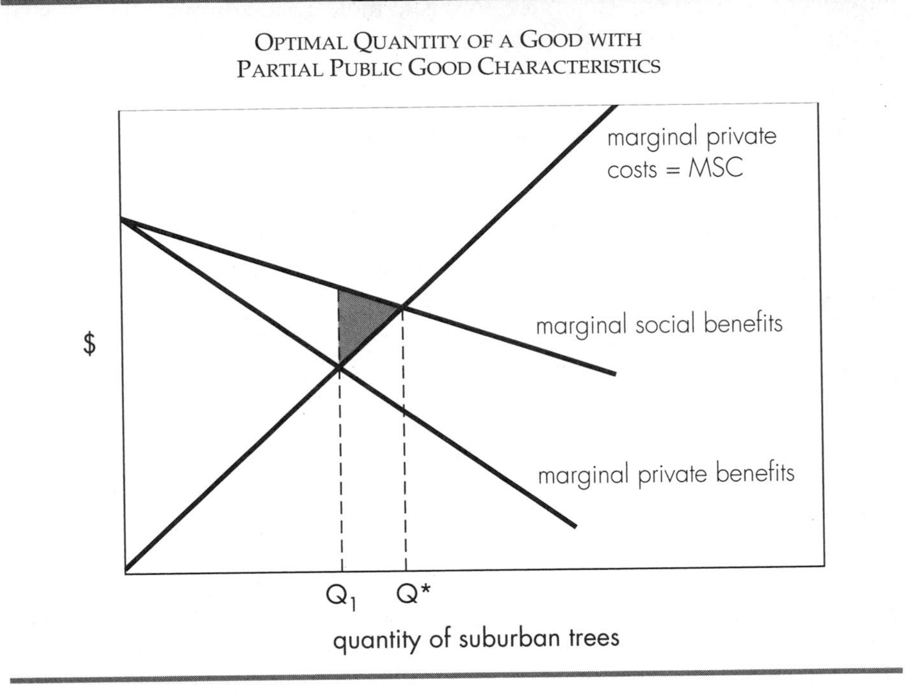

An externality is present whenever some individual's (say A's) utility or production relationships include real (that is nonmonetary) variables, whose values are chosen by others (persons, corporations, governments) without particular attention to the effects on A's welfare.

Baumol and Oates have chosen the words in this definition very deliberately. The key points are production and utility, real variables, and unintended effects.

The first point to examine is that Baumol and Oates are talking about unintended effects. If you were to intentionally blow cigar smoke in someone's face, that is not an externality, as you are doing it with the purpose of lowering that person's welfare. However, if your cigar smoke drifts from your restaurant table to another, then this would be an externality, as you are disregarding the utility of the affected people rather than making your decisions based upon your effects on their utility.

The second point to examine is that they are talking about real variables, not prices. This rules out an unintended price change as an externality. For example, assume that the demand for blue jeans increases, which increases the price of cotton (from which blue jeans are made). This will increase the price of flannel shirts (also made of cotton), which hurts people who like to wear flannel shirts. Note, however, that nothing has been done to interfere with the ability to produce or enjoy flannel shirts; they are simply more expensive.

Although this type of price effect is not regarded as an externality, it is generally called a pecuniary externality. The adjective "pecuniary" indicates that it has to do with money variables and not real variables. A pecuniary externality is not considered an externality despite the term "externality" appearing in its name. Although this jargon is unnecessarily confusing, it is widely used in the literature, so new terms will not be developed in this textbook.

The third point to examine is the focus on the effects on production and utility relationships. If we examine air pollution, which we suspect might be an externality, several effects on production and utility functions can be specified. First, certain types of air pollution lead to reduced yields of agricultural crops. If there is air pollution in a cotton area, that would imply that it now takes more resources (land, labor, fertilizer, and so on) to grow cotton than it did in the absence of the pollution. This reduced capability to grow cotton can be contrasted with the mere price increase of the above blue jean/flannel shirt example. In addition to interfering with the production of goods, air pollution may interfere with the production of utility. For example, a person will get less utility from an outdoor sport, such as jogging or tennis, in a polluted environment than in a clean-air environment. This type of externality (which affects production of goods or utility) is called a technological externality.

An ideal way to distinguish between a technological externality and a pecuniary externality is to examine their differential effects on the production possibilities frontier. The production possibilities frontiers in Figure 2.7 are based on the assumption of an economy based on only two goods, in this case, cotton and steel. The production possibilities frontier labeled p1 in Figure 2.7 shows the set of all feasible production points. These feasible points are all the combinations of levels of cotton and steel that it is possible to produce with the economy's endowment of resources, and they include all the points on or below the production possibilities frontier p1. The production possibilities frontier p1 is also based on the assumption that there are no externalities.

The actual combination of goods that are produced is determined both by the production possibilities frontier and consumer preferences. The production possibilities frontier determines what is possible to produce, while consumer preferences determine which of the possible combinations is actually produced. For example, if the current combination of cotton and steel leaves consumers unsatisfied and wanting more cotton, the price of cotton will be bid up as consumers try to buy more cotton, which will encourage more production of cotton and less production of steel. Let's assume that the

Shift in production possibilities frontier due to pollution.

FIGURE 2.7

TECHNOLOGICAL EXTERNALITY

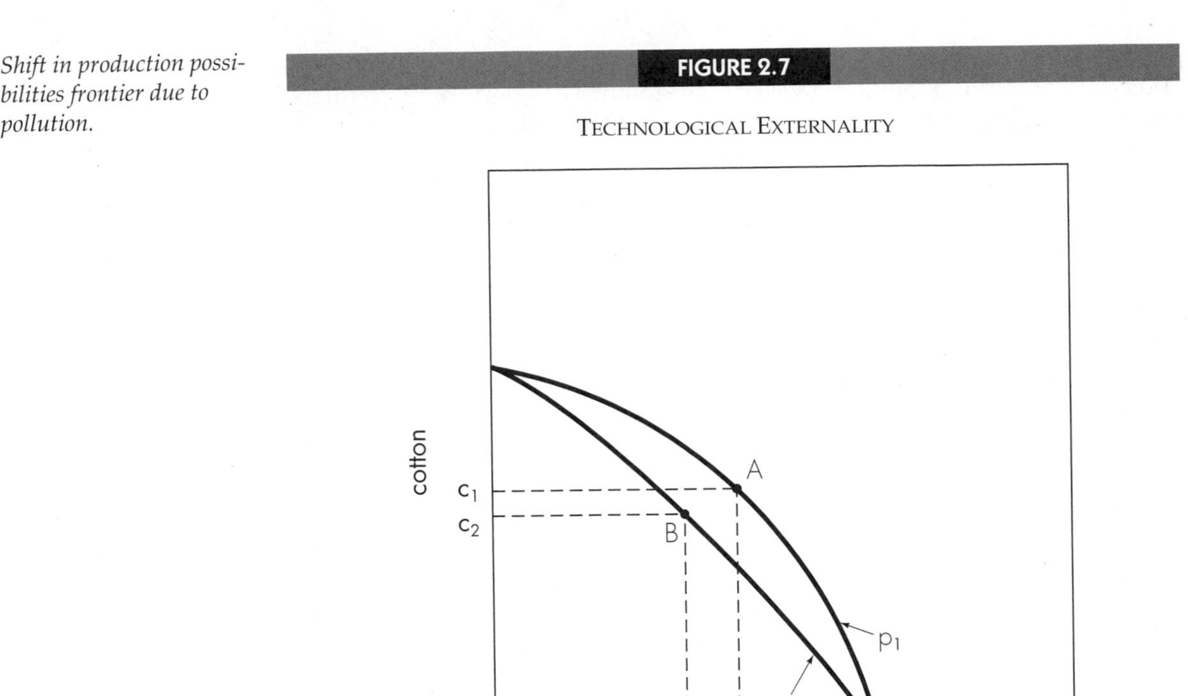

level of steel and cotton which maximizes society's well-being is represented by point A.[4]

Now, instead of assuming a world of no externalities, let's assume that the production of steel generates air pollution that reduces the yield per acre of cotton. This means that in order to produce as much cotton as before, more resources (land, labor, energy, and so on) will have to be devoted to growing cotton. If these resources are devoted to cotton, they can't be devoted to steel, which means that steel production must fall. Alternatively, some resources could be devoted to reducing pollution so the effects on cotton would be mitigated, but this would mean that there would still be fewer resources available for either cotton or steel. The end result is that since the pollution adversely affects the production of cotton, the economy can no longer produce either as much cotton as before, or as much steel as before, or as much

[4]A more comprehensive understanding of this point can be gained by indifference curve analysis, which models the consumer preference side of the adjustment process. Indifference curve analysis is a topic that is generally covered in intermediate microeconomic courses.

of both as before. The only exception to this, as pointed out by Baumol and Oates, is when the economy specializes in the production of only cotton or only steel. If the economy is only producing cotton, then there is no pollution from steel production, so as much cotton can be produced as before. If the economy is producing only steel, then there is pollution, but nothing to be adversely affected by pollution (remember that in this model there are only two goods in the economy), so as much steel can be produced as before.

The negative effect of pollution on the production possibilities of an economy is represented by the downward shift of the production possibilities frontier from p_1 to p_2 in Figure 2.7.[5] This will result in a new equilibrium at a point such as B. In this example, point B is clearly inferior to point A, since point A is associated with more of both goods.[6]

In contrast, a pecuniary externality represents a movement along a production possibilities frontier rather than a shift of it. Figure 2.8 shows a shift in preferences toward blue jeans which is represented by the equilibrium shifting from point A to point B. The number of blue jeans demanded and produced increases from j_1 to j_2, and the number of flannel shirts produced decreases from s_1 to s_2. It is important to note that this new allocation of resources has not decreased society's welfare. The change in prices results in a transfer from one segment of society to another. It is important to understand the difference between a true externality (technological externality) and a price effect (pecuniary externality). Figure 2.9 summarizes the important differences between a pecuniary externality and a technological externality.

Externalities as Public Goods. Many externalities have public good characteristics. These are called nondepletable externalities and are characterized by the public good property of nonrivalry in consumption. This means that one person's consumption of the externality does not reduce the amount of the externality available to others to consume. The pollution of drinking water supplies is a good example of a nondepletable externality, as one person's consumption of the water pollution does not reduce the amount of the water pollution to which other people are exposed.

MARKET FAILURE AND PROPERTY RIGHTS

Although we have discussed how externalities are generated by a disparity between social costs and social benefits, the reasons for the existence of this

---

[5]In this simple example, utility has been lowered because the pollution has reduced society's ability to produce manufactured outputs. Obviously, pollution can have other types of effects on society's welfare, such as loss of ecosystem productivity, biodiversity, recreational opportunities, and so on.

[6]A beneficial externality would shift the production possibilities frontier upward instead of downward and move society to a higher indifference curve and a higher level of utility.

Movement along production possibilities frontier due to changes in preferences.

FIGURE 2.8

PECUNIARY EXTERNALITY

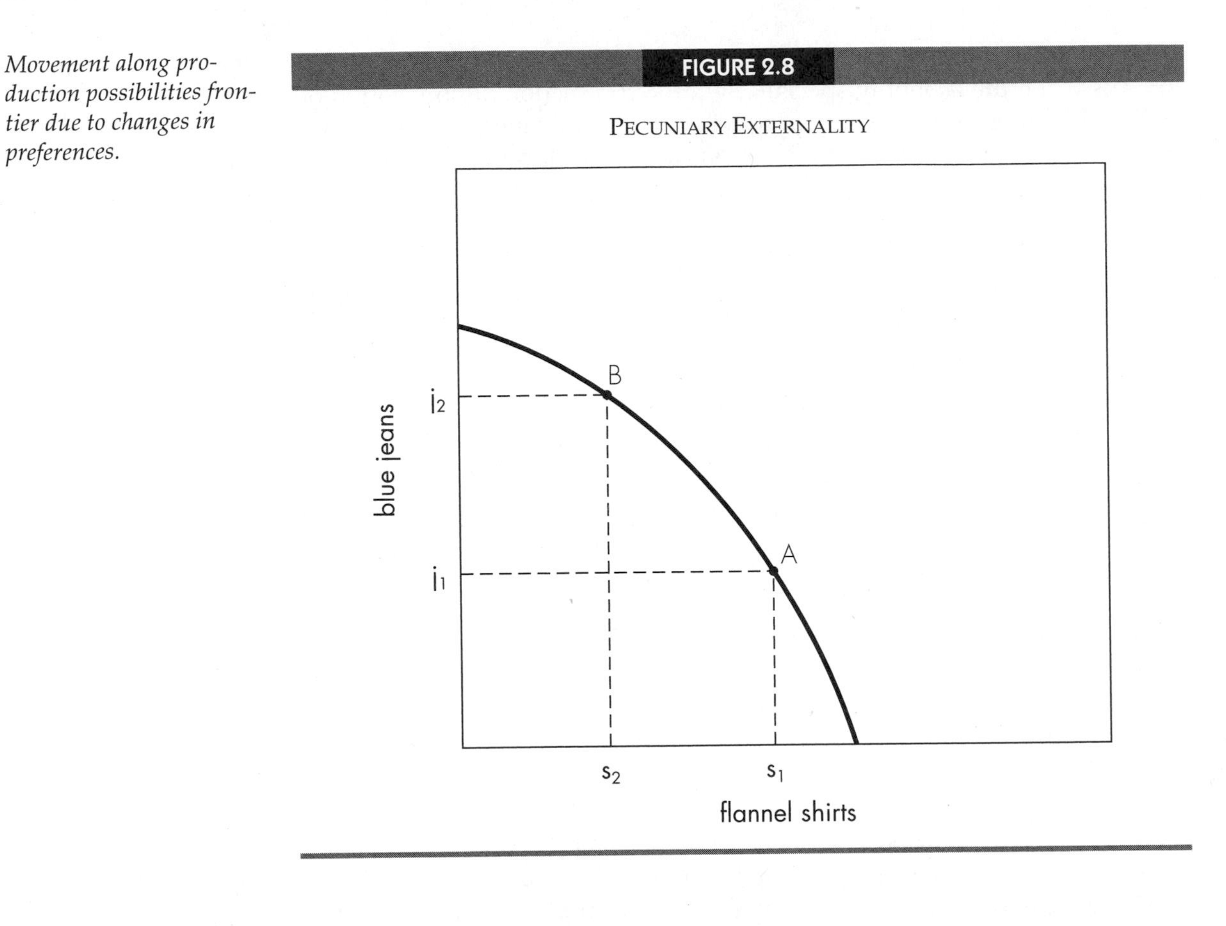

FIGURE 2.9

SUMMARY OF DIFFERENCES BETWEEN PECUNIARY AND TECHNOLOGICAL EXTERNALITIES

TYPE OF EXTERNALITY	TYPES OF VARIABLES AFFECTED	EFFECT ON PRODUCTION POSSIBILITIES FRONTIER	EFFECT ON SOCIAL WELFARE
Pecuniary externality [not a real externality]	prices	movement along frontier	transfer from one segment of society to another
technological externality	ability to produce goods or utility	shift of frontier (downward in the case of detrimental externalities)	net change in welfare (loss in the case of detrimental externalities)

disparity have not been examined. One of the important reasons has to do with the definition of property rights.

For example, no one has property rights to clean air, unless environmental legislation specifically defines these property rights. Consequently, a person who suffers from air pollution has no legal recourse to prevent someone from polluting the air or to collect damages from someone who does pollute the air. In contrast, a homeowner has the legal right to prevent another person from using the homeowner's front yard as a garbage disposal site. The difference between the two situations is that property rights to front yards are well defined, whereas property rights to air are not well defined.

A special class of externality that is generated by a lack of property rights (or an inability to enforce property rights) is the open-access externality. An open-access externality exists when property rights are insufficient to prevent general use of a resource and when this uncontrolled use leads to destruction or damage of the resource. For example, if access is not controlled to a fishery, then anybody who wants to can pursue the fish, diminishing the availability of future fish and increasing the cost of harvesting fish (see Chapter 10).

Even if property rights exist, a lack of ability to enforce them might lead to destruction of the resource. For example, many rain forest areas are communally owned by indigenous people who manage the forests in a sustainable fashion. However, even though the indigenous people have property rights that may be legally recognized by the national government, the indigenous people may not have the ability to enforce the property rights when outsiders migrate to the region. Since the outsiders have no legal right to the forest, they have no incentive to preserve it and destructive deforestation results (see Chapter 12).

THE INVISIBLE HAND AND EQUITY

It has been shown that, absent of market failure, the invisible hand of the market is an efficient allocator of resources. As discussed at the beginning of the chapter, "efficient" means maximizing the difference between social benefits and social costs. However, nothing has yet been said about how those costs and benefits are distributed among the members of society.

Many critics of the market system criticize it on these distributional grounds rather than for efficiency reasons. There is nothing inherently superior in how the market distributes costs and benefits across society. The market distribution of net benefits is only one possible distribution out of many possible distributions. The "best" distribution depends on what view of equity and fairness is held. There will be extensive discussion of equity issues in Chapters 4 and 5, as well as the application chapters.

THE INVISIBLE HAND AND DYNAMIC EFFICIENCY

The discussion in the first half of this chapter showed how the market can be an efficient allocator of resources and under what circumstances the market fails to do this. This discussion, while appropriate for most markets, is not a complete discussion of efficiency as it has been examining efficiency in a static sense.

A static analysis is an analysis that is only concerned with one time period. While one might initially think that a one-period analysis has limited basis in reality, this is certainly not the case. This is because most multiperiod analyses are actually a series of independent one-period analyses. This will be true as long as the optimizing decisions in one period are independent of the optimizing decisions in other periods. If the optimizing decisions are independent across periods, then a static analysis is sufficient. However, if the optimizing decisions of one period depend on the optimizing decisions of past or future periods, then a dynamic analysis must be employed.

The difference between the types of markets that can be examined statically, and the types that must be examined dynamically can be best illustrated by example. Let's look at a small farmer who has 100 acres of cropland. He or she looks at market conditions today and decides that corn is the best crop to grow this year. Then, next year at planting season, the farmer is faced with the same decision as last year, but the answer may turn out to be something different than corn if market conditions have changed. The point is that each year the farmer makes choices that are, for the most part, independent of the decisions that were previously made.

However, let us now assume that the farmer has 100 acres of forest instead of 100 acres of cropland. If the farmer decides to cut the trees today and plant corn, then in the next period the farmer cannot decide to have trees again.

An important and often cited example of dynamic optimization is the investment decision. Today's decision of how much output to consume and how much output to invest will affect these decisions in future periods because investment determines the capital stock that is used to produce output. Social and private efficiency of possible time paths of investment are examples of one area where dynamic, or intertemporal, considerations must be taken into account.

DYNAMIC EFFICIENCY AND EXHAUSTIBLE RESOURCES

Since decisions today can influence the quality of the environment and the stock of a resource far into the future, one would expect dynamic considerations to be important in the study of environmental and resource economics. For example, the decision of how much oil to take out of the ground today affects our ability to take oil out of the ground in the future. As will be illustrated below, the decision of how much oil to take out of the ground today is also dependent on how much people are willing to pay for oil in the future.

In order to examine the ability of the market to generate intertemporal efficiency, let's assume that you are an oil producer who has to decide how much of your oil to sell today and how much to sell in the future. As you might expect, efficiency requires that you continue to sell oil today until the point where marginal cost equals marginal revenue. For the time being, we will assume that there are no market failures associated with oil, such as imperfect information, imperfect competition, externalities, public goods, or inappropriate government intervention. This is not a realistic assumption, as will be documented in Chapter 8, but it will allow us to examine to what extent the market is capable of generating a dynamically efficient allocation of resources.

As an individual owner of oil, you will maximize your profits by setting private marginal cost equal to private marginal benefits, which in the absence of the market failures discussed above also generates marginal social cost equal to marginal social benefit. So far, nothing is different from the static case, and we have done nothing to show how the market takes intertemporal considerations into account. The way in which the future is considered by the market is that individual owners of oil, in deciding whether to sell their oil today or wait and sell it in the future, incorporate an additional opportunity cost in their decision making. The opportunity cost of not having the oil available in the future is sometimes known as user cost or rent.

As one might expect, the interest rate has an important influence in the process of intertemporal allocation of oil. This is because the individual oil owner has two choices for producing income into the future. First, the oil producer can sell all his or her oil and invest the money and earn interest income, or the oil producer can hold the oil and sell it in the future at a higher price. The oil producer will make a choice that will maximize the sum of the present values of the earnings potentially received in each period. (See Appendix 2.A if you need review on the concepts of discounting and present value.)

To illustrate the intertemporal allocation process, let's assume that this particular oil producer, as well as the other oil producers, believes that the future price will be too low, so he or she is better off selling the oil now. This will mean that more oil is sold in the present, so less will be available for the future, so the expected future price will rise. The sale of more oil in the present will cause present prices to fall. The combination of the higher expected future prices and the lower present prices will make holding oil for future sale a more attractive alternative.

On the other hand, if the initial conditions are such that the current price of oil is perceived as low relative to the future price, oil producers might initially view selling in the future to be the more attractive option. Production in the present will be withheld, which will drive up the present price. At the same time, the expected future price will fall as oil owners and oil consumers realize that too much oil is being held for future use. The combination of the rising present price and the falling future price will cause selling in the present to be a more attractive option.

As long as one option (selling in the present versus selling in the future) appears to be a more attractive option than the other, prices will adjust. This process of price adjustment will continue until owners of oil are indifferent between the options of selling in the present and selling in the future.

Since the present price is dependent on the future price, and the future price is dependent on the present price, there is an opportunity cost for using a barrel of oil at any particular point in time. This opportunity cost is the cost associated with not having the barrel of oil available in another time period. Note that this opportunity cost would not exist if there was an unlimited amount of oil.

This opportunity cost is a social cost, but at the same time, it is a private cost to the owner of the barrel of oil. This particular type of opportunity cost is often referred to as user cost. As a private cost, it is incorporated into price, and so the intertemporal dimension does not necessarily introduce a market failure, although in Chapter 8 we will discuss how the intertemporal dimension can cause a market failure in conjunction with imperfect information about the future. However, for the present we will assume perfect information about the future, which means that the marginal user cost of a barrel of oil is known.

The price of oil, at any particular time *t*, can be represented by equation 2.1.[7] In this equation, MUC refers to marginal user cost, and MEC refers to marginal extraction cost.

$$P_t = MUC_t + MEC_t \qquad 2.1$$

The most important thing to be noted about this equation is that one can predict changes in the price of oil by predicting changes in the marginal extraction cost and the marginal user cost. In particular, looking at marginal user cost can be especially informative. For example, one can explain how the Persian Gulf War generated changes in the price of oil by looking at marginal user cost.

When Iraqi forces invaded Kuwait in early August of 1990, and it looked as if Iraq might also gain control of (or destroy) Saudi oil fields, the price of a barrel of oil immediately rose about 50 percent. While many people attributed this to "price gouging," it can also be explained by changes in marginal

---

[7]A short mathematical and graphical derivation is presented in Appendix 2.B. For a complete mathematical discussion of the notion of user cost and how the time path of the price of an exhaustible resource is determined, three references are suggested, in order of level of difficulty: J. M. Griffen and H. B. Steele, *Energy Economics* (New York: Academic Press, 1980). Dasgupta and Heal, *Economic Theory and Exhaustible Resources* (London: Cambridge University Press, 1979). H. Hotelling, "The Economics of Exhaustible Resources," *Journal of Political Economy* 39 (1931): 137–175.

user cost. Marginal user cost rose dramatically as the taking of Kuwait and threats to Saudi Arabia dramatically changed the opportunity cost of using a barrel of oil today. However, when the United States and coalition forces began their air attacks on January 16, 1991, and rapidly achieved air superiority and then air supremacy, it became apparent that Iraq could not damage Saudi oil fields or affect Persian Gulf exports in the long run. This lowered people's perceptions of the opportunity cost of using a barrel of oil today (lowered marginal user cost), and the price of oil fell almost as rapidly as it had risen. By the beginning of the ground campaign (late February 1991), the price of oil was approximately at its pre-war levels. It is easy to see how marginal user cost serves as a vehicle for other phenomena to affect price. For example, if you wake up tomorrow morning and read in the newspaper that an engineer has developed an inexpensive and efficient cell for converting sunlight into electricity, this would lower the opportunity cost of using a barrel of oil today, and the price would fall.

Two other important observations can be made with respect to marginal user cost. First, the existence of marginal user cost implies that price will be different from marginal extraction cost. This means that the fact that price is greater than the marginal cost of extraction is not in itself an indication of the existence of monopoly profits. Although marginal user cost accrues to the owner of the resource, it is a scarcity rent rather than a monopoly profit. Second, in order for an owner of oil to be indifferent as to the period in which he or she sells the oil, the present value of the marginal user cost of oil must be the same in all periods. This means that, *ceteris paribus,*[8] the marginal user cost of an exhaustible resource will increase at the discount rate.

OTHER EXAMPLES OF DYNAMIC PROBLEMS IN ENVIRONMENTAL AND RESOURCE ECONOMICS

Market failures can also have a dynamic dimension. For example, the decision of how much to pollute has intertemporal dimensions when pollutants accumulate in the environment without breaking down. These are referred to as persistent pollutants, chronic pollutants, or stock pollutants. Some examples of this include heavy metals (such as lead, mercury, and cadmium) certain classes of pesticides, radioactive wastes, PCBs, and chlorofluorocarbons. Since today's decision to generate a stock pollutant (such as radioactive waste) has an effect on environmental quality far into the future, the present value of future social costs must be considered when deciding upon today's pollution standards.

---

[8]*Ceteris paribus* is a Latin phrase meaning "everything else remaining the same." It is often used in economics to mean that all the other factors that could influence the outcome are being held constant so that they do not influence the outcome. In this example, factors that could influence the values of marginal user costs in the future include factors such as unforeseen changes in demand (such as those caused by technological innovation in solar energy) or unforeseen changes in supply (such as unexpected discoveries of large amounts of low cost oil).

SUMMARY

Although the focus of this book is environmental and natural resources, which often are nonmarket goods, we have spent the bulk of this chapter focusing on markets, how they work, and their significance in the promotion of social welfare. This focus on markets is an important first step in the understanding of environmental and natural resource issues, as the primary reason we need to study environmental and natural resource economics is that the market often fails with respect to generating the socially optimal level of these resources.

For market goods, the market can be very effective in allocating resources, since the market can equate marginal social costs and marginal social benefits. This is true even when one period's production decisions affect the production decisions in future periods, as marginal user cost serves to equilibrate the market and maintain this equality. However, the market mechanism fails to efficiently allocate resources when private costs are not equal to social costs, or private benefits are not equal to social benefits. When marginal private cost is not equal to marginal social cost, or when marginal private benefit is not equal to marginal social benefit, then the market is characterized by failure. Market failures are caused by phenomena such as externalities, public goods, imperfect competition, imperfect information, and inappropriate government intervention.

One of the most important types of market failure, from the perspective of the study of environmental and resource economics, is the externality. An externality occurs when one person (or firm or agency) chooses values of variables in another person's (or firm's or agency's) production or utility function. Externalities are the source of most of our most important environmental problems. For example, when persons use a fossil fuel such as oil or coal, they make their decision of how to use the fuel based on a comparison of private costs and private benefits, and they do not consider the social costs associated with the fuel, such as its impact on air quality, acid rain and global warming.

This chapter has highlighted the sources of market failure with a focus on the disparity between marginal social cost and marginal private cost that a market failure such as an externality can create. In the next chapter is a discussion of the different strategies with which society can mitigate the social losses associated with market failure. With the critical emphasis that we have placed here on market failure and the disparities between marginal social cost and marginal private cost, it will not be surprising when in Chapter 3 we discuss policy tools which can help to eliminate that disparity.

REVIEW QUESTIONS

1. When does the invisible hand fail to maximize net social benefits?
2. Place the following goods on the two-dimensional privateness/publicness spectrum of Figure 2.4:
   a. elementary education
   b. secondary education
   c. undergraduate education
   d. a metropolitan opera company
   e. health care
   f. Yosemite National Park
   g. indoor air quality
   h. outdoor air quality
   i. the ozone layer
   j. biodiversity
   k. urban parks
   l. hiking trails
3. Distinguish between a pecuniary and technological externality.
4. Assume that the demand curve for a particular good is fully coincidental with the marginal social benefit function and can be described by $MSB = MPB = 24 - 2q$, where q refers to the quantity of the good. Assume that the marginal private cost function can be described by $MPC = q$, and that marginal social costs are always double the marginal private cost. Graph the functions and algebraically determine the market level of output and the optimal level of output.
5. Assume that a dam costs $20 million to build in one year and that, beginning in the second year, the dam yields net benefits of $2 million per year for 30 years. If the discount rate is equal to five percent, what is the net present value of the dam?
6. Define user cost.

QUESTIONS FOR FURTHER STUDY

1. Discuss the process by which the invisible hand efficiently allocates resources. List the market failures that may lead to inefficiency. Show how each market failure causes a loss in social benefits.
2. Derive the "rule of 70" discussed in the appendix.
3. Discuss the process by which an exhaustible resource is efficiently allocated over time.
4. If the price of oil is observed to be greater than the marginal extraction cost, can one infer that monopoly profits are being generated in the oil industry? Why or why not?
5. Do you believe that markets promote social welfare? Why or why not?

SUGGESTED PAPER TOPICS

1. Write a paper that discusses the development of economic thought on externalities. Start with early work by Pigou, Coase, Bator, and their contemporaries and move forward to the work by Fisher and Peterson, Baumol and Oates, and others.
2. Trace the development of the concept of user cost. Start with the work of Ricardo, who was one of the first to be concerned with the scarcity of exhaustible resources, and examine the work by Hotelling and more contemporary authors. Textbooks by Fisher, Dasgupta and Heal, and Griffen and Steele may be useful. Also, consult the survey article by Peterson and Fisher.

WORKS CITED AND SELECTED READINGS

1. Bator, F. M. "The Anatomy of a Market Failure." *Quarterly Journal of Economics* 72(1958): 351–379.
2. Baumol, W. J. and W. E. Oates. *Theory of Environmental Policy.* London: Cambridge University Press, 1988.
3. Coase, R. H. "The Problem of Social Cost." *Journal of Law and Economics* 3(1960): 1–44.
4. Dasgupta and Heal. *Economic Theory and Exhaustible Resources.* London: Cambridge University Press, 1979.
5. Dorfman, R. and N. S. Dorfman. *Economics of the Environment: Selected Readings.* New York: Norton, 1993.
6. Fisher, A. C. and F. M. Peterson. "The Environment in Economics: A Survey." *Journal of Economic Literature* 14(1976): 1–33.
7. Griffen, J. M. and H. B. Steele. *Energy Economics.* New York: Academic Press. 1980.

8. Hotelling, H. "The Economics of Exhaustible Resources." *Journal of Political Economy* 39(1931): 137–175.
9. Meade, James E. "The Theory of Economic Externalities." Geneva: Institute Universitaire de Haustes Etudes, 1973.
10. Peterson, F. M. and S. C. Fisher. "The Exploitation of Extractive Resources: A Survey." *Economic Journal* 87(1977): 681–721.
11. Pigou, A. C. *The Economics of Welfare.* London: MacMillan and Company, 1938.
12. Ricardo, D. *Principles of Political Economy and Taxation.* London: Everyman, reprint 1926.
13. Smith, Adam. *The Wealth of Nations, vol. 1,* 1776. Chicago: University of Chicago Press, reprint 1976.

APPENDIX 2.A

Discounting and Present Value

The concepts of discounting and present value are based on a type of behavior called time preference, which suggests that people prefer to realize benefits sooner rather than later (and realize costs later rather than sooner). Since people prefer benefits in the present, the individual is not indifferent between one dollar of benefits today and one dollar of benefits sometime in the future.

Discounting is a procedure by which dollars of benefits in different periods can be expressed in a common metric. The common metric is called present value, whereby all future values are converted to a value in today's dollars with the conversion constructed so that the individual is indifferent between the dollars in the future and the present value of those dollars today.

One can obtain an informal idea of the process of constructing present values by playing the following game:

Suppose that you are given the option of buying a bond for $100 with complete certainty of payoff one year from now. What is the minimum amount of money that the bond would have to pay one year from now in order to make you willing to buy it today?

Let's say that you said $110 as your answer. That means that you are indifferent between $100 today and $110 one year in the future or that you view the present value of $110 one year from now to be $100. From your response to the question, we can observe that your rate of time preference, or your discount rate, is ten percent.

This process of obtaining present values by discounting future values can be formalized into the mathematical expression of Equation 2a.1 below, where FV refers to the value that occurs t periods into the future, r is the discount rate, and PV is the equivalent present value.

$$PV = (1 + r)^{-t} \times FV \qquad 2a.1$$

In our example above, $t = 1$, $FV = \$110$, and $r = 0.1$. Performing the calculations indicated by Equation 2a.1, we see that:

$$\$100 = \left(\frac{1}{1 + 0.1}\right) \times \$110 \qquad 2a.2$$

The formula can be better understood by looking at another example. What is the present value of $1,000 payable in ten years, if the discount rate is eight percent (0.08 in decimal terms)? Using our formula above, with the values of $t = 10$, $r = 0.08$, and $FV = \$1{,}000$, the present value can be calculated according to Equation 2a.3:

$$\begin{aligned} PV &= (1 + 1.08)^{-10} \times \$1000 \\ &= (0.463) \times \$1000 \qquad 2a.3 \\ &= \$463 \end{aligned}$$

This indicates that the present value of $1,000, payable ten years from today, is $463. In other words, a person with a discount rate of 8 percent is indifferent between receiving $463 dollars today and $1,000 ten years in the future.

In our examination of environmental and resource economics, it will be necessary to examine slightly more complex formulations. For example, let's say that we wanted to examine the desirability of a project to upgrade New York City's sewage treatment plant to protect area beaches and other marine resources. In each year of the project, there will be costs and benefits. In some years, the costs might exceed the benefits, and in other years the benefits might exceed the costs. In order to determine if the project is a good idea, one must take the present value of the benefits and costs in each year and then determine if the present value of the benefits minus the present value of the costs (present value of the whole time stream of net benefits) is positive or negative. If we let B_t refer to the benefits (measured in dollars of year t), C_t be the corresponding measure of costs in each year, and T be the number of periods in which the project will be yielding either costs or benefits, then the present value of the total net benefits of the project are given by Equation 2a.4:

$$TPV\ of\ NB = \sum_{t=1}^{T} (B_t - C_t)(1 + r)^{-t} \qquad 2a.4$$

An analogous process can be used to show the extent to which amounts in the present will grow through some future date. Note that if Equation 2a.1 was solved for FV, then the result would be:

$$FV = (1 + r)^t \times PV \qquad 2a.5$$

For example, if $100 was invested for ten years at an annually compounding interest rate of ten percent, then the future value of this $100, ten years in the future could be calculated by substituting the appropriate values into Equation 2a.5.

$$\begin{aligned} FV &= (1 + r)^t \times PV \\ &= (1 + 0.1)^{10} \times \$100 \\ &= (2.59) \times \$100 \\ &= \$259 \end{aligned} \qquad 2a.5$$

In the above example, the interest was compounded annually, meaning that after each year, the interest would be earned on the previous year's interest. While this is an adequate representation of some types of interest growth, it is not an adequate representation of many types of growth, either economic or biological. For many assets, interest compounds instantaneously, which means there is no waiting period for earning interest on the interest previously earned. This is certainly the case for measuring the growth of the populations of most organisms, including human beings. Equation 2a.6 illustrates a growth function with continuous compounding, while Equation 2a.7 represents the analogous discounting process. Equation 2a.8 is the continuous analog of Equation 2a.4.

$$FV = PV \times e^{rt} \qquad 2a.6$$

$$PV = FV \times e^{-rt} \qquad 2a.7$$

$$PV\ of\ NB = \int_{t=1}^{T} (B(t) - C(t))e^{-rt}dt \qquad 2a.8$$

One convenient mechanism for looking at growth and discounting without the aid of a calculator is the "rule of 70." The rule of 70 states that if you take the growth rate, multiply it by 100, and divide it into 70, the answer is the number of years it will take the sum to double when growing at that rate (or the number of years it would take to shrink to half its value when discounted at the given rate). For example, if your money was invested at ten percent (0.1 in decimals) interest, it would double every $70/(0.1 \times 100) = 7$ years.

The rule of 70 can be derived from Equation 2a.6. Let the quantity that is growing be represented by the variable X. Then the length of time it would take for X to grow to $2X$ would represent the doubling time. Let $2X$ represent FV in Equation 2a.6 and X represent PV. Then Equation 2a.6 can be rewritten as Equation 2a.9 and then solved for t.

$$2X = Xe^{rt} \qquad 2a.9$$

divide both sides by X

$$2 = e^{rt}$$

take the natural logarithm of both sides

$$\ln(2) = rt$$

$$0.693 = rt$$

$$\frac{0.693}{r} = t$$

$$\frac{69.3}{100r} = t$$

$$t \text{ is approximately} = \frac{70}{100r}$$

APPENDIX 2.B

Dynamic Efficiency

The concepts of present value that have been developed in Appendix 2.A can be used to look more precisely at the question of the dynamically efficient allocation of an exhaustible resource. Assume that there are 100 tons of coal, that marginal extraction costs are zero, and that time consists of two periods, period 1 and period 2. This two-period model is a convenient way of collapsing continuous time into a tractable model. Period 1 can be viewed as the present, and period 2 can be viewed as the rest of time.

Assume that the demand curve in each period is equal to $P = 500 - .5q$. In any period, people are willing to purchase the whole 100 tons (at 100 tons, the price people would be willing to pay is $450). How, then, will the 100 tons be allocated over the two periods?

The answer is that the owners of the coal would try to maximize the present value of the income that they would receive in each period. The income that they would receive in period 1 is equal to the price that they would receive multiplied by the quantity that is sold, or $(500 - .5q_1)q_1$. The income that they would receive in period 2 is equal to $(500 - .5q_2)q_2$. Since $q_1 + q_2$ must equal 100, the income received in period 2 can be written as $[500 - .5(100 - q_1)](100 - q_1)$. If the discount rate is equal to five percent, the present value of the income received in period 2 can be rewritten as $[500 - .5(100 - q_1)](100 - q_1)/1.05$ or as $[476.2 - .4762(100 - q_1)](100 - q_1)$.

Present value of the income from the two periods can be computed as:

$$PV = (500 - .5q_1)q_1 + [476.2 - .4762(100 - q_1)](100 - q_1)$$

To maximize PV, set $dPV/dq_1 = 0$ and solve for q_1.

The solution is that $q_1 = 60.97$ and $q_2 = 39.02$,

with $p_1 = 469.5$ and $p_2 = 480.5$.

Several things are important to note:

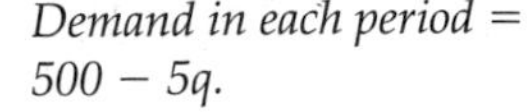
*Demand in each period =
500 − 5q.*

FIGURE 2.10

ALLOCATION OF 100 TONS OF COAL IN A 2-PERIOD MODEL

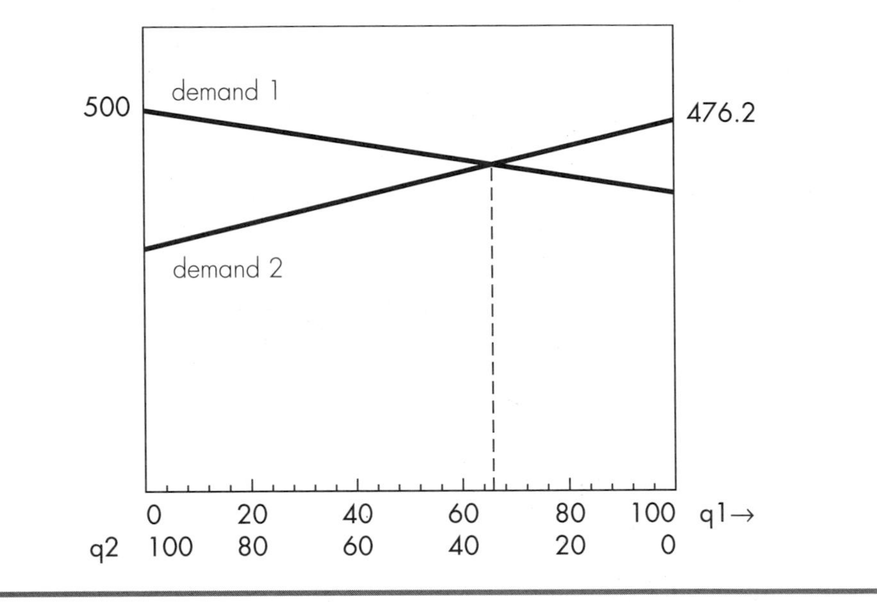

1. Although marginal extraction cost is equal to zero, price in both periods is positive. The price is composed exclusively of user cost, the opportunity cost of not having the coal available in another period.
2. Price is higher in period 2 than in period 1; the difference between the two reflects the positive rate of time reference.

Figure 2.10 contains a graphical analysis of this issue. The horizontal axis of this graph is equal to 100 tons of coal. As one reads from left to right, the quantity in period 1 gets larger. As one reads from right to left, the quantity in period 2 gets larger.

Demand 1 is the demand curve in period 1. Demand 2 is the present value of the demand in period 2 (which is read from right to left). The present value of total income is maximized where the two demand curves intersect. The reason for this is that to the left of the intersection, *PV* can be increased by allocating more coal to period 1, since the price that people are willing to pay in period 1 is greater than the present value of the price in period 2. To the right of the intersection, present value can be maximized by allocating more coal to period 2, since the present value of the price that people are willing to pay in period 2 is greater than the price in period 1.

Government Intervention in Market Failure

Since its creation two decades ago, EPA has made great strides in protecting the environment. For the most part, these environmental improvements were made through the use of command-and-control regulation; that is, promulgation of uniform, source-specific emission of effluent limits.

It is becoming increasingly clear that reliance on the command-and-control approach will not, by itself allow EPA to achieve its mission....

To maintain momentum in meeting environmental goals, we must move beyond prescriptive approaches by increasing our use of policy instruments such as economic incentives. Properly employed, economic incentives can be a powerful force for environmental improvement.[1]

INTRODUCTION

The above quote from former U.S. Environmental Protection Agency (EPA) Administrator William K. Reilly and the actual existence of an agency such as the EPA are based on the supposition that government intervention to correct environmental externalities improves society's well-being. The issue to which the quote relates is not whether the government should intervene, but what form government intervention should take. In most high income countries, the necessity for environmental policy is accepted by the majority of the citizens, and the focus of the debate is how policy should be formulated. The major thrust of this chapter is a discussion of how policy should be structured and the relative merits of alternative policy instruments. However, this chapter will begin by taking a step back to examine the intellectual arguments that would support or refute the necessity of government intervention.

In the previous chapter, market failures and their implications for social welfare were discussed. It was shown that market failures lead to inefficient allocations of resources, as marginal social costs and marginal social benefits are not equal. Since the existence of market failures implies losses in social

[1]William K. Reilly, Administrator's Preface, in Alan Carlin, *The United States Experience with Economic Incentives to Control Environmental Pollution,* USEPA, 1992, 230-R-92-001.

welfare, the natural question to investigate is whether the mitigation of market failures can lead to improvements in social welfare. In other words, can we make ourselves better off by correcting the problem?

Since externalities are such an important component of environmental market failures, our discussion of whether to intervene to correct market failures will focus on externalities and leave the discussion of monopoly, for example, to industrial organization courses. Specifically, should society, through collective action, actively intervene to correct market failures associated with such environmental externalities as pollution?

This question is one that has been addressed by many economists in past writings, with arguments on both sides. The early work of greatest significance arguing for government intervention was by A. C. Pigou. The counterpoint argument was provided by Ronald Coase, who argued that government intervention was not only unnecessary, but counterproductive. The next section of this chapter will focus on their arguments and on critical analysis of their work by subsequent authors.

Once the issue of whether the government should intervene is resolved, then the question of how it should intervene must be examined. Different sets of policies will have different effects on efficiency and on how social welfare is distributed among the members of society. These issues are analyzed in the remaining sections of the chapter.

SHOULD THE GOVERNMENT INTERVENE TO CORRECT ENVIRONMENTAL EXTERNALITIES?

A. C. PIGOU

The discussion of this issue begins with the work of A. C. Pigou (1938) who was among the first to recognize the existence of externalities and the associated divergence between private and social costs and benefits. Pigou argues that the externality cannot be mitigated by contractual negotiation between the affected parties and recommends either direct coercion on the part of the government or the judicious uses of taxes against the offending activity. These externality taxes are often referred to as Pigouvian taxes, after the economist who first proposed them.

The basic principle behind the use of externality taxes is that the tax eliminates the divergence between marginal private cost (MPC) and marginal social cost (MSC). This is illustrated in Figure 3.1, which reproduces the graph contained in Figure 2.2. In this graph, MPC′ represents the marginal private cost function, Q1 represents the market equilibrium (where MPC = MPB), and Q* represents the optimal level of output (where marginal social cost [MSC] = marginal social benefit). If an externalities tax equal to the divergence (measured at Q*) between MPC and MSC was charged, it would

FIGURE 3.1

AN EXTERNALITY TAX ON OUTPUT

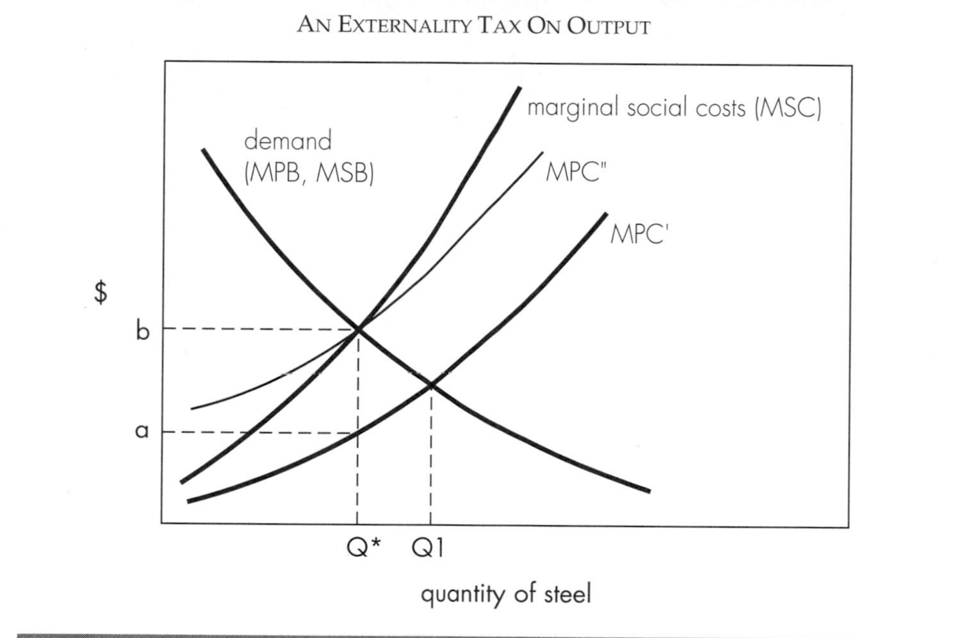

raise the steel firms' private costs, because they would have to pay the tax on each unit of output. This tax would be equal to the vertical distance from *a* to *b* in Figure 3.1. The tax would shift the marginal private cost curve by a corresponding amount, from MPC′ to MPC″. This new higher level of marginal private costs would force marginal private costs equal to marginal social costs, and thus the market would arrive at the optimal equilibrium of Q*. This process is sometimes known as internalizing the externality. Subsequent examination of externalities taxes have shown that the tax should not be placed on the output (such as steel) but on the externality (such as emissions of sulfur dioxide). This is because the output should not be directly discouraged, because the output yields benefits.[2] It is the externality that generates the additional social costs, and so it is the externality that should be taxed. The important differences between a tax on emissions and a tax on output will be discussed in more detail when pollution taxes are more thoroughly analyzed later in the chapter.

---

[2]This statement would not be true if the only way of reducing the externality was to reduce the output.

TABLE 3.1

RANCHER'S BENEFITS

NUMBER OF CATTLE	TOTAL BENEFITS	MARGINAL BENEFITS
1	$10	$10
2	19	9
3	27	8
4	34	7
5	40	6
6	45	5
7	49	4
8	52	3
9	54	2
10	55	1
11	54	-1
12	52	-2
13	49	-3

RONALD COASE

Ronald Coase (1960) argues that not only is such a tax unnecessary, but it is often undesirable. The reason for this, Coase believes, is that a market for the externality will develop. Coase makes two major arguments: The market will automatically generate the optimal level of the externality, and this optimal level will be achieved regardless of the definition of property rights. The definition of property rights refers to whether the generator of the externality has the legal right to generate the externality or the victim of the externality has the legal right to be free from exposure to the externality. In other words, do steel mills have the right to dump their emissions in the air, or do citizens have the right to clean air? Coase's supposition that the market will generate the optimal level of the externality, regardless of the definition of property rights, has come to be known as the Coase Theorem. The Coase Theorem is often the point of embarkation for discussion of the necessity of government intervention to correct externalities.

Our discussion of the Coase Theorem will begin with a discussion of his primary example. The numbers have been modified to allow for a more straightforward examination, but they model the same behavior as Coase, and they lead to the same results.

Coase's example, which forms the primary conceptual evidence for the Coase Theorem, is based on the interaction of a cattle rancher and a crop farmer. The interaction occurs because the cattle occasionally leave the rancher's property, venture to the farmer's property, and damage the farmer's crops. Table 3.1 contains the marginal and total benefits that the

TABLE 3.2

TOTAL AND MARGINAL COSTS (DAMAGES)
TO THE FARMER FROM THE RANCHER'S CATTLE

NUMBER OF CATTLE	TOTAL DAMAGES	MARGINAL DAMAGES
1	$ 2	$ 2
2	5	3
3	9	4
4	14	5
5	20	6
6	27	7
7	35	8
8	44	9
9	54	10
10	65	11
11	77	12
12	90	13
13	104	14

rancher realizes by having a herd of a given size. These benefits are net of internal ranching costs, such as the cost of cattle feed, veterinary services, and so on. Notice that the numbers in this example imply declining marginal product of cattle, although this is not crucial to Coase's argument.

If the rancher were acting in a vacuum, he or she would choose to have a herd of 10, as that would maximize his or her total benefits from ranching. If there were no externalities, this would also be the socially optimal level of cattle.

However, as mentioned earlier, the cattle ranching generates an externality as the cattle destroy the neighboring farmer's crops. Thus, in order to find the socially optimal level of cattle, we must consider the damages that the cattle generate for the farmer. These damages are presented in Table 3.2. Note that in this example, marginal damages are assumed to be increasing.

The socially optimal level of cattle is the level that maximizes the difference between the rancher's benefits and the farmer's costs. These benefits and costs are compared in Table 3.3.

As can be seen in Table 3.3, the level of cattle that maximizes society's net benefits is a herd size of five. This maximizes the difference between the rancher's benefits ($40) and the farmer's costs ($20) to yield a maximum net social benefit of $20.[3] This optimal level could also have been computed by

[3]In an example such as this, where quantity levels must be integers (e.g. you can't have 5.32 cows) there will often be two quantity levels (4 and 5 in this example) which have the same maximum benefits. However, in a continuous number example, there will only be one maximum level which will occur at the point where marginal benefit equals marginal cost.

TABLE 3.3

DERIVATION OF THE OPTIMAL NUMBER OF CATTLE

NUMBER OF CATTLE	RANCHER'S TOTAL BENEFITS	FARMER'S TOTAL DAMAGES	SOCIAL NET BENEFITS
1	$10	$ 2	$ 8
2	19	5	14
3	27	9	18
4	34	14	20
5	40	20	20
6	45	27	18
7	49	35	14
8	52	44	8
9	54	54	0
10	55	65	-10
11	54	77	-23
12	52	90	-38
13	49	104	-55

looking at the marginal benefit column of Table 3.1 and the marginal damage column of Table 3.2 and observing that the rancher's marginal benefits equals the farmer's marginal damages (both are equal to six dollars) at a herd size of five cattle.

The important question that Coase addresses is the question of whether market forces automatically generate a herd size of five or if a different herd size will arise. An examination of Table 3.1 might cause one to think that a herd size of ten would arise, as this is the level that maximizes the rancher's net benefits. This is a logical conclusion, as one would think that the profit-maximizing rancher would ignore the effects on the farmer. However, Coase says that this is not the case, that market forces will automatically generate a herd size of five cattle.

The rationale behind Coase's argument is that the farmer and the rancher will negotiate a mutually profitable arrangement. Since five cattle is the herd size with the greatest net social benefits, five cattle will maximize their mutual profits. Let us examine how this negotiation process works. First, we will examine the negotiation process assuming that the rancher has the right to let his or her cattle wander and is not responsible for the damages the cattle generate. In the second case, we will assume that the farmer has the right to untrampled crops, and the rancher is responsible for damages. These rights are called property rights since they signify what restrictions society places on the use of private property. The Coase Theorem asserts that it does not matter in whose favor property rights are defined, the optimal level of the externality will still be generated by the model.

If the property rights are defined in favor of the rancher, then the rancher might initially choose his or her seemingly privately optimal herd size of ten. However, the farmer will react to the rancher's choice. The farmer will look at the herd and see that the 10th animal is causing $11 of damage to his or her crops. Since the benefit to the rancher of the 10th animal in the herd is only $1, the farmer can offer the rancher a payment of more than $1 but less than $11 to reduce the herd size by one animal. Such a transaction with a payment greater than $1 but less than $11 would make both the rancher and the farmer better off. When the farmer thought about the ninth animal, a similar conclusion would be reached. The cost of the ninth animal to the farmer is $10 but the benefit to the rancher is only $2. Again, the farmer can pay the rancher to get rid of an animal, and as long as the payment is greater than $2 but less than $10, they both will be better off. This negotiation process will continue until the herd size is equal to five, which is the level at which the farmer's willingness to pay to reduce herd size by an additional animal is exactly equal to the benefits that the rancher receives from an additional animal. The herd size would not be reduced below that level, as the farmer's marginal willingness to pay for a reduction would be smaller than the rancher's foregone benefits.

A similar negotiation process would take place if the farmer had property rights defined in his or her favor. If property rights were so defined, then the farmer could prohibit the rancher from having any cattle. However, the rancher would observe that a herd size of one animal would generate benefits of $10 for the rancher, but costs of only $2 to the farmer. The rancher could then pay the farmer some amount greater than $2 but less than $10 for the right to have one animal, and both the farmer and the rancher would be better off. This process would continue until the herd size increased to the socially optimal level of five animals.

The point behind this exercise of Coase's is to show that a negotiation process will develop, regardless of the direction of the definition of property rights, that leads to the optimal level of cattle. Coase also shows that if some other strategy, such as building a fence, was optimal under one property right regime, it would be optimal under the other property right regime.

Since Coase believes that the market automatically generates the correct level of an externality, he argues against the imposition of interventions such as the externalities tax Pigou suggested. In fact, he argues that a Pigouvian tax is not only unnecessary but counterproductive, as it will encourage people to locate in the vicinity to collect the compensation. Before addressing this issue, let us examine whether Coase's primary assertion (the market will automatically generate the optimal amount of the externality) is correct.

One critical assumption that Coase makes is that transactions costs are insignificant. Transactions costs are those costs that are borne by the victim and the generator of the externality in negotiating an agreed upon level of the externality, with compensation to one party or the other as part of this agreement. In the case of the rancher and the farmer, this is an appropriate

FIGURE 3.2

DIFFICULTY OF COMMUNICATION

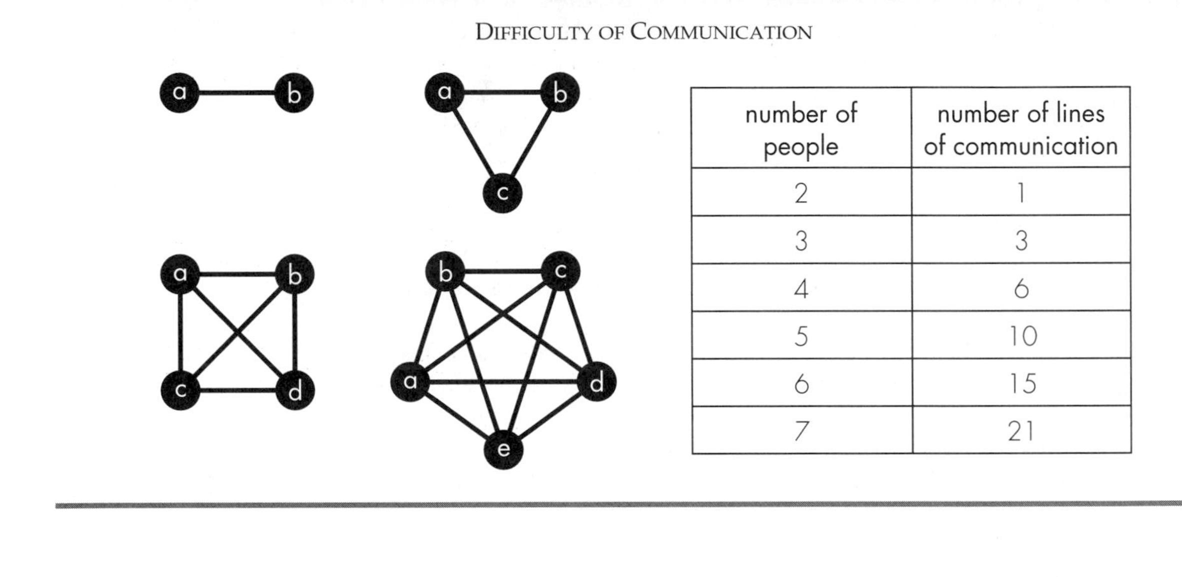

number of people	number of lines of communication
2	1
3	3
4	6
5	10
6	15
7	21

assumption, as the two can get together over a cup of coffee to discuss who is going to make an adjustment and at what level of compensation. However, if the externality in question is sulfur dioxide emissions in North America, there are hundreds of millions of generators of the externality (everybody who burns a hydrocarbon-based fuel) and hundreds of millions of victims of the pollution. In addition, the pollution generated in one part of the continent affects environmental quality in other parts of the continent, with significant amounts of pollution migrating across the U.S.–Canadian border. Under these conditions, transactions costs will be significant.

The polar examples of the rancher–farmer interaction and the sulfur dioxide pollution problem illustrate the relevance of the number of participants to the importance of transactions costs. It is important to note that not only are transactions costs likely to be positively related to the number of participants, but they are likely to be increasing at an increasing rate. This is illustrated in Figure 3.2, where the schematic drawings show the number of participants and the lines of communication among the participants. As shown in the accompanying chart, as the number of participants increases by a constant of one, the number of lines of communication among the participants increases according to the progression {1, 3, 6, 10, 15, 21, ...}.

One way to reduce transactions costs is to appoint an agent who acts in behalf of a large number of people. For instance, the Sierra Club acts as an agent for many thousands of its members in making their positions known (or potentially negotiating in their behalf) on environmental externalities such as sulfur dioxide pollution. However, not everybody who desires lower sulfur dioxide levels contributes to Sierra Club for this goal. This phenomenon may be due to what economists term the *free rider problem*. The free rider problem occurs in the case of a public good, where *an individual may decline to share in the costs* knowing that he or she *cannot be precluded from sharing the benefits*. If the Sierra Club is successful in reducing sulfur dioxide emissions, all people who prefer cleaner air benefit, not just the dues-paying members of the Sierra Club. A similar problem can occur in terms of taking collective action in a Coasian negotiation framework.

Although recognizing the relevance of transactions costs, in the 1969s and early 1970s, many economists still argued the general applicability of the Coase Theorem, as most of the externalities that economists discussed were of the one generator/one victim format of Coase's example. However, Baumol and Oates derive a result which shows that the Coase Theorem breaks down even if zero transactions costs are assumed. Their result is that optimality (maximizing social welfare) requires that the generator of a negative externality be charged a price, while the victim remains uncompensated. Since optimality requires a different price for the victim and the generator (asymmetric prices), an efficient market cannot develop, as a market requires the same price for the buyer and the seller.

The intuition behind the need for this asymmetric pricing is that since the victim does not control the amount of the externality to which he or she is exposed, an externality price is superfluous to that individual's decision-making process. The inability of an efficient market to develop is a critically important flaw in the Coase Theorem. This flaw exists regardless of the presence or absence of transactions costs.

One of the problems with the Coase example, (and the same argument can be applied to the writings of many of Coase's contemporaries) is that if the example only has one victim of the externality, then the externality may quickly cease to be an externality. The reason for this is that in the one-versus-one case, the total amount of the externality is identical to the amount to which the victim is exposed, and the negotiation process therefore allows the victim control over how much of the externality to which he or she is exposed. It is very tenuous logic to develop an example for a one-versus-one situation and then generalize to large numbers.

Another problem with the generalization from the one generator/one victim example to the large-numbers case is that focusing on one generator also misses an important dimension of the problem. This is because focusing on this one-on-one interaction masks the importance of entry and exit. The entry and exit of firms into or out of the market will affect both the number of generators and the number of victims. If the ranchers have the right to let their cattle roam without worrying about the damages to farmers, then there will be more ranchers than in a society where they are responsible for their damages. An analogous statement can be made about the number of farmers. If this is the case, then in a multiple generator/multiple victim world, the definition of property rights will have a critically important impact on the outcome, as it will have an impact on the relative numbers of generators and victims.

Finally, property rights might matter because there may be important differences between the victim's willingness to pay for reducing the detrimental externality and the victim's willingness to accept compensation to allow increases in the level of the externality. Part of the difference may be due to the existence of *income effects*.

The existence of income effects would imply that the *higher an individual's standard of living, the greater his or her marginal valuation of a normal good or service*. Since being exposed to detrimental externalities such as pollution lowers an individual's standard of living, and being free from a detrimental externality raises an individual's standard of living, the marginal valuation of the damages of an externality depends on the definition of property rights. If the value of the externality depends on the definition of property rights, then the way property rights are defined must have an effect on the negotiated level of the externality.

In summary, the Coase Theorem suggests that there is no need for government intervention to correct market failures due to externalities. However, the Coase Theorem suffers from four important flaws:

1. The Coase Theorem assumes zero or insignificant transactions costs, an assumption inappropriate for environmental externalities, because it is likely that there are large numbers of generators and victims.
2. Even if one assumes zero transactions costs, the Theorem cannot generate an efficient market, since efficiency requires asymmetric prices.
3. The definition of property rights will have an effect on the outcome since this definition affects the number of potential participants in the market for the output and the market for the externality associated with the output.
4. The definition of property rights may be important due to the existence of income effects, which may affect the marginal value of the externality.

RESOLUTION OF THE ISSUE OF GOVERNMENT INTERVENTION

In Chapter 2, it was shown that market failures, such as environmental externalities, can lead to losses in social welfare. In the economic literature, many economists, such as A. C. Pigou, suggest that government intervention to correct market failure would lead to improvements in social welfare. Other economists, most notably Ronald Coase, argue that government intervention is unnecessary, because a market for the externality would develop that would lead to the socially optimal level of the externality being chosen. However, it was shown in the previous section that under most circumstances, and certainly under the circumstances surrounding most environmental externalities, the Coase Theorem is not applicable, and consideration must be given to the option of government intervention. Of course, the costs of government intervention must be compared to the benefits of government intervention before deciding that government intervention is desirable. This comparison cannot be done for externalities in general, but must be done on a case-by-case basis for the externalities of greatest concern. Also, it should be recognized that the costs and benefits of government intervention are a function of the type of intervention. For example, one way to control the externalities associated with littering is to execute all people who are found littering. However, most people would regard such a policy as having costs far in excess of its benefits.

TYPES OF GOVERNMENT INTERVENTION

There are five broad classes of government intervention to correct market failures associated with externalities. These are:

1. moral suasion
2. direct production of environmental quality
3. pollution prevention
4. command and control regulations
5. economic incentives

Each of these classes of intervention or policy instruments represents a different philosophy toward the role of government in society, generates different behavioral incentives for those who are generating externalities, and leads to different levels of costs and benefits. These different instruments should not be viewed as mutually exclusive methods, as an appropriate environmental policy may have elements of all five classes of policy instruments.

MORAL SUASION

Moral suasion is a term that is used to describe the governmental attempts to influence behavior without actually stipulating any rules that constrain behavior. Governmental leaders, usually the chief executive of the jurisdiction, make public announcements that detail the way in which they would like people to behave. In order for this to be effective, the statements of the government officials must convince the public that the benefits of behaving in the desired fashion are substantially greater than the costs. Such federal government programs as Woodsy Owl's "Give a hoot, don't pollute" and

Smokey Bear's "Only you can prevent forest fires"[4] are examples of relatively successful moral suasion programs aimed at environmental problems. Currently, many levels of government as well as corporate and not-for-profit organizations are engaged in moral suasion programs aimed at reducing the volume of waste and increasing recycling.

As mentioned above, the effectiveness of moral suasion programs depends upon the extent to which the people (household, firm or organization members) who are being asked to change their behavior believe that it is in their individual and collective interests to do so. A change in their behavior may be in their individual or collective interests because of the relative magnitudes of the direct costs and benefits of the program. This change may also be in their private interests if the people believe that if they do not voluntarily adopt the behavior advocated by the moral suasion program, more severe restrictions on their behavior will be adopted as permanent law.

Moral suasion can be an effective method for generating environmental improvement, as the success of voluntary recycling programs indicates; however, it may not be practical in many circumstances. In particular, the free rider problem (defined on page 47) can inhibit the effectiveness of moral suasion.

DIRECT PRODUCTION OF ENVIRONMENTAL QUALITY

The direct production of environmental quality is another way in which the government can mitigate environmental market failures. At first, it might seem unrealistic to say that government programs could undo environmental degradation, but planting trees, stocking fish, creating wetlands, treating sewage, and cleaning up toxic sites are all examples of this type of activity. As one might suspect, government production of environmental quality is largely an ameliorative action, and in many cases it would have been better for society if the environmental degradation had been prevented in the first place.

Although both moral suasion and direct production of environmental quality are important features of the policy arsenal, they have limited applicability to environmental problems in general. In particular, it is unlikely that such pressing problems as air pollution, global warming, water pollution, and the depletion of the ozone layer are likely to be adequately addressed by either of these policies. As a consequence, we must consider more stringent types of government intervention.

POLLUTION PREVENTION

Pollution prevention programs are not designed to control the externality itself, but to address a related market failure of imperfect information. These

---

[4]Although the Smokey the Bear program has been effective in reducing accidental forest fires, many forest ecologists believe that fire is a natural element in the forest, and that fire reduction programs are counter-productive in the long run.

programs are partnerships of business and government agencies designed to increase the profitability of reducing pollution by developing technologies that are both more profitable and cleaner. The basic premise underlying pollution prevention programs is that the knowledge to develop these cleaner and more profitable technologies is beyond the capability of an individual firm to develop. However, the combined efforts of government agencies, national laboratories, universities and private firms can lead to the development of these innovative and beneficial technologies. These programs emphasize being proactive in reducing pollution, rather than waiting to react to new regulations. A proactive policy can lead to lower costs of abatement in the long run.

Command and Control Regulations

Command and control regulations are a class of policy instruments that have greater ability to modify environmentally degrading behavior. In many textbooks and articles, they are also referred to as direct controls. Command and control regulations are distinguished from other policy instruments, as they place constraints on the behavior of households and firms (and any other generators of externalities). If behavior remains within these boundaries, then the household or firm is behaving lawfully. However, if behavior violates these boundaries, then the firm or household is behaving illegally and suffer penalties specified by the rule or law that established the direct control. These constraints generally take the form of limits on inputs or outputs to the consumption or production process. Examples of command and control restrictions that constitute restrictions on inputs would include requiring sulfur-removing scrubbers on the smokestacks of coal-burning utilities, requiring catalytic converters on automobiles, and banning the use of leaded gasoline. Command and control regulations that take the form of restrictions on outputs include emissions limitations on the exhaust of automobiles, prohibitions against the dumping of toxic substances, and prohibitions against littering.

Economic Incentives

Economic incentives are based on a different philosophy than command and control regulations. Rather than defining certain behaviors as legal or illegal and specifying penalties for engaging in illegal behavior, economic incentives simply make individual self interest coincide with the social interest. Examples of economic incentives include pollution taxes, pollution subsidies, marketable pollution permits, deposit–refund systems, and bonding and liability systems.

CHOOSING THE CORRECT LEVEL OF ENVIRONMENTAL QUALITY

Whether one employs command and control techniques or economic incentives, a crucial issue involves the determination of the desirable level of pollution or environmental degradation. At first, this might seem to be an illogical question, as the desirable level of something that is bad (like pollution) should be zero. However, some reflection will reveal that this is not likely to be true because the reduction of pollution will have opportunity costs. In actuality, a zero level of pollution is impossible to achieve due to a principle of physics known as the law of mass balance.

The law of mass balance is based on the proposition that an activity cannot destroy the matter in the reaction; it can only change its form.[5] The law of mass balance states that the mass of the outputs of any activity are equal to the mass of the inputs. For example, if 10 pounds of wood are burned in a fireplace, 10 pounds of matter are not destroyed. The 10 pounds of matter still exist, although the form of the matter may have been altered to smoke or ash. The law of mass balance directly indicates that any production or consumption activity will be associated with waste. Eliminating all pollution means eliminating all production and consumption activities, as all consumption and production activities must produce waste. A society whose sole economic activity consisted of organic agriculture would still be a polluting society.

The above discussion suggests that some pollution is inevitable and that zero pollution is neither desirable nor achievable. The discussion in Chapter 2 suggested that the unregulated market level of pollution is likely to be excessive, since the market failure implies that social welfare is lower than it could be with less of the externality. If zero pollution is likely to be too costly or unattainable, and the unregulated level is excessive, then some level of pollution between zero and the unregulated market level is desirable.

The desired level of pollution will be a function of the social costs that are associated with the pollution. There are two categories of social costs associated with pollution. The first of these is the damage that pollution creates by degrading the physical, natural, and social environment. These damages would include effects on flora and fauna, human health effects, damage to human-made structures, and aesthetic effects. The second type of cost is the cost of reducing pollution and includes the opportunity costs of the resources used to reduce pollution and the value of any foregone outputs.

THE MARGINAL DAMAGE FUNCTION

The damages that pollution generates by degrading the environment are modeled in the marginal damage function of Figure 3.3. The emissions of pollution are specified on the horizontal axis, and dollar measures of the damages pollution generates are specified on the vertical axis. Many people question the ability to derive dollar measures of the damages from pollution

---

[5]Of course, fusion and fission reactions are exceptions to this, where the mass of the materials that are inputs to the reaction is different than the mass of the materials after the reaction. The difference in mass becomes converted to energy.

FIGURE 3.3

THE MARGINAL DAMAGE FUNCTION

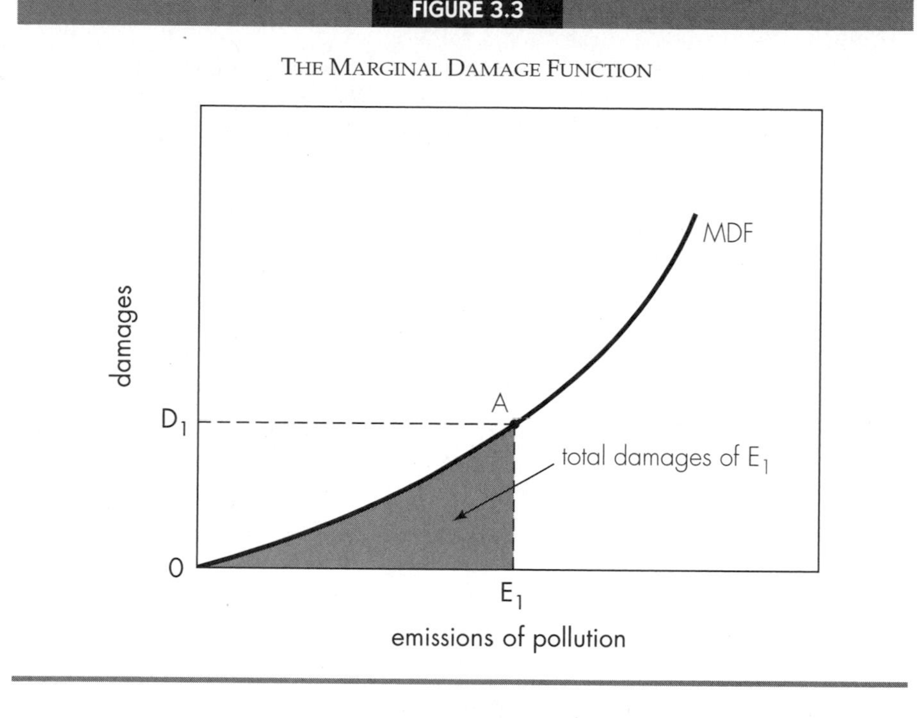

(such as human health or ecosystem effects). However, the investigation of this issue will be postponed, and for now it will be assumed that these effects are quantifiable. The issues associated with the measurement of the value of environmental resources and the damages from pollution will be examined in Chapter 4.

It should be noted that the damage function presented in Figure 3.3 is a marginal damage function which specifies the damages associated with an additional unit of pollution. For example, if the level of emissions is equal to E_1, then the damage of an additional unit of emissions is equal to D_1. The total damages generated by a particular level of pollution can be examined by looking at the area under the marginal damage function. For example, the total damages associated with E_1 units of pollution are equal to the shaded area OAE_1. Several other characteristics of the marginal damage function in Figure 3.3 warrant discussion. These characteristics concern the slope and the intercept of the marginal damage function. The marginal damage function in Figure 3.3 has a slope that increases at an increasing rate. This means that as the level of pollution becomes larger, the damages associated with the marginal unit of pollution becomes larger, and that the rate at which they become larger is increasing. While there has not been enough study of marginal damage functions to conclude that this is true for pollution in general,

ALTERNATIVE SHAPED MARGINAL DAMAGE FUNCTION:
INCREASING AND THEN DECREASING

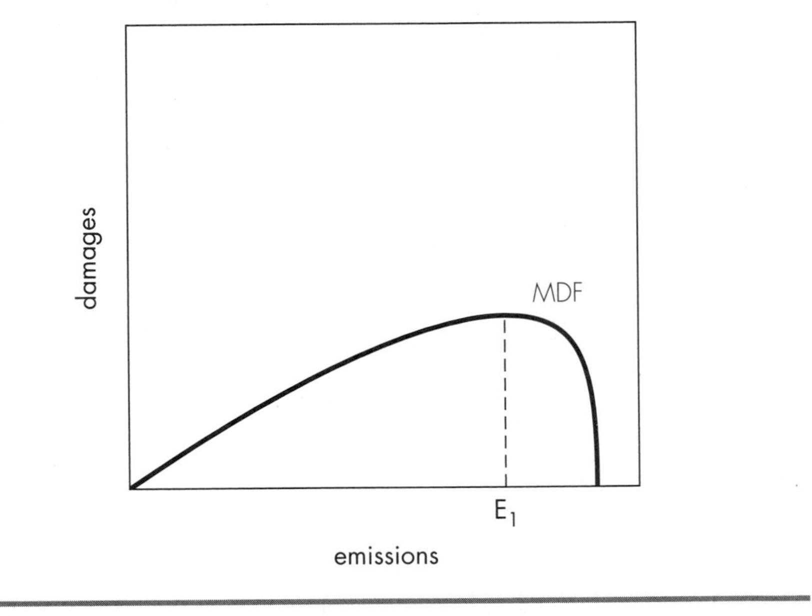

the author of this book believes this to be likely for many types of pollution, particularly those that have human health effects or those that have global effects.

However, it is possible for particular marginal damage functions to increase at a decreasing rate (as in Figure 3.4a) and even to have a negatively sloped range if the initial levels of pollution cause so much damage and environmental degradation that additional amounts of pollution can have little or no additional negative impact on the resource. An example of this may occur when pollution to a lake is so severe that all the life in the lake dies, and the lake is too foul to swim in, use as a source of drinking water, or even be in proximity of. If this is the case, additional units of pollution would have diminishing marginal damages, as the quality of the lake is already so degraded that further pollution has limited impact.

The marginal damage function in Figure 3.3 is drawn with the origin as its vertical intercept. This means that marginal damages approach zero as emissions approach zero. This assumption is probably good for all but the most toxic and long-lived pollutants, or pollutants for which small emissions could have a strong localized impact. An example of the second situation would occur if small levels of pollution were emitted into a small pond and

FIGURE 3.4B

ALTERNATIVE SHAPED MARGINAL DAMAGE FUNCTION: POSITIVE INTERCEPT

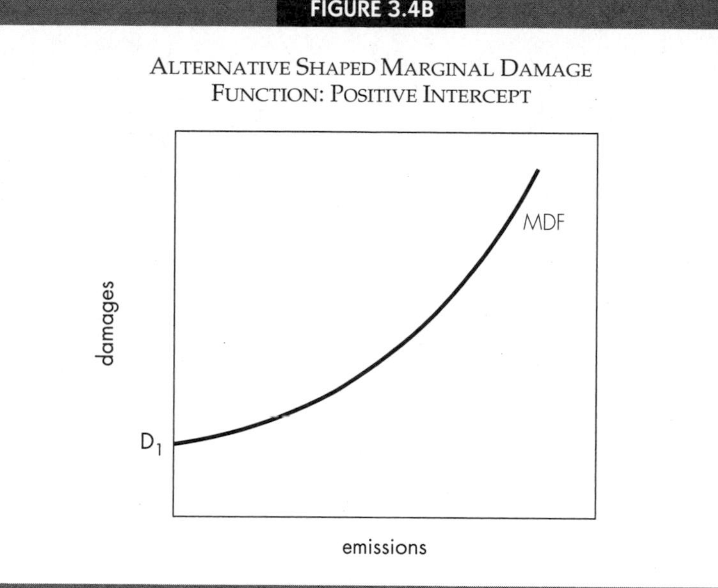

caused the demise of a species of fish that was found only in that pond. A marginal damage function of this nature is illustrated in Figure 3.4b, where the marginal damages of the first unit of pollution is equal to D_1.

Occasionally, marginal damage functions will be presented as step functions, as in Figure 3.4c. The intuition behind a function of this shape is that marginal damages remain constant for a range of values of emissions and then become larger and remain constant over another range of values. For example, in Figure 3.4c marginal damages are equal to zero when emissions are between 0 and E_1, equal to D_1 when emissions are between E_1 and E_2, equal to D_2 when emissions are between E_2 and E_3, and equal to D_3 when emissions are greater than E_3. This shape implies that there are thresholds at which new impacts occur.

Although some people believe that the step function is a convenient way to conceptualize the processes that generate the damages, such thresholds are unlikely to occur. For instance, thresholds are often postulated and reflected in standards that regulate people's exposure to pollutants. For example, the levels of PCB (a toxic substance) in striped bass at which the striped bass is still deemed safe for human consumption is 2 parts per million. 2.01 parts per million is legally defined as unsafe. Yet even if 2.01 parts per million but not 2 parts per million generated a cancer risk in the typical person, 2 parts per million might generate a cancer risk for individuals who were more sensitive to PCB or who had a greater genetic tendency to contract cancer. Thus, even if a pollutant was associated with thresholds for a particular type

FIGURE 3.4C

A MARGINAL DAMAGE FUNCTION WITH THRESHOLD EFFECTS

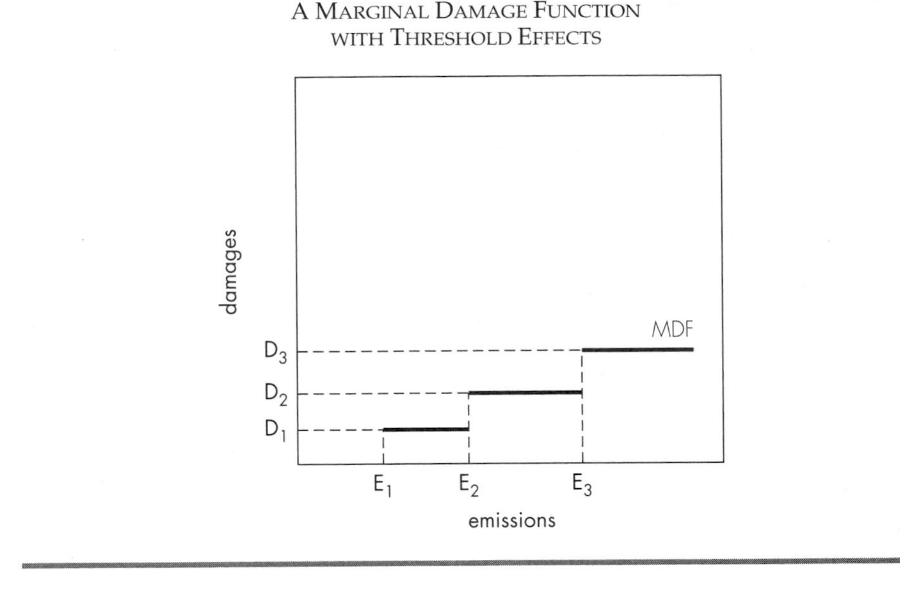

of individual, the broad range of individuals in society (different ages, overall levels of health, exposure to other carcinogens, and so on) tends to suggest that the relationship between pollution and human health risks should be continuous, rather than discrete. Since similar arguments can be made with respect to flora and fauna, it is unlikely that marginal damage functions are step functions, such as the function depicted in Figure 3.4c.

Although thresholds such as those depicted in Figure 3.4c are unlikely to occur, thresholds may be very important when measuring the relationship between emissions and damages. Typically, we conceptualize the damage function as marginal increases in emissions which yield marginal increases in damages. However, there may be a case where thresholds exist because of positive feedback effects. This case occurs when an increase in emissions not only increases damages, but also causes a large and fundamental change in the environment. For example, carbon dioxide emissions increase global average temperature through the greenhouse effect (see Chapter 6 for a complete discussion of global warming). Small increases in carbon dioxide concentrations in the atmosphere lead to small changes in average global temperature. However, many scientists believe that global warming gradually will increase to the point where a small increase in carbon dioxide emissions will lead to a huge increase in mean global temperature. The reason for this is that the small initial amount of warming associated with the increase in emissions will cause the temperature to rise to the point where the permafrost in the arctic regions melts. This melting will lead to rapid anoxic decay of previously frozen organic material, releasing large quantities of methane, which itself has a strong global warming effect. Thus the small initial increase in carbon dioxide emissions can lead to a snowballing feedback effect which would dramatically and perhaps irreversibly increase mean global temperature.

THE MARGINAL ABATEMENT COST FUNCTION

As mentioned above, damages are only part of the social costs associated with pollution. The other cost is the cost of abating (reducing) pollution to a lower level so that there are fewer damages.[6] Such costs include the costs of the labor, capital, and energy needed to lessen the emissions of pollution associated with particular levels of production or consumption; or the costs may take the form of opportunity costs from reducing the levels of production or consumption.

Figure 3.5 contains a marginal abatement cost function, where E_u represents the level of pollution that would be generated in the absence of any government intervention. At this level of emissions, the marginal cost of abatement is zero, as polluters have not yet taken any steps to reduce the level of pollution below the level they each privately regard as optimal. As pollution is reduced below E_u, the marginal cost of abatement (the cost of reducing pollution by one more unit) increases, as the cheapest options for reducing pollution are the first to be employed. As the cheaper alternatives are exhausted, more expensive steps must be taken to further reduce pollution, so marginal abatement costs rise as the level of emissions moves from E_u to the left. At E_1, the marginal abatement cost is equal to C_1, and the total abatement cost is equal to the shaded area under the marginal abatement cost function between E_u and E_1.

As was the case with the marginal damage function, it is important to discuss the nature of the slope and intercept of the marginal abatement cost function. The marginal abatement cost function in Figure 3.5 has a slope that decreases at a decreasing rate. This means that as one abates emissions from a starting point of E_u (as one moves in the direction of zero pollution), the costs of further reducing pollution increases at an increasing rate. This is consistent with an extremely high vertical intercept or the case where the marginal abatement cost function approaches an asymptote as in Figure 3.5. The extreme magnitude of the intercept implies that the cost of eliminating the last few units of pollutants is extremely high. This would most likely be

---

[6]An additional cost is the cost of avoiding pollution damages. For example, some of the damages from air pollution can be avoided by filtering air as it passes though air conditioning and heating systems. The marginal damage function presented here can be viewed to be incorporating this type of avoidance cost.

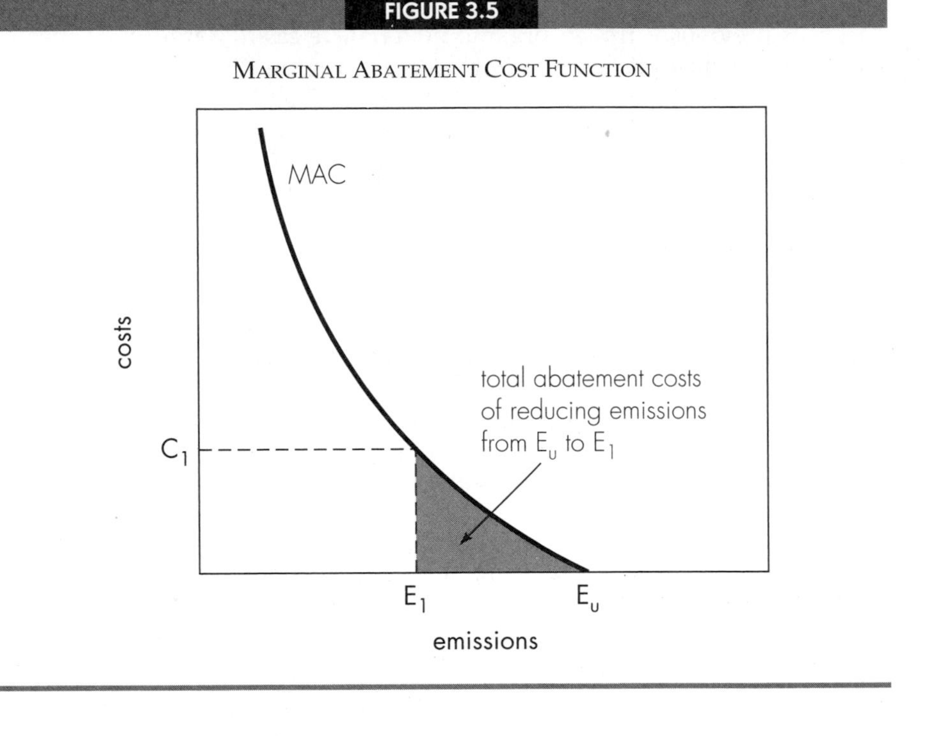

FIGURE 3.5

MARGINAL ABATEMENT COST FUNCTION

the case for pollutants, such as sulfur dioxide and nitrogen oxides, that result from the burning of fossil fuels, such as coal, oil, and gasoline.

On the other hand, there are many pollutants for which the slope and intercept would have the shape of the marginal abatement cost function in Figure 3.6. This function is decreasing at an increasing rate, implying that the costs of reducing pollution increase at a decreasing rate as one abates from the unregulated level of emissions (E_u). The marginal abatement cost function of Figure 3.6 also has a relatively low intercept, indicating that the cost of eliminating the last few units is not that much higher than the cost of eliminating the first few units. The release of lead emissions into the air through automobile exhaust is a good example of this type of pollution.[7] Lead was added to gasoline to reduce refining costs, as the presence of lead allows gasoline to be made with fewer of its higher octane components. Simply stated, removing lead means more of other inputs need to be used in producing gasoline. The cost of not adding lead to the last few gallons of

[7]Lead is a heavy metal that is extremely toxic (especially to children), which was extensively used as an additive in gasoline and emitted in automobile exhaust. Although small amounts of lead are still added to gasoline, recent regulations have reduced the use of lead in gasoline to a trace of past levels.

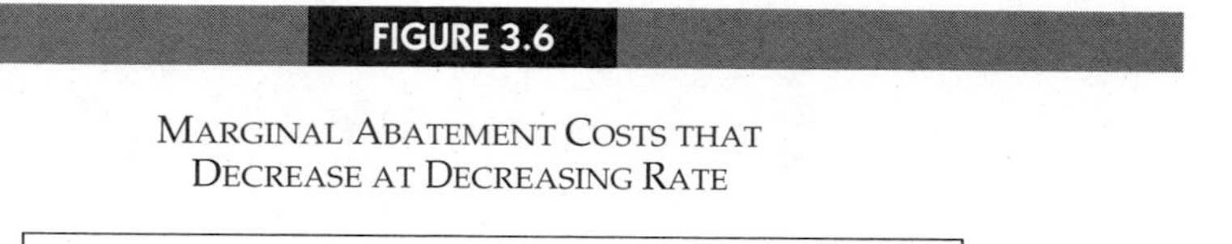

MARGINAL ABATEMENT COSTS THAT
DECREASE AT DECREASING RATE

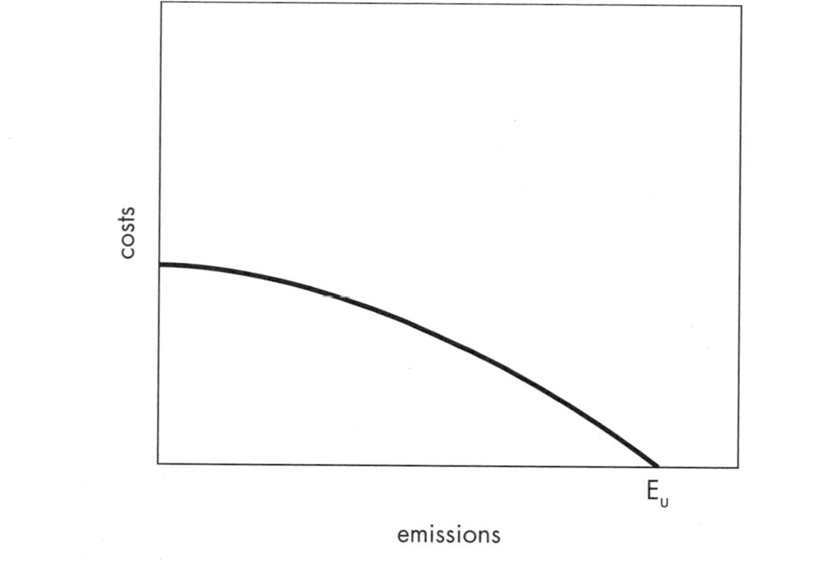

gasoline is not likely to be significantly larger than the cost of not adding lead to the first few units of gasoline. In fact, these costs would even be decreasing if economies of scale could be realized in refining.

MARGINAL DAMAGES, MARGINAL ABATEMENT COSTS, AND THE OPTIMAL LEVEL OF POLLUTION

Now that the two costs of pollution (abatement costs and damages) are fully understood, it is possible to determine the optimal level of pollution emissions. The optimal level of emissions is the level that minimizes the total social costs of pollution, which is the sum of total abatement cost and total damages. This level occurs at the point where marginal abatement costs are equal to marginal damages, which is E_1 in Figure 3.7.

It is relatively easy to demonstrate that equality of marginal abatement cost and marginal damages is the condition for determining the optimal level of emissions. If the level of emissions is less than E_1 (to its left), then marginal abatement costs are greater than marginal damages. This means that the cost of eliminating the marginal unit of pollution is greater than the damages the unit of pollution would have caused. Hence, it would be

FIGURE 3.7

THE OPTIMAL LEVEL OF POLLUTION

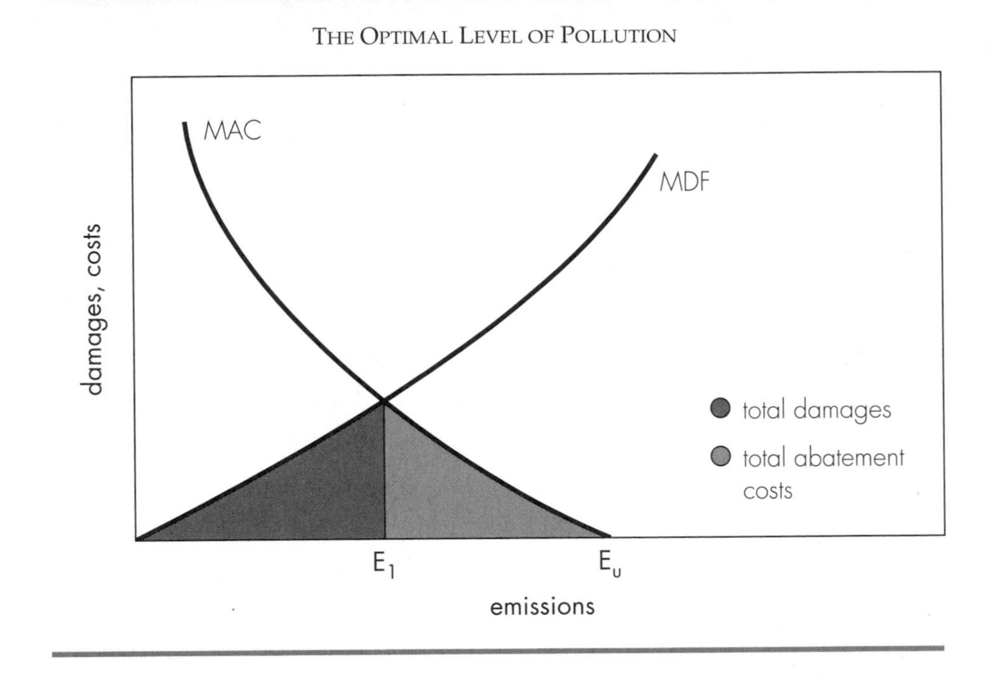

beneficial to have more pollution. This statement would be true as long as the level of pollution is less than E_1. On the other hand, if the level of emissions is greater than E_1, then marginal damages are greater than marginal abatement costs. Since the damages caused by the marginal unit of pollution are greater than the costs of eliminating that unit of pollution, then society is better off eliminating it. This will be true as long as the level of emissions is greater than E_1.

At E_1, where marginal abatement costs are equal to marginal damages, the total social costs of pollution cannot be lowered by changing the level of emissions. This minimized total social cost is shown as the shaded triangular area in Figure 3.7. Figures 3.8a and 3.8b more graphically illustrate the excess social costs associated with a level of pollution different from E_1. In Figure 3.8a, the actual level of emissions (E_2) is above the optimal level (E_1). In the range of emissions between E_1 and E_2, marginal damages are greater than marginal abatement costs, generating excess social costs of the area abc. Similarly, in Figure 3.8b, the actual level of pollution (E_3) is lower than the optimal level (E_1), leading to excess social costs of area ade. The optimal level of pollution need not be static but may change over time. For example, if an engineer developed a new device for cars that enabled them to get substan-

FIGURE 3.8A

SOCIAL COSTS WHEN POLLUTION IS GREATER THAN OPTIMAL

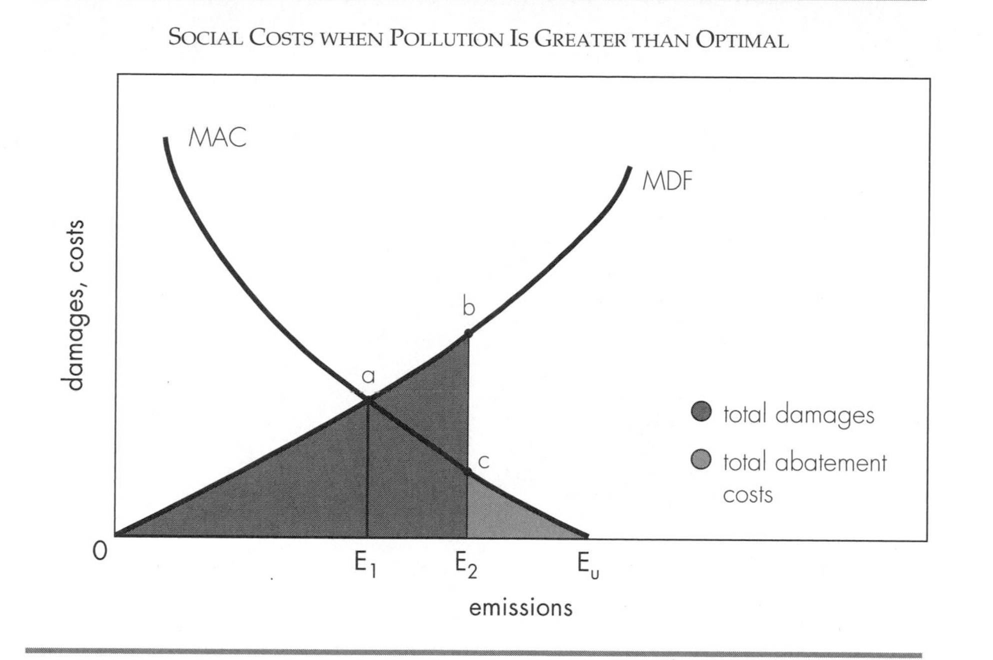

tially better gas mileage (which would reduce the pollution per mile), then the marginal abatement cost function would shift downward, as in Figure 3.9. The optimal level of pollution would decrease from E_1 to E_2, and the unregulated level of pollution would decrease from E_u' to E_u''. It may appear strange that the unregulated, or market level, of pollution would decline, but many people would desire the new device on their cars—independent of environmental regulations—as better gas mileage reduces their operating expenses.

It should be noted that knowledge of the marginal abatement cost function and marginal damage function is not always available to policy makers. If this is the case, then a target level of pollution is chosen that usually corresponds to some sort of legislative goal, such as a level of air pollution that is consistent with "protecting the public health" or a level of water pollution that is consistent with "fishable/swimmable waters." The implications of imperfect information for the development of environmental policy will be discussed later in this chapter and in many of the applications chapters, such as the chapters on Global Environmental Change (Chapter 6), Acid Deposition (Chapter 7), and Fossil Fuels (Chapter 8).

FIGURE 3.8B

SOCIAL COSTS WHEN POLLUTION IS LESS THAN OPTIMAL

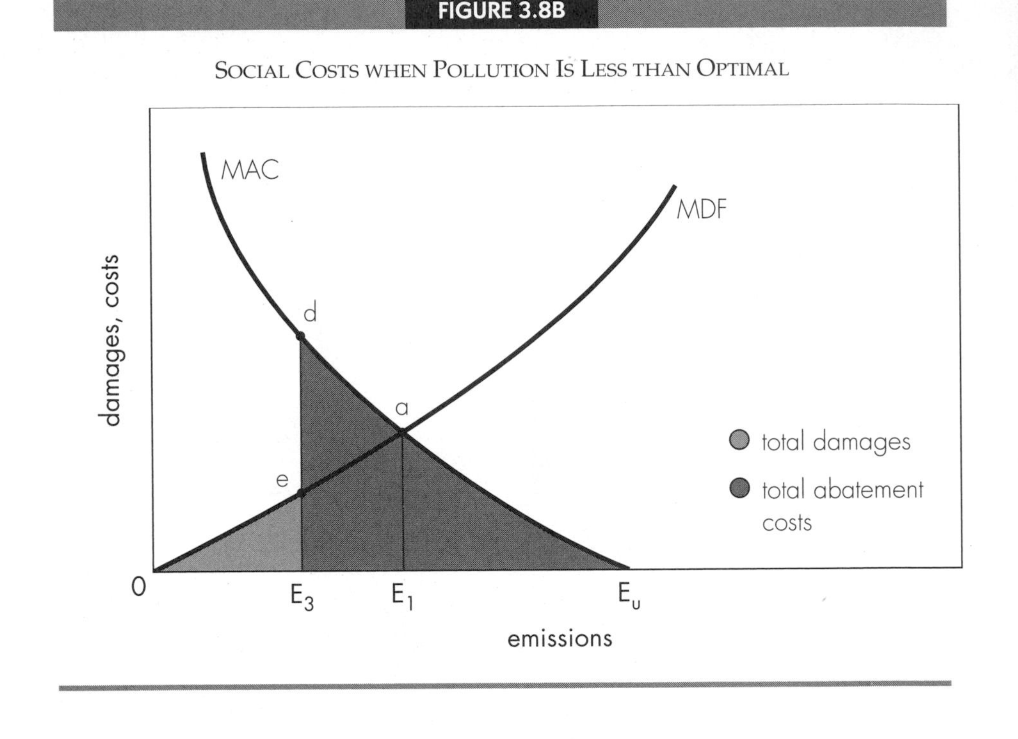

PURSUING ENVIRONMENTAL QUALITY WITH COMMAND AND CONTROL POLICIES

One potential way of achieving the optimal level of pollution is through use of command and control polices. If the optimal level of pollution is known to be E_1, then E_1 units of pollution could be allocated among all the polluters. For example, if the optimal level of pollution was equal to half of the unregulated level, then a regulation could be imposed that would require each polluter to reduce his or her level of pollution by 50 percent.

COMMAND AND CONTROL POLICIES AND EXCESS ABATEMENT COST

Command and control regulations have been criticized as generating more abatement costs than necessary to achieve a given level of emissions. This point can be best illustrated by assuming that there are only two polluters in society and then looking at the marginal abatement cost functions of the individual polluters. Figure 3.10 contains the marginal abatement costs for two polluters and for society as a whole. The aggregate marginal abatement

FIGURE 3.9

TECHNOLOGICAL INNOVATION THAT LOWERS ABATEMENT COSTS

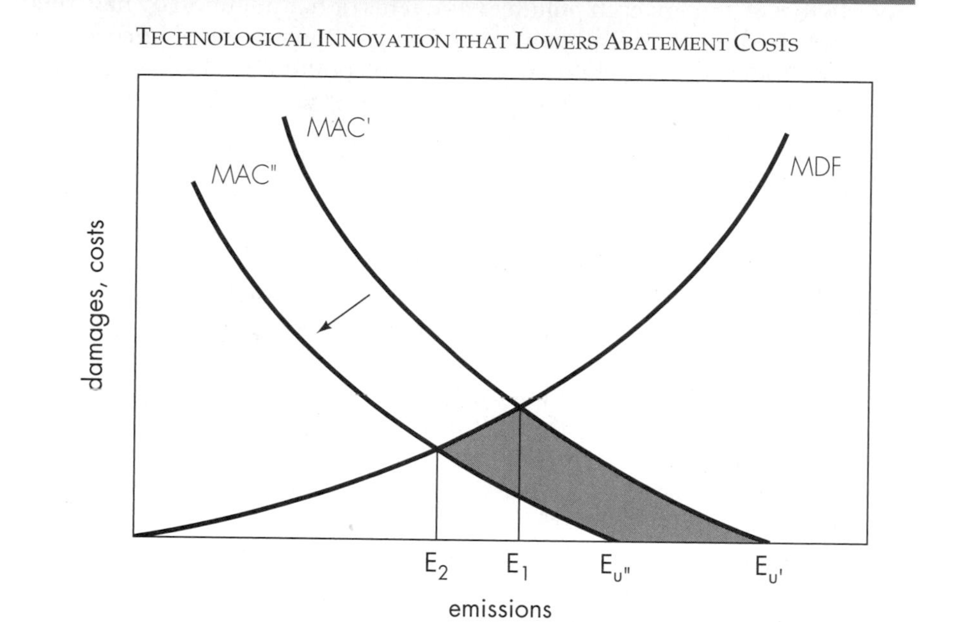

cost function (societal marginal abatement cost function) is derived by horizontally summing the marginal abatement cost functions for all polluters. In this example, there are just two polluters, polluter 1 and polluter 2. For example, if there are no environmental regulations, then polluter 1 will emit 8 units of pollution, and polluter 2 will emit 6 units. The total for society, therefore, will be 14 units, which becomes the horizontal intercept for the aggregate marginal abatement cost function. Other points on the aggregate marginal abatement cost function may be derived in similar fashion. For example, if each polluter reduced his or her level of pollution to the point where marginal abatement costs were $2, then polluter 1 would be emitting 4 units, and polluter 2 would be emitting 4 units, so when marginal abatement costs are equal to $2, society is emitting a total of 8 units. Notice the kink that occurs in the aggregate abatement cost function at a marginal abatement cost of $4, since polluter 1 has eliminated all emissions at this level of marginal abatement costs.

The effects of a command and control regulation that requires all polluters to reduce pollution by 50 percent can now be analyzed with the aid of a similar graph, which is contained in Figure 3.11. In this example the unregulated level of pollution is 10 units for polluter 1 and 6 units for polluter 2. If pollution is cut by 50 percent for each polluter, then polluter 1's pollution declines from 10 to 5 units, polluter 2's from 6 to 3, and society as a whole's from 16 to 8. Although each polluter has cut his or her pollution by half, their marginal abatement costs are not equal. Polluter 2 incurs a higher cost (than polluter 1) for reducing the marginal unit of pollution. This represents a misallocation of resources from society's point of view, as the total costs of obtaining any particular level of emissions will occur when the marginal abatement costs are equal across all polluters.

This point is illustrated in Figure 3.12, which shows how society's total abatement costs can be lowered by keeping the total emissions constant, but changing the allocation of emissions across the two polluters so that one polluter pollutes more, the other pollutes less, and the total remains constant. Since polluter 2 has higher marginal abatement costs, polluter 2 will be allowed to emit more, and polluter 1 will be required to pollute less. In this case, polluter 1 reduces pollution by one half unit (to 4 1/2) and polluter 2 increases pollution by one half unit (to 3 1/2).

The heavily shaded areas in Figure 3.12 show the changes in each polluter's abatement costs, which arise from the marginal reallocation of one-half unit of pollution. Polluter 1's abatement costs increase as a result of the reallocation, while polluter 2's abatement costs decrease. Since the decrease in polluter 2's abatement costs is greater than the increase for polluter 1's, the reallocation reduces costs for society as a whole. Only when the marginal abatement costs are equal for each polluter will there be no possible cost-saving reallocations of emissions. The principle that total abatement costs are minimized when marginal abatement costs are equalized across polluters is fundamental to understanding the differences among pollution control policies.

It has been shown that command and control regulations that have the property of assigning pollution levels to different polluters are not likely to result in achieving the minimum abatement costs by obtaining the target levels of pollution, unless, of course, the assignment was made in such a fashion as to equate these marginal abatement costs across all polluters. This assignment is only feasible under two conditions. The first condition is the unlikely condition that all polluters face the same marginal abatement cost function. Then, equalizing emissions across polluters (or requiring each polluter to reduce emissions by the same percentage amount) will equalize marginal abatement costs. However, since many different types of producers and consumers emit the same pollutants, it is highly unlikely that all polluters will have the same marginal cost functions. For example, steel factories, electric utilities, and automobile owners all emit sulfur dioxide pollution, yet they have very different production or consumption technologies. The second condition is if the regulating authority knew the marginal abatement cost function for each polluter, then it could choose an allocation across polluters in which the level of marginal abatement costs were equal for all polluters. However, the same technological diversity that makes the first condition unlikely makes the second condition prohibitively expensive.

FIGURE 3.10

AGGREGATING MARGINAL ABATEMENT COST FUNCTIONS

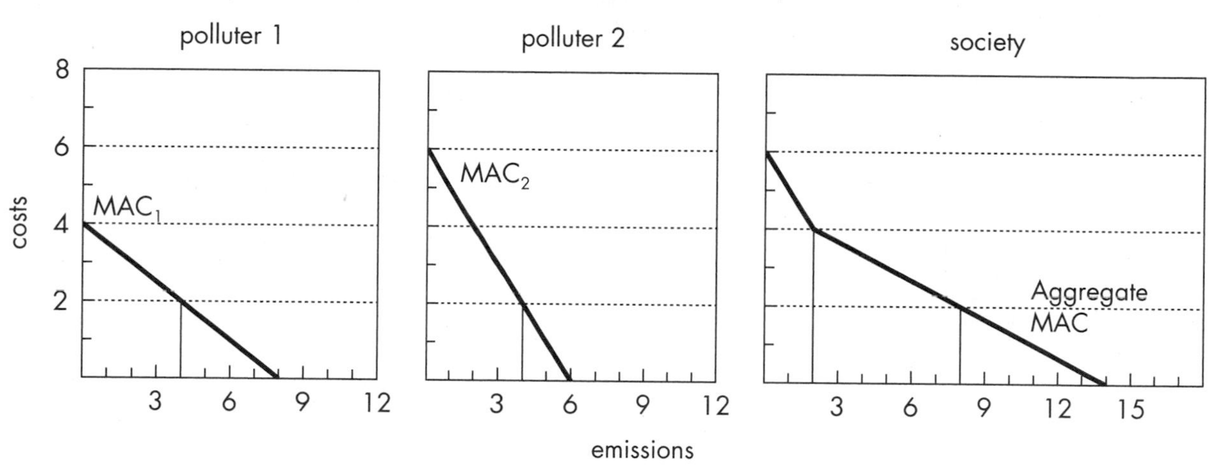

FIGURE 3.11

REDUCTION IN POLLUTION THROUGH EQUAL PERCENTAGE REDUCTION BY EACH POLLUTER

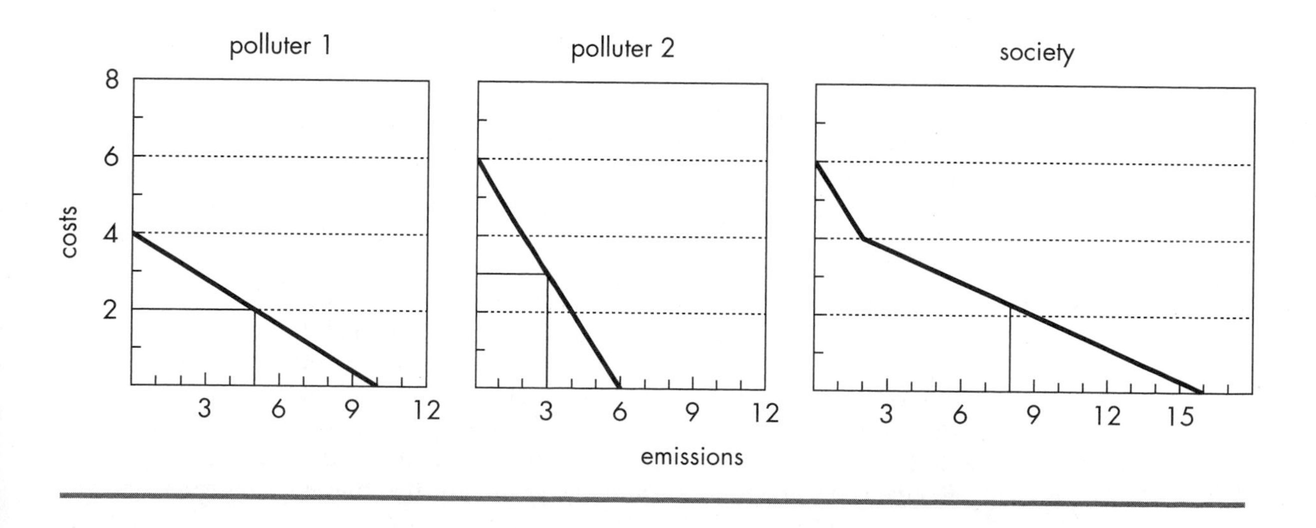

FIGURE 3.12

MINIMIZING TAC BY EQUATING MAC

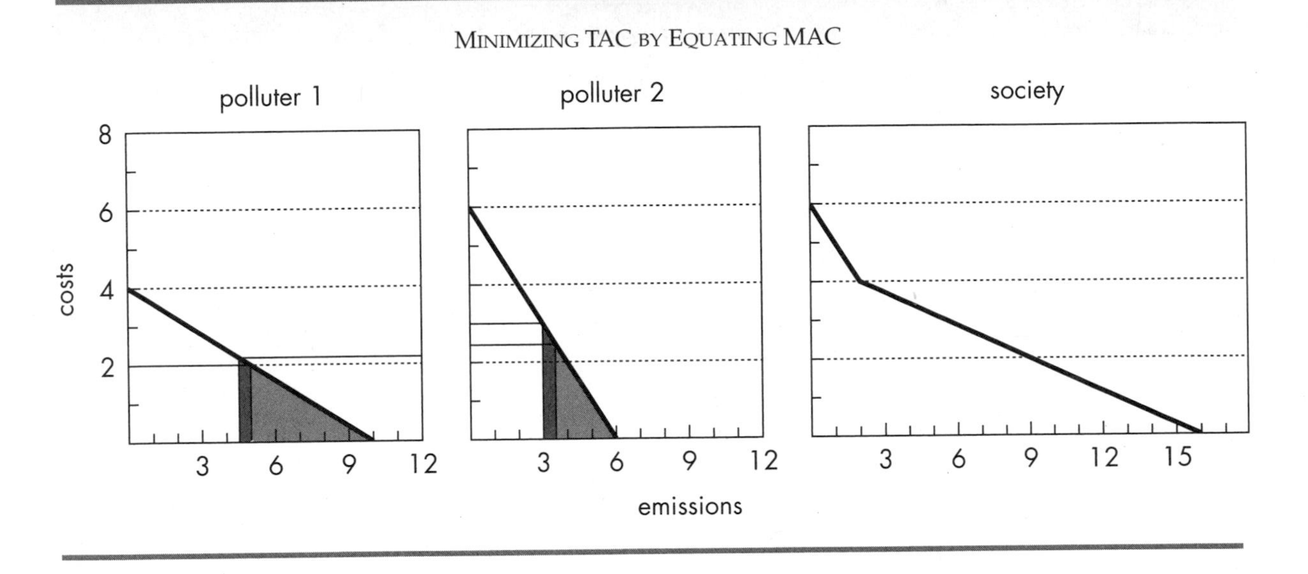

Since there are literally thousands of types of polluters for the more common pollutants, it would be prohibitively expensive to develop even rough estimates of the marginal abatement cost function of each polluter.

Since an assignment of pollution levels to the various polluters cannot generally equate the marginal abatement costs across the various polluters, this type of command and control policy will result in higher abatement costs than necessary, for any target level of pollution. Other types of command and control regulations that have the property of specifying the specific abatement technologies that must be employed may be even worse. For example, the best way to deal with sulfur dioxide pollution from electric utilities may be to burn a low-sulfur fuel. However, regulations that require more expensive scrubber stacks do not allow firms to use lower cost technologies. Even worse from an efficiency standpoint, such regulations reduce the profit motive for research and development of lower cost methods of reducing pollution.

Many critics of U.S. environmental policy believe that the expenditures on pollution control are excessively high. Other critics of U.S. environmental policy argue that current levels of pollution are much higher than optimal. In what might appear to be a paradox, both may be right.

What makes this apparent paradox non-paradoxical is that U.S. environmental policies, which are primarily of the command and control method, move society onto a higher marginal abatement cost curve than is necessary. For example, in Figure 3.13, MAC_2 represents the lowest cost means for abat-

FIGURE 3.13

EXCESS SOCIAL COSTS FROM INEFFICIENT REGULATIONS

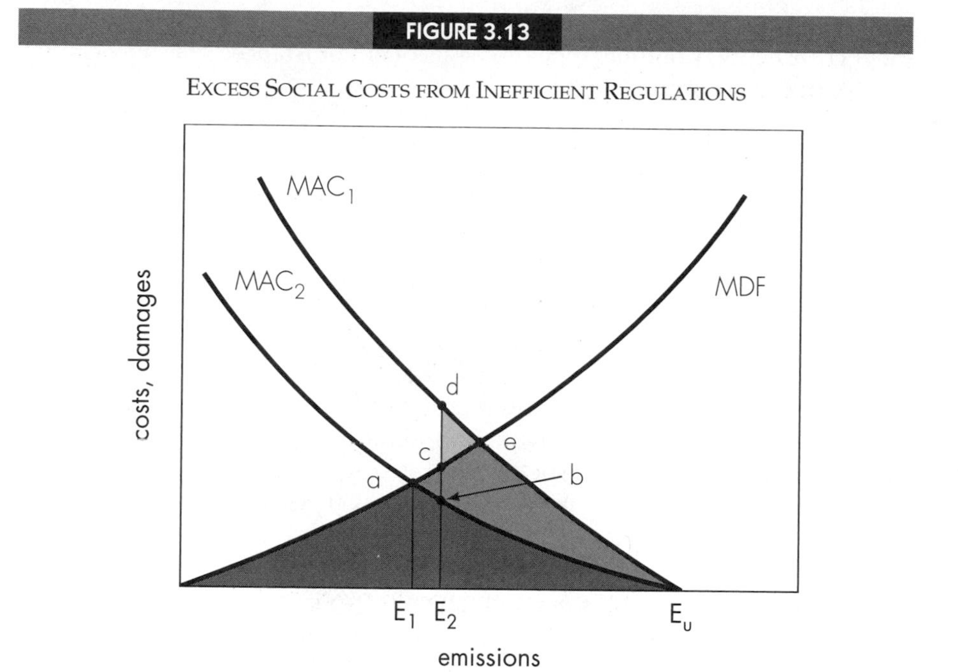

ing pollution, but MAC_1 is the higher cost function that inefficient and ill-advised policies force upon society. Because of the high abatement costs, it is likely that the target level of pollution chosen will be greater than the optimal level of pollution (E_1).

Let us assume that E_2 is the target level of pollution. The excess social costs that inefficient regulations force upon society are diagrammed in Figure 3.13. Area $E_u$ecb plus area cde represent the extra costs associated with using inefficient policies to reduce pollution from E_u to E_2. Area acb represents the excess social costs from having a level of pollution that is higher than optimal.

THE ROLE OF COMMAND AND CONTROL POLICIES

In the next section of this chapter, it will be shown that economic incentives such as pollution taxes and marketable pollution permits do equate the marginal abatement costs across polluters. Since marginal abatement costs will be equated across all polluters with these policies, they will achieve the target level of pollution with the minimum total abatement costs for society. Before beginning our discussion of these policies, we will discuss several circumstances for which command and control regulations may still be the most desirable policy instrument, despite their inability to equate marginal abatement costs across polluters. The three sets of circumstances that may call for the use of direct controls are:

1. when monitoring costs are high
2. when the optimal level of emissions is at or near zero
3. during random events or emergencies that change the relationship between emissions and damages

Many types of command and control regulations, particularly those that set allowable levels of pollution for each polluter, require as extensive monitoring as pollution taxes or marketable pollution permits. In all these cases, the pollution control authority must know the exact (or approximate) amount of pollution each polluter emits. However, consider an environmental problem such as littering, where the optimal amount of emissions is some nonzero amount. For the sake of the example, let us say that the optimal level is 1,000 units of litter per week in a society of 1,000 people. One could achieve this level through a direct control that specified that each person could litter 1 unit per week, or one could set a tax designed to achieve this. However, in order to implement either policy, one would need "litter police" constantly observing the actions of all individuals so they could measure the amount of litter being released into the environment. Obviously, these monitoring costs would be so high that the policies would not be workable.

An alternative would be to institute a command and control policy that only requires sporadic, rather than continuous, monitoring. This could be done by making all littering illegal and stipulating that anyone caught littering must pay a punitive fine. The fine multiplied by the probability of being caught is the expected private cost of littering. Potential litterers would compare the expected private cost of littering with the private benefits of littering (that is, not having to carry the dirty trash around in their pocket) when deciding whether to litter. The higher the fine and the greater the probability of being caught, the greater the expected cost of littering; thus, less littering will take place. If monitoring is relatively easy, so that the probability of being caught is near one, the fine could be very low and still generate compliance. Under these circumstances the system more resembles a per unit pollution tax. However, if the cost of monitoring is extremely high, then the best way to maintain the expected cost is with a relatively high fine. Care must be taken, however, that the fine is not so high that authorities are reluctant to support it. For example, a fine of $100,000 for throwing a gum wrapper on the ground is not likely to be enforced, so it does not constitute a credible threat.

The second set of circumstances where direct controls are advantageous is when the optimal level of release into the environment is at or near zero. When one casually lists pollutants for which the optimal level seems to be at or near zero, extremely dangerous pollutants—such as heavy metals and radioactive waste—immediately come to mind. These are pollutants for

OPTIMAL LEVEL OF EMISSIONS AT OR NEAR ZERO

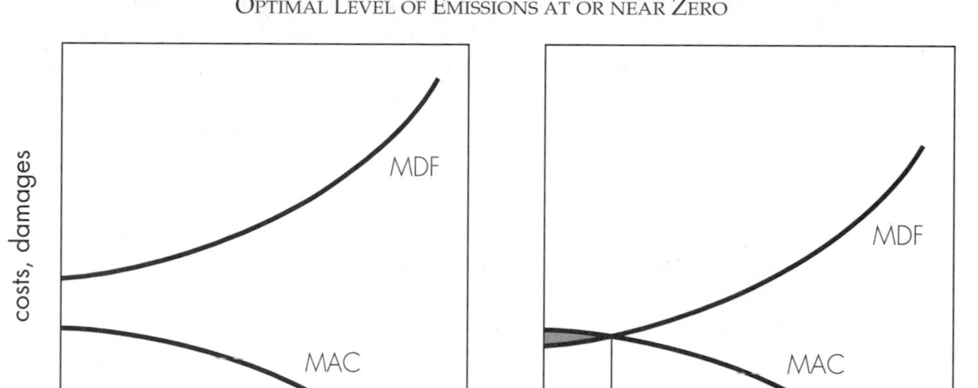

which initial marginal damages are likely to be quite high relative to marginal abatement costs, because the damages associated with these pollutants are quite severe (and because the pollutants persist in the environment without decomposing to more innocuous substances). There is another class of pollutants for which the initial marginal damages are likely to be quite high relative to marginal abatement costs; these are pollutants for which the costs of abating the last few units of the pollutant are quite low. This will generally be true for all pollutants for which there are substitutes which damage less. For example, chlorofluorocarbons (CFCs), which are used as propellants in spray cans (such as hair spray or deodorant), deplete the ozone layer (see Chapter 6). The marginal cost of reducing CFC pollution from this source is likely to be extremely low, as there are extremely good substitutes for CFC as a propellant in toiletries (roll-on deodorants, mechanical pumps, and so on). Figure 3.14a contains representations of marginal abatement and marginal damage functions for which the optimal level of emission is zero, as the marginal damages are always greater than the marginal abatement costs. Although pollution could always be reduced to zero with an appropriately high pollution tax, it is simpler and less expensive to simply ban it. For example, the use of CFCs as a propellant in toiletries has been banned.[8]

---

[8]Chapter 6 discusses the Montreal Protocol, an international agreement for reducing and eventually eliminating the use of CFCs.

Another example is banning the use of lead shot (steel is the substitute) for hunting ducks and geese.

A ban may also be cost effective when the optimal level of pollution is near zero, as in Figure 3.14b. Here, a positive level of pollution, E_1, is optimal. If emissions were reduced to zero, excess social costs in the shaded area would be generated. However, these excess social costs may be small relative to the costs of implementing a government program (such as pollution taxes) designed to bring society to the optimal level of E_1. Since the costs associated with reducing pollution from the optimal level to zero are relatively small, the cost minimizing action may be to ban the pollution entirely and not expend the resources needed to achieve the optimal level. This scenario may describe the situation for lead additives in gasoline, which were recently banned by the U.S. Environmental Protection Agency.

The last set of circumstances in which direct controls may be optimal is in dealing with emergency situations such as smog alerts or droughts. The problem generated by these events is that they temporarily change the nature of the relationship between emissions and damages. For example, if there is a drought that reduces the volume of water in a river, then each unit of pollution that is emitted into the river is less diluted and causes greater damages. Similarly, when there is a thermal inversion over southern California, which traps pollutants in the lower atmosphere, the damages associated with each unit of air pollution increases. In the case of these events, which may occur in a random or unpredictable fashion, taxes may not provide enough flexibility to adequately deal with the new set of circumstances. The nature of the relationships between emissions and damages has changed so much as a result of random events that immediate and drastic action is often warranted. For example, during smog alerts, the Los Angeles area prohibits single-person commuting, requiring carpooling or the use of mass transit. Also, some factories and schools are required to close until the emergency passes.

PURSUING ENVIRONMENTAL QUALITY WITH ECONOMIC INCENTIVES

Command and control policies may be preferable when monitoring is particularly difficult,[9] when the optimal level of pollution is near zero, or during unpredictable emergency events. However, most economists advocate general policies based on economic incentives for two primary reasons:

---

[9]It should be noted that for most types of conventional pollution, such as air pollution from factories, the monitoring costs associated with command and control are the same as for economic incentives. When the pollution is sporadic, nonstationary, or emitted by large numbers of small polluters, monitoring becomes more difficult.

FIGURE 3.15

POLLUTER BEHAVIOR IN THE PRESENCE OF A POLLUTION TAX

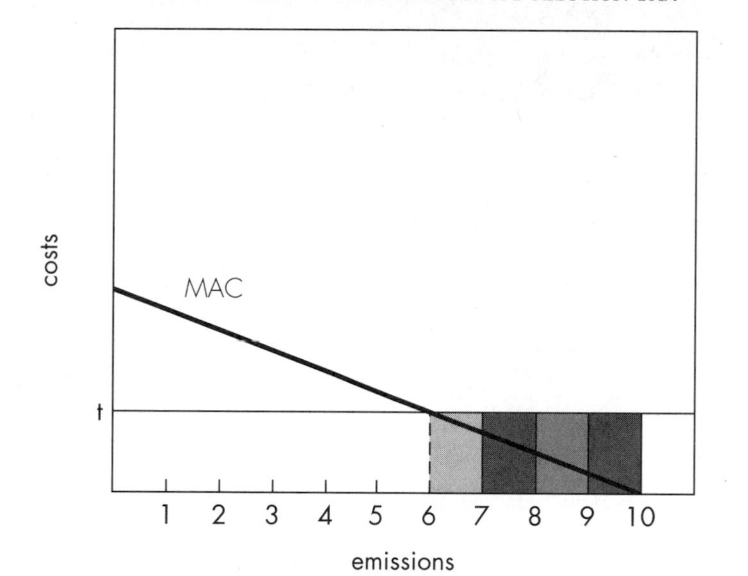

1. Economic incentives minimize total abatement costs by equating the marginal abatement costs across polluters and encouraging a broader array of abatement options.
2. Economic incentives encourage more research and development into abatement technologies and alternatives to the activities that generate the pollution.

ECONOMIC INCENTIVES AND MINIMIZED TOTAL ABATEMENT COST

Several times it has been mentioned that economic incentives can equate the marginal abatement costs across all polluters, but this assertion has not yet been demonstrated in this text. The proof is actually quite simple and can be easily visualized by referring to Figure 3.15.

Assume that a polluter is polluting at the unregulated level of 10 units, and the government imposes a tax equal to t dollars. If the polluter continues to pollute 10 units, he or she will be liable for 10t dollars in total taxes. However, the polluter will not continue to pollute 10 units, as he or she will realize that costs can be reduced by reducing the level of pollution. For example, the polluter will consider whether to reduce pollution to 9 units. The cost of the tax on that tenth unit can be visualized graphically as the shaded

CIGARETTE AND ALCOHOL TAXES AS EXTERNALITY TAXES

The Clinton administration has discussed a set of policies that would pay some of the costs of health care reform. A major component of the financing would be increased federal cigarette and alcohol taxes. Cigarette and alcohol taxes (often called "sin taxes") are often advocated as revenue sources since the demand functions of both goods are inelastic. This inelasticity implies that if price increases, the associated reduction in the quantity demanded is proportionately smaller. Since quantity demanded is relatively insensitive to price increases, increasing the tax will lead to an increase in revenue.

Although these taxes are often viewed simply as revenue raising devices, they also can be interpreted as externality taxes. Cigarette smoking generates externalities associated with secondhand smoke and increased health care costs. Alcohol is associated with many externalities, including drunk driving, which is responsible for tens of thousands of traffic fatalities each year. Since both alcohol and cigarettes are associated with external costs, it makes sense to tax them to reduce the external costs. However, since the demand for these goods is inelastic, large increases in taxes will be necessary to reduce the use of these goods.

rectangle with its base between 9 and 10, and its height equal to t. Notice that the cost of abating the tenth unit is the area underneath the marginal abatement cost function between 9 and 10. As visual inspection clearly shows, this abatement cost is less than the cost of the tax, so the polluter saves money by reducing pollution from 10 units to 9 units. This will be true as long as the per unit pollution tax is greater than the marginal abatement costs. Conversely, if the tax is less than the marginal abatement cost, the polluter can reduce costs by increasing pollution and paying the lower tax rather than the higher abatement costs. The only level of pollution for which a change cannot reduce total costs (total abatement costs plus tax payments) is where the tax is equal to the marginal abatement costs.

Since any individual polluter will adjust his or her level of emissions to the point where marginal abatement cost is equal to the per unit tax, all polluters will adjust to the point where their marginal abatement costs are equal to the tax. This means that as long as all polluters face the same tax, a tax will equate the marginal abatement costs across polluters. Thus, a tax on each unit of pollution will minimize the total cost of obtaining the target level of pollution.

ECONOMIC INCENTIVES AND THE CERTAINTY OF ATTAINING THE TARGET LEVEL OF POLLUTION

While a tax can modify polluters' actions so that the target level of pollution is achieved at minimum total abatement costs, the question of what tax level will induce the target level of pollution is a much more difficult problem to address. If the aggregate marginal abatement cost function is known, the problem becomes trivial. If E_1 in Figure 3.16 is the target level, then t_1 is the

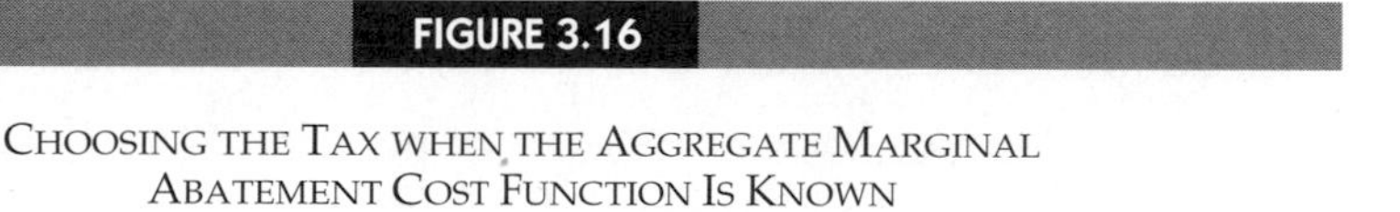

CHOOSING THE TAX WHEN THE AGGREGATE MARGINAL ABATEMENT COST FUNCTION IS KNOWN

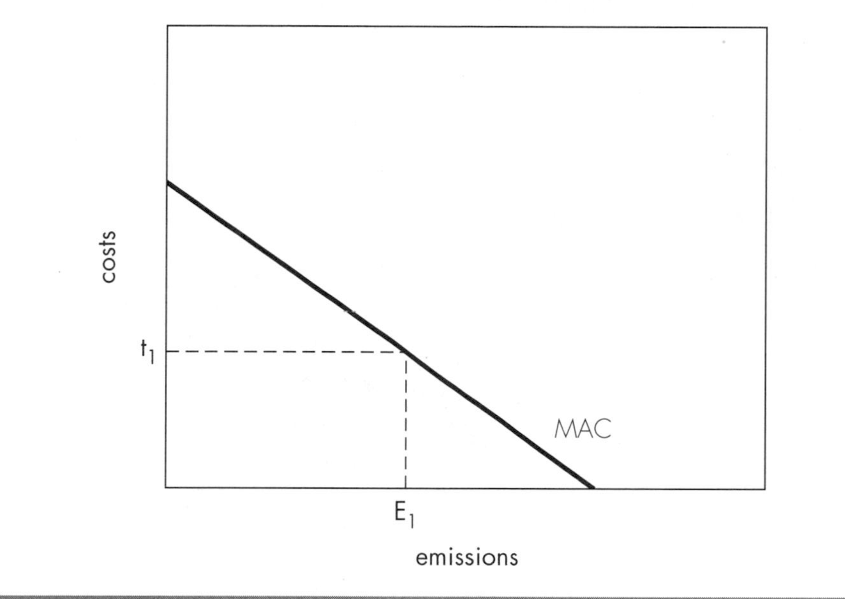

tax that will achieve the target level. The problem is determining what tax level should be chosen when the aggregate marginal abatement cost function is not known.

This problem is examined in Figure 3.17, where it is assumed that evidence suggests that the true marginal abatement cost function lies between the upper and lower bounds of MAC_u and MAC_1. Let us further assume that policy makers choose a target level of pollution of E_1, based on this "fuzzy" knowledge of the position of the marginal abatement cost function. However, let us assume that policy makers think that the true marginal abatement cost function is MAC_1 and, therefore, choose a tax of t_1, which would generate a level of emissions of E_1 if MAC_1 were the true marginal abatement cost function. However, if MAC_t describes how polluters actually respond to the tax, the level of pollution that t_1 would generate would be E_2. Notice that E_2 is higher than the true optimal level of pollution (E_1) and would generate the excess costs of area abc. In reality, policy makers would have chosen the higher tax (t_2) if they had known how polluters would react to the tax. In an interesting (but somewhat technical) article, Martin Weitzmann (1974) shows that the flatter the marginal abatement cost function, the greater the disparity between the "arrived at" level of pollution and the target level of pollution.

CHOOSING THE TAX WHEN THE AGGREGATE MAC IS NOT KNOWN

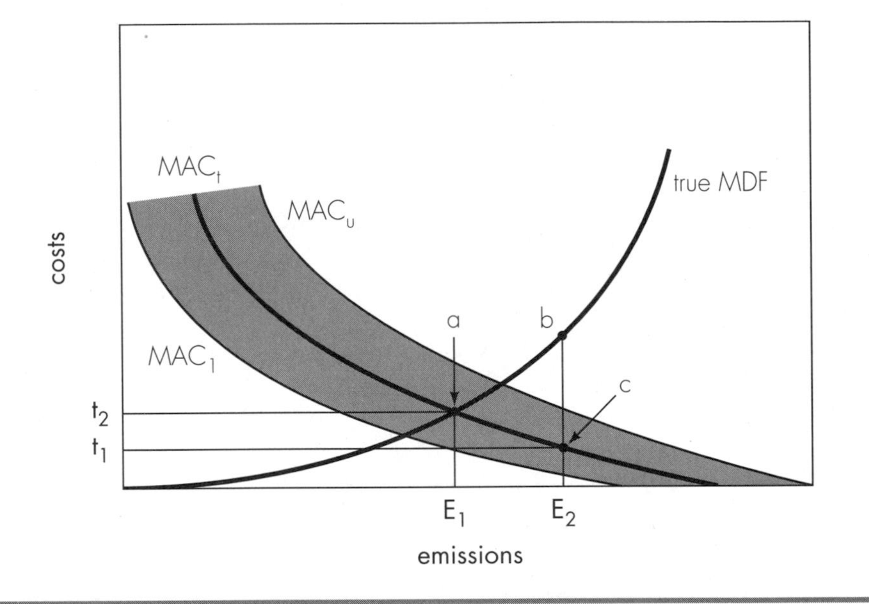

Weitzmann also shows that the steeper the marginal damage function, the greater the social losses associated with the disparity between the "arrived at" level of pollution and the target level of pollution. Weitzmann suggests that with appropriate slopes of the marginal abatement cost and marginal damage functions, quantity restrictions (command and control) may be preferable to price restrictions (pollution taxes). This represents another apparent argument in favor of command and control policies. Command and control techniques can achieve the target level of pollution with greater certainty than can taxes, since policy makers seldom have perfect prior information on the responses of polluters to the taxes.

In the past, it has been argued that errors in the choice of the pollution tax are not important, as one can iterate toward the correct tax rate: If a tax is chosen that leaves the level of emissions too far above the target level, simply raise the tax in the next period. If a tax is chosen that reduces the level of pollution too far below the target level of pollution, simply lower the tax in the next period. This argument may seem to have some appeal, but it should be realized that pollution control is not generally achieved by changing the level of easily variable inputs such as labor, materials, or energy. In fact, pollution control is relatively capital intensive. Pollution abatement and production technologies that are optimal under one set of taxes would not be chosen under a different set of taxes.

For example, if a carbon tax (to control greenhouse gas emissions) was instituted, the price of gasoline would increase. Let us assume that the price increased from $1.20 to $2.20 per gallon. Consumers would be willing to pay more money for cars that offered better gas mileage, and this would be reflected in their purchases. However, if the pollution control authority then decides that there is still too much carbon dioxide pollution and increases the tax so the price of gasoline rises to $3.20 per gallon, then consumers would want even more efficient cars. However, they are stuck with their purchasing decisions of the past period. If they sell their existing cars, they will suffer losses on resale value; if they keep driving these cars, they will pay high taxes. Their ability to react to the new situation in a cost minimizing fashion is constrained by their previous purchase of a capital good. Consumers would have preferred to know about the higher taxes before they made their purchases. The same thing would be true for firms in respect to their capital purchases.

MARKETABLE POLLUTION PERMITS

A summary of our discussion of potential policy instruments would suggest that:

1. Pollution taxes are preferable to command and control techniques since pollution taxes minimize abatement costs and provide other desirable incentives.
2. Under certain conditions, pollution taxes are less proficient than command and control techniques in achieving the target level of pollution.

Since pollution taxes are preferable on one count, and command and control techniques are preferable on the other, the choice between the two appears to be a difficult one. However, it is a choice that does not have to be made, as an instrument exists that combines the desirable properties of both.[10] This instrument is known as the marketable pollution permit (sometimes referred to as transferable discharge permit or a transferable emissions permit).

Marketable Pollution Permits and Efficiency. The institution of a system of marketable pollution permits begins with the determination of the target level of pollution. For the purposes of this discussion, it will be assumed that there are 100 polluters, and the target level of pollution is 1,000 units. The

---

[10]Command and control techniques will still be an essential part on environmental policy when monitoring is difficult, the optimal level of emissions is at or near zero, or in the presence of unpredictable emergency events.

next step is to determine the allocation of pollution across polluters. For example, one could authorize each polluter to pollute 10 units. So far, this is exactly the same as a command and control policy that specifies the legal amounts of pollution for each polluter. However, the major difference between a command and control system and a system of marketable pollution permits is that once the initial allocation of pollution is made, polluters are free to buy and sell the rights to pollute.

At first, this ability to buy and sell the rights to pollute may seem to be unimportant, but it is this feature that equates the marginal abatement costs across polluters. Marketable pollution permits (MPPs) equate the marginal costs across polluters because each polluter compares his or her marginal abatement costs with the price of a permit in deciding whether to reduce (or increase) pollution by 1 unit or to sell (or buy) another permit. For polluters whose marginal abatement costs are greater than the price of a permit, total costs can be reduced by buying more permits and polluting more. For polluters whose marginal abatement costs are less than the price of a permit, profits can be increased by selling permits and polluting less. Polluters will have an incentive to buy or sell permits as long as the price of permits is different than the individual polluter's marginal abatement costs. Since our knowledge of competitive markets tells us that there can be but one equilibrium price for the permits, marginal abatement costs for all polluters will be equated when the market for permits is in equilibrium. It is important to note that this result is independent of the method that is used to initially distribute the permits. The permits could be given to existing polluters proportional to historic pollution levels, auctioned to the highest bidder, distributed by lottery, or allocated by some other scheme or combination of schemes. As long as the permits are marketable, polluters' attempts to minimize their total pollution costs (the cost of abatement for pollution that is eliminated plus the cost of permits for pollution that is still emitted) will result in marginal costs being equated across all polluters and the minimization of the total abatement costs of achieving the target level of pollution.

Although the method for the initial allocation of MPPs is irrelevant from an efficiency viewpoint, it is controversial because there are important equity considerations. If permits are auctioned, this creates a substantial initial cost for polluters (and revenue for the government), but if permits are allocated based on historical levels or lotteries, this creates an asset for polluters.

Marketable Pollution Permits and Geographic Considerations. So far, the discussion of pollution policy has proceeded as if a unit of pollution has the same importance regardless of the location at which it is emitted. For some types of pollution, such as the emission of greenhouse gases or chemicals which deplete the ozone layer, this statement is true. However, for other types of pollution, the geographic location of the emissions can have a profound effect on the damages the pollution generates. For example, air pollution generated in the vicinity of large numbers of people will

Marketable Pollution Permits and the 1990 Clean Air Act Amendments

The United States has had a very limited experience with marketable pollution permits. Except for a very small system developed for water pollution in the Fox River in Wisconsin (Erhard F. Joeres and Martin Heidenhaim David, *Buying a Better Environment: Cost-Effective Regulation through Permit Trading,* Madison: The University of Wisconsin Press, 1983), there have been no systems set up prior to 1990. The Fox River system actually did not even constitute a true marketable permit system, as the system was so restrictive that trades could not take place.

However, the 1990 Clean Air Act Amendments established a large-scale system of marketable pollution permits. This system is authorized in Title IV, which is designed to control sulfur dioxide emissions that generate acid rain (see Chapter 7). The system is designed to allow trades among the eastern electric power generators who are responsible for a large portion of sulfur dioxide emissions. Although most economists are heartened to see a system of economic incentives adapted on a large scale, the system does not deal with geographic variability in the damages of emissions. Although there may not be much variation in the effect of location on the acid rain contribution of the sulfur dioxide emissions, there may be other local pollution effects that need to be accounted for in a permit system. It will be interesting to follow this system as it is implemented, track the cost–savings generated by trading, and see if any local "hotspots" develop as a result of the permit system.

generate greater health effects, as more people are exposed to the pollution. On the other hand, air pollution emitted downwind from population centers will have very little effect on the health of people in the population centers.

Central to the importance of the location of the emissions is the manner in which the pollution disperses when it enters the environment. For example, air pollution will tend to disperse in all directions when emitted from a smokestack, but there will be a tendency for the pollution to move from west to east, which is the prevailing wind direction in much of North America.

As mentioned above, geographic location is not relevant for all types of emissions. Pollutants that generate their damages in the upper atmosphere (greenhouse gases and chemicals that deplete the ozone layer) have the same effect regardless of the location of emissions. This makes the establishment of a system of marketable emissions permits more straightforward, as only one market for emissions need be established.

In contrast, the release of carbon monoxide (from automobile exhaust) in California does not affect air quality in Manhattan, although the release of carbon monoxide in northern Virginia does affect air quality in Washington, D.C. A unit of carbon monoxide released in Raleigh, North Carolina, would create more damages than a unit released in Wilmington, North Carolina (where most of it would blow out over the ocean). In order for any set of pollution controls to be effective, they must take into consideration the

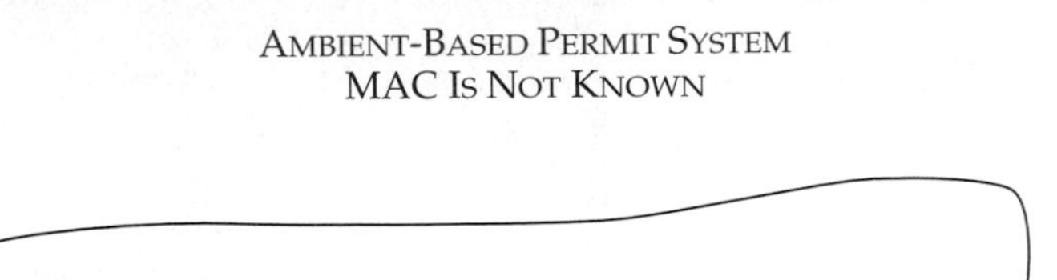

FIGURE 3.18

AMBIENT-BASED PERMIT SYSTEM
MAC IS NOT KNOWN

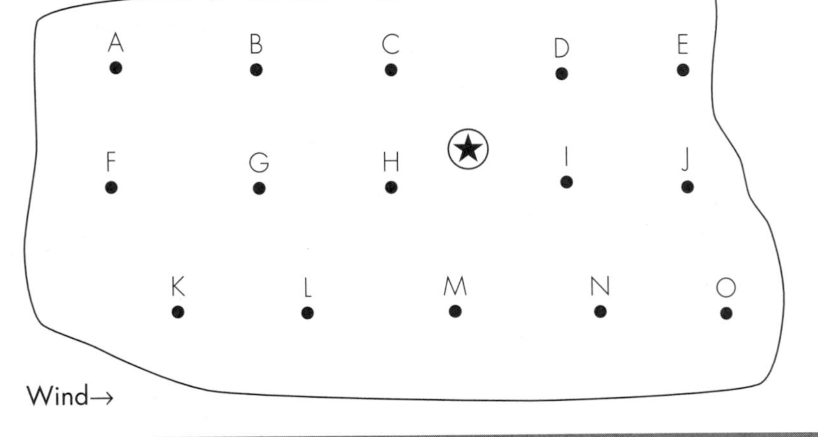

geographic variation in the effect of pollution on society. This consideration is necessary whether one is using taxes or marketable pollution permits.

A tax system would take this variation into account by charging higher taxes in areas where emissions are more damaging. One would expect higher taxes in metropolitan areas, such as Los Angeles, Atlanta, St. Louis, Chicago, and so on, and lower taxes in more rural areas like Wyoming, Alaska, or South Dakota. The policy would consider the effects of one region's pollution on the welfare of other regions. For example, sulfur dioxide taxes in the Midwest should be set with consideration of the effects of these emissions on acid rain in the Northeast.

It is slightly more complex to adapt a marketable pollution permit system to deal with the geographic variability in the effects of emissions. Generally, this must be done by dividing the overall region into subregions. Then, the subregions can be used to account for geographic variability in one of two ways. First, receptors (pollution measurement locations) can be located across the region. These are denoted by the lettered points in Figure 3.18, where the location of a particular polluter is denoted by the circled star. Under this type of receptor-based system (which is often called an ambient-based system), the polluter would need to purchase marketable pollution permits based on the effect of his or her pollution on each receptor in the region. Since receptor "I" is relatively close to, and downwind from, the polluter, the polluter would need to purchase more "I" permits than "A"

FIGURE 3.19

Emission-Based Permit System

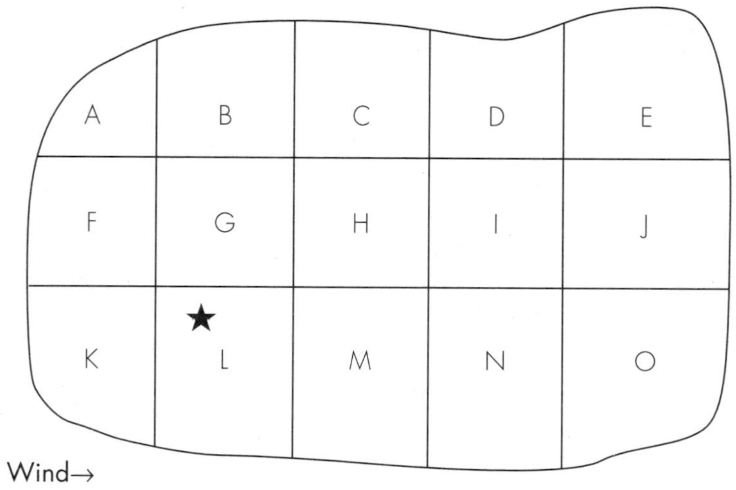

permits, which is farther away and upwind. The effect of a polluter's emissions on the concentration of pollution receptors is determined by dispersion coefficients. Dispersion coefficients are derived from mathematical models of wind, topography and other factors, and measure the way pollution spreads from its source to other locations. These dispersion coefficients can be used to help in defining the terms of trade in a marketable pollution permit market. For example, for each unit of pollution that the polluter at the circled star emits, he or she may need to purchase one "I" permit, but only one quarter of an "A" permit. The polluter would have to buy 15 different types of permits (one for each receptor, A through O). Although this system does a good job in dealing with the geographic variability in the effect of pollution, it has high administrative and transactions costs associated with it, due to the many different markets in which each polluter must participate.

An alternative to the ambient-based system is to establish separate markets for each subregion, where the total number of permits available in each subregion is set with consideration of the effect of pollution from the subregions on the overall region. The polluter need only purchase permits for the subregion in which he or she is located. This is called an emissions-based system and is illustrated in Figure 3.19. A polluter located at the star would only have to buy permits for subregion L. Although this system greatly reduces transactions costs by limiting the number of markets in which the polluter must participate, it does have an important shortcoming. This shortcoming is that a polluter located in one subregion cannot trade with a polluter located in another subregion. This restriction is bothersome, because there could be polluters in one subregion that have a relatively low abatement cost function and polluters in another subregion with a relatively high marginal abatement cost function. Even though it would be in both the public and private interest for the high cost polluter to buy permits from the low cost polluter (and for the high cost polluter to pollute less and the low cost polluter to pollute more), this would not be permissible under an emissions-based system. More trades could take place if each subregion constituted a larger fraction of the overall region, but if the subregions become larger, then the division into subregions does not do as good a job accounting for the geographic variability in the effect of emissions.

Alan Krupnick, Wallace Oates, and Eric Van de Verg (1983) suggest a compromise system that allows more trades but with lower transactions cost. Many different receptors would be defined, as in the ambient-based system. However, there would only be one type of permit for the overall system, and any polluters could trade emissions (buy and sell marketable pollution permits), provided that their trade does not result in the ambient air quality standards being violated at any receptor point.

OTHER TYPES OF ECONOMIC INCENTIVES

Deposit–refunds. Many students are familiar with deposit–refunds as a means of controlling environmental externalities, as many states now have deposit–refund regulations in place for beverage containers (primarily beer and soda). Deposit–refund systems are a good way of employing economic incentives when monitoring costs are high.

For example, the monitoring costs of making sure that everyone properly disposed of their soda cans are extremely high. The social costs associated with improper disposal are not likely to exceed several cents a can. In most cases, the private benefits of improper disposal are likely to be several cents per can as well. Figure 3.20 shows the hypothetical social cost associated with improper disposal, as well as the hypothetical private benefits of improper disposal. Let us assume that by proper disposal it is meant that the beverage containers are collected for recycling. In this case, the social costs could include aesthetic values associated with reduced littering, as well as the costs of landfill space, and the environmental externalities associated with the use of energy to produce containers from new metal, glass, or plastic. The private benefits of improper disposal are the inconvenience and other costs that the consumer avoids with improper disposal. In this illustration, the optimal amount of improper disposal is about 40 percent of total beverage containers.

One way to achieve the optimal level of improper disposal is to set a tax equal to five cents for each container that is not disposed of properly. Since the marginal private benefits of improper disposal are less than five cents for 60 percent of the containers, the consumers would suffer the lost benefits

FIGURE 3.20

COSTS AND BENEFITS OF IMPROPER DISPOSALS

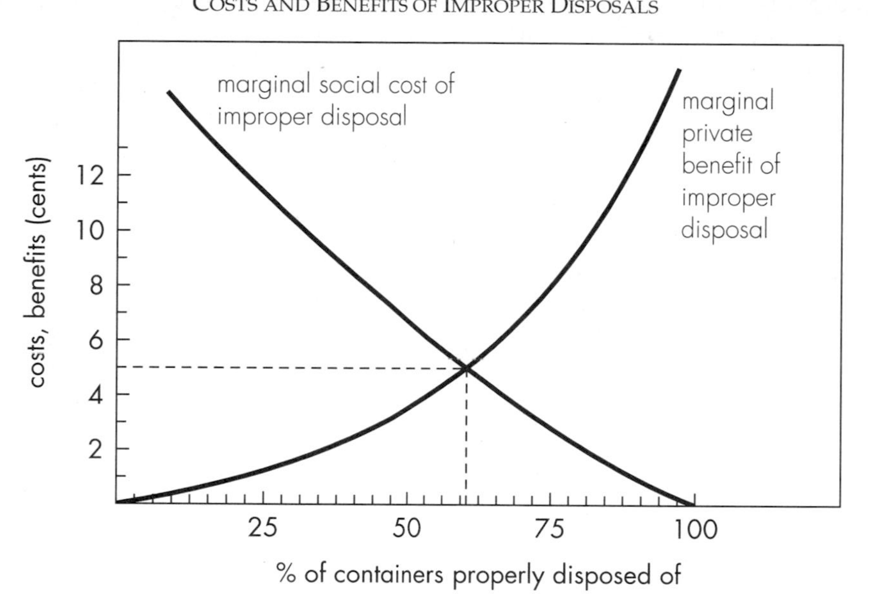

rather than pay the tax, and 60 percent of the containers would be disposed of properly. However, a tax such as this would be difficult to enforce because there are virtually unlimited opportunities for improper disposal, which would make monitoring costs high.

A deposit–refund system is similar to a tax, but instead of making the individual pay for undesirable acts as they occur, the individual pays up front and then is rewarded if he or she acts properly. A five cent deposit would be collected upon purchase, which is refunded when the container is returned. Although the most apparent example of the deposit–refund system is beverage containers, it has been used on cars in Scandinavia and suggested for use with tires, batteries, CFCs in refrigerator coils (see Chapter 6), and containers of toxic household products (such as pesticides or paints; see Chapter 15).

Bonding Systems. Bonding systems are closely related to deposit–refund systems. With bonding systems, a potential degrader of the environment is required to place a large sum of money in an escrow account. The money will be returned if the environment is left undamaged (or returned to its original condition) and will be otherwise forfeited. The size of the bond should be large enough to provide appropriate safeguards by those posting the bond, or large enough that the government can use the funds to clean up damage if it occurs. Bonds are employed in strip mining areas, where mining companies forfeit the bond if the land is not returned to its original condition. Bonds have also been suggested for companies that have leases to cut trees in public forests (see Chapter 12) and for companies that transport oil or toxic substances (see Chapters 8 and 15).

Liability Systems. Liability systems are based on defining legal liability for the damages caused by certain types of pollution discharges and facilitating the collection of these damages. The Comprehensive Environmental Response, Compensation and Liability Act of 1980 (CERCLA) defines legal rights to natural resources for local, state, and federal governments and specifies methods by which damages may be measured. The provisions of this act provide a means for facilitating the incorporation of the expected social cost of spills into the private cost calculation by potential polluters or internalizing the expected damages of spills. The legislation increases the probability that the firm will have to pay the social cost of its spills, so the firm is more likely to take appropriate safety measures.

A related system that has also been suggested is to define legal liability as above and then require potential polluters to obtain full insurance against any damages they might generate. The idea behind this suggestion is that the insurance industry would then require appropriate safety measures on the part of potential polluters. This type of system has been suggested for generators, haulers, and disposers of toxic waste.

Pollution Subsidies. The last type of economic incentive to be discussed is the per-unit pollution subsidy. This type of incentive pays the polluter a fixed amount of money for each unit of pollution that is reduced. This is depicted in Figure 3.21, where s is the amount paid per unit of pollution that is reduced. In this example, the polluter pollutes 10 units in an unregulated environment. If the subsidy system is imposed, the polluter will evaluate his or her tenth unit of pollution and discover that the cost of abating it is less than the payment that would be received for abating it. Thus, the polluter would make a net profit by reducing pollution from 10 units to 9 units. This would be true as long as the per-unit subsidy is greater than the cost of abatement. The polluter will reduce pollution to the point where the subsidy is equal to the marginal cost of abatement. Remember that under the tax system, pollution was reduced until the point where the tax was equal to marginal abatement cost. Therefore, if the amount of the per-unit subsidy is equal to the amount of the per-unit tax, the two systems will generate identical behavior on the part of an individual polluter.

For many years, economists argued that tax and subsidy systems had equivalent efficiency effects, although different distributional effects—the most notable being that the subsidy system transfers resources toward polluters, since they are receiving a payment rather than making a payment. Other potential problems with subsidies include lack of political acceptability (poli-

FIGURE 3.21

POLLUTER BEHAVIOR IN THE PRESENCE OF A PER-UNIT POLLUTION SUBSIDY

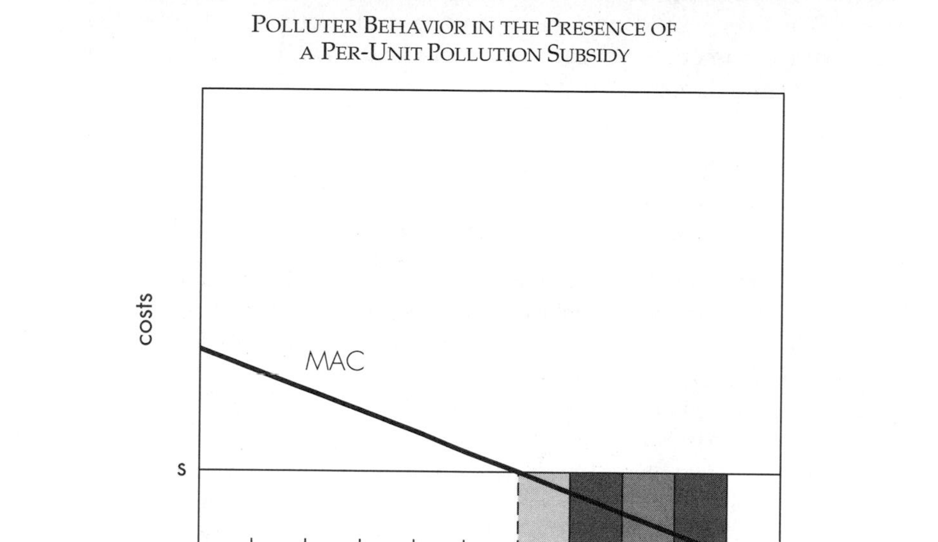

ticians would be reluctant to support a system that paid polluters, due to potential adverse reaction by voters) and the possibility of strategic behavior (increasing the initial amount of pollution before the system goes into place, so that a larger subsidy can be obtained). There is an additional problem that has only recently been articulated (Baumol and Oates, pp. 211–234, 1988), and that is that while taxes and subsidies have equivalent effects on an individual polluter, they have quite different effects on the number of polluters. Quite simply stated, subsidies make firms more profitable, so there will be more firms under a subsidy system than under a tax system. Although both reduce the amount of pollution per polluter, there will be more polluters under a subsidy system than under a tax system. A subsidy system that reduced the amount of pollution per polluter could actually result in an increase in the amount of pollution if enough new firms entered the market due to the increased profitability generated by the subsidy. The reader will note that this is a very similar situation to the one discussed with the Coase Theorem, where the definition of property rights had an important effect on the number of polluters.

Since subsidies have the perverse effect of increasing the number of polluters, it is the author's opinion that they should not constitute an important part of the policy arsenal to control environmental externalities. While there may be some situations in which the political power of the polluters is so strong that they would not accept any control method but subsidies, it is believed that the other policy instruments that have been described in this chapter are much more effective.

SUMMARY

The market failures associated with environmental externalities generate losses in welfare. It has been shown that despite the arguments advanced by Ronald Coase, government intervention may be necessary to reduce the magnitude of these social losses.

Command and control policies form the basis of current U.S. environmental policy. Although these policies may be preferable under certain conditions (the presence of high monitoring costs, the possibility of unpredictable emergency events, or the case when the optimal level of release is at or near zero), command and control policies do not equate the marginal abatement costs across polluters and also restrict the options which are available to control pollution. Consequently, command and control policies do not minimize the costs of obtaining the target level of environmental quality.

Economic incentives, such as taxes and marketable pollution permits, do equate the marginal abatement costs across polluters and therefore minimize the costs of obtaining the target level of environmental quality. Taxes have a shortcoming in that the response of polluters to the taxes is not known in advance, so the level of pollution induced by the taxes may differ substantially from the target level of pollution. The more elastic the marginal abatement cost function, the more likely it is to be a problem. However, marketable pollution permits do not suffer from this problem and generate the target level of pollution with much more certainty. In addition, economic incentives create additional motivation for technological innovation in developing better ways to reduce pollution.

In conclusion, this chapter has demonstrated important social benefits which can be realized by the use of economic incentives in environmental policy, but current environmental policy (both in the United States and in other countries) tends to be based almost entirely on direct controls. Although this chapter has reached its conclusion, our examination of these issues has not. As we examine problems, such as the use of fossil fuels, global warming, the depletion of the ozone layer, water pollution, deforestation, loss of biodiversity, depletion of our marine resources, and so on, we will discuss these problems in the context of the different environmental types of policy instruments. In particular, we will highlight how economic incentives and command and control policies are currently used to control these problems and how they can be used more effectively.

REVIEW QUESTIONS

1. What types of environmental problems are best handled with deposit–refund systems?
2. What are the five broad classes of policy instruments that are available to the government to correct market failures associated with externalities?
3. Describe how the various policy instruments give differential incentives for research and development of new production and abatement technologies.
4. What is the Coase Theorem? Does it have important implications for environmental policy?
5. Show how an economic incentive system can minimize total abatement costs.
6. Assume that a society is composed of two polluters, with the marginal abatement costs of polluters 1 and 2, respectively, equal to:

   $$MAC_1 = 18 - E_1$$
   $$MAC_2 = 12 - 2E_2$$

   where MAC_1 refers to the marginal abatement costs of polluter 1, and E_1 refers to the level of emissions of polluter 1. What is the unregulated level of pollution for each polluter? Find the total level of emissions that would be generated if a per-unit pollution tax of four dollars were imposed. Perform the same exercise for taxes of six dollars and eight dollars.
7. Given the same two marginal abatement cost functions in question 6, find the market price of a marketable pollution permit if pollution was limited to eighteen units through the issuance of marketable pollution permits.
8. Given a societal marginal abatement cost function of:

   $$MAC = 100 - 3E$$

   and a societal marginal damage function of:

   $$MD = 2E,$$

   find the optimal level of pollution and the per-unit pollution tax that would achieve it.
9. MAC_1 and MAC_2 are two different societal marginal abatement cost functions. Which one is more likely to be associated with an optimal level of pollution that is at or near zero? Why?

   $$MAC_1 = 10 - 0.2E$$
   $$MAC_2 = 1/E$$

QUESTIONS FOR FURTHER STUDY

1. What issues must be considered in the development of marketable pollution permit systems?
2. How would you counter the argument that economic incentive systems (such as pollution taxes and marketable pollution permit systems) are immoral, as they sell the right to do something that is bad for society?
3. Use what you have learned about environmental policy to develop a policy for the parking problem on your campus.
4. Use what you have learned about environmental policy to develop a policy for the problem of large numbers of students being closed out of classes. (Assume that you cannot hire more professors and cannot make professors teach more or larger classes.)
5. Under what conditions would per-unit pollution taxes be preferable to marketable pollution permits?
6. Why are pollution subsidies not given serious attention as environmental policy instruments?
7. What issues must be addressed when establishing a system of marketable pollution permits?
8. Are economic incentives likely to be effective in controlling environmental degredation arising from non-market activities, such as subsistence farming or fishing in developing countries?

SUGGESTED PAPER TOPICS

1. Choose a local or regional pollution problem and develop a system of controls (economic incentives, command and control policies, or a combination of both) that best deals with the problem. Make sure to look at the literature (*Journal of Environmental Economics and Management, Land Economics, Journal of Environmental Management)* to see what others have suggested or done in analogous circumstances. Discuss the implications of your choices and justify them over alternatives. Check the local newspapers, environmental groups, or your professors for suggested problems. Develop background information from newspapers and local authorities. Then, develop your plan by comparing your problem to others which have been analyzed in the literature.

2. Analyze the marketable pollution permit provisions of Title IV (Acid Precipitation) of the 1990 Clean Air Act Amendments. Look at issues such as how markets are defined, who participates, the initial allocation, and local emissions effects. Suggest changes that could improve the way the market is established. Check the references to Chapter 7 for more information on these amendments. Also, look at recent issues of the *Journal of Environmental Economics and Management, Land Economics,* and the *Rand Journal of Economics.* Search bibliographic databases using *pollution trading, sulfur dioxide, acid rain,* and *1990 Clean Air Act Amendments* as keywords.
3. Write a survey paper about the use of deposit–refund systems to control environmental externalities. The articles by Porter and the book by Bohm (listed below) are a good place to start your search for references.
4. Investigate the role of property rights in mitigating environmental externalities. Search for references in the *Journal of Economic Literature* and in the bibliographic databases, using *environment, pollution,* and *property rights* as keywords.

References and Selected Readings

1. Anderson, Frederick, et al. *Environmental Improvement through Economic Incentives.* Washington, D.C.: Resources for the Future, 1977.
2. Baumol, W. J., and W. E. Oates. *The Theory of Environmental Policy.* London: Cambridge University Press, 1988.
3. Bohm, Peter. *Deposit–Refund Systems: Theory and Applications to Environmental Conservation and Consumer Policy.* Washington, D.C.: Resources for the Future, 1981.
4. Burrows, Paul. *The Economic Theory of Pollution Control.* Cambridge, MA: MIT Press, 1980.
5. Coase, R. H. "The Problem of Social Cost." *Journal of Law and Economics* 3 (1960): 1–44.
6. Dales, J. H. "Land, Water and Ownership." *Canadian Journal of Economics* 1968, reprinted in R. Dorfman, and N. S. Dorfman, *Economics of the Environment.* New York: Norton, 1993.
7. Freeman, A. M. "Depletable Externalities and Pigouvian Taxes." *Journal of Environmental Economics and Management* 11 (1984): 173–179.
8. Hahn, R. W. "Economic Prescriptions for Environmental Problems: How the Patient Followed the Doctor's Orders." *Journal of Economic Perspectives* 3 (1989): 95–114.
9. Joeres, Erhard F., and Martin Heidenhain David. *Buying a Better Environment: Cost–Effective Regulation through Permit Trading.* Madison: The University of Wisconsin Press, 1983.
10. Krupnick, A., W. E. Oates, and E. Van de Verg. "On Marketable Pollution Permits: The Case for a System of Pollution Offsets." *Journal of Environmental Economics and Management* 10 (1983): 233–247.
11. Oates, W. E. "Markets for Pollution Control." *Challenge* (1984): 11–17.
12. Pigou, A. C. *The Economics of Welfare.* London: Macmillan and Company, 1938.
13. Porter, R. C. "A Social Benefit–Cost Analysis of Beverage Containers: A Correction." *Journal of Environmental Economics and Management* 10 (1983): 191–193.
14. Portney, Paul R., and Roger C. Dower. *Public Policies for Environmental Protection.* Washington, D.C.: Resources for the Future, 1990.
15. Tietenberg, Thomas H. *Emissions Trading: An Exercise in Reforming Pollution Policy.* Washington, D.C.: Resources for the Future, 1985.
16. Weitzman, M. L. "Prices vs. Quantities." *The Review of Economic Studies* 41 (1974): 477–499.

Valuing the Environment for Environmental Decison Making

The same thing is good and true for all men,
but the pleasant differs from one and another.

Democritus (c. 460–370 b.c.)[1]

INTRODUCTION

The measurement of both the value of environmental resources and changes in the level of environmental quality is a critical step in the development of the objectives of environmental policy. This kind of information is essential in determining the benefits of environmental policy, which must be compared to the costs of obtaining these environmental goals.

For example, in the late 1980s, there were substantial problems resulting from debris washing ashore on Atlantic Ocean beaches, particularly in the New York/New Jersey area. Beaches closed, marine recreational activity dropped off precipitously, and seafood sales fell as people were afraid to eat seafood. As it turned out, a major source of the debris problem was—and still is—the New York City combined sanitary and storm sewer system, which overflowed and bypassed the treatment facilities whenever there was significant rainfall.

In order for there to be significant improvement in beach and water quality, New York City must overhaul its sewer system at a cost that is likely to exceed $10 billion. Is it worth it? Not only is it important to measure value for the evaluation of a particular project, it is important to measure value when comparing alternative projects. For example, it has been estimated that the costs of completely cleaning one former weapons facility in eastern Washington (Hanford) is roughly $50 billion. Should we spend $50 billion to clean up the coastal environment all along the Atlantic Coast, or should we spend $50 billion to clean up a contaminated, but isolated, former weapons facility, or should we do neither? Is it worth higher energy costs to have less air

[1]Taken from Henry Spiegel, *The Growth of Economic Thought* (Durham: Duke University Press, 1971).

pollution? Should we restrict the use of automobile air conditioners to protect the ozone layer? As might be expected, questions of how much we should spend on improving environmental quality, and what areas of environmental quality should be emphasized, cannot be resolved unless the value of clean beaches, clean water, and all the other environmental resources is clearly understood.

The measurement of value is particularly important in two facets of environmental economics. One is the determination of the optimal level of pollution. In Chapter 3, the concepts of the marginal damage and marginal abatement cost functions were introduced. Obviously, damages and costs must be monetized in order to perform this analysis. The second area in which the measurement of value is important is in cost–benefit analysis, where the choice is among several options rather than the choice of a continuum of levels. For example, a site that is contaminated by toxic waste (such as the previously mentioned weapons site) can be left alone, the waste can be contained at the site, or the site can be restored to a near pristine state. The choice among these alternatives involves comparing the costs and benefits of each option. Both the marginal damage function approach and cost–benefit analysis will be discussed later in the chapter. However, before this is done, the whole notion of value must be addressed.

WHAT IS VALUE?

The definition of value can be approached from many different perspectives. In fact, it is a subject about which whole books have been written. However, a discussion of these different perspectives will be postponed until later in the chapter so that we can focus on the way economists define value. Toward the end of the chapter, some alternative philosophies such as deep ecology will be examined.

The first point that distinguishes the economic view of value from other perspectives is that it is an anthropocentric concept. In the economic perspective, value is determined by people and not by either natural law or government. While government representatives may have their own values and may incorporate these values into policy, their values do not necessarily reflect society's values.

The second point is that value is determined by peoples' willingness to make trade-offs. This can best be seen with market goods, where the willingness to make trade-offs is reflected in peoples' willingness to pay a monetary price for the good. First, we will examine how value is measured for market goods, and then we will examine the measurement of value for nonmarket goods, with an emphasis on environmental resources and environmental quality.

FIGURE 4.1

MARGINAL AND TOTAL WILLINGNESS TO PAY

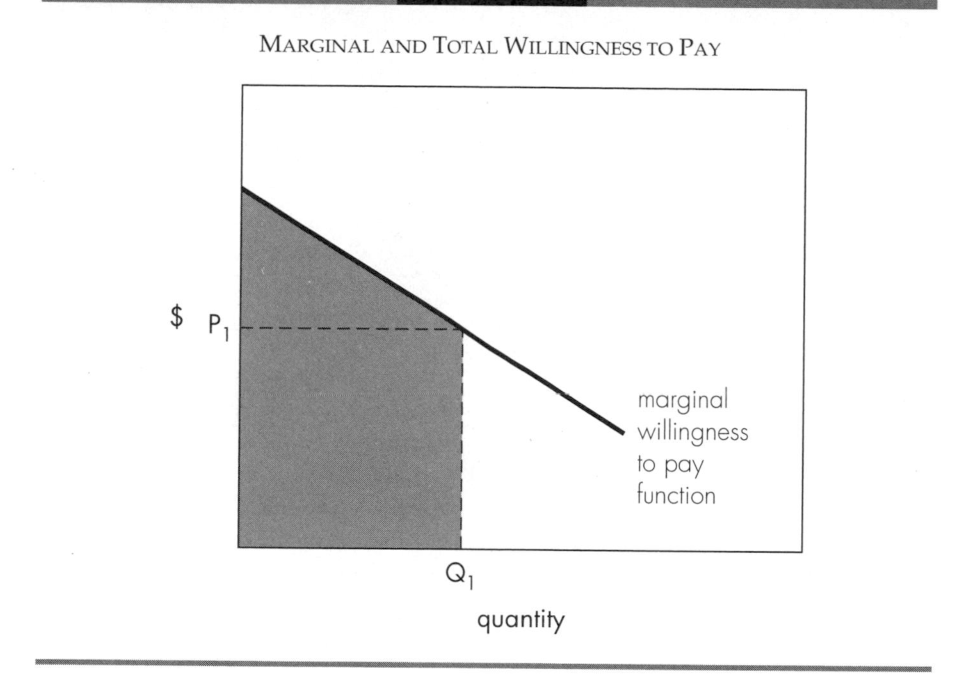

For market goods, the inverse demand curve represents a marginal willingness to pay function. For example, in Figure 4.1, P_1 represents how much people are willing to pay for an additional unit of the good, given that Q_1 units are already being consumed. The total willingness to pay for Q_1 units is represented by the shaded area. While the shaded area represents the total value or the total benefit associated with the Q_1 units of the good, total benefit is not usually an appropriate measure of the contribution of the good to society's well-being, as the cost of producing the good is not taken into account. Since the resources that are used to produce this good could be used to produce other goods that would benefit society, these resource costs must be subtracted from total value to yield net value.

Total resource costs can be examined with the aid of the marginal cost function. In Figure 4.2, the relevant resource cost associated with Q_1 units of the good is given by the area under the marginal cost function, or the shaded triangle OBQ_1. This notion of cost may be somewhat confusing to students who are aware that total revenue is equal to area OP_1BQ_1. Since total revenue must equal total cost in a perfectly competitive market, it seems as if total costs should be area OP_1BQ_1 and not OBQ_1. The resolution to this apparent contradiction is that area OBP_1Q_1 contains producers' surplus (sometimes referred to as economic rent), which represents the benefit gained by society

FIGURE 4.2

OPPORTUNITY COST, CONSUMERS' SURPLUS, AND PRODUCERS' SURPLUS

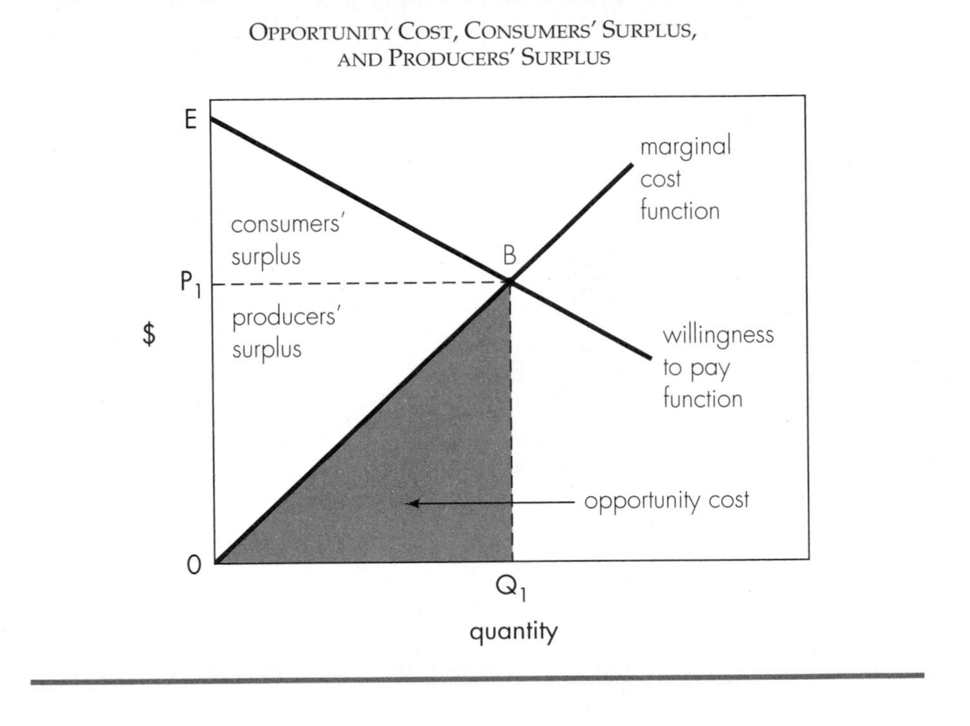

from having these resources utilized in their most productive application. Only the opportunity cost (the productivity of these resources in their next most productive application) is subtracted from total value. Opportunity cost is represented by the area under the marginal cost function, which is the shaded area in Figure 4.2. Net value is therefore equal to area OBE, which for market goods is equal to the sum of consumers' surplus (EP_1B) and producers' surplus (P_1OB). Analogous measures can be developed for nonmarket goods.

VALUE AND NONMARKET GOODS

It should be noted that although money is one thing people give up or trade off to obtain goods, it need not be the only thing that people trade off. Time or other opportunities are sacrificed to obtain both market and nonmarket goods,[2] and an examination of these trade-offs can serve as a basis for valuing goods that have no market price.

[2]Market goods are goods for which the invisible hand of the market generates a market price that is paid by consumers to suppliers. Nonmarket goods are not traded and have no corresponding market price. Nonmarket goods include environmental resources, outdoor recreation, and many other amenities.

Nonmarket goods may have both direct use and indirect use values. Direct use values are those associated with tangible uses of environmental resources, such as when clean water is used as an input in the manufacturing process, when environmental resources are used in recreational activities, or when environmental quality affects human health. Indirect use values are those associated with more intangible uses of the environment, such as aesthetic benefits or the satisfaction derived from the existence of environmental resources. Indirect use values are also called passive use and nonuse values. In addition to existence value, indirect use value would also include bequest value, altruistic value, option value, and the value of ecological services.

Option value is the closest of these preservation values to use value, and it refers to the fact that an individual's current value includes the desire to preserve the option to use a resource in the future. For example, a 20-year-old student may have no current desire to visit the Grand Canyon but may wish to preserve the option to take his or her future children to visit the Grand Canyon.

Bequest value refers to the fact that an individual values having an environmental resource or general environmental quality available for his or her children and grandchildren to experience. It is based on the desire to currently work to raise the well-being of one's descendants. Existence value refers to the fact that an individual's utility may be increased by the knowledge of the existence of an environmental resource even though the individual has no current or potential direct use of the resource. For example, an individual may have a positive willingness to pay to preserve whales even though that individual may get so hopelessly seasick that he or she would never want to go on a whale-watching trip under any circumstances.

Altruistic value occurs out of one individual's concern for another. A person values the environment not just because that person benefits from environmental quality but because the person values the opportunity for other people to enjoy high environmental quality.

None of the values are mutually exclusive. A person who has a direct use value for preservation of old-growth forests (such as a backpacker) may also have option, bequest, altruistic and existence values for old-growth forests.

The value of ecological services (which include nutrient cycling, atmospheric processes, carbon cycling, clean air, clean water, and biodiversity, among other services) are unique among indirect use values in that people do not always know that these services positively affect their well-being. For example, although people recognize the value of clean water, ecological services such as biodiversity and nutrient cycling affect their well-being through a less obvious path than many other types of indirect use values. Since these paths are less obvious, it makes the valuation process more difficult for these very important contributions of the environment to social welfare.

V. Kerry Smith (1993) reiterates and argues persuasively the point made by Charles Plourde (1975) and Kevin McConnell (1983) that these indirect use values are not really values that are unique to environmental resources. Indirect use values are the values from the pure public good aspects of environmental resources, while direct use values are the private good or mixed good (part public and part private) values of environmental resources (see Chapter 2 for a discussion of private and public goods). Smith's 1993 article provides an accessible and insightful perspective on the measurement of direct use and indirect use values associated with environmental resources.

TECHNIQUES FOR MEASURING THE VALUE OF NONMARKET GOODS

There are two major classes of techniques for measuring the value of nonmarket goods: revealed preference and stated preference techniques. Revealed preference approaches look at decisions people make regarding activities that utilize or are affected by an environmental amenity, to reveal the value of the amenity. These approaches focus on measuring direct use value and are not particularly useful in measuring indirect use value. Stated preference methods elicit values directly from individuals, through survey methods. These techniques can be used to measure both direct use and indirect use values.

REVEALED PREFERENCE APPROACHES

At first, one might expect it to be very difficult, if not impossible, to measure the value of an environmental amenity. How can one put a value on an unpriced resource such as clean air? In fact this can be done, and a process for doing it can be illustrated by the example discussed below. Imagine a neighborhood composed of identical houses. Assume that the neighborhood is featureless (it has no attractive facilities, such as parks, and no noxious facilities, such as slaughterhouses). Further, assume that everyone works in his or her home. Under these conditions, all the houses will have identical prices.

Now, imagine that a factory is built in the eastern part of the neighborhood, which emits air pollution and creates smog over the eastern half of the city. Now, people will prefer to live in the cleaner western part of the city. As people try to move from the eastern part of the city to the western part of the city, housing prices will fall in the east and increase in the west, which reflects the increase in demand for houses in the cleaner west and the fall in demand for the dirtier east. There will always be upward pressure on western prices and downward pressure on eastern prices as long as people prefer living in the west over living in the east. The price movement will cease when the price differential is large enough to make people indifferent between living in the clean west and in the dirty east. Since the only difference between the two areas is the higher air pollution in the east, the price differential reveals people's willingness to pay to avoid the air pollution.

Housing is just one area in which people's willingness to pay for environmental quality can be observed. There are many other types of behavior that can reveal an individual's willingness to pay for environmental quality. For example, recreationists will travel further for higher quality outdoor recreational sites (cleaner beaches, better fishing, and so on). Other types of observable behavior include the choice of location among cities, the choice of jobs, and the choice of consumer goods.

Hedonic Pricing Techniques. The prior discussion of air pollution is an example of a hedonic pricing technique, which is one methodology for revealing willingness to pay. Hedonic price techniques are based on a theory of consumer behavior that suggests that people value a good because they value the characteristics of the good rather than the good itself. According to this theory, an individual would not value a car because the car directly gives him or her utility, but because the characteristics of safety, operating cost per mile, luxury, comfort, and status provide utility. If this is the case, an examination of how the price of a car varies with changes in the levels of these characteristics can reveal the prices of the characteristics.

To illustrate this process in an environmental context, let us return to the housing price and air pollution model. Housing prices will be related to a variety of characteristics including attributes of the house itself (number and size of rooms, size of lot, number of bathrooms, quality of construction, and so on) and attributes of the neighborhood (distance to employment centers, level of crime, quality of schools, air quality, and so on).

For the time being, let us assume that all characteristics of houses and neighborhoods are the same throughout the city, except air pollution, which varies with location. Then, houses in the areas with higher air quality will have higher prices. This is reflected in Figure 4.3, where each dot represents the housing price and air quality levels associated with each individual house in the city.

The general upward sloping constellation of these points suggests a positive relationship between air quality and housing prices, which can be formalized through regression analysis. Regression analysis is a statistical process for fitting a line through the cluster of points. In this case, the line would describe the fashion in which improvements in air quality lead to increases in housing prices. For example, an ordinary least-squares regression fits the line by minimizing the sum of the squared values of the distances between each point and the line, as in Figure 4.3.

If H represents the housing price, and Q represents air quality, then the regression will draw the line by choosing the intercept and slope of Equation 4.1:

$$H = a + bQ \tag{4.1}$$

FIGURE 4.3

THE RELATIONSHIP BETWEEN
AIR QUALITY AND HOUSING PRICE

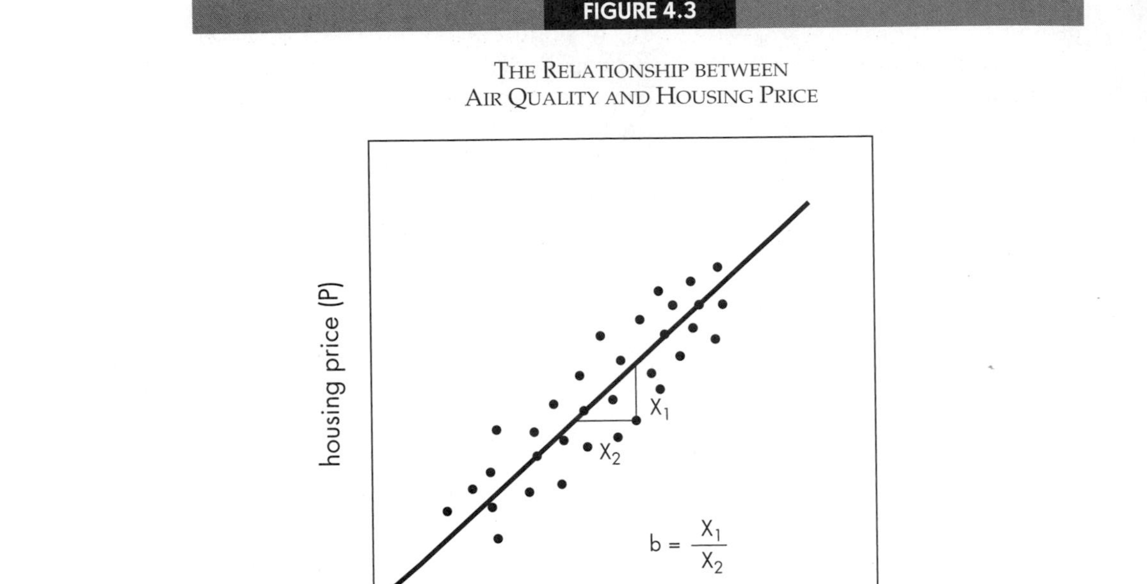

The estimate of b tells the researcher how many units H will increase for each unit of increase in air quality. The estimate b can be interpreted as the slope of the regression line.

Of course, in reality the prices of houses are dependent upon many different characteristics which the researcher needs to consider. For example, the size of the house (S) is likely to have an effect on the price of housing, so Equation 4.2 might better explain the variation in housing prices that one observes when looking at the spectrum of housing prices in a city.

$$H = a + bQ + cS \tag{4.2}$$

A regression analysis chooses a, b, and c by fitting a surface to a cluster of points in three-dimensional space, with each point corresponding to the price, air quality, and size of a particular house. It is conceptually identical to the process depicted in Figure 4.3, but performed in three-dimensional space.

Although it is difficult for a person to conceptualize more than three dimensions, the formulae for computing distance in multidimensional space are relatively simple. The regression analysis can be expanded to include many right-hand side variables, such as the characteristics of the house and the characteristics of the neighborhood. The regression can also be estimated in a fashion that represents a nonlinear relationship among the variables. For

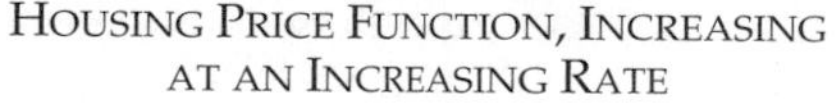

HOUSING PRICE FUNCTION, INCREASING AT AN INCREASING RATE

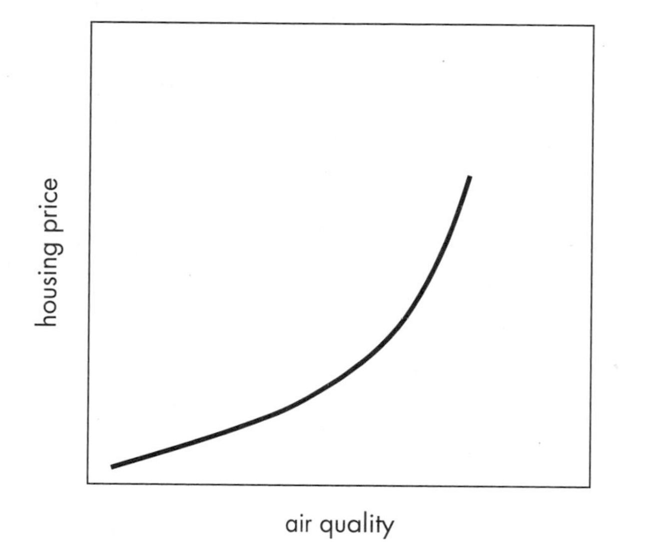

example, housing price is likely to increase with air quality but not necessarily at the constant rate of a linear relationship. Housing prices could increase with increased air quality at either an increasing rate (Figure 4.4) or a decreasing rate (Figure 4.5).

Table 4.1 contains some examples of studies that have related housing prices to air quality to measure the value of air quality improvement. The air quality variable examined is listed in the second column and the willingness to pay for improvement in air quality is contained in the third column.

Hedonic Wage Studies. An analogous technique to the hedonic housing price approach is the hedonic wage approach. The hedonic wage approach is based on the idea that an individual will choose the city in which he or she resides in order to maximize his or her utility. The individual will consider the wage that can be earned in a particular city, as well as a host of other factors including negative characteristics (such as crime, pollution, high cost of living, extreme climate, and congestion) and positive characteristics (such as educational opportunity, recreational opportunity, fine arts, social life, sports, and mild climate). One might question why these market wages will reveal information about people's valuation of nonmarket goods; however, a simple example will show how wages adjust to compensate people for

FIGURE 4.5

HOUSING PRICE FUNCTION, INCREASING AT A DECREASING RATE

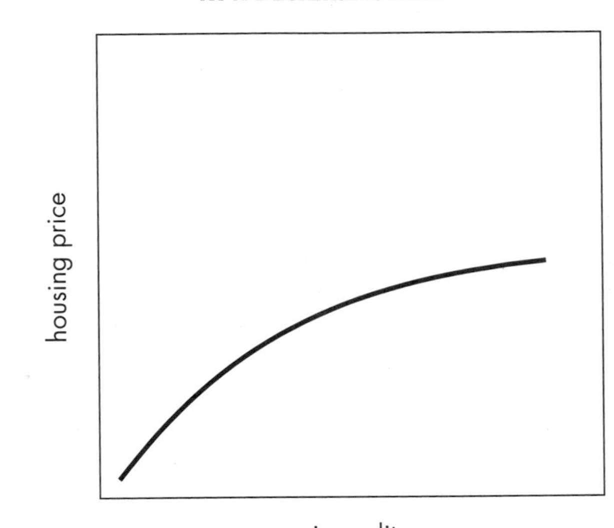

TABLE 4.1

HEDONIC ESTIMATES OF THE WILLINGNESS TO PAY FOR REDUCTIONS IN AIR POLLUTANTS

STUDY	POLLUTANT	HOUSEHOLD WILLINGNESS TO PAY PER UNIT OF AIR QUALITY CHANGE
Ridker and Henning (1967)	sulfur particles	$9.8 per mg/m^3
Nelson (1978)	average summer ozone concentration	$15.17 per ppm
Nelson (1978)	average monthly particulate concentration	$69.34 per $\mu g/m^3$
Harrison and Rubinfield (1978)	NO_X	$1,613 per ppm
Harrison and Rubinfield (1978)	particulate concentration	$18 per ppm
Brookshire, et al. (1982)	NO_2	$45.92 for poor to fair $59.02 for fair to good[A]
Bloomquist, et al. (1988)	particulates	$360 per mg/m^3

[A]poor > 11 parts per hundred million (pphm), fair = between 9 and 11 pphm, good < 9 pphm differential city characteristics. This compensation will be positive for negative characteristics or disamenities and negative for positive characteristics or amenities.

In this example, assume that a person has two job offers, one in a cold weather city (such as Buffalo) and one in a warm weather city (such as Atlanta). Let us also assume that in both cities, the person is offered a job with the same characteristics and the same pay. Which job will the person accept? If we assume that Buffalo and Atlanta are identical in all aspects but weather, then the person will choose Atlanta if he or she prefers warm weather to cold weather. If everybody prefers warm weather to cold weather, then everyone could make themselves better off by moving to warm weather cities such as Atlanta. However, this will increase the supply of labor in warm weather cities and decrease the supply of labor in cold weather cities. Wages will fall in warm weather cities and increase in cold weather cities until people can no longer make themselves better off by moving. The difference between the wage in a warm weather city and the wage in a cold weather city will be just sufficient to compensate people for the disutility associated with the colder weather. Notice that although some people may prefer cold weather to warm weather, there will still be a positive differential associated with cold weather if the marginal worker prefers warm to cold. As mentioned above, many different characteristics will affect people's choice among cities.

Many city characteristics are associated with city size. Disamenities such as crime, pollution, congestion, and cost of living tend to increase as city size increases. Similarly, some amenities are associated with city size, such as cultural opportunities, spectator sports, and social activities.

The hedonic wage method can also be used to gain information about the value of human life. If two jobs are identical in all respects except the risk of accidental death, then the wage in the riskier job must be higher to induce people to accept that job. The estimation of a hedonic wage function, with the degree of risk as an explanatory variable, can be used to quantify this relationship between risk and the willingness to accept higher wages to be exposed to greater risk.

If a study revealed that an individual must be compensated $1,000 per year to accept an additional 0.1 percent (1 in 1,000) annual risk of dying, then the value of saving a life can be computed in the following fashion:

1. Each individual is willing to receive $1,000 to accept an increased 1/1000 risk of dying.
2. The size of the population for which one would expect exactly one person to die is 1,000 people.
3. These 1,000 people have a collective willingness to accept $1 million ($1,000 multiplied by 1,000 people) to be exposed to this risk, which would be expected to lead to the death of 1 individual.
4. The collective willingness to be compensated to accept a loss of 1 life is, therefore, equal to $1 million.

FIGURE 4.6

ESTIMATION OF TRAVEL COST DEMAND CURVE

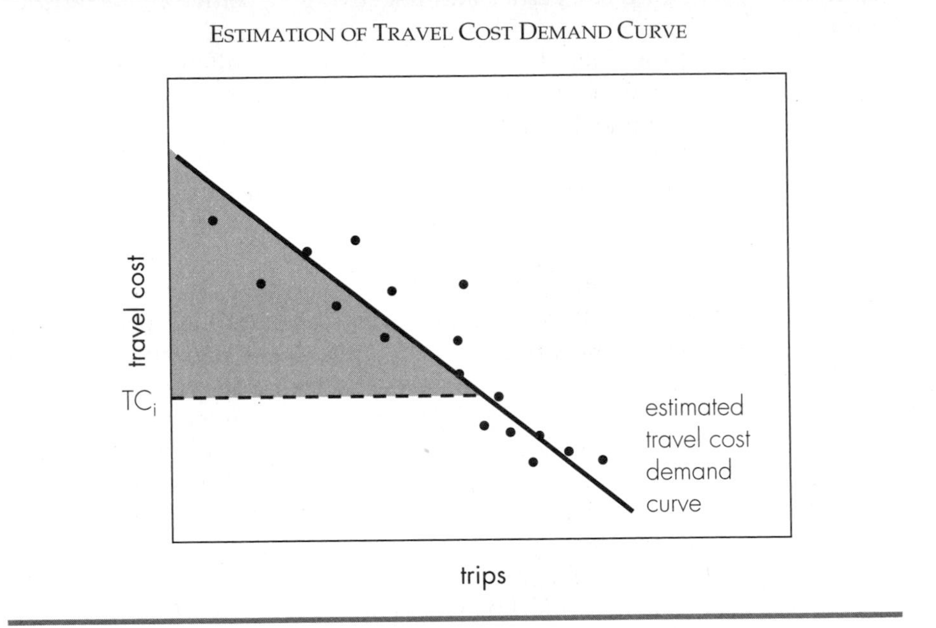

In empirical studies of willingness to pay for reduction in risk, there is considerable variation in the estimates. An informal examination of these studies suggests an average willingness to pay of about $3 million per life saved. However, caution must be exercised when utilizing this type of measure of the value of a human life, because it measures the value of saving a statistical life and not the life of a specific person. If a specific person was faced with the certain prospect of death, that person and his or her family and friends would be willing to pay a near infinite sum to keep the specific person alive. This discrepancy also explains why we as a society are willing to pay tens of millions of dollars to save the life of a specific little girl who has fallen into a well or the life of a specific coal miner who is trapped in a mine, but we will not pay the same (or even lesser amounts) for highway improvements that would save the lives of more (but unidentified) individuals.

Both the hedonic housing price model and the hedonic wage model are powerful tools in estimating the health costs associated with air and water pollution. As long as people are aware of the health consequences of living in polluted areas, housing prices and wages should reflect their willingness to pay to avoid the health risk. Even if this information requirement is not met for some pollutants, one can value the benefit of a reduction in these pollutants by looking at an analogous risk for which people are better informed.

Travel Cost Model. The travel cost model is a method for valuing environmental resources associated with recreational activity. The travel cost method was first proposed by Harold Hotelling in 1947. The basic premise behind the model is that the travel cost to a site can be regarded as the price of access to the site. If recreationists were asked questions about the number of trips they take and their travel cost to the site, enough information would be generated to estimate a demand curve.

Figure 4.6 contains a collection of such observations on travel cost and number of trips. In this figure, each point represents the combination of travel costs and number of trips reported by an individual. Using regression analysis, a line is fitted through these points, and that line represents an individual demand curve. The value of the site to an individual can be estimated by computing the consumer surplus for each individual in the survey, averaging the consumers' surplus, and then multiplying by the estimated number of recreationists. Consumers' surplus is the area under the demand curve and above the price (the travel cost) that the individual actually incurred. For example, if TC_i was the travel cost paid by the individual, the individual consumer's surplus would be equal to the shaded triangle in Figure 4.6. The travel cost demand curve is often expanded to include other explanatory variables incurred, such as age, income, family size, educational level, and other socioeconomic variables.

Table 4.2 lists annual consumers' surplus associated with different types of recreational activities for a selected group of studies. For a comprehensive discussion of how and why consumers' surplus estimates vary across studies, see Smith and Kaoru (1990a).

The studies in the following table give the value of the recreational activity or the value of the sites that provide the opportunities. They do not give information about the value of changing the quality of the activity, which is the information one would need to know to determine the value of environmental improvement. For example, increased water quality will increase fish populations, which will increase recreational fishers' utility. How can the travel cost method be used to measure their willingness to pay for this increased water quality?

Two travel cost demand curves are contained in Figure 4.7. The lower one (D_1) represents an individual's willingness to pay for recreational fishing trips at the current lower level of water quality. The upper one (D_2) reflects willingness to pay for recreational fishing trips with an improved level of water quality. If P_1 represents the travel cost actually incurred by the typical recreational fisher, then the area ABCE represents the value of the improvement in water quality to that recreational fisher. Note that the improvement in water quality would generate an increase in the number of trips from Q_1 to Q_2.

In order to measure how the travel cost demand curve shifts as environmental quality shifts, the travel cost demand curve must be estimated with quality as an explanatory variable. For example, in the recreational fishing context, the travel cost demand curve could be estimated using the number

TABLE 4.2

CONSUMERS' SURPLUS ESTIMATES FROM SELECTED STUDIES
(1980 DOLLARS UNLESS OTHERWISE INDICATED)

	STUDY	REGION AND ACTIVITY	CONSUMERS' SURPLUS
SALTWATER FISHING	Bockstael, Hannemann, and Strand	Chesapeake Bay and Ocean City, MD fishery	$2,054 to $2,564 (1978 dollars)
	Bockstael, McConnell, and Strand	Southern California private boat fishery	$2,703 to $4,148 (1983 dollars)
	Bell, Sorenson, and Leeworthy	Florida resident saltwater fishing	$401 to $1,033
FRESHWATER FISHING	Kealy and Bishop	Lake Michigan fishery (Wisconsin side)	$625 (1978 dollars)
	Kahn	Great Lakes fishing	$300 to $700
HUNTING	Balkan and Kahn	U.S. deer hunting	$1,043
BEACH USE	Bockstael, Strand, and Hannemann	Chesapeake beaches	$1,080 to $2,757
	Bell and Leeworthy	Florida tourist beach users	$203 (1984 dollars)
		Florida resident beach users	$286
OTHER	Rockel and Kealy	Nonconsumptive wildlife-associated recreation	$1,282 to $2,404

SOURCE: Kahn, 1991.

of trips as the dependent variable, and travel cost, socioeconomic variables (age, income, family size, and so on), and the average catch per day at the fishing site or sites as the explanatory variables. Then, fishery biologists would be asked to help establish the relationships among water quality, catch, and fish populations to establish the links between water quality and value. Table 4.3 contains information from empirical studies of recreational activities, which shows how consumers' surplus varies with quality.

Although the travel cost procedure has been widely used and holds great promise for further understanding the way that environmental quality affects the value of recreational activities, many methodological issues remain unresolved. These include:

1. how to incorporate the opportunity cost of travel time into the measure of travel cost;
2. how to properly account for substitutes (multiple sites) in estimating a travel cost demand curve;

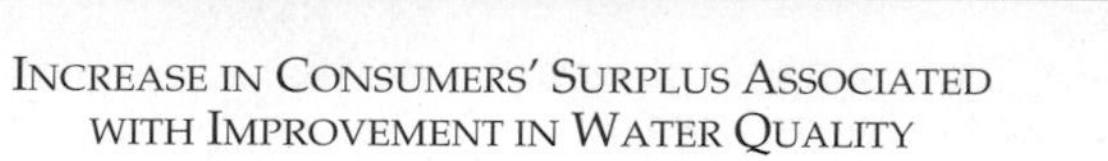

FIGURE 4.7

INCREASE IN CONSUMERS' SURPLUS ASSOCIATED WITH IMPROVEMENT IN WATER QUALITY

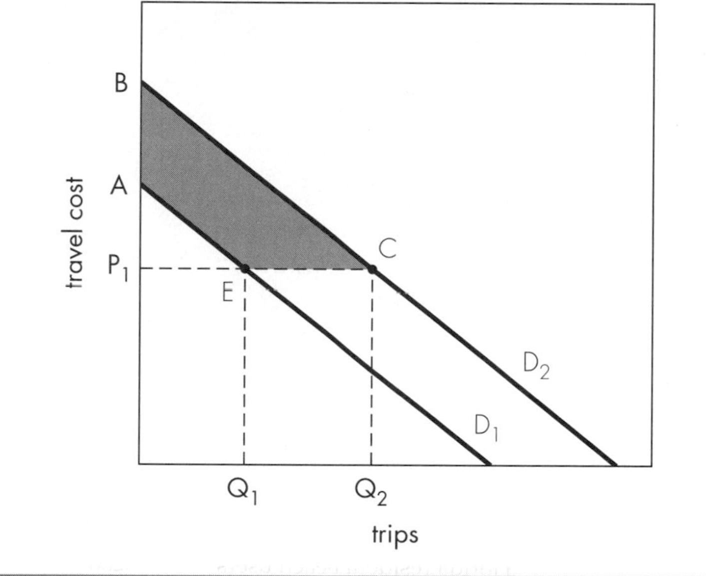

3. how to account for a variety of sampling biases that arise when data is collected by interviewing recreationists at recreation sites;
4. how to properly measure recreational quality and relate recreational quality to environmental quality.

STATED PREFERENCE TECHNIQUES

Stated preference valuation techniques are somewhat different than the revealed preference techniques already discussed. The indirect techniques are designed to measure the value of environmental resources by looking at how people's actual behavior (choice of city, house within a city, job and/or recreational activities) changes as the level of environmental quality changes. Stated preference techniques do not have this link to actual behavior. Instead, they solicit value measures directly by asking individuals hypothetical questions.

Contingent Valuation. The most widely used stated preference valuation technique is the contingent valuation method. The contingent valuation method (CVM) ascertains value by asking people their willingness to pay for a change in environmental quality. One of the reasons many researchers are

TABLE 4.3

CROSS STUDY COMPARISON OF BENEFIT ESTIMATES OF CHANGES IN SPORT FISH CATCH RATES

STUDY	DATE	LOCATION	SPECIES/TYPE	MODEL[A]	CATCH RATE % VALUE
Bockstael, McConnell, and Strand (1988b)	Nov/Dec 1987	Florida (Atlantic Coast)		RUM Nested Logit	+(20%)
			Big Game		$1.56
			Small Game		$0.33
			Nontargeted Small Game		$0.32
			Bottom Fish (from boat)		$1.27
Smith and Palmquist (1988)	Summer 1981 and 1982	Albemarle/ Pamlico Estuary	All Species	RUM Nested Logit	+(25%)
					(per choice occasion)
		All Sites	All Species		$ 2.43
		Closest 4 Sites			$ 0.60
		Outer Banks Range Across Most Sites	All Species	CTC	$ 1.74 to $ 38.07
		Pamlico Range		CTC	$ 1.99 to $ 14.24
Huppert (1988)	July 1985– June 1986	San Francisco Bay and Ocean Area	Anadromous Fish Runs	CVM	−(50%) (per year)
			(Salmon and Striped	EV	$ 37.12
			Bass)	CV	$ 83.71 +(100%)

enthusiastic about contingent valuation is that it enables them to measure values that are not related to a direct use of the resource.

The contingent valuation method is based on asking an individual to state his or her willingness to pay to bring about an environmental improvement, such as improved visibility from lessened air pollution, the protection of an endangered species, or the preservation of a wilderness area. Questions have been asked in a variety of ways, using both open-ended and close-ended questions. In open-ended questions, respondents are asked to state their maximum willingness to pay. In close-ended questions, respondents are asked to say whether or not they would be willing to pay a particular amount to preserve the environmental resource.

In addition to the hypothetical question of willingness to pay, the contingent valuation procedure must specify the mechanism by which the payment will be made. In order for the question to be effective, the respondent must

TABLE 4.3

CONTINUED

				CV CTC EV CV CV	$ 43.05 −(50%) $126.76 $127.34 +(100%) $ 69.58
Morey and Shaw	1976/77	Adirondacks (Saranac, Raquette, Lake Placid, and Piseco)	Brook and Lake Trout (Average)	SHR	+(25%) (per season) $ 8.68 +(50%) $ 15.51
Bockstael (1988b)	1980	Chesapeake Bay	Striped Bass	CTC	+(20%) (per trip) $ 2.00 to $ 6.30
Milon (1988b, 1988c)	1981 and 1986	Gulf of Mexico	King Mackerel Average for 4 Sites	RUM	+(25%) (per trip) $ 5.40 +(50%) $ 16.70

SOURCE: Brown, et al., 1990.

[A]The guide to symbols is RUM for Random Utility Model; CTC for Conventional Travel Cost (includes Simple Travel Cost and Varying Parameter/Regional Travel Cost Models); CVM for Contingent Valuation Method (EV for Equivalent Variation and CV for Compensating Variation); and SHR for Share Model.

believe that if the money was paid, whoever was collecting it could effect the specified environmental change. An example, of such a question, which specifies a payment mechanism (or vehicle, as it's known in the contingent valuation literature), follows.

> *Water quality in the Tennessee River is adversely affected by problems of contamination from combined sewage outflow. This problem is caused by the connection between storm sewers and sanitary sewers, which causes the sewer system to overflow when there are heavy rains and spill untreated sewage into the Tennessee River. If the storm runoff was handled through a separate system, it would eliminate the spillage of untreated sewage, significantly improving water quality. The water, which is currently too polluted to swim in, would become safe for swimming. Would you be willing to pay an additional $X per year in your monthly sewer and water bill if the money was used to correct this problem?*

NATURAL RESOURCE DAMAGE ASSESSMENT AND VALUATION METHODOLOGIES

The Comprehensive Environmental Response, Compensation, and Liability Act of 1980 (CERCLA) establishes the legal right of the trustees of natural resources (local governments, state governments, the federal government, and Native American nations) to collect damages from firms or individuals who release hazardous substances that then damage or destroy environmental resources.

These liability cases are often settled out of court, but many times the cases go to trial. In order to collect damages in court, the trustees must establish the existence of damages. The federal government, under the provisions of CERCLA, has established guidelines for the methodologies that have legal standing in these proceedings. Both the travel cost technique and the contingent valuation method have attained this status, and the use of these methodologies may not be challenged by the parties to the dispute (although the specifics of how they are used may certainly be challenged in court). One of the most prominent cases in which these techniques were used was the measurement of damages in the Exxon Valdez spill, which was eventually settled out of court. Since this case was settled out of court, many of the estimated values have not been made public.

Since there is such controversy surrounding the use of contingent valuation in natural resource damage assessment, the National Oceanic and Atmospheric Administration commissioned a "Blue Ribbon" panel of experts to investigate contingent valuation and make recommendations as to the best ways it could be employed in assessing the monetary value of damage from oil spills and other compensable environmental accidents. These findings have been published in the Federal Register (see Department of Commerce, 1993).

Several aspects of the structure of this question are important. First, information was provided about the cause and significance of the environmental problem. Second, the payment vehicle was appropriate to the particular problem. The payment question was structured as a closed-ended question, where different values of $X are specified for different individuals. Although these close-ended questions only let us know that the individual is willing to pay at least the particular amount, appropriate statistical techniques exist to derive willingness to pay measures from these responses.

Care must be taken that the contingent valuation exercise does not become a referendum on the payment vehicle. For example, in some communities high property taxes are a subject of political contention. If a researcher was to use a property tax payment vehicle in a contingent valuation questionnaire, respondents' stated values might be biased by the fact that people already think property taxes are inappropriately high, and they may react to the tax rather than the environmental issue.

Although the contingent value method has been widely used for the past two decades, there is considerable controversy over whether it adequately measures people's willingness to pay for environmental quality. These arguments are based on informational issues and on the fact that people may be indicating their value of something other than the particular environmental issue that is the subject of the questionnaire.

The informational issue revolves around the fact that with market goods, people have practice making choices, so their purchasing decisions are likely to reflect their true willingness to pay. Consumers buy clothes frequently, so when they see a clothes item for a given price, they are likely to know if their true willingness to pay is greater than the price. However, we do not have practice valuing threatened species, so if persons are asked if their willingness to pay to save the black-footed ferret from extinction is greater than $40 per year, they are not certain that their true willingness to pay is greater than $40. Information is a big problem as well, as persons may not know why black-footed ferrets are important to biodiversity or how biodiversity affects their utility function.

The second issue revolves around the fact that the expressed answers to a willingness to pay question in a contingent value format may be biased because the respondent actually answers a different question than the surveyor intended. For example, the respondent may state a positive willingness to pay, because the respondent wants the surveyor to think he or she is a good person. Alternatively, the respondent may express a willingness to pay because he or she feels good about the act of giving for the social good, and the social cause itself is unimportant. Finally, a respondent may just be trying to signal that he or she places importance on improved environmental quality by stating a positive willingness to pay.

Another problem is that the respondent may make associations between environmental goods that the researcher had not intended. For example, if the question asks for the willingness to pay for improved visibility through reduced pollution, the respondent may actually answer based on the health risks that he or she associates with dirty air. This may seem appropriate; however, a big problem of double counting may develop if health risks are examined separately. Similarly, researchers have found that there is very little difference between stated willingness to pay for a more encompassing set of environmental resources and a less encompassing set. For example, researchers have found that the willingness to pay for improved air quality in a system of Rocky Mountain parks is not too different from the willingness to pay for improved air quality in one Rocky Mountain park (William Schultze, 1991). This may occur for some of the reasons stated previously, or because people do not believe that it is possible to increase air quality in one park without increasing air quality in another park.

Some researchers argue that there is a fundamental difference in the way that people make hypothetical decisions relative to the way they make actual decisions. A series of experiments looked at hypothetical versus actual decision making. Some experiments (M. S. Kealy, M. Montgomery, and J. F. Dovido, 1990; Ronald Cummings and Glen Harrison, 1992) ask the hypothetical willingness to pay for a private good (candy, electric juicer, calculator, etc.) and then offer respondents the opportunity to purchase at the stated willingness to pay. Many of the survey respondents do not elect to purchase at that price. Similar experiments have been conducted with public goods with similar results. (J. W. Duffield and D. A. Patterson, 1992; Kalle Seip and Jon Strand, 1992).

Although these experiments provide some evidence that hypothetical values exceed actual values, one cannot conclude that this result is definitely true. So few experiments have been conducted that it is difficult to determine if the differences between hypothetical and actual values are a result of the experimental method or if they truly reflect the way people behave. As with many issues in nonmarket valuation, the answers await further research.

It should be emphasized that the significance of the problems mentioned above is highly debated. While some researchers hypothesize the existence and insurmountability of these problems, other researchers feel that these are technical issues that can be resolved by proper survey design. Contingent valuation remains an important technique in the researcher's valuation arsenal. However, many of these issues need to be resolved in order for researchers and policy makers to have more confidence in the value measures resulting from these techniques. Research has been proceeding at a rapid pace, and many studies have taken great care to minimize the magnitude of the problems discussed above. However, many questions still must be answered before contingent valuation receives general acceptance among environmental economists, scholars from other disciplines, and environmental decision makers.

Conjoint Analysis. Conjoint analysis is a technique employed by researchers in marketing, transportation, and psychology for determining individual preference across different levels of characteristics of a multiattribute choice. For example, a marketing study might ask potential consumers to state which of two hypothetical cars they prefer, with each car having a stated level of different characteristics such as price, roominess, reliability, safety, fuel economy, power, and so on. These choices can be made in a pair-wise fashion or by ranking a number of alternatives, although economists generally prefer the pair-wise choices. Statistical techniques are then used to establish a relationship between the characteristics and preference. As long as one of the characteristics is price, it is possible to use the preference function to derive the willingness to pay for changes in the levels of the other characteristics. It should be noted that conjoint analysis can be viewed as a hedonic price technique, where hypothetical price substitutes for actual market price.

The applications of conjoint analysis to valuing other types of environmental resources can be seen immediately. For example, if one was trying to value forest quality, one could define characteristics such as the age of trees, the diversity of trees, the diversity of other organisms, soil productivity, water quality, and so on. Then one could use conjoint analysis to determine the importance of the various characteristics.

Alternatively, one could use conjoint analysis to directly value alternative national environmental policy scenarios. For example, respondents can be asked to compare alternative scenarios with different levels of clean-up of toxic waste sites, different levels of acid precipitation, different levels of global warming, different levels of ambient air quality, different levels of old-growth forests, and most importantly, different tax burdens for the respondent. The respondents' preference ordering will then allow a determination of the willingness to be taxed in order to obtain different levels of the environmental variables.

One advantage of conjoint analysis over contingent valuation is that it does not ask respondents to make a trade-off directly between environmental quality and money. Rather, it asks respondents to state a preference between one bundle of environmental characteristics (at a given tax burden) and another bundle of environmental characteristics at a different tax burden. Although environmental characteristic/money trade-offs can be statistically derived from conjoint analysis, respondents do not perceive the ranking as a direct trade-off between environmental quality and money. More importantly, conjoint analysis presents a type of choice with which respondents are more familiar. In contingent valuation, respondents are asked to state their willingness to pay for a nonmarket good. As mentioned earlier, this is the type of decision with which the individual is not typically familiar. However, conjoint analysis asks respondents to choose between two states of the world, which have different levels of both market and nonmarket goods. This is a choice process with which we are all familiar and practiced. We engage in this type of choice process when we are faced with voting Democratic or Republican, getting married or staying single, choosing a major in college, or choosing which college to attend. Since conjoint analysis is based on a more familiar choice process, it should, therefore, reduce problems with protest bids, signaling, and some of the other sources of potential bias associated with contingent valuation.

Although conjoint analysis has great potential for valuing environmental resources, there has not been much experience with it in this application. More research needs to be done with this technique, particularly comparison studies with contingent valuation.

THE USE OF VALUE MEASURES IN DETERMINING ENVIRONMENTAL POLICY

Although many questions need to be answered to fine tune the various valuation methodologies, these methods can be very effective in measuring many types of values associated with environmental change. These measures can be used to help formulate environmental policy in two contexts: cost–benefit analysis and marginal damage analysis.

USE OF CONJOINT ANALYSIS TO VALUE RECREATIONAL FISHING QUALITY

Recreational fishermen and fisherwomen would be presented with cards with different characteristics of recreational fishing trips printed on the cards; for example, they might be asked: Which fishing trip would you prefer?

Trip A, which has the following characteristics:

1. The water in the river is sufficiently clear that you can see the bottom in water of three feet or less.
2. You see a pair of bald eagles during the trip.
3. Other fishers get in your way four times during the day.
4. You catch three largemouth bass of over three pounds each and two under two pounds each.
5. You are required to pay an entrance fee of $11.

Trip B, which has the following characteristics:

1. The water in the river is sufficiently clear that you can see the bottom in water of one foot or less.
2. Other fishers get in your way five times during the day.
3. You catch one largemouth bass of over three pounds and six under one pound each.
4. You are required to pay an entrance fee of $5.

This type of question would be repeated several times for each fisher in the sample, and one would obtain a sample of at least several hundred anglers. The level of the characteristics (including price) would be changed to obtain the variation needed to statistically determine the relationship between each characteristic and preference.

Cost–benefit analysis looks at a decision, such as the decision to build a dam, and determines whether the benefits are greater than the costs. If the benefits exceed the costs, then the dam is built. If the costs are greater than the benefits, then the decision is made not to build the dam. The important thing to note is that cost–benefit analysis is used when the policy maker has a discrete decision to make. Either build the dam or not. It is not a decision to choose a level out of a continuous spectrum but to choose among a finite number of alternatives, although there may be more than two alternatives. For example, the choice may be to build the dam at site A, build the dam at site B, or not build the dam at all. The alternative with the largest net benefits should be chosen.

Marginal analysis is used when the choice must be made from a continuous spectrum of alternatives, rather than a discrete set of choices. For example, the choice in setting air quality is not between zero sulfur dioxide emissions and a particular level of sulfur dioxide emissions, but what level of sulfur dioxide (out of a continuous spectrum of possible levels) is optimal. As pointed out in Chapter 3, this decision is made by choosing the level of sulfur dioxide emissions for which the marginal damage of another unit of pollution is equal to the marginal cost of abating the pollution.

POTENTIAL PARETO IMPROVEMENTS AND ENVIRONMENTAL POLICY

In both benefit–cost analysis and marginal analysis, benefits are being compared to costs with the idea of improving society's well-being. Although many criteria have been suggested as to what constitutes an improvement in social welfare, we shall use the criterion most employed by economists—the Hicks-Kaldor criterion, or the criterion of a potential Pareto improvement. A Pareto improvement is said to occur when resources are reallocated in a fashion that makes some people better off and no one worse off. In a complex real world, any reallocation is likely to hurt someone, so economists have adopted the criterion of a potential Pareto improvement. A potential Pareto improvement is a reallocation where the gain of people who are helped is larger than the losses of those who are made worse off. It is called a potential Pareto improvement because if the gainers would compensate the losers for their losses, the gainers would still be better off. After compensation, the potential Pareto improvement is an actual Pareto improvement.

Economists believe that public policies should be pursued provided that these policies improve the social welfare, with the potential Pareto improvement viewed as the operational definition of improved social welfare. Note that no compensation need take place in order for the change to be deemed desirable.

It also should be noted that this is not the only criterion on which policy should be based. For example, equity, or fairness, is also an important criterion. Cost–benefit analysis examines the magnitude of costs and benefits, but the question of who receives the costs and benefits, as well as the fairness of this allocation, is important as well. For example, most people would argue that a policy that benefitted billionaires at the expense of minimum wage earners was not desirable even if it meets the criterion of being a potential Pareto improvement. Equity and other additional criteria for environmental decision making will be further discussed later in the chapter.

Cost–Benefit Analysis

Cost–benefit analysis is one of the most used— and abused tools—available to policy makers. It is often abused by policy makers who use it subjectively to justify the positions they take. How can an objective tool like cost–benefit analysis be applied subjectively? The practitioner of cost–benefit analysis must make many subjective decisions before analysis, such as what costs and benefits must be included in the analysis, how the costs and benefits should be measured, and what discount rate should be used to compute the present value of future costs and benefits. The focus of this section will be to explain how these decisions should be made in order to generate an objective analysis and to recognize improper assumptions and methods so that potentially biased cost–benefit studies can be recognized.

The best way to illustrate the ins and outs of cost–benefit analysis is with an example. Imagine there is a river for which the construction of a dam is proposed. The first step in the process is to list the potential costs and benefits. This step is done in Table 4.4, to which we will refer in the next several sections.

TABLE 4.4

POTENTIAL COSTS AND BENEFITS OF BUILDING A DAM

POTENTIAL COSTS	POTENTIAL BENEFITS
1. Costs of construction (years 1 through 7)	1. Electrical generation (years 8 through life of dam)
2. Costs of operation and maintenance (years 8 through 50)	2. Flood control (years 8 through life of dam)
3. The opportunity cost of the land that is inundated (years 8 through infinity)	3. Reservoir recreation (years 8 through life of dam)
4. Loss of the services of a free-flowing river	
a. loss of river recreation (years 8 through infinity)	
b. loss of riverine ecosystem (years 8 through infinity)	

Discount Rates. One important aspect of understanding cost–benefit analysis is understanding the time horizon of the various costs and benefits. The time horizon (the period in which the costs and benefits are realized) is listed for each cost and benefit in Table 4.4. Since this is just a hypothetical dam, we will assume that it takes seven years to build the dam, and that the economic life of the hydroelectric facility is 50 years. These figures, particularly the duration of construction, would vary substantially for different real world hydro-projects.

Of course, since many of the costs and benefits occur in future years, they must all be converted into present value in order to make the cost–benefit comparison. The mathematical process of discounting was explained in the first appendix to Chapter 2. To reiterate, if the present value is denoted *PV*, the future value is denoted *FV*, the discount rate is denoted *r*, and the length of time into the future is denoted *t*, then:

$$PV = FV\left[\frac{1}{(1+r)^t}\right] \tag{4.3}$$

This formula can be extended to a project that has costs and benefits every year over some period of time. If v_i is equal to the net benefits in any year *i*, then the present value of the project can be described as:

$$PV = \sum_{i=0}^{t}\left[\frac{v_i}{(1+r)^i}\right] \tag{4.4}$$

A careful examination of the above equation will show that the higher the discount rate, the less important are the future costs and benefits; therefore, the present value of the net benefits of a project will be critically dependent upon the choice of discount rates. For example, a nuclear power plant generates benefits in the first few decades of its existence, and then it generates costs over the next few millennia since long-lived radioactive wastes must be safely stored for tens of thousands of years. A high discount rate will give less weight to those future costs and make it more likely that a proposed nuclear power plant would pass a cost–benefit test. Richard Porter (1982) provides an excellent discussion of the sensitivity of the net benefits of the economic development of an environmental resource to the choice of discount rates.

As one might expect, the choice of discount rates is the subject of much controversy. There are two basic positions. One position is that the social discount rate should be set equal to the risk-free real market rate of interest. The other is that the social discount rate should be less than the risk-free real market rate of interest.

What Is the Risk-Free Real Market Rate of Interest? Before presenting arguments as to why many economists feel that the social discount rate should be equal to the risk-free market rate of interest, it should be made clear exactly what is meant by this rate of interest. Any market rate of interest (such as the prime rate, mortgage rates, or credit card interest rate) has three components: an inflation component, a risk component, and a real interest rate.

The inflation component is relatively easy to understand. The higher inflation, the higher the rate of interest must be in order to compensate the lender for the loss of purchasing power generated by inflation before the loan is repaid. Since future inflation is not very predictable, cost–benefit analysis adjusts for inflation and expresses everything in inflation-adjusted dollars (also called constant dollars or real dollars). A real rate of interest is simply a market rate of interest with the inflation rate taken out. For example, if a nominal (unadjusted for inflation) market rate of interest is eight percent and the inflation rate is five percent, then the real rate of interest is three percent. Since cost–benefit analysis accounts for the loss of purchasing power by adjusting costs and benefits for inflation and expressing these values in real values, a real rate of interest must be employed in discounting.

The other component of the market rate of interest that must be netted out before discounting is the risk premium. In general, the higher the risk of default on a loan, the higher the interest rate the lender will demand in order to make the loan. However, economists argue that as a society, we should ignore the risk component, because society's portfolio of investments is so diversified that the risk is eliminated. Economists in the United States use the interest rate on long-term U.S. Treasury bonds as a proxy for the risk-free rate, as the probability of the U.S. government defaulting on these obligations is close to zero. Of course, this rate has an inflation component that must be subtracted from the nominal interest rate to obtain the real rate. The current 30-year Treasury bond rate may be found in the Credit Market section of *The Wall Street Journal*.

The Risk-Free Real Market Rate of Interest as the Social Discount Rate. The basic argument behind using the risk-free market rate of interest as the social discount rate is that it reflects the aggregation of individual rates of time preference. It is easy to show that the market rate of interest reflects the aggregate social rate of time preference with the following example. If the market rate of interest is higher than most people's rate of time preference, then these people will try to lend money, and more money will be supplied than demanded. This would put a downward pressure on the market rate of interest. Conversely, if most people's rate of time preference is higher than the market rate of interest, more people will try to borrow money than to lend it. This situation would put an upward pressure on the market rate of interest. As can be seen in the above discussion, the market rate of interest is determined by the forces of supply and demand, which are in turn driven by the aggregation of individual rates of time preference.

Another reason for choosing the risk-free real market rate of interest has to do with the opportunity cost of government projects or activities. The opportunity cost of devoting resources to a project (or causing firms or households to expend resources due to new rules or regulations) is the cost of using these resources in a private investment. A firm that contemplates making an investment compares the return from the investment to the cost of borrowing (or the opportunity cost of not lending their own funds to someone else). Thus, at the margin, the market rate of interest reflects the rate of return on private investment, which is the opportunity cost of government investment.

An additional reason for choosing the risk-free market rate of interest as the social discount rate has developed over the past ten or fifteen years. This reason also has to do with the opportunity cost of government expenditure of resources. If the government does not expend resources on a particular project, these funds could be used to retire a portion of the national debt. The cost of not retiring a portion of the debt is the interest that must be paid upon it. Since the debt is generally financed by the sale of long-term bonds, the interest rate paid on the national debt is, in fact, the risk-free market rate of interest.

Should the Social Discount Rate Be Lower than the Risk-Free Real Market Rate of Interest? Arguments that the social discount rate should be lower than the risk-free real market rate of interest revolve around the idea that citizens feel society as a whole should have a lower rate of time preference than people have as individuals. In other words, they feel that society as a whole should be more "future oriented" than individual members of society.

One reason for this opinion is that individual rates of time preference reflect the trade-off of current consumption for future consumption within an individual's lifetime—or sometimes extending to that individual's children or grandchildren, but seldom beyond that. Generations beyond that are not generally part of the present generation's planning horizon, and future generations might wish that the current generation had a lower rate of time preference and had more adequately considered the future.

The question of intergenerational equity is an extremely important one, since many people feel that discounting is unfair to future generations. A project that yields small net benefits to the current generation and large net costs to all future generations could pass a present value-based cost–benefit test with virtually any positive discount rate.

For this reason, many people advocate a different criterion, such as giving each generation one vote. In order to operationalize such a criterion, the current generation must predict how each future generation would vote. This procedure, which would eliminate the need for discounting, is advocated by many noneconomists.

However, if one uses the criterion of a potential Pareto improvement as a means for allocating resources, then one particular discount rate may be most appropriate when considering future generations. This rate is the average real rate of growth of GDP (Gross Domestic Product), which has historically been between two percent and three percent.[3]

Why should one use the real rate of growth of GDP in discounting? The answer is that if we want to know whether future generations could be compensated for the costs they incur, we would have to be able to determine if the present generation's benefits could grow to be larger than the future generations' costs. Since no component of GDP could grow faster than the entire GDP for long periods of time, the real rate of growth of GDP will determine whether compensation could potentially take place.[4]

The use of discounting to determine the existence of intergenerational potential Pareto improvements does not eliminate the intergenerational equity issue. For example, future generations never get a chance to agree on the fairness of the potential Pareto criterion. However, from a fairness perspective, the use of the rate of growth of GDP as the social discount rate has the advantage of treating reallocations of resources across generations in the same fashion as reallocations within a generation. Despite the fact that the use of this rate is more consistent with intergenerational equity, further consideration of intergenerational equity and other equity issues will generally be required.

Measuring Costs in Cost–Benefit Analysis. We will return to discussion of the hypothetical dam (see Table 4.4) to give some insights into how to measure

---

[3]GDP is a measure of national income. When measuring the growth rate of GDP, it is imperative that GDP be measured correctly. See Chapter 5 for a discussion of measurement issues associated with GDP.

[4]For example, if GDP equaled $1 trillion and grew at two percent a year, a $10 million investment (which is equal to the approximate cost of a typical university classroom building) that grew at five percent a year would exceed GDP in less than 400 years. This illustrates the fact that with exponential growth, no part can grow faster than the whole for long periods of time.

costs. The guiding principle is to identify all the opportunity costs associated with the dam.

The costs of construction are rather straightforward. Market prices of steel, equipment, labor, and so on, reflect their opportunity costs, with the exception of the pollution costs associated with the manufacture of materials such as the steel and cement. However, most cost–benefit studies do not attempt to include the pollution externalities associated with the construction materials, for two reasons. First, the pollution costs are likely to be small relative to the costs of the materials. Second, since pollution abatement is required of industries, such as steel and cement manufacturers, some of the social costs of the externality are incorporated into market price.

The costs of operation and maintenance are also relatively straightforward to measure. Again, market prices of material and labor tend to reflect opportunity cost. These operation and maintenance costs are incurred every year the plant is in operation. Even after the plant ceases to produce electricity, maintenance will still be needed to ensure the safety of the dam.

It should be realized that the lack of concern about pollution externalities associated with the operation of the hydroelectric facility is not a general proposition and is certainly not applicable for all electric plants. For example, pollution costs will be a major component of the operating costs of a coal-burning power plant. Of course, there are corresponding environmental costs associated with the hydroelectric plant, including the inundation of land and the alteration of the river, but these costs will be analyzed separately from the conventional operating costs.

When the dam is built, a large amount of land will be inundated by the reservoir. The opportunity cost of this land must be included as a cost of the project. This opportunity cost of the land is not necessarily the price that the dam builders pay to obtain title to this land.

If the land that will be under the future reservoir is farmland, its opportunity cost can be computed relatively easily by looking at the value of similar land in the same region. The costs of relocating the dislocated families should also be factored into the analysis.

If the land that will be under the future reservoir is natural environment or is land associated with important cultural significance, then the market price of the land will not reflect its opportunity cost. The types of valuation techniques discussed earlier in this chapter must be employed to measure the opportunity costs of the land in this case. Similar techniques can measure the losses of the ecological services of the river.

Finally, the costs of the loss of river recreation must be measured. Canoeing, white-water rafting, boating, tubing (floating on truck tire inner tubes), fishing, and hiking are important recreational activities associated with free-flowing rivers. Of particular importance is the effect of dams on anadromous fish (fish such as salmon, steelhead, and striped bass that live in the ocean but return to rivers to spawn). These recreational values are likely to be important because there are few free-flowing (undammed) rivers left in most regions of the United States. In parts of Canada and Alaska, there are more free-flowing rivers.

Measuring Benefits in Cost–Benefit Analysis. The major benefit of a hydroelectric facility is the electricity that is produced. Since the cost of producing the electricity and the environmental impacts will be measured separately, one would think that the benefit of the electricity could be measured as the price of electricity multiplied by the amount that is produced. However, electricity is produced by a regulated monopoly, and price does not always equal marginal social benefit. Also, since the electricity could be produced from a different plant, the real benefit is the difference between the cost of producing electricity in the hydroelectric facility under consideration and the next cheapest alternative.[5] Of course, measuring the cost of the next best alternative should include all the social costs, such as air pollution from a coal-burning plant or the costs of storing radioactive wastes generated by a nuclear power plant.

Measuring the dam's flood control benefits is a little more difficult because floods are not completely predictable. However, climatologists and hydrologists can estimate the annual probability of floods of a given intensity. The annual probability of a given intensity of flooding can be multiplied by an estimate of the damages that would occur if a flood of this intensity happened. The results would then be summed over the different intensity levels to give the expected damages in a given year. These damages would include loss of property and potential loss of human life. Of course, the greater the number of existing dams in a given river system, the lower the flood control benefits associated with a new dam.

The final benefit of the project is from recreation in the reservoir. The reservoir can be used for fishing, boating, waterskiing, and other recreation. In general, the benefits of the reservoir recreation are going to be smaller than the value of the lost river recreation. The reason for this discrepancy is that in many regions of the country, there are many more reservoirs than free-flowing rivers. A basic principle of economics suggests that (holding everything else constant) the more (and closer) substitutes for a good, the less its value.

Since many of the participants in recreation at the proposed reservoir would have recreated at already existing reservoirs, the recreational value of the new reservoir has to be computed relative to other reservoirs. The recreational benefit of the new reservoir is actually equal to the reduced travel costs incurred by those who now have a closer reservoir to visit and to the reduction in congestion at existing reservoirs. An excellent example of this type of analysis can be found in Krutilla and Fisher, who conducted a

---

[5]The alternative power plant could be located in almost any location because of the interconnectiveness of the electric power grid. However, the further the distance from the plant to the consumption location, the more electricity is dissipated in the transmission process.

cost–benefit analysis of a proposed ski resort in a national forest in California. They also used this type of analysis to show that there were no reservoir recreational benefits at a proposed dam site at Hell's Canyon. In that case, the costs of making the reservoir suitable for recreation (access roads, parking lots, boat launching ramps, toilets, and so on) were more than the recreational benefits from the reservoir, primarily because there were existing underutilized reservoirs in closer proximity to population centers.

Are Jobs Benefits? The above discussion of benefits makes no mention of something most noneconomists feel to be an important benefit associated with a project such as a hydroelectric facility. This benefit is the creation of jobs. However, as will be demonstrated below, the jobs associated with a particular project are seldom a social benefit, because the jobs are generally transferred from another area of the economy and are not new jobs.

For example, many people will be employed building the dam and generation facility. However, if the money is not spent building the hydrofacility, it could be spent building roads or sewers or some other project. Alternatively, if the taxpayers' money is not spent on this project, taxes could be lower; people would be employed making more private goods to satisfy the increased demand associated with higher after-tax incomes.

If new industry moves into the area to take advantage of cheaper electricity, the jobs associated with the new industry are not generally included in net social benefits. The industry has moved to this location from some other location, so one region's gain is another region's loss.

Although jobs always generate benefits for the individuals who are hired, they only generate social benefits under one of two conditions. The first condition is when the project creates jobs that would not otherwise exist in the economy. The generation of social benefits through the creation of jobs could take place because the project increases the productivity of the economy or because it increases the economy's comparative advantage relative to international competition.[6] The second condition is when the project creates jobs in areas of high structural unemployment, where barriers to mobility prevent potential workers from relocating to other areas where unemployment is lower. Projects that bring jobs to inner-city areas are an example of projects for which jobs should be counted as benefits. The above arguments are applicable to developed countries with strong fully employed economies such as the United States and Canada. These arguments generally do not hold in developing countries where structural employment is at a much higher level.

Cost–Benefit Analysis and Sensitivity Analysis. In any cost–benefit study, the analyst makes many choices with regard to which valuation methodologies to employ, the fashion in which to employ them, and a host of parameters

---

[6]See Chapter 5 for a discussion of this point.

TABLE 4.5

PRESENT VALUE OF NET BENEFITS OF A HYPOTHETICAL PROJECT AS A FUNCTION OF RESEARCH CHOICES (VALUES ARE EXPRESSED IN MILLIONS OF 1993 DOLLARS)

	REAL DISCOUNT RATE			
	2%	4%	6%	8%
1% annual growth in demand for recreation	2.0	2.2	2.4	2.6
2% annual growth in demand for recreation	0.0	0.1	0.2	0.3
3% annual growth in demand for recreation	−2.0	−2.2	−2.4	−2.6

which might affect the analysis. These parameters might include the discount rate, the value of human life, or the rate of growth of certain economic and social variables, such as population, GDP, or the demand for a good such as wood or electricity. Consequently, two economists may perform a cost–benefit analysis and come up with very different answers. Neither one is right or wrong, as their analyses are correct based on the assumptions and choices that were made. This potential for differing answers does not mean, however, that cost–benefit analysis is meaningless but that one should be very skeptical when a cost–benefit analysis presents a single number. The most useful way to conduct such a study is to allow the treatment of assumptions to change and see how the different treatments affect the results. Similarly, one can allow the values of parameters, such as the discount rate and the value of life, to change and see how sensitive the results are to changes in these values.

At a minimum, the researcher should present both upper and lower bound results, but even more useful is a matrix that illustrates the sensitivity of the results to key methodological choices. The rows and columns of Table 4.5 represent different choices the researcher might make, and the elements of the matrix represent the net benefits of the project associated with a particular set of choices. Under one set of choices, the project has positive net benefits; under another set of choices, they are essentially zero; and under a third, they are negative. Although it may seem as if the matrix presents confusing results to the policy maker, it actually presents enlightening results, as it allows the policy maker to be aware of the ramifications of the researchers' methodological choices.

When Some Benefits or Costs Cannot Be Measured. Very often a researcher will be in a position where measuring certain aspects of benefits or costs is impossible or would require more resources (time and money) than have

TABLE 4.6

HYPOTHETICAL HYDROELECTRIC PROJECTS

	NET MARKET BENEFITS	ENVIRONMENTAL COSTS
Project A	$10,000,000	smaller, but unquantified
Project B	$ 8,000,000	larger, but unquantified
Project C	$10,000,000	larger, but unquantified
Project D	$ 8,000,000	smaller, but unquantified

been allocated for the analysis. For example, computing existence values or the values of lost biodiversity may be extremely difficult, yet these values may be important components of total costs or benefits. If a value cannot be measured, can anything meaningful come out of an incomplete cost–benefit analysis? The answer to this question is sometimes yes.

For example, let us assume there is a river that floods frequently, and a political decision has been made to put in a dam for flood control. It could be located in an upstream location (location A) or a downstream location (location B). Let us assume it is relatively easy to measure the benefits of flood control in either location, and the benefits are higher for the upstream location because it protects more of the river basin. Let us also assume that natural scientists find that the upstream location has lower environmental costs, since more of the river remains below the dam and fewer fish spawning areas are blocked by the dam. As can easily be seen in Table 4.6, location A is preferable to location B because it has both greater economic benefits and lower environmental costs than location B. A decision can be made without actually measuring the value of the alternative environmental impacts.

However, sometimes one alternative will not dominate the others. In Table 4.6, alternative C has higher economic benefits and greater environmental costs than alternative D. It is not possible to say which alternative is better without measuring the economic value of the environmental costs.

Krutilla and Fisher (1985) used this type of comparison in analyzing the alternative pipelines for taking oil from Prudhoe Bay (Arctic Coasts of Alaska and Canada) to markets to the south. They show that a "Trans-Canadian" pipeline running southeast to Winnepeg (where it would join an existing pipeline network) had both greater economic benefits and lower environmental costs than the "Trans-Alaskan" route that ran south to the Alaskan coastal town of Valdez, where the oil was loaded on tankers to be delivered to the lower 48 states. (Incidentally, the Trans-Alaska Pipeline was built, and some of these potentially higher environmental costs were realized with the Exxon Valdez oil spill.)

An incomplete analysis can also be useful if one looks at the difference between the measured benefits and measured costs. If measured costs are greater than measured benefits, there is generally no need to measure the environmental costs. However, if measured benefits are greater than measured costs, one can still make a conclusion if one can construct a logical argument about whether the unmeasured environmental costs are likely to affect the differential between costs and benefits.

For example, let us say that one is considering a potential dam project, which results in the loss of river recreation, but the value of the river recreation has not been measured due to a lack of data. Let us assume that the present value of measured costs exceeds the present value of measured benefits by $1 million and that there are no benefits from the dam that remain unmeasured. Under these conditions, the dam would fail a cost–benefit test if the present value of river recreation exceeded $1 million. However, if the present value of river recreation was less than $1 million and there were no other unmeasured costs (such as an indirect use value associated with the riverine ecosystem), then the dam would pass a cost–benefit test. Notice that it requires less information about unmeasured costs for the dam to fail the cost–benefit test than for the dam to pass the test.

The present value of river recreation that exactly equates benefits and costs is $1 million. What ability do we have to make a reasonable estimate of river recreation benefits when there is insufficient data (or insufficient time and money to obtain the data) required for travel cost or contingent valuation studies?

The first step toward resolving this issue is to convert the present value of $1 million into an equivalent stream of annual benefits. This can be done according to the formula of Equation 4.5, where X is the unknown annual benefits for which one solves the equation:[7]

$$\$1{,}000{,}000 = X + X\left[\frac{1}{1+r}\right]^1 + X\left[\frac{1}{1+r}\right]^2 + X\left[\frac{1}{1+r}\right]^3 + X\left[\frac{1}{1+r}\right]^4 + \dots + X\left[\frac{1}{1+r}\right]^n \quad (4.5)$$

If $R = 0.1$, then X will approximately equal $90,000.
Now we have transformed the question from "Is the present value of river recreation benefits likely to exceed $1 million?" to "Are the annual river recreation benefits likely to exceed $90,000?" The two questions are mathematically identical, but the latter question is easier to conceptualize.

[7]The solution of this equation is based on an important geometric series, which takes the form:

$$A = X(B) + X(B^2) + X(B^3) + \dots + X(B^n).$$

If n approaches infinity and $0 < B < 1$, then $A = X[1/(1 - B)]$. The convergence of this geometric series has many important uses in economics. It is used to derive the multiplier in macroeconomics, and it can show the extent to which recycling can extend the effective reserves of a mineral such as copper.

How can one estimate annual recreation benefits without a travel cost or contingent valuation procedure? One way is through a process called benefits transfer, which is to take per trip value from studies of other recreational activity (such as river recreation in another region) and transfer those benefits to the problem at hand. One must be careful which studies one uses in such a transfer so that the activities, the recreational site, and the population of recreationists correspond between the study one is transferring from and the activity one is examining.

Other extrapolation methods can be used to develop a general impression of the unmeasured environmental benefits of preserving the river and compare them to the cost differential between measured costs and measured benefits. Many times these general impressions will be much larger or much smaller than this differential, so that even though this general impression is imprecise, it can help make a determination whether total costs are greater than total benefits or vice versa. Other times, however, the general impression of unmeasured benefits will be roughly the same order of magnitude as the differential, and the imprecision becomes critically relevant. In these circumstances, it is not possible to determine whether total benefits are greater than total costs or vice versa.

Cost–Benefit Caveats. Although cost–benefit analysis, with its reliance on numbers and economic theory, may seem as if it might be a precise science, this is not the case. Cost–benefit analysis is only as good as the men or women who perform the analysis, and it is critically dependent upon both the assumptions that are made and the quality of the data that are employed. A reader should be extremely cautious about accepting the results of a cost–benefit analysis that does not explicitly diagram the assumptions underlying the study, that does not explain how the data are generated, or that does not conduct a sensitivity analysis.

MARGINAL DAMAGE FUNCTIONS

Marginal damage functions were discussed in Chapter 3, where it was shown that the optimal level of pollution (E^*) occurs at the level where marginal damages are equal to marginal costs. This optimal level is shown in Figure 4.8. Although one requires information on both abatement costs and damages in order to identify (or approximate) the optimal level of pollution, this chapter will focus on the process for identifying the damage functions.

As can be seen in Figure 4.8, a marginal damage function specifies a relationship between an incremental unit of emissions and the damages that the emissions generate. As one might expect, the relationship between emissions and damages is actually a complex series of cause and effect relationships. These are diagramed in Figure 4.9.

The first set of relationships depicted in Figure 4.9 are the relationships between emissions of pollution and concentrations of pollutants. Emissions disperse to different locations and are transformed into other substances,

FIGURE 4.8

OPTIMAL LEVEL OF POLLUTION

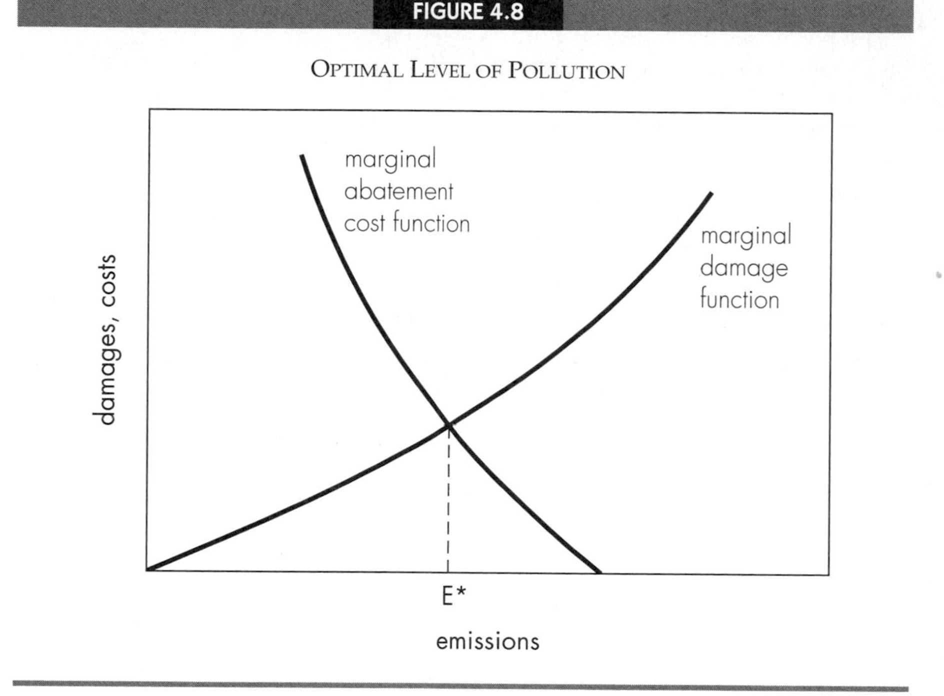

some of which are harmful and some of which are benign. Elaborate computer models that take into account the chemical properties of the pollutants, the geographic features of land, and wind and weather patterns are used to predict how emissions of one type of pollution turn into concentrations of different types of pollution.[8]

The second set of relationships identifies the exposure to pollution. What populations (human and nonhuman) are exposed to the pollution and what is the level of exposure?

The next set of relationships in Figure 4.9 models how exposure to pollution gives rise to physical effects. These are often termed dose–response relationships, as they show how physical effects respond to changes in exposure to pollution. These dose–response relationships are estimated using either laboratory or field data. For example, laboratory data might be used to discover how acidity affects fish physiology. Alternatively, this could be examined in the field by looking at a sample of lakes and streams of different

[8]For example, sulfur dioxide pollution is transported over large distances and turns into sulfuric acid, sulfate aerosols, and other types of pollution that generate damages (see Chapter 7 on Acid Precipitation).

FLOW DIAGRAM OF A DAMAGE FUNCTION

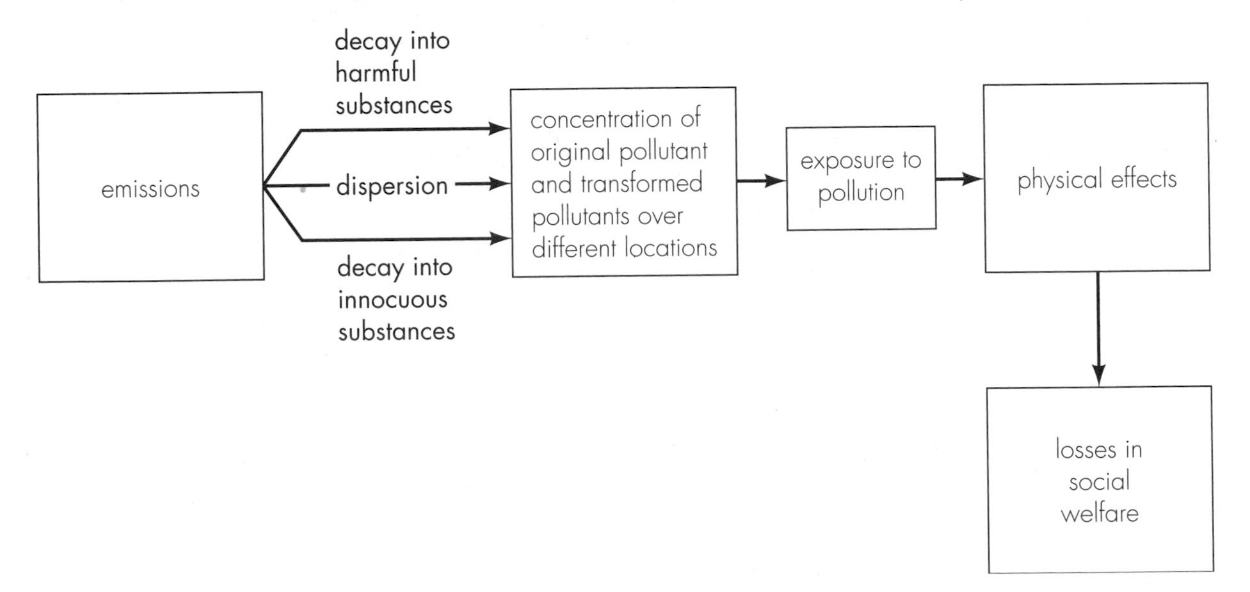

acidity levels and noting how fish populations differ. With laboratory tests, researchers have the advantage of more easily being able to hold constant other factors that might affect fish physiology. Field studies give researchers the advantage of deriving "real world" data. Good studies usually rely on both field and laboratory data.

Once the physical effects are known, then the effects on social welfare can be estimated by utilizing the revealed and stated preference valuation techniques that were presented earlier in the chapter.

OTHER APPROACHES TO VALUE

The economic approach to value and to valuation of the environment is but one perspective into incorporating environmental values into social decision making. As indicated earlier, it is a particularly anthropocentric approach. In contrast, several philosophical perspectives stress a biocentric perspective on value.

In North America, the earliest of these perspectives was the Native American ethic, which existed prior to the arrival of settlers from the "Old

World." Although there were many different tribes and cultures in North America, and it is difficult to generalize across these cultures, some perspectives on the value of environmental resources were shared by many Native American cultures. First, Native Americans believed that although their actions affected the natural world, the human role was not to dominate or control nature. Native Americans typically felt that they were a component of the natural world and no more important than any other component. As a consequence, Native Americans viewed the land and all plant and animal species as intrinsically valuable, where value was not defined in terms of the satisfaction of human needs (Suzuke and Knudtson, 1992; Callicott, 1989).

Biocentric approaches developed within the non-Native American culture as well. One of the most important developers of this approach was Aldo Leopold (Flader and Callicott, 1991), who developed an approach to value based on a land ethic. The basic thrust of his approach was that the components of nature were linked through the balance of the ecosystem, and therefore, their value was based on ecological roles, which was independent of other sources of value.

A more recent development has been deep ecology (Devall, 1985), also a biocentric perspective. Deep ecology was developed by a Norwegian, Arne Ness, and it rejects a management approach toward the environment. Environmental resources are not viewed as existing for human benefit; in fact, it is unethical to use environmental resources to satisfy any human need except vital human needs. All components of the ecosystem, as well as the ecosystems themselves, have an intrinsic value and an inherent right to exist in unaltered form.

Another example of a different perspective on value (which is neither purely biocentric nor purely anthropocentric), is the energy theory of value, which is advocated by a group of applied ecologists (Odum and Odum, 1976). This theory suggests that the one thing common to, necessary for, and scarce in both economic systems and ecological systems is energy. Thus, energy can be viewed as a "currency of exchange," and the value of any component of an economic system or ecological system is based on the energy embodied in that component. Economists reject this concept because it places too little emphasis on the scarcity of the component. Under this system, a barnyard goose could be viewed as more valuable than a whooping crane (an endangered species), and a big painting by the author of this textbook could be viewed as more valuable than a small painting by Picasso.

The economic perspective on value, with its anthropocentric orientation, varies substantially from both biocentric and energy approaches. This book will proceed based on the economic perspective on value. This approach should not be interpreted as a rejection of the importance or validity of these other approaches. The choice of this paradigm reflects the orientation of the book as an environmental economics textbook.

ADDITIONAL CRITERIA FOR EVALUATING POLICY

The preceding sections have looked at methods for comparing benefits and costs as means of comparing alternative environmental policies and looking for those policies which maximize net benefits. While this is an extremely important evaluation criterion, it is not the only criterion that should be used. In recent years, there has been an increasing focus on equity (sometimes referred to as environmental justice), sustainability, and process as criteria against which alternative environmental policies should be judged.

PROCESS

Recent experience with the implementation of environmental policy has dramatically illustrated that people are interested not only in the outcomes of environmental policy (such as how much risk to human health is reduced), but also in the process by which the policies and outcomes are generated. People feel that since they are affected by the outcomes, they should participate in the process which generates the outcomes. As a consequence, public participation in the environmental policy process has increased, especially at the local level. In general, environmental problems which are more local in nature (such as what to do about the toxic site near the elementary school or where to site the nuclear waste facility) generate a greater need for public participation. Many environmental decision processes legally require that public participation be a part of the decision-making process.

A good public participation process begins by identifying the stakeholders, or shareholders, in the decision. For example, if the environmental problem revolved around agricultural pollution of an estuary, stakeholders would include farmers, agrochemical suppliers, county extension agents, recreational fishers, commercial fishers, marina owners, boat owners, environmental groups, people living near the estuary and other citizens who were concerned with the outcome. Productive processes seldom begin in a high school gymnasium full of shouting people. Rather, representatives of the different positions must be selected in order to create a group of workable size. The next step is to develop a process by which the different stakeholders are able to express their concerns, and these concerns help define the decisions which are made. Skilled moderating is needed to help people understand that everyone has a stake in the outcome and that confrontation does not lead to resolution. In addition, one can move away from the stakeholder process, select citizens who are not charged with representing a particular group, and have them work together to develop recommendation. One way of doing this is roleplaying, where the citizens step into the shoes of the various stakeholders and discuss the issues from each particular perspective, with the idea of everyone developing an understanding of all the different viewpoints. A slightly different alternative is to appoint a citizen jury (Bruce Tonn, 1994), which weighs the evidence (the cost, benefits, fairness, etc.) of alternative policies and makes a recommendation.

In the early days of the public participation process, government officials would often hold public hearings only after many of the important decisions had been made. Stakeholders commented on the plans already developed, but many stakeholders would have preferred to consider options that had been excluded from consideration. Stakeholders felt they were being dictated to because they did not have a role in helping to select among the many available alternatives to deal with the problem. Consequently, most environmental agencies now begin the public participation process early in the decision-making process.

EQUITY

Although most economists tend to focus on efficiency, the equity implications of environmental change are extremely important to the policy process. These equity implications mean that environmental change and environmental policy may have different impacts among members of society. For example, a particular environmental change may benefit the well-off members of society more than the less well-to-do. Equity considerations are particularly important when the change exacerbates the quality of life differential between the well-off and the not so well-off. There are several important dimensions to the equity issues. Policy makers need to be concerned about equity across generations, across countries, and across different groups within a country.

EQUITY ACROSS GENERATIONS AND SUSTAINABILITY

Equity across generations, or intergenerational equity, is an extremely important consideration in environmental policy. Because today's decisions may generate important environmental costs for future generations, and because future generations cannot participate in making current decisions, future generations may potentially pay the price for the actions of the current generation. For example, current emissions of greenhouse gases which cause global warming, current conversion of habitat which causes loss of biodiversity, and current generation of toxic and radioactive wastes all cause irreversible environmental change which lowers the well-being of future generations.

Many people feel that the unfair treatment of future generations is made worse by the process of discounting, which makes future costs and benefits insignificant in comparison to present costs and benefits (see above discussion of discounting). For this reason, they advocate a criterion of sustainability rather than efficiency (maximizing present values) for guiding policy.

The concept of sustainability is closely related to the concept of intergenerational equity. The idea behind sustainability is that it is desirable to attempt to improve the condition of the current generation, but this improvement must not compromise the ability of future generations to meet their needs and improve their quality of life. Actions which create benefits in the present but large environmental costs in the future could pass a cost–benefit test because of the impact of discounting when measuring the present value of future costs. However, such actions are unlikely to pass a sustainability test, which does not look at present values, but examines barriers to future improvement in the quality of life which are created by today's actions. In other words, policies that benefit the current generation but diminish the prospects of future generations would not be allowed under this criterion. The criterion of sustainability is very important when examining environmental problems such as deforestation, global climate change, nuclear power and biodiversity. The economic theory of sustainability will be more fully discussed in Chapters 5 and 17.

Equity Across Countries

Equity across countries is also extremely important to environmental policy. Some environmental problems, such as global warming, require international cooperation. There is also a demand in wealthy countries to preserve the environmental resources in the lower-income countries. In either case, the lower-income countries have to bear costs to generate benefits for higher-income countries. Obviously, many people, particularly those in developing countries, think that this burden is inherently unfair and that prosperous countries should compensate the lower-income countries for the costs they incur in improving the global environment. Another dimension of environmental equity that is often overlooked is the lower level of environmental quality in poorer countries, which inhibits their ability to improve their standard of living. These points are discussed in great detail in Chapter 17.

Intra-country Equity

Both environmental degradation and environmental policy create consequences for reducing or increasing the differential in the quality of life that may exist across subgroups of a country's population. Initially, many economists argued that improving the quality of the environment would actually widen this quality gap. Environmental quality was perceived to be a luxury, with the improvement in environmental quality benefitting wealthier people more than the poor. Because stricter environmental policy may result in a loss of jobs in polluting industries, and in higher product prices, it was often also argued that the poor bore a disproportionate share of the cost of improving environmental quality.

This view is not as widely held now, for several reasons. First, in developing countries the poor are primarily agrarian, dependent on the environment and suffering the most from the effects of environmental degradation (soil erosion, deforestation, contaminated drinking water, etc.). Also, urban dwellers in these countries suffer from many environmental externalities including contaminated drinking water, severe air pollution, and exposure to toxic waste.

The same thing is true in developed countries such as the United States. Air quality tends to be the worst in our largest cities, especially in the inner-city areas where low income people reside. Also, low income and minority populations may face greater exposure to toxic waste and to other environmental hazards (see Chapter 15).

SUMMARY

This chapter represents a basic introduction to the concept of value, how to measure value, and how to use value in decision making. It should be stressed that this discussion occurs from an economist's perspective, where value is defined by peoples' willingness to make trade-offs.

Even though the economist's definition of value is not universally accepted, it can make a substantial contribution to the decision-making process. Revealed preference approaches such as hedonic pricing techniques and the travel cost method can help measure direct use values associated with environmental change. Stated preference valuation techniques such as the contingent valuation method and conjoint analysis can estimate indirect use values, although the measurement of indirect use values remains quite.

One important use of value estimates is in cost–benefit analysis, where a limited number of alternatives are compared. For example, cost–benefit analysis can be used to help choose among alternative locations of a waste disposal site. Value estimates also are important in estimating marginal damage functions, which can be used to identify the target level of emissions of a particular pollutant.

Although economic valuation cannot provide exact answers to all our resource allocation decisions, it can provide substantial insight to help us make informed decisions. Other criteria such as sustainability, equity and the outcome of public participation processes should also help shape the final form of environmental policy.

Review Questions

1. What is the difference between marginal damage function analysis and cost–benefit analysis?
2. What reasons suggest that the social discount rate should be less than the risk-free real market rate of interest?
3. Is the creation of jobs a societal benefit?
4. List the potential biases associated with contingent valuation.
5. Assume that you have data that suggest that if
   a. travel cost is greater than or equal to $15, no trips are taken.
   b. travel costs are zero, 100 trips are taken.

   Draw a travel cost demand curve based on these data. Calculate ordinary consumer's surplus for the individual whose travel costs are equal to $5.

6. What can you say about the relative preferability of three projects that have the following costs and benefits (compare A to B, A to C, and C to B)?

Project	Measured Costs	Measured Benefits	Unmeasured Environmental Costs
A	$10,000,000	$10,000,000	insignificant
B	$ 5,000,000	$10,000,000	insignificant
C	$10,000,000	$15,000,000	large

7. If construction workers are willing to accept a 1/2,000 annual risk of death if their income increases by $3,000 per year, what is the collective willingness to be compensated to accept the loss of one life?

QUESTIONS FOR FURTHER STUDY

1. What are the relative strengths and weaknesses of indirect and direct valuation techniques?
2. Can human life be valued?
3. What valuation techniques would you use to measure the value of improving water quality in a local lake or river? Why?
4. What valuation techniques would you use to measure the value of preserving tropical rain forests? Why?
5. Why is it possible to look at wages or housing prices and draw a conclusion about the value of environmental quality?
6. Is cost–benefit analysis objective?
7. What criteria should be used to develop global environmental policy, such as limiting greenhouse gas emissions?

SUGGESTED PAPER TOPICS

1. Use the hedonic wage model to estimate willingness to pay for an environmental characteristic. The basic data set for estimating the hedonic wage model is the Public Use Micro-Data Subset (PUMS) of the Census of Population. (Most large universities will have this data set on computer tape or CD-ROM.) This data set contains income and socio-economic data on individual households. Since the data set also lists the city in which the household lives, all one needs to do is develop appropriate environmental variables (concentration of various air pollutants, drinking water quality, proximity of Superfund sites [toxic hot spots], etc.). The USEPA, Department of Health, and Department of Energy collect and publish this type of data. See Clark and Kahn; Epple; Kask; and Gegax, Gerking, and Schulze for references to this approach.
2. Conduct a contingent valuation study for a local environmental quality problem using students as your sample. Look to Mitchell and Carson and the NOAA "Blue Ribbon Panel Study" (Department of Commerce, 1993) for guidance as to how to structure the questions. Make sure to ask your professor if you need approval from your university's "Human Subjects Review Committee."
3. Choose a national or local environmental problem and outline the steps for conducting a cost–benefit analysis. Look at related studies in the literature (*Journal of Environmental Economics and Management, Land Economics, Ecological Economics, American Journal of Agricultural Economics)* for estimates of the value of closely related resources.
4. Write a critique of a valuation method, such as contingent valuation. Start with the most recent issues of the journals listed above (as well as the appropriate references at the end of this chapter) for critical discussion of the method you choose.
5. Many university libraries have the 1991 (or 1985) U.S. Fish and Wildlife Survey of Fishing, Hunting and Wildlife Associated Recreation on CD-ROM or mainframe computer tapes. If you have good computer skills, you might enjoy taking data from this survey and estimating a travel cost demand curve. Look at Brown, et al. and Smith and Kaoru as initial references.

REFERENCES AND SELECTED READINGS

1. Ajzen, Icek, and George L. Peterson. "Contingent Value Measurement: The Price of Everything and the Value of Nothing?" In *Amenity Resource Valuation: Integrating Economics with Other Disciplines,* George L. Peterson, B. L. Driver, and Robin Gregory, 65–76. State College, PA: Venture, 1988.
2. Balkan, Erol, and James R. Kahn. "The Value of Changes in Deer Hunting Quality: A Travel Cost Approach." *Applied Economics* 20 (1988): 533–539.
3. Bell, F. W., and V. R. Leeworthy. *An Economic Analysis of the Importance of Saltwater Beaches in Florida.* Sea Grant Report 47, FSU-Tallahassee, 1986.

4. Bell, F.W., P.E. Sorenson, and V.R. Leeworthy. *The Economic Valuation of Saltwater Recreational Fisheries in Florida.* Sea Grant Report 47, FSU-Tallahassee, 1982.
5. Berger, M. C., G. C. Blomquist, D. Kenkel, and G. S. Tolley. "Valuing Changes in Health Risks: A Comparison of Alternative Measures." *Southern Economic Journal 53*, 4 (April 1987): 967–984.
6. Blomquist, G. C., M. C. Berger, and J. P. Hoehn. "New Estimates of Quality of Life in Urban Areas." *American Economic Review* 78 (1988): 89–107.
7. Bockstael, N. E., W. M. Hannemann, and Ivar Strand. "Benefit Analysis Using Indirect or Imputed Methods." *Measuring the Benefits of Water Improvements Using Recreation Demand Models 2*, Environmental Protection Agency, Washington, D.C., 1986.
8. Bockstael, N. E., K. McConnell, and I. E. Strand. *Benefits from Improvement in Chesapeake Bay Water Quality,* EPA Contract 811043-01, USEPA, 1988.
9. Bockstael, N. E., Ivar Strand, and W. M. Hannemann. "Time and the Recreation Demand Model." *American Journal of Agricultural Economics* 69 (1987): 293–302.
10. Brookshire, D. S., M. A. Thayer, W. W. Schultze, and R. L. d'Arge. "Valuing Public Goods: A Comparison of Survey and Hedonic Approaches." *American Economic Review* 72 (1982): 165–177.
11. Brown, Gardner, et al. "Methods for Valuing Acidic Deposition and Air Pollution Effects." *Acid Deposition: State of Science and Technology,* IV, 27: NAPAP, Washington, D.C.: U.S. Government Printing Office, 1990.
12. Brown, Thomas C. "The Concept of Value in Resource Allocation." *Land Economics* 6 (1984): 232–245.
13. Callicott, J. Baird. *In Defense of the Land Ethic: Essays in Environmental Philosophy*. Albany: SUNY Press, 1989.
14. Clark, D. E., and J. R. Kahn. "The Two-Stage Hedonic Wage Approach: A Methodology for the Valuation of Environmental Amenities." *Journal of Environmental Economics and Management* 16 (1989): 106–120.
15. Coursey, D. L., W. D. Schulze, and J. L. Hovis. "The Disparity between Willingness to Accept and Willingness to Pay Measures of Value." *Quarterly Journal of Economics* (August 1987): 679–690.
16. Cummings, Ronald G., and Glenn W. Harrison. *Identifying and Measuring Nonuse Values for Natural and Environmental Resources: A Critical Review of the State of the Art*. Final Report, U.S. Environmental Protection Agency, Washington, D.C., April 1992.
17. Cummings, R. G., David Brookshire, and William D. Schultze, eds. *Valuing Public Goods: A State of the Arts Assessment of the Contingent Valuation Method*. Towato, NJ: Rowan and Allanheld, 1986.
18. Devall, Bill. *Deep Ecology*. Salt Lake City: Peregrine Smith Books, 1985.
19. Duffield, J. W., and D. A. Patterson. "Field Testing Existence Values: An Instream Flow Trust Fund for Montana Rivers." Paper presented at AERE meetings, New Orleans, January, 1992.
20. Epple, D. "Hedonic Prices and Implicit Markets: Estimating Demand and Supply Functions for Differentiated Products." *Journal of Political Economy* (1987): 59–80.
21. Flader, Susan L., and J. Baird Callicott, eds. *Aldo Leopold: The River of the Mother of God and Other Essays*. Madison: The University of Wisconsin Press, 1991.
22. Freeman, A. M., III. *The Benefits of Environmental Improvement: Theory and Practice*. Baltimore: Johns Hopkins University Press, 1979.
23. Gegax, D., S. Gerking, and W. Schulze. "Perceived Risk and the Marginal Value of Safety." *The Review of Economics and Statistics 73*, 4, (November 1991): 589–596.
24. Gramlich, E. M. *A Guide to Benefit–Cost Analysis*. New York: Prentice Hall, 1990.
25. Harrison, D., and D. L. Rubinfeld. "Hedonic Housing Prices and the Demand for Clean Air." *Journal of Environmental Economics and Management,* 5 (1978): 81–102.
26. Hotelling, Harold. Letter to National Park Service in *An Econometric Study of the Monetary Evaluation of Recreation in the National Parks,* dated 1947. U.S. Department of Interior, NPS and Recreational Planning Division, 1949.
27. Huppert, D. D. "Two Empirical Issues in Recreational Fishery Economics: Mail Survey Self-Selection Bias and Divergence Between WTP and WTA" in Fourth Annual AERE Workshop: Marine and Sport Fisheries Economic Valuation and Management, June 1988, Seattle, EPA230-08-88-034, USEPA, Washington, D.C.

28. ICF Incorporated. "Pollution Prevention Benefits Manual." U.S. Environmental Protection Agency, Washington, D.C., October 1989.
29. Johnson, Lawrence E. *A Morally Deep World: An Essay on Moral Significance and Environmental Ethics*. Cambridge: Cambridge University Press, 1991.
30. Kahn, J. R. "The Economic Value of Long Island Saltwater Recreational Fisheries." *New York Economic Review* 21 (1991): 3–23.
31. Kask, S. B. "Long-Term Health Risk Valuation: Pigeon River, North Carolina." In *Benefits Transfer: Procedures, Problems and Research Needs*. Proceedings of the 1992 Association of Environmental and Resource Economists Workshop, Snowbird, UT.
32. Kealy, Mary Jo, Mark Montgomery, and J. F. Dovido. "Reliability and Predictive Validity of Contingent Valuation: Does the Nature of the Good Matter?" *Journal of Environmental Economics and Management* 19 (1990): 244–263.
33. Kealy, M. J., and Mark Rockel. "The Value of Nonconsumptive Recreation in the United States." Unpublished manuscript, 1990.
34. Krutilla, John V. "Conservation Reconsidered." *American Economic Review* 47 (1967): 777–786.
35. Krutilla, John and A. C. Fisher. *The Economics of Natural Environments*. Washington, D.C.: Resources for the Future, 1985.
36. Louviere, Jordan J. "Conjoint Analysis Modelling of Stated Preferences: A Review of Theory, Methods, Recent Developments and External Validity." *Journal of Transport Economics and Policy* (1988): 93–119.
37. McConnell, K. E. *Existence and Bequest Value, in Managing Air Quality and Scenic Resources at National Parks and Wilderness Areas*, eds. R. D. Rowe and L. G. Chestnut, Boulder, CO: Westview Press, 1983.
38. Milon, J. W. "Travel Cost Methods for Estimating the Recreational Use Benefits of Artificial Marine Habitat." *Southern Agricultural Economics* (1988): 87–101.
39. ———. *Estimating Recreational Angler Participation and Economic Impact in the Gulf of Mexico Mackeral Fishery.* Prepared for Southeast Regional Office, NMFS, NA86wc-h-06616, RAS/CC31, 1988.
40. Mishan, E. J. *Cost–Benefit Analysis*. New York: Praeger Publishing, 1976.
41. Mitchell, R. C., and R. T. Carson. *Using Surveys to Value Public Goods—The Contingent Valuation Method*. Washington, D.C.: Resources for the Future, 1989.
42. Morey, E. R., and W. D. Shaw. "An Economic Model to Assess the Impact of Acid Rain: A Characteristics Approach to Estimating the Demand for and Benefits from Recreational Fishing." In V. K. Smith and A. D. Witte, eds., *Advances in Applied Microeconomic Theory,* 8. Greenwich, CT: JAI Press, Inc., 1992.
43. Nash, Roderick Frazier. *The Rights of Nature: A History of Environmental Ethics*. Madison: The University of Wisconsin Press, 1989.
44. Nelson, Jon. "Residential Choice, Hedonic Prices and the Demand for Urban Air Quality." *Journal of Urban Economics* 5 (1978): 357–369.
45. Odum, H. T., and E. C. Odum. *Energy Basis for Man and Nature*. New York: McGraw-Hill, 1976.
46. Plourde, Charles. "Conservation of Extinguishable Species." *Natural Resources Journal* 15 (1975): 791–797.
47. Porter, R. C. "The New Approach to Wilderness Preservation through Benefit-Cost Analysis." *Journal of Environmental Economics and Management* 9 (1982): 59–80.
48. Ridker, R. G., and J. A. Henning. "The Determination of Residential Property Values with a Special Reference to Air Pollution." *Review of Economics and Statistics* 49 (1967): 246–257.
49. Rosen, Sherwin. "Wage-based Indexes of Urban Quality of Life." *Current Issues in Urban Economics,* P. Mieskowski and M. Straszheim, eds. Baltimore: Johns Hopkins University Press, 1979.
50. Schulze, W. D., and G. H. McClelland. "Valuing Winter Visibility Improvement in the Grand Canyon," unpublished paper presented at Allied Social Sciences Association meetings, New Orleans, January 1991.
51. Schulze, W. D., G. H. McClelland, and D. L. Coursey. "Valuing Risk: A Comparison of Expected Utility with Models from Cognitive Psychology." Unpublished manuscript, University of Colorado, Boulder, CO, 1986.
52. Seip, Kalle, and Jon Strand. "Willingness to Pay for Environmental Goods in Norway: A Contingent Valuation Study with Real Payment." *Environmental and Resource Economics* 2 (1992): 91–106.

53. Smith, V. Kerry. "Nonmarket Valuation of Environmental Resources: An Interpretive Appraisal." *Land Economics* 69 (1993): 1–26.
54. Smith, V. Kerry, and Yoshiaki Kaoru. "What Have We Learned Since Hotelling's Letter? A Meta-Analysis." *Economic Letters* 32 (1990a): 267–272.
55. ———. "Signals or Noise: Explaining the Variation in Recreation Benefit Measurement." *American Journal of Agricultural Economics* 72 (1990b): 419–433.
56. Smith, V. K., and R. B. Palmquist. *The Value of Recreational Fishing on the Albemarle and Pamlico Estuaries*, USEPA, CX814569-01, 1988.
57. Suzuki, David, and Peter Knudtson. *Wisdom of the Elders*. Bantam Books, 1992.
58. U.S. Department of Commerce. "Natural Resource Damage Assessments Under the Oil Protection Act of 1990." 15 CFR Chapter IX, National Oceanic and Atmospheric Administration. *Federal Register* 58(10), January 15, 1993, Proposed Rules, 4601–4614.
59. Viscusi, W. K., W. Magat, J. Huber. "Pricing Environmental Health Risks: Survey Assessment of Risk-Risk and Risk-Dollar Tradeoffs for Chronic Bronchitis." *Journal of Environmental Economics and Management* 21, 1 (July 1991): 32–51.

The Macroeconomics of the Environment

The conflict between environmental protection and economic competitiveness is a false dichotomy.

Michael Porter[1]

INTRODUCTION

Chapters 1 through 4 examine microeconomic aspects of the environment. While developing an understanding of the microeconomic issues is critically important to understanding environmental economics, it is not sufficient. Microeconomic environmental issues involve choosing the optimal level of an environmental resource (this could be a quantity or quality choice). Environmental resources are viewed as goods that provide services, and the optimal level of these services is found by equating marginal social benefits with marginal social costs in a manner that is identical to all other types of goods. In addition to choosing the optimal level of resources from a microeconomic perspective, it is also essential to look at the more aggregated economic issues surrounding the environment. These issues take two basic forms. First is the question of how environmental policy affects the economy as a whole. For example, what is the effect of environmental policy on international competitiveness, which Porter refers to in the controversial assertion that begins this chapter? Although it might seem artificial to separate microeconomic environmental issues from macroeconomic environmental issues, some distinctions may be drawn. The second issue revolves around the interaction of the environmental system and the economic system.

What is the effect of the macroeconomic growth on the environment, and what is the effect of the environment on the macroeconomy? Specifically, does improving environmental quality reduce the Gross Domestic Product

[1]Michael Porter, "America's Green Strategy," *Scientific American* (1991): 168.

(GDP)[2] and reduce the number of jobs, or does it improve the economy? Also, does a growing economy imply that environmental quality must diminish? These questions are becoming increasingly important for several reasons. As we begin to experience more system-wide environmental changes, such as the loss of biodiversity, acid precipitation, global warming, and ozone depletion, these relationships between the environment and the macroeconomy become more important. In addition, political discussions on future directions of environmental policy are increasingly focused on the effect of environmental improvement on jobs and GDP.

Although much of the current focus of environmental economics is on more microeconomic issues, the conceptual discussion of the interrelationship between the environment and the macroeconomy began early in the development of the field of environmental economics. Boulding (1966), who made the analogy of the earth to a spaceship, stressed that our activities were constrained by our endowment of resources and the ability of environmental systems to assimilate wastes. Rather than living in a "frontier" society (where one can always expand the frontiers when currently utilized land areas are depleted, or resources are spoiled by waste), we live in a "spaceship" society where our inputs are finite, and we are trapped with our wastes. Although Boulding did not develop formal models of the relationship between the economy and the environment, his article was central in specifying the idea that the overall economy was constrained by the environment and that the economy impacts the environment and may change the nature of that constraint.

CONCEPTUAL MODEL OF THE ENVIRONMENT AND THE ECONOMY

Although this chapter differs from earlier chapters in its more macroeconomic focus, it is similar in that the focus is still on how environmental and economic interactions affect social welfare. Figure 5.1 illustrates several ways in which interactions between the environment and the macroeconomy affect social welfare. In the conceptual model illustrated in Figure 5.1, the health of the environment, the health of the economy, the health of the (human) population, and social justice all affect social welfare, both independently

[2]Gross Domestic Product, or GDP, is a measure of national income. It is related to Gross National Product, or GNP, with which readers may be more familiar. The primary difference between GDP and GNP is that GNP includes only the income of factors of production which are owned by citizens of the country, while GDP includes that income of all factors of productions that are employed in the country. For example, the income of a Japanese engineer or a Peruvian Major League Soccer player who works in the United States is considered part of GDP but not part of GNP. Because of the increased globalization of national economies, GDP has become the preferred measure of national income.

FIGURE 5.1

AREAS OF SOCIAL POLICY AND SOCIAL WELFARE

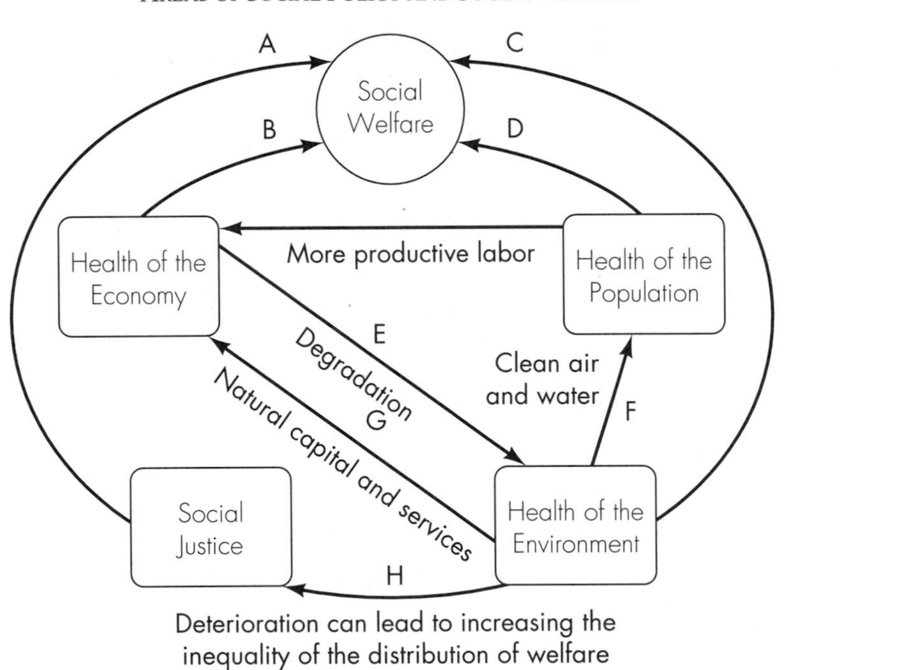

and through their interaction with each other. Other areas of social policy, such as education, are important as well, but they have been excluded from the diagram in the interests of simplicity.

Environmental quality impacts social welfare in a variety of ways. People benefit from improved environmental quality if it beneficially affects their health (arrow F in Figure 5.1), if it improves social justice (H), if it improves the economy (G), or simply because they view their quality of life as higher if environmental quality is higher (C). Even more indirect effects can be important as well. For example, improved environmental quality will increase the health of the population, which will increase the marginal productivity of labor, which in turn will increase the health of the economy.

While Figure 5.1 is useful for illustrating some of the important paths by which the environment and the economy can affect each other and consequently affect social welfare, in some senses this diagram has a limited perspective. The reason is that Figure 5.1 depicts the economy and the environment as inherently separate systems.

FIGURE 5.2

The Economic System as a Component of the Environmental System

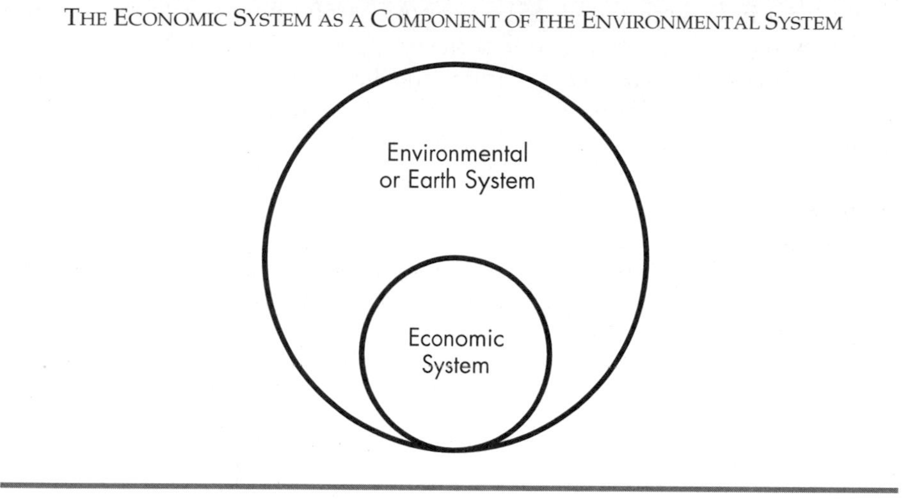

This depiction of the environment and the economy as inherently separate systems with a finite set of linkages is very useful for highlighting some of the most important linkages between the two. However, it is also important to realize that the two systems are not distinct, since the economic system is contained within the larger environmental (or earth) system, as illustrated in Figure 5.2. This perspective is extremely important since it implies that the properties of the economic system are fundamentally constrained by the properties of the larger earth system.

Which perspective is the correct one? The answer to this question depends on what issue we are looking at. If we are focusing on issues such as the impact of a cleaner environment on the productivity of the environment, of the impact of environmental regulations, then the first perspective is sufficient. However, if we are looking at issues such as how much growth is possible and what steps do we need to take to assure sustainability, then the latter perspective is the one that should be taken.

The Impact of the Environment on Economic Productivity

Although the relationships between the economy and the environment are quite complex, there are three basic mechanisms by which the environment and environmental policy affect economic productivity. The first has to do with the impact of environmental policy on productivity (a negative impact), and the last two have to do with the impact of environmental quality on economic productivity (positive impacts).

The Negative Impact of Environmental Policy on Economic Productivity. Environmental policy forces firms to make decisions that are in the social interest, but not in their private interest. This constraint necessarily reduces the efficiency with which firms produce their outputs. For example, if firms devote resources such as energy, capital and labor to reducing emissions of pollutants, the cost of producing their outputs increases. More importantly, these resources used to reduce pollution cannot be used to produce economic outputs. Thus, as we use resources to reduce pollution, Gross Domestic Product must necessarily fall.

The same arguments can be made with respect to land use. If land use is restricted in order to protect the environment, the outputs produced in conjunction with that land will now require additional resources to produce the same level of outputs.

The Positive Impacts of Environmental Quality on Economic Productivity. While the use of resources to reduce emissions can have a negative impact on the economy, an increase in environmental quality also would be expected to have a positive impact on the economy. This positive impact occurs in two ways. First, environmental resources are an input to production processes. For example, clean water is needed to produce many products. The cleaner the water is to begin with, the fewer resources that are needed to further clean the water.

Second, environmental quality affects the productivity of other inputs. For example, the cleaner the environment, the healthier and more productive the labor force.[3] Another example of this type of impact is the positive impact of reduced air pollution (such as tropospheric ozone) on agricultural productivity. Reduced air pollution increases yield per acre of many crops, increasing the productivity of land and other agricultural impacts.

Important benefits to the economy also can be found through the impact on health and health care. The impact of improved environmental quality on the marginal productivity of labor was discussed above, but there is another very important impact. Health care consumes a very large proportion of the national income in most countries, but particularly so in the United States. If the population is healthier because they have cleaner water to drink, cleaner air to breathe, and less exposure to toxic substances, then less resources need be devoted to health care. If health care is an industry characterized by increasing average cost,[4] then increasing environmental quality not only will reduce the resources necessary to treat environmentally related illness, but it also could reduce health costs generally by reducing the average cost of treatment. This effect is illustrated in Figure 5.3, where reducing the number of cases of treatment from N_1 to N_2 reduces the average cost of treatment

---

[3]For example, children who are exposed to lead suffer permanent neurological damages, including a permanent reduction in IQ.

[4]In this case, increasing average costs would mean that for each additional case of illness, the cost of treating the illness is higher than for previous cases.

FIGURE 5.3

POTENTIAL REDUCTION IN AVERAGE COST OF HEALTH CARE FROM AN IMPROVEMENT IN ENVIRONMENTAL QUALITY

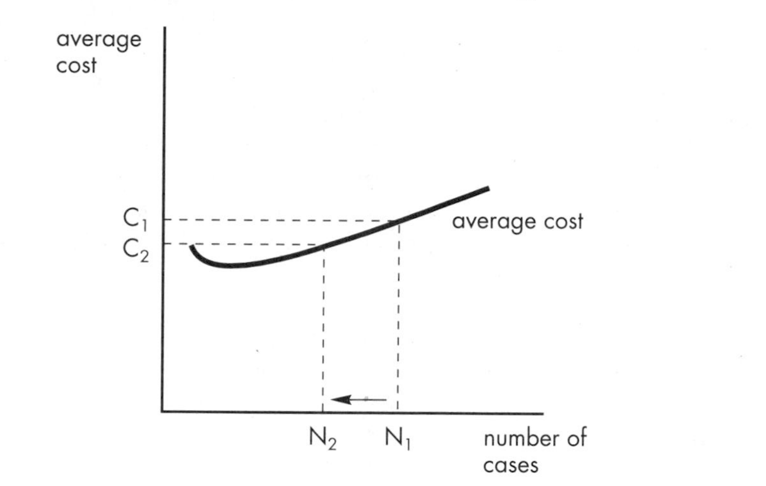

from C_1 to C_2. This reduction in the average cost of general health care will occur only if the average cost of health care is increasing. While there is some evidence to suggest this is true, this relationship is very difficult to measure due to the heterogenous nature of disease and health care.

In summary, cleaner air, less exposure to toxic waste, cleaner water, more forests, more fish, greater levels of ecological services (such as biological diversity), and greater protection from ultraviolet radiation make the economy more productive. This impact on the productivity of the economy is in addition to the other beneficial aspects of environmental improvement such as aesthetic, recreational, health, and ecological benefits.

Which Effect Dominates? The above sections describe both positive and negative economic impacts of improving environmental quality. This discussion leads immediately to the question of determining the net effect of these two opposing impacts. Unfortunately, this question is very difficult to answer. Theoretically, one would want to estimate an equation[5] of the following form:

$$GDP = f(L_1, K_1, EQ[GDP, L_2, K_2]) \tag{5.1}$$

[5]Students who are more familiar with economic and mathematical modeling will recognize this equation as a "reduced form," which represents a more complicated system of several equations. It should also be noted that in this equation, environmental quality is a function of GDP. The impact of GDP on environmental quality will be discussed later in the chapter.

where L_1 is the amount of labor directly used to produce GDP, K_1 is the amount of capital directly used to produce GDP, EQ is the level of environmental quality, L_2 is the amount of labor used to reduce emissions, and K_2 is the amount of capital used to reduce emissions. This equation has the potential to show both the negative and positive impacts of environmental change on GDP. The negative impact would be seen as an increase in the amount of labor or capital devoted to abating emissions reduces the amount of labor and capital that can be used to produce GDP. The equation also allows for the positive impact as an increase in environmental quality (EQ) can increase GDP for the reasons discussed in the sections above.

Unfortunately, none of the empirical studies really has attempted to measure a relationship of the nature of Equation 5.1. The primary reason is that it is very hard to determine how to measure environmental quality because of all the potential measures that can be used. Air quality, water quality, the health of ecosystems (such as forests, wetlands, aquatic ecosystems), toxic contamination, biodiversity, the density of the ozone layer, emissions of greenhouse gases, and many other measures of environmental quality could be defined. Which are most important, and how can they be combined into a single measure or a small set of measures? Of even more serious concern is the fact that in order to estimate an equation of the nature of Equation 5.1, it would be necessary to have an environmental measure not only for today, but for the past twenty or thirty years so one would have a sufficient number of data points to estimate the equation. Different possibilities for measuring environmental quality in this context will be discussed later in the chapter.

The lack of ability to include a comprehensive environmental quality measure has substantially shaped the empirical studies of the relationship between the environment and the economy. It has caused people to focus on the negative impact (resources used to reduce emissions can't be used to produce GDP) and has precluded them from measuring the positive impact (environmental quality is an input to producing GDP and makes other inputs more productive). For example, Jorgenson and Wilcoxen (1990) estimate that the impact of environmental regulation on the economy is to reduce GDP by 2.592 %. Hazilla and Kopp (1990) find a cumulative impact through 1990 of a reduction of nearly 6%. The *EPA Retrospective Study*[6] found qualitatively similar results. However, the economic model (computable general equilibrium model) which they used only allowed them to measure negative impact, and they did not attempt to measure the positive impact, presumably because of a lack of data.

More recently, Gillis et al. (1996)[7] have attempted to rectify this problem by using the same type of computable general equilibrium model but

---

[6]US Environmental Protection Agency "The Benefits and Costs of the Clean Air Act, 1970 to 1990," (Draft, October 1996).

[7]T. Gillis, A. McGartland, D. Nestor, C. Pasurka, and L. Wiggins, *The Social Benefits of Air Quality Management Programs: A General Equilibrium Approach* (1996).

also by incorporating the impact of changes in environmental quality on economic activity. They incorporate the health benefits of the air quality regulations, including reduced medical costs, increased productivity as well as impacts on agriculture and household soiling. Gillis et al's estimation of the combined negative and positive impacts is that GDP is approximately 2% higher than it would be in the absence of air quality regulations.

Additional Potential Impacts of Environmental Policy— The Porter Hypothesis

In the above section on negative impacts of environmental policy on the economy, it was argued that stricter environmental policy will have a negative direct impact on the economy as resources that could be used elsewhere in the economy are devoted to production. However, in recent years, Porter (1990, 1991), Gore (1992), and others argue that it is not clear that stricter environmental policy reduces overall resource availability and increases firms' costs. Since stricter environmental policy requires firms to cut wastes, increase the efficiency with which energy is used, and choose newer, more efficient production technologies, costs may be reduced. *World Resources 1992–93* cites several studies that show that firms forced to reduce their release of pollution and wastes actually reduce their production costs. Porter further argues that the reduction in production costs associated with strict environmental policy not only directly stimulates the domestic economy, but also increases the competitiveness of the domestic economy relative to foreign economies, bolstering exports and reducing imports. This hypothesis that regulation can reduce production costs is essentially an argument that there are cost savings opportunities available that firms are not seizing. Many economists refer to these unseized opportunities as analogous to five dollar bills laying on the sidewalk that no one bothers to pick up.

Are there Five Dollar Bills on the Sidewalk? Simpson (1993) points out that the most logical way to attempt to explain both the Porter hypothesis and the cases of cost reduction that have been observed is to differentiate between variable and fixed costs. Stricter environmental policy forces firms to change their production technologies, which increases their fixed costs in the short run as they install new equipment. However, since the cleaner technologies reduce the firms' needs for inputs (such as energy) and produce less waste, which is expensive to treat or dispose, variable costs will be reduced after the new technology is installed. Advocates of the Porter hypothesis would argue that the reduction in variable costs exceeds the increase in fixed costs, so total costs decline.

However, many economists do not believe that there are cost-reducing options available that firms do not voluntarily pursue. In order for government environmental policy to generate opportunities that would otherwise be unavailable to firms, there must be institutional barriers that prevent the firms from pursuing these cost-reducing opportunities.

FIGURE 5.4

THE INVERSE U-SHAPED RELATIONSHIP BETWEEN POLLUTION AND INCOME
URBAN CONCENTRATIONS OF SULFUR DIOXIDE

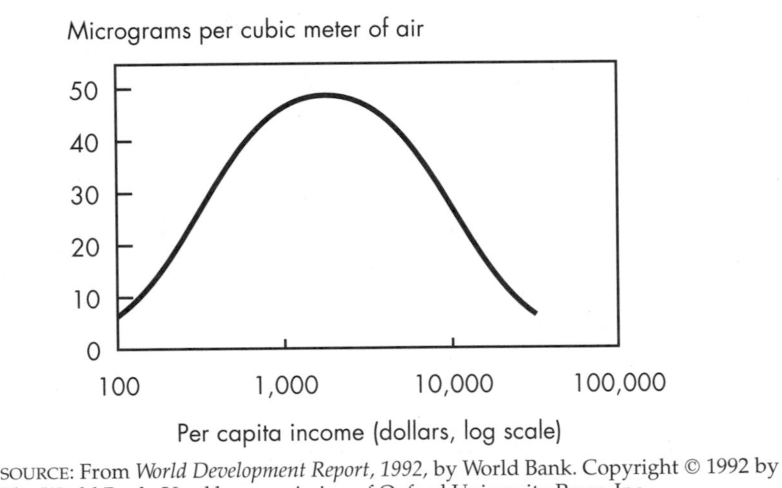

SOURCE: From *World Development Report, 1992,* by World Bank. Copyright © 1992 by The World Bank. Used by permission of Oxford University Press, Inc.

Palmer and Simpson (1993) investigate the types of barriers that might prevent firms from pursuing these cost-reducing activities. These barriers include the inefficiency and shortsightedness of the firm and the difficulty of appropriating the benefits from research and development into cleaner technologies.

The explanation that firms may be stupid (or ignorant) and simply fail to take advantage of beneficial opportunities is not very satisfying. Although individual firms certainly have bypassed many golden opportunities, Palmer and Simpson suggest that it is difficult to imagine that this oversight would occur on a systematic basis, particularly since our free market system encourages survival of the most efficient firms. However, deviations from cost-minimizing behavior could occur in industries where there was little competition from either domestic or foreign producers (such as electric utilities). Also, if firms are shortsighted and inadequately consider the future benefits of current investments, then they might not undertake investments that lower costs in the long run. However, the free market system also should discourage this. In the long run, firms that are farsighted should replace firms that are shortsighted, as Darwinian survival of the fittest will reward firms that engage in better planning.

Another reason firms may bypass cost-reducing opportunities is that the difficulty of appropriating the benefits from research and development prevents firms from engaging in the socially optimal level of research and development into technologies that are less polluting. If a potential technology is difficult to protect with patent laws (either because foreign competitors do not honor U.S. patents, or because it is easy to modify the technology and get around the patent), then firms have little incentive to invest in such technologies. However, it is then difficult to argue that government policies that encourage such research and development would increase competitiveness relative to foreign competitors (such as Porter argues). If the technologies are easy to imitate, then it is not likely that their promotion will create a cost advantage for domestic producers.

A further reason firms might not undertake these potentially beneficial investments is explored by Farmer, Kahn, and McDonald (1996). This reason has to do with the asymmetric nature of the potential rewards and penalties of adopting new technologies. Let us assume that there is an equal probability that a potential technology will either increase or reduce costs. Let us also assume that the technology is easy to copy. Under these circumstances, firms have no incentive to innovate. If they adopt the new technology and it proves to reduce costs, the cost advantage will be short-lived as the technology will soon be copied by the firm's competitors. If the new technology proves to be a bust and increases costs, then the firm's competitors will not follow the innovating firm. Under these circumstances, it always pays to be a follower and not a leader, so cost-reducing technologies might not be implemented.

THE IMPACT OF THE ECONOMY ON THE ENVIRONMENT

The impact of environmental policy and changes in environmental quality on the economy has been discussed above. However, the earlier discussion covers only one set of interactions between the economy and the environment. The other set involves the impact of economic growth on the environment.

There has been much discussion of this issue in the environmental economics and economic development literature. One perspective argues that as the economy grows and per capita income becomes greater, environmental quality will increase. The argument used to support this perspective says that as income increases, the demand for environmental quality will increase as well. The World Bank (1992) looks at the data and finds empirical evidence which supports this hypothesis with respect to some environmental problems but not others. The World Bank looks at data across countries and charts environmental quality as a function of per capita income. Their work shows that an inverse U-shaped relationship exists for air pollution (see Figure 5.4), but not for other measures such as per capita municipal waste and carbon dioxide emissions. That is to say, as income initially increases, pollution increases, but then it reaches a point where increases in per capita income cause a decline in pollution. Although the World Bank does not find this proposition (that environmental quality will improve with increases in income) to be true in general, many people interpret the air pollution results to be true for environmental degradation in general. Based on this unwarranted generalization, it is often argued that an improvement in environmental quality is a natural outcome of increases in per capita income.

Unfortunately, the relationship depicted in Figure 5.4 does not really present convincing evidence that increasing income generates increasing environmental quality. One important problem is that emissions of air pollution in itself is not a particularly good measure of general environmental quality. First, conventional air pollutants, such as sulfur dioxide, do not represent an environmental problem whose impacts accumulate over time. If impacts accumulate over time—even if the incremental environmental degradation becomes smaller as income increases—the cumulative environmental impact will continue to grow larger. This effect is one of the reasons that most ecologists feel that land use issues and not emissions issues are the most pressing environmental problems. Related to this problem of using air pollution as a proxy for overall environmental change is that air pollution is not as degrading of ecological services as such environmental problems as wetlands conversion or deforestation or soil loss.

More importantly, if environmental quality is an input into producing income, income may never increase. For example, if current environmental degradation causes the desert to move into food producing areas, income will not be able to grow in the future, so the impact of income increasing environmental quality could never be realized. This point will be discussed more fully in Chapter 17.

An additional problem with the argument that increasing income leads to environmental improvement is that the measures we use for income may be flawed in that they do not measure the loss of environmental capital that is used to produce income. This point is discussed more fully below.

ENVIRONMENTAL DEGRADATION AND THE MEASUREMENT OF INCOME

GDP, which measures the market value of the goods and services produced in a society in a year, is almost universally used as a measure of national income. It has become a measure highly associated with the standard of living of a society, but it has several important shortcomings.

First, it does not include certain types of economic activities, such as home production. The value of the services provided by a parent that stays home and raises a child is not included in GDP, but if the services were provided by a day care provider, they would be included in GDP. More serious, however, is the fact that GDP does not include other aspects of the quality of life. Chief among these omissions is environmental quality. The higher the environmental quality, the higher social welfare; but this is not measured in GDP. Third, and most important from the point of view of the discussion of the impact of environment on the economy, it does not measure how current economic activity reduces future income-producing capabilities by depleting environmental resources. Although the first two flaws of GDP are well recognized by economists and presented in virtually every principles of economics textbook, the third flaw is not as well discussed and needs more explanation.

Many economists argue that GDP, and the National Income and Product Accounts (see any principles of macroeconomics textbook for a discussion of the National Income and Product Accounts) that underlie GDP, should be modified to take the environment into account. In particular, Daly (1977, 1991), Peskin (1976), and Repetto (1989) have argued that disastrous consequences can occur when macroeconomic policy is based on promoting the growth of GDP without taking the environment into account. They argue that not only does this omission ignore other aspects of the quality of life, but that GDP is seriously flawed as a measure of economic progress. This flaw has to do with the fact that measures of NDP (Net Domestic Product, which subtracts depreciation from GDP and is a better indicator of economic progress) subtract the depreciation of human-made capital, but do not subtract the depreciation of natural capital. Thus, when a machine is worn-out, the loss in income-producing ability is subtracted from the measure of current income.

The consumption of human-made capital is subtracted from GDP to give a more accurate measure of the current economic well-being of a nation. However, when a forest is clear-cut, soil degraded, or stocks of minerals depleted in order to produce current income, a similar debit is not made. Although one can argue that this issue is just a definitional, and that all definitions are arbitrary in nature, there are serious implications when national economic policy is based on this flawed measure of NDP. If increasing current NDP is a primary policy goal, then natural capital and its ability to produce future income (or other services) may be expended even if it is detrimental to producing future social welfare. Although this problem is crucial in developed countries such as the United States, it is perhaps even more important in developing countries, where pressing needs to increase current income have caused catastrophic deforestation, pollution, soil erosion, and desertification (see Chapter 17). As Repetto and others indicate, this difference in the treatment of natural capital and human-made capital

> *. . . reinforces the false dichotomy between the economy and the "environment" that leads policy makers to ignore or destroy the latter in the name of economic development. It confuses the depletion of valuable assets with the generation of income. Thus, it promotes and seems to validate the idea that rapid rates of growth can be achieved and sustained by exploiting the resource base. The result can be illusory growth and permanent losses in wealth.*

Therefore, Repetto argues that the depreciation of natural capital should be factored into NDP in a fashion analogous to the depreciation of human-made capital. Repetto recomputes Indonesia's national income and product accounts, making corrections for deforestation, soil erosion, and depletion of oil reserves. Repetto found that the measured 7.1 percent annual growth rate of GDP is actually only 4 percent when these corrections are made.

GDP, Sustainability and Environmental Quality

The central argument used by those who advocate adjusting the measure of GDP to incorporate the loss of future income earning capacity is that some growth patterns are inherently unsustainable. If the soil is made barren in pursuit of raising current income levels, then it is argued that these income levels are unsustainable and that future income levels must decline. In actuality, the question of the sustainability of economic growth and quality of life is much more complex and must be examined in further detail. In this examination of sustainability, we will focus on sustainability of GDP, although we recognize that GDP is not a complete measure of social welfare or quality of life. The reason for focusing on GDP is that the major purpose of this chapter is to examine the interaction between the macroeconomy and the environment. In Chapter 17, which examines the relation between the environment and development in developing nations, we will focus on the more multidimensional nature of sustainable development.

What patterns of growth of GDP are sustainable and what patterns are not? The answer to this question generally hinges on the accumulation of capital which accompanies growth. In general, if the growth of income is associated with an accumulation of capital, the growth is sustainable. However, if the growth of income is associated with a decline in the amount of capital, the income level is unsustainable and will fall in the future. Discussions of the feasibility of sustainable growth are critically dependent on a broad view of capital and particular assumptions about the substitutability of different types of capital.

Economists who are optimistic about the prospects for sustainable growth of GDP define three types of capital: human-made capital (such as factories, roads, etc.), human capital (the quantity and quality of the labor force) and natural capital. These optimistic economists view the three types of capital as near perfect substitutes, so if a society fuels its growth in income by consuming natural capital, the growth will be sustainable—provided that a sufficient investment in human and human-made capital is made to compensate for the loss of natural capital. However, these economists tend to view natural capital as the stock of extractable resources, such as oil, coal, iron, etc. If natural capital is composed solely of extractable resources, then it is reasonable to argue that human-made capital and human capital are good substitutes for natural capital. For example, as oil reserves become depleted in the generation of the growth of income, society can substitute away from this scarcity by developing more fuel-efficient vehicles, adding insulation to buildings, and developing better solar energy technology.

Unfortunately, this perspective on natural capital is relatively limited. Natural capital should be divided into two categories: extractable resources such as discussed above, and environmental resources. Environmental resources are those natural resources which produce ecological services such as nutrient cycling, production of oxygen, carbon fixation, biodiversity and ecosystem resilience and stability. It is unrealistic to think that human capital or human-made capital can be an effective substitute for environmental resources in the production of ecological services. However, ecological services are essential to the productivity of the economy. This implies that growth in income that reduces the ability of environmental resources to produce ecological services is inherently unsustainable, and that a focus of sustainable development policies should be the conservation of environmental resources and ecological services.

The Measurement of Environmental Quality

One theme which echoes through this entire chapter is that in order to measure the impact of changes in environmental quality on the economy and the changes in the economy on the environment, one must be able to develop measures of the state of the environment that can reflect environmental change, and which are relevant to the state of the economy. Typical data collected, such as concentrations of specific air pollutants, have an important flaw in that they are not informative of the environmental system's ability to produce ecological services. This lack of data is the primary reason why economic models are so incomplete in incorporating positive feedbacks of the health of the environment on the health of the economy.

This lack of indicator variables has been recognized at many levels, and there has been a flurry of activity of estimating ecological indicators and indicators of sustainability. One large program is EPA's Environmental Assessment and Monitoring Program (EMAP), which develops indicators of the health of different types of ecosystems (forests, estuaries, wetlands, etc.). Although this system is relatively good at describing the properties of each individual system, when looking across systems there are thousands of individual indicators, and it is difficult to aggregate these indicators into a small set of variables that can be used in statistical analysis of the relationship between the economy and the environment.

Environmental Analogues to GDP. In a separate paper, the author of this text (Kahn, 1996) argues that effective social policy requires the development of a separate indicator of the health of the environment that is comparable to the measures we have for guiding other areas of social policy. For example, we use GDP, the unemployment rate, and the inflation rate as measures of the health of the economy.[8]

---

[8]The measure of GDP should include corrections that show how environmental degradation reduces our ability to produce income.

It is important to develop a corresponding set of indicators of a healthy environment. These indicators need not be flawless, as existing indicators in other areas of policy are not flawless. For example, the official unemployment rate does not give a complete picture of unemployment because it does not include discouraged workers who are not actively seeking work. The unemployment rate also does not adequately reflect employment opportunities for all segments of the labor force (for example, inner-city youth). However, if the people who base economic policy on indicators such as the unemployment rate understand the measure's limitations, the measure can be effectively used to make policies that improve society's well-being.

In his 1996 paper, Kahn suggests developing an index of the health of the environment by surveying people and presenting them with alternative states of the environment, as defined by a number of measurements of the physical characteristics of the states of the environment. They would be asked to determine which of a set of choices they prefer, and then a statistical analysis, such as conjoint analysis (see Chapter 4), would be used to estimate a preference function. This preference function would show the influence of each characteristic on the choice process. The coefficients of this function could be used as weights to aggregate the environmental characteristics into a set of indices or a single index. These measures would then be indices where physical characteristics are aggregated based on weights that are derived from people's willingness to make trade-offs. This is exactly analogous to GDP, which is also a trade-off weighted index of physical quantities. In the GDP case, the physical quantities are the quantities of market goods, and the weights are market prices. In the environmental index case, the physical quantities are the levels of environmental characteristics, and the weights are the preference weightings determined through survey research. Although this environmental index would not be expressed in monetary terms, it is completely analogous to GDP.

This index can be used in analysis in two ways. First, it can inform policy makers if a change leads to an improvement or decline in overall environmental quality. Second, it can be used in empirical analysis of the relationship between environmental quality and the health of the economy as a variable which explains the health of the economy, such as EQ in Equation 5.1 (page 137).

ENVIRONMENTAL TAXATION AND MACROECONOMIC BENEFITS

Within the last several years, an energetic debate has developed surrounding the potential existence of macroeconomic benefits associated with environmental taxation. These macroeconomic benefits might arise if the revenue

FIGURE 5.5

LABOR MARKET LOSSES ASSOCIATED WITH AN INCOME TAX

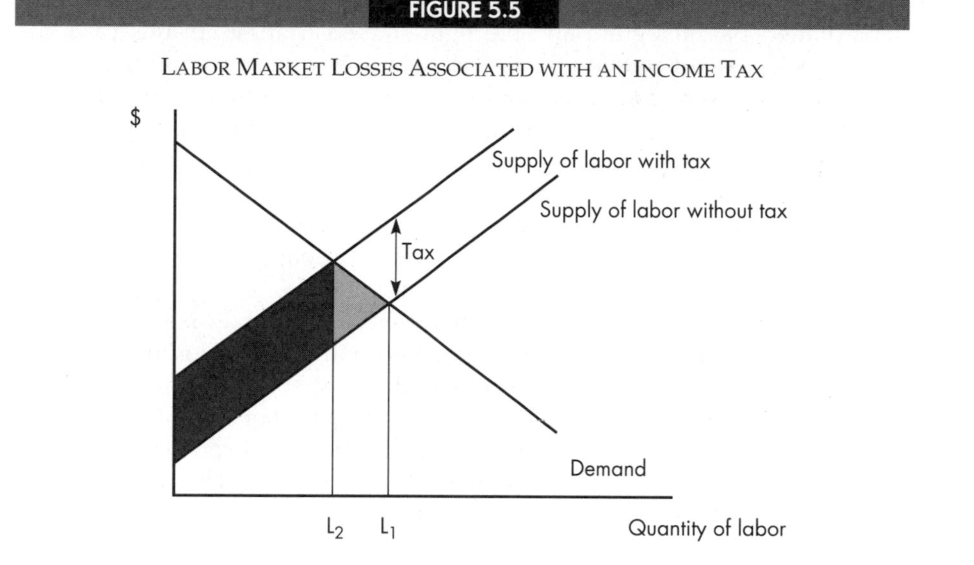

associated with an environmental tax was used to reduce the level of income tax (or property or sales tax).[9] The above statement immediately raises the question of why tax dollars from one type of tax might be more beneficial than tax dollars from another source. The answer to this question is that environmental taxes (per unit pollution taxes) correct a market distortion, while income taxes create a market distortion.

In Chapters 2 and 3, it was shown how the existence of an unregulated environmental externality could generate an inefficiently high level of both the externality and the externality-generating activity (see Figures 2.2 and 3.5). The tax reduces these levels toward their socially optimal level. In contrast, a tax on income creates a market distortion by generating an inefficiently low level of labor. This contrast is illustrated in Figure 5.5, which depicts the marginal benefit and marginal cost functions for labor.[10] L_1 is the amount of labor that market behavior would generate in the absence of a tax, but if an income tax is imposed, the supply curve will shift up by an amount equal to the tax, as the individual's return to providing labor declines by the amount of the tax. The new level of labor will be L_2, which will be associated

---

[9]The following discussion will take place in the context of income taxes, but similar arguments could be constructed for sales and property taxes.

[10]In this discussion, we will assume that there are no externalities or public good characteristics associated with labor.

with a reduction in consumers' and producers' surplus of the sum of the dark-shaded parallelogram and the light-shaded triangle. In this case, the cost to society is not equal to the loss in consumers' and producers' surplus, as part of the loss to consumers and producers is not a loss to society. Since the tax payment is used to produce public services or make payments to other members of society, it is not a loss to society. The total tax payment is equal to L_2 units of labor, multiplied by the amount of the tax (equal to the area of the dark-shaded parallelogram). That leaves a net loss to society of the area of the light-shaded triangle. This loss of benefits arising from the tax is often called the deadweight loss of the tax.

In summary, if one were to substitute environmental taxes for income taxes, there would be two benefits associated with the environmental taxes. The first benefit would be the correction of the market failure associated with the pollution externalities, which would lead to a higher level of social welfare.[11] The second benefit is the reduction of the market distortion in the labor market, which also will increase the level of social welfare. These two benefits are often called the "double dividend" associated with an environmental tax (see Pearce, 1991). A very intuitive explanation for the existence of the double dividend is that environmental taxes penalize something that is bad for society (pollution) while income taxes discourage something that is good for society (people working). If a government were to implement a pollution tax and use the revenue to reduce the income tax, it must increase the level of social welfare.

Although the above arguments for the existence of an environmental taxation double dividend have intuitive appeal, there has been much controversy over its actual existence. Several noted economists (Bovenberg and de Mooij, 1994) argue that an environmental tax can actually lead to greater distortion in the labor market, since the combination of an income tax and higher prices for goods (the higher price is caused by the environmental tax) will lead to an even lower level of labor than with the income tax alone. Although the mathematical complexity of the arguments underlying the opposing viewpoints is beyond the scope of this textbook, the potential existence of a double dividend remains an unresolved controversy. Developing a better understanding of whether a double dividend exists should be a primary focus of research in environmental economics since the resolution of this question has very important implications for environmental policy. If the double dividend exists, then environmental taxation will make the macroeconomy stronger; this effect would be a powerful argument for using pollution taxes instead of other policy instruments such as marketable pollution permits or direct regulations.

---

[11]Of course, if the environmental tax was too high, it could reduce the level of the externality below the level that was socially optimal. See Chapter 3 for a discussion of this point.

THE ENVIRONMENT AND INTERNATIONAL ECONOMIC ISSUES

There are several significant environmental issues intertwined with international economic issues.[12] These include the global public good nature of many environmental resources, transfrontier pollution, the effect of environmental policy on international competitiveness, and the effect of international trade policy on environmental policy.

GLOBAL PUBLIC GOODS

Many environmental resources have global public good characteristics. That is to say, the environmental resources in one country generate public good benefits for people in other countries, or environmental resources span the borders of several or all countries. Rain forests, wetlands, and biodiversity are examples of environmental resources contained in one country that create public good benefits for citizens of other countries. Oceans and atmospheric processes are examples of resources that span international borders.

The difference between global public goods and local or national public goods is that with global public goods there will be international separation between those who benefit from the environmental resources, and those who bear the costs of preserving the environmental resources. For example, the whole world benefits through the preservation of Brazil's rain forests, but Brazil bears the entire cost of preserving the rain forest. Thus, what is optimal from the global perspective is unlikely to be generated by the policies of a particular country, so global optimality cannot be generated without negotiating an international agreement. The difficulty of generating a negotiated agreement across countries is made more complex because it is often the lower income countries who bear the cost of preserving the resources and the higher income countries who receive the public good benefits from preserving them.

International policies that shift some of the benefits of preserving these environmental resources toward the countries who preserve them need to be developed. In other words, these potential Pareto improvements must be turned into actual Pareto improvements.[13] Policies for accomplishing this transformation vary across resources. They are discussed in more detail in

---

[12]A good source of information on international environmental issues is the U.S. State Department's Oceans, Environment, and Science home page available at http://www.state.gov/www/global/oes.

[13]A potential Pareto improvement is a change where those who benefit gain more than the losses of those made worse off. If the winners compensated the losers it would turn the situation into an actual Pareto improvement, where some people are made better off and no one is made worse off. See Chapter 4 for futher discussion of this point.

Chapter 12 (Tropical Forests), Chapter 13 (Biodiversity), and Chapter 17 (The Environment and Economic Growth in Third World Countries).

TRANSFRONTIER POLLUTION

Transfrontier pollution is pollution generated in one political jurisdiction that creates damages in another political jurisdiction. The problems generated by transfrontier pollution become more complex when the boundaries the pollution crosses are international boundaries rather than state, provisional or municipal boundaries. The country that generates the pollution does not consider all the social costs when devising environmental policy, because some of the social costs are borne by other countries. This leads to an environmental policy that would generate a level of emissions higher than optimal from a global point of view.

International negotiation must be conducted in order to generate the appropriate level of emissions. In Europe, this process tends to be easier since the shared agencies of the European Union provide a structure within which the negotiations may take place. In North America, the process is more difficult, as the past controversy between Canada and the United States over acid rain attests (see Chapter 7). The newly created North American Free Trade Agreement (NAFTA) could provide a framework for resolving these issues. However, NAFTA is in its beginning stages, and it is too early to see what type of effect it will have on environmental quality in North America (Canada, Mexico, and the United States). In the coming years, there is a high probability that NAFTA will merge with MERCUSOL, the South American free trade organization.

ENVIRONMENTAL POLICY AND ITS EFFECTS ON INTERNATIONAL TRADE

The effect of environmental policy on international trade is a highly contested issue. There are three major ways that environmental policy can affect international trade. First, if environmental policy increases the cost of production in an environmentally strict country, that country's goods could be more expensive relative to those produced in environmentally lax countries. Firms also may choose to relocate to countries where production costs are lower because there is less environmental regulation. Alternatively, if strict environmental policy reduces the cost of producing goods (as the previously discussed Porter hypothesis suggests), then the goods of the environmentally strict country will be less expensive relative to the goods produced in environmentally lax countries. The potential effect of environmental policy on the cost of producing goods was discussed at the beginning of the chapter. However, these impacts are likely to be small, since intercountry differences in the cost of environmental regulation tend to be very small in comparison with intercountry differences in the cost of labor and other factors of production.

The third way in which environmental policy can affect international trade is if environmental policy encourages the development of less polluting or "green technologies," and the green technologies are exported to other countries. Although research into this area is beginning, we do not know enough to be able to determine the potential of increasing exports in this fashion.

THE IMPACT OF INTERNATIONAL TRADE POLICY ON THE ENVIRONMENT AND ENVIRONMENTAL POLICY

Although there was much discussion concerning the potential of the North American Free Trade Agreement (NAFTA) to reduce environmental quality in the United States and Mexico, the potential effect of NAFTA on environmental quality was probably misunderstood by most of the U.S. public. The primary fear was that NAFTA was going to lead to increased movement of U.S. industrial activity to Mexico, which would not only reduce the number of jobs in the United States, but would increase pollution in Mexico, since Mexico has more lax and less enforced environmental regulations than the United States. An additional fear was that this pollution would spill across the border as it moved through rivers, wind currents, and ocean currents and negatively affect environmental quality in the United States.

Obviously, if industrial activity increases in northern Mexico, this increase has the potential to adversely affect environmental quality in both Mexico and the United States. However, Mexican environmental law requires U.S. companies operating in Mexico to adhere to the same environmental standards as the companies would in the United States. Critics of NAFTA, however, argue that this provision is often unenforced. There is little doubt that unless environmental protection increases, industrial growth in northern Mexico will lead to reduced environmental quality in both the United States and Mexico.

The important issue is whether NAFTA will exacerbate or mitigate this process. Critics of NAFTA argue that NAFTA will lead to much more industrial activity and that the existence of NAFTA removes the incentives for Mexico to agree to improved environmental standards. NAFTA proponents argue that industrial activity will continue to increase in northern Mexico and that NAFTA provides a negotiating framework for increasing environmental standards in Mexico. This controversy cannot be resolved through theoretical or logical arguments, and only time will tell if NAFTA will have a negative or positive impact on environmental quality in Mexico and the United States.[14]

Although environmentalists have focused on NAFTA as a potential environmental problem, a different international agreement poses a much more

---

[14]This discussion has focused on environmental quality in the United States and Mexico and has not focused on U.S.–Canadian issues. The reason is that Canada and the United States have much more similar environmental standards, and several bilateral commissions have been established to develop joint environmental and resource policy.

serious threat to environmental quality. This international trade agreement is through the auspices of the World Trade Organization (WTO) which is an organization created by international treaty, which establishes procedures that define "fair trade practices" and establishes procedures by which a country can legally retaliate against a country that engages in unfair trade practices against it.

The critical element of the WTO with respect to the environment is that a country cannot establish import barriers for a good based on the way the good is produced. This provision was designed to prevent countries from developing artificial reasons for keeping foreign goods out. For example, a heavily unionized country might say that only goods produced with union labor can be imported, which would reduce import competition. Another example of an import barrier based on production is a requirement that food and beverages that are imported be free of preservatives. This requirement effectively eliminates competition from foods and beverages from distant locations. The problem for environmental quality is that the effect of production methods on environmental quality cannot be used as a means for discriminating against imports. For example, when the United States required domestic canners of tuna to produce "dolphin safe" tuna, it attempted to ban imports of Mexican canned tuna which were not dolphin safe. However, WTO (which was then known as the General Agreement on Trade and Tariffs (GATT) regarded this as discrimination based on the method of production and ruled that the United States was in violation of GATT. The United States could not even force Mexican tuna to be labeled in order to state that it was not dolphin safe, since that would also be unfair discrimination, according to these rules.

The environmental provisions of WTO are hotly debated. Free trade enthusiasts argue that environmental standards must be coordinated at the minimum level in order to maximize free trade. Those who are more concerned with the environment argue that the environmental implications of production methods should be a legitimate reason to discriminate against imported products. Since both free trade and global environmental quality benefit the people of the earth, some sort of compromise must be established. One potential compromise is to allow environmental discrimination, but to establish an international panel to review each case so that countries do not use insignificant environmental reasons as a phony excuse to protect their own goods at the expense of other countries. This panel should represent the diversity of countries as well as contain representatives with both environmental and business backgrounds.[15]

---

[15]The World Trade Organization Web page can be found at http://www.wto.org. Information on current environmental issues before the WTO can be found by searching on "World Trade Organization" and "environment."

SUMMARY

Both microeconomic and macroeconomic methodologies are important in helping to guide the development of environmental policy. This chapter has focused on macroeconomic issues to complement the microeconomic methods that were discussed in Chapters 2 through 4.

The environment and environmental policy can have both a negative and positive impact on the economy. The negative impact occurs when resources that could be used elsewhere in the economy are devoted to reducing pollution. The positive impact occurs because environmental quality is an input to production processes and can positively impact the productivity of other inputs such as labor and agricultural land.

Many previous modeling attempts have concluded that increasing environmental quality will reduce the productivity of the economy, but these models did not allow for the existence of the positive impact. A new study by EPA that measures just a small subset of these positive impacts shows that the air quality laws have had a positive impact on the economy.

One of the primary problems with measuring the relationship between the economy and the environment is the lack of a suitable measure of environmental quality. However, a set of trade-off based operational indicators can be developed that are consistent with operational indicators that measure the success of the pursuit of social policy goals. Although this process would require a substantial research effort, there are significant potential gains in terms of developing objective measures that can be used to shape environmental policy in a direction that most improves social welfare.

This chapter has just scratched the surface in highlighting some of the models that have been used to understand the effect of the environment on the macroeconomy. In the application chapters that follow, both microeconomic and macroeconomic theory will be applied in order to understand the environmental issues that are the focus of the chapters. As an example, in this chapter, we have not completely examined the issue of whether environmental regulations and the pursuit of better environmental quality will lead to a loss of jobs. This will be examined in the context of the focus of each chapter, where appropriate. For example, in the chapter on fossil fuels (Chapter 8), the macroeconomic impact of air quality regulations will be addressed. Similarly, the employment impacts of alternative forestry policies are addressed in Chapter 11.

As a final note, it should be reemphasized that although macroeconomic implications are important, they should not be the sole determinant of environmental policy. Our target levels of environmental quality should not be determined solely by the effect of environmental quality on GDP or NDP, but on the effect of environmental quality on social welfare, to which both GDP and environmental quality contribute.

REVIEW QUESTIONS

1. What is the Porter hypothesis? Are there economic justifications for the effects Porter describes?
2. How can input–output models examine the interaction between the environment and the economy?
3. How does WTO constrain environmental options?
4. Why should GDP be modified to include certain types of environmental degradation?
5. How do the environment and economy interact?
6. What is meant by the double dividend?

QUESTIONS FOR FURTHER STUDY

1. What is the effect of environmental policy on international trade?
2. What is the international significance of global public goods?
3. Can environmental policy increase international competitiveness?
4. How would you handle environmental issues in the latest reformulation of GATT?
5. Why is the macroeconomic analysis of environmental issues important?

SUGGESTED PAPER TOPICS

1. Look at the Porter hypothesis and determine whether stricter environmental policy can lead to more jobs. Start with the references by Simpson, Palmer and Simpson, and Oates, Palmer, and Portney, and survey current journals for more recent papers.
2. Look at the impact of WTO on the environment. Start with the book by Pearce and Warford and search bibliographic databases and the Web on WTO and the environment.
3. Choose an environmental problem such as global warming and examine the equity issues associated with it. Pay particular attention to intertemporal and North/South equity issues. Search bibliographic databases using the specific environmental problem and *equity* as keywords.

REFERENCES AND SELECTED READINGS

1. Ayres, R. U., and A. V. Kneese. "Production, Consumption, and Externalities." *American Economic Review* 59 (1969): 282–297.
2. Baumol, William, and Wallace Oates. *The Theory of Environmental Policy.* London: Cambridge University Press, 1979.
3. Bovenberg, A. Lans and Ruud A. de Mooij. "Environmental Levies and Distortionary Taxation," *American Economic Review,* September 1994, 1085–1089.
4. Boulding, Kenneth. "The Economics of the Coming Spaceship Earth." In *Environmental Quality in a Growing Economy,* edited by Henry Jarrett. Baltimore: Johns Hopkins University Press, 1966.
5. Farmer, Amy, James Kahn, and Judith McDonald. "Strict Environmental Policy: Microeconomic and Macroeconomic Justifications." Unpublished draft paper, University of Tennessee, 1994.
6. Daly, Herman E. *Steady-State Economics: The Economics of Biophysical Equilibrium and Moral Growth.* San Francisco: Freeman, 1977.
7. ———. "On Economics as a Life Science." *The Journal of Political Economy* 76 (1969b): 392–406.
8. ———. "Towards an Environmental Macroeconomics." *Land Economics* 67 (1991): 255–259.
9. Hazilla, Michael and Raymond J. Kopp (1990), "Social Cost of Environmental Quality Regulations: A General Equilibrium Analysis," *Journal of Political Economy,* 98, No. 4, pp. 853–873.
10. Jorgenson, Dale W. and Peter Wilcoxen (1990), "Intertemporal General Equilibrium Modeling of U.S. Environmental Regulation," *Journal of Policy Modeling,* 12, No. 4, pp. 715–744.
11. Kahn, James R. "Trade-off Based Indicators of Environmental Quality: An Environmental Analogue to GDP." Unpublished draft paper, University of Tennessee, 1994.
12. Knoester, Anthonie and Jarig van Sinderan, "Taxation and the Abuse of Environmental Policies," in *Environmental Economics* (G. Boera and A. Silberston, eds.), New York, St. Martins Press, 1995.
13. Malthus, Thomas Robert. "An Essay on the Principle of Population," In *An Essay on the Principle of Population, Text, Sources and Background, Criticism,* edited by Phillip Appleman. New York: Norton, 1975.
14. Oates, Wallace, Karen L. Palmer, and Paul R. Portney. "Environmental Regulation and Competitiveness: Thinking about the Porter Hypothesis." Discussion Paper 94–02. Washington, D.C.: Resources for the Future, 1993.

15. Palmer, Karen L., and David R. Simpson. "Environmental Policy as Industrial Policy." *Resources* 112 (1993): 17–21.
16. Pearce, David, "The Role of Carbon Taxes in Adjusting to Global Warming," *Economic Journal*, July 1991, 938–48.
17. Pearce, David W., and Jeremy J. Warford. *World Without End.* Washington, D.C.: Oxford University Press for the World Bank, 1993.
18. Peskin, Henry M. "A National Accounting Framework for Environmental Assets." *Journal of Environmental Economics and Management* 2 (1976): 255–262.
19. Porter, Michael A. *The Competitive Advantage of Nations.* New York: Free Press, 1990.
20. ———. "America's Green Strategy." *Scientific American* (1991) 168.
21. Repetto, Robert, et al. *Wasting Assets: Natural Resources in the National Income and Product Accounts.* Washington, D.C.: World Resources Institute, 1989.
22. Simpson, David R. "Taxing Variable Cost: Environmental Regulation as Industrial Policy." Discussion Paper ENR 93–12. Washington, D.C.: Resources for the Future, 1993.
23. World Bank. *World Development Report 1992: Development and the Environment.* New York: Oxford University Press, 1992.
24. World Commission on Environment and Development. *Our Common Future (The Brundtland Report).* New York: Oxford University Press, 1987.
25. World Resource Institute. *World Resources 1992–93.* New York: Oxford University Press, 1992.

Exhaustible Resources, Pollution, and the Environment

Part II examines the role that the production and disposal of exhaustible resources has on the economy and the environment. Energy figures prominently in this discussion due to its central role in economic processes and its role in the emission of greenhouse gas and acid deposition, as well as a host of other environmental problems.

Global Environmental Change: Ozone Depletion and Global Warming

The greenhouse effect itself is simple enough to understand and is not in any real dispute. What is in dispute is its magnitude over the coming century, its translation into changes in climates around the globe, and the impacts of those climate changes on human welfare and the natural environment.

Thomas C. Schelling[1]

INTRODUCTION

Very few environmental changes have generated as much anxiety, discussion, and confusion as "the greenhouse effect" and "the hole in the ozone layer." Although these are separate environmental problems, they are linked in this chapter because of their commonalities. These commonalities include the following:

1. Both problems are the result of pollutants modifying basic atmospheric chemistry, thereby altering atmospheric processes and functions.
2. Both problems are caused by stock pollutants that persist in the atmosphere for long periods (up to hundreds of years) after their emission into the atmosphere.
3. In both cases, the pollutants are global in the sense that their contributions to environmental problems are independent of the location of the emissions. It does not matter if the pollutants are emitted in New York, London, Tokyo, or Nairobi, since the ultimate effect on atmospheric processes is identical.

[1] Thomas C. Schelling, "Some Economics of Global Warming," *American Economic Review* (1992).

4. In both cases, there is the potential for significant global environmental change and significant impacts on social, economic, and ecological systems.

Each of these commonalities is a characteristic that differentiates these types of pollutants from conventional pollutants, such as those that form smog. Consequently, these global, atmosphere-altering pollutants require a substantially different treatment in the development of policy.

The fact that the basic effect of the pollutants is the alteration of the atmosphere (its chemistry, processes, and functions) is extremely important. It makes it much more difficult to estimate a damage function than it is to estimate one for conventional pollutants, such as those that form smog. For example, to test the relationship between smog and agricultural productivity, one can observe agricultural plots in high smog areas and low smog areas and (controlling for other influences) determine the effect of smog on agricultural yield. Similarly, one can measure how human respiratory ailments are influenced by smog levels. Since there are lots of agricultural plots and lots of people, it is possible to make enough observations to test whether smog affects people and plants. However, there is only one atmosphere, and it cannot be subdivided into smaller portions to increase the number of observations to provide a large enough sample to conduct statistical tests. The only way to make multiple observations is to look at the atmosphere, its characteristics, and the pollutants as they change over time. However, since we have only recently begun collecting all of the appropriate data, it is difficult to separate natural variation from pollution-induced variation. This problem of attributing cause and effect is probably more pronounced for global warming than it is for ozone depletion, as there appears to be a greater consensus that certain classes of chemicals cause the depletion of stratospheric ozone.

Both the pollutants that deplete ozone and the pollutants that are linked to global warming are stock pollutants, meaning that they have long lifetimes in the environment. Ozone depleting chemicals may last up to 100 hundred years in the stratosphere, while some of the greenhouse gases (gases that cause global warming) last even longer. A portion of the carbon dioxide released into the environment today may still be present 500 years into the future. The persistence of these pollutants is important, because when one assesses the damages associated with these pollutants (such as in estimating a marginal damage function), one needs to calculate the damages that current emissions will generate in the future. As the chapter-opening quotation from Schelling indicates, uncertainties involving the damages are the central scientific and policy questions.

The global nature of these pollutants makes them very different from other types of pollutants. For example, the emissions of nitrogen oxides and other pollutants in the southern California area increase the smog problem in that area. If the emissions are moved from southern California to New

York, smog will increase in New York and decrease in California. The effects of the pollutants tend to be local and regional. In contrast, the effects of global pollutants are completely independent of the location of their emissions. This locational equivalence both complicates and simplifies the policy making process. It simplifies the process because there is no need to account for geographic variability in the effects of emissions into taxes or marketable pollution permits (see Chapter 3 for a discussion of this need for geographic variability in pollution control instruments). It complicates the policy making process because the problems cannot be dealt with by one country, since the emissions of all countries matter. One country cannot unilaterally solve these problems. International cooperation is required to deal with these global externalities.

A final reason for looking at these two problems in the same chapter is that the processes that generate the depletion of the ozone layer also have both mitigating and intensifying effects on global warming. The pollution that causes the depletion of the ozone layer also has a strong greenhouse effect, which intensifies global warming. On the other hand, the depletion of the ozone layer may have a cooling effect that partially offsets global warming. Since the way ozone depletion occurs has an effect on global warming (and global warming has no effect on ozone depletion), we will first examine ozone depletion and then focus on global warming.

THE DEPLETION OF THE OZONE LAYER

CAUSES OF THE DEPLETION OF THE OZONE LAYER

The term ozone layer refers to the presence of ozone in the stratosphere. The stratosphere, the outer layer of the atmosphere, is separated from the troposphere (lower atmosphere) by the tropopause. An important characteristic of the atmosphere is that, throughout the troposphere, the air becomes colder the further the distance from the earth's surface. This temperature gradient changes at the tropopause, where the lowest layer of the stratosphere is warmer than the highest level of the troposphere. The significance of this thermal relationship is that there is a layer of warm air sitting on top of a layer of cold air. Since hot air rises, there is little mixing of air across this temperature inversion. Also, once pollutants make their way into the stratosphere, they tend to remain there, since they are above the rain and other mechanisms that can remove them from the atmosphere. Figure 6.1 contains a diagram of the atmosphere, which is not drawn to scale. If it was drawn to scale, the thickness of the atmosphere would be less than the thickness of the line used to depict the earth's surface.

The pollutants that most adversely affect the ozone layer are fluorocarbons, particularly those that contain chloride or bromide. Of these, those

FIGURE 6.1

TROPOSPHERE, TROPOPAUSE AND STRATOSPHERE

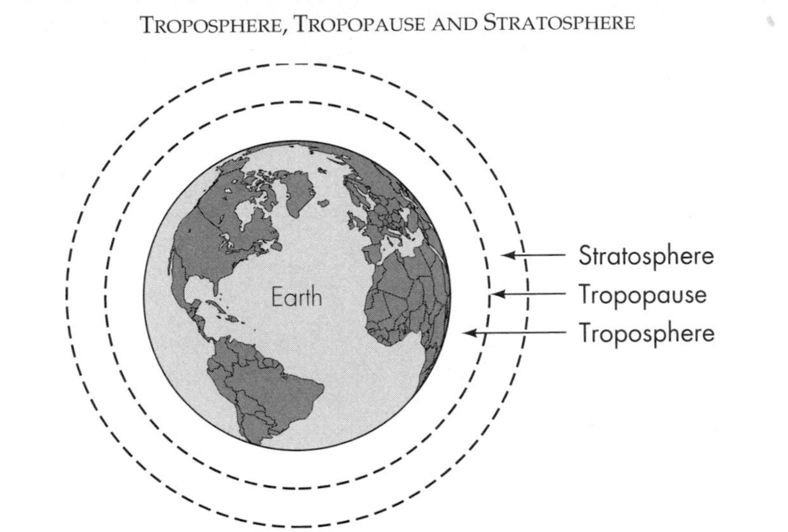

containing chloride (chlorofluorocarbons, or CFCs) are responsible for most of the depletion of the ozone layer.

These chemicals deplete the ozone layer by serving as a catalyst in a chemical reaction that converts ozone (0_3 and 0_1) to oxygen (0_2). This reaction is accelerated by the presence of ice crystals in the stratosphere, which is one of the explanations for the large loss in ozone over the Antarctic (50–60 percent loss, World Resources Institute, 1992, p. 200). Since the CFCs act as a catalyst, they are not consumed in the reaction that converts ozone to oxygen, but remain in the stratosphere and continue the destruction of ozone for many decades.

CONSEQUENCES OF THE DEPLETION OF THE OZONE LAYER

As discussed in greater detail in Chapter 8, ozone in the lower atmosphere causes a variety of health, ecological, and agricultural damages. Based on that fact, one might wonder why there is so much excitement about the depletion of stratospheric ozone. It should be emphasized that while ozone in the lower atmosphere (troposphere) causes damages, ozone in the upper atmosphere (stratosphere) does not come into contact with living organisms and fulfills a critical function: blocking the penetration of ultraviolet light and limiting the amount of ultraviolet light that strikes the earth's surface.

Consequently, as the ozone layer becomes depleted, the amount of ultraviolet radiation striking the earth's surface increases. Since ultraviolet radiation causes living cells to mutate, this increase has profound consequences for human and ecological health. In October 1991, a panel of international scientists said that there has been a 3 percent reduction in stratospheric ozone (World Resources Institute, 1992, p. 200). This 3 percent reduction in the protective layer of stratospheric ozone could lead to a 6 percent increase in the amount of ultraviolet radiation striking the earth's surface. This increase in radiation could lead to an additional 12 million cases of skin cancer in the United States over the next 50 years, with potentially bigger problems in the southern temperate countries (Australia, New Zealand, Argentina, Chile, and South Africa) where the ozone depletion is even greater than in the Northern Hemisphere (World Resources Institute, 1992, p. 200).

In addition to humans, other organisms also are adversely affected by increased ultraviolet radiation. Increased ultraviolet radiation damages most plants and could significantly reduce agricultural yields. In addition to altering agriculture, the increased ultraviolet radiation may have very important impacts on ecosystems, since the increased radiation may change competitive interactions among plant species in ecosystems (Caldwell, Teramura and Tevini, 1989). Perhaps the most important nonhuman effect is on the phytoplankton layer in the ocean. Phytoplankton form the foundation of the oceanic food web. In addition, the larval and juvenile stages of many fish live and feed in the phytoplankton layer. These organisms undergo metamorphosis several times before achieving adult form and are very vulnerable to the mutagenic effects of increased ultraviolet radiation.

Additionally, increased ultraviolet radiation accelerates the deterioration of materials. Synthetic materials, such as plastics and nylon, are particularly susceptible to deterioration from ultraviolet radiation.

USES OF CFCS AND OTHER OZONE DEPLETING CHEMICALS

CFCs and other ozone depleting chemicals are valued in a variety of applications. These include their use in refrigeration and air conditioning systems, as propellants in spray cans, as propellants for manufacturing foam products, and as solvents to degrease and clean machine parts and electrical components. One of the properties of these chemicals that makes them so useful is that they do not react with other chemicals (this property is called inertness). Unfortunately, this inertness also is what makes them persist so long in the environment, as their disaffinity for chemical reaction means they do not readily break down into simpler and less harmful substances.

POLICY TOWARD OZONE DEPLETION

When the problem of ozone depletion was discovered in the 1970s, alternative solutions to the problem were proposed. Many economists favored traditional economic incentives. In particular, economist Peter Bohm constructed a powerful argument for the use of deposit–refund systems to recapture CFCs in refrigerator coils (see boxed example). Other economists argued that CFCs represented a perfect case for the use of marketable pollution permits or pollution taxes. Since CFCs are a global pollutant and the damages generated are independent of the location of the emissions, taxes or marketable pollution permits would not have to be modified to take into account the geographic location of the emissions.

The first policy the United States adopted toward ozone depleting chemicals was a command and control type of policy, a 1977 ban on the use of CFCs as a propellant in spray cans of deodorants, hair sprays, and other consumer products. At first glance, one might think this is another case of abatement cost-minimizing economic incentives being rejected for more costly command and control techniques, a process that has happened repeatedly in the formulation of environmental policy.

However, in the case of these uses of CFCs, a ban is probably socially efficient. The reason is that very good substitutes exist for CFCs as propellants in spray cans. In other words, the cost of eliminating these emissions is low relative to the damages the emissions create. For example, substitutes for CFC-propelled deodorant spray include the use of more benign gases as propellants, mechanical (thumb) pumps, stick deodorants, roll-on deodorants, and more frequent showers. Figure 6.2 contains a graph of marginal abatement costs and marginal damages that relate positions of costs and benefits for CFC emissions from personal hygiene and other consumer spray products.

If the costs of abatement are low due to the availability of substitutes, then this is a case where the optimal level of emissions is at or near zero. Although an efficient level of emissions can be achieved with economic incentives, why incur the costs of such a program when a ban is so much cheaper in terms of administrative costs? (See Chapter 3 for a complete discussion of this point.)

In the 1970s and the 1980s, it was not apparent that the optimal level of CFC emissions from other sources (solvents, foam manufacture, refrigeration, and air conditioning) was equal to zero. However, one thing becoming increasingly clear is that further ozone depletion policy could not be developed in the U.S. (or any other country) in isolation. Since ozone depletion is a global pollution problem, other countries' emissions of CFCs were just as important as U.S. emissions. Hence, effective policy must be developed in the context of an international agreement.

The discovery of the hole in the ozone layer above the Antarctic, and the evidence of continued depletion at mid-latitudes, spurred the development of an international agreement on chemicals that deplete the ozone layer. This agreement is based on an internationally shared belief that the emissions of CFCs and other ozone depleting chemicals generate damages far in excess of abatement cost, with the optimal level of emissions of CFCs equal to zero.

In 1987, the Montreal Protocol on Substances that Deplete the Ozone Layer was signed by most developed and developing countries. This international agreement was amended in 1990 to speed up the elimination of

FIGURE 6.2

MARGINAL ABATEMENT COSTS AND MARGINAL DAMAGES FOR CFC EMISSIONS FROM PERSONAL HYGIENE SPRAY CANS

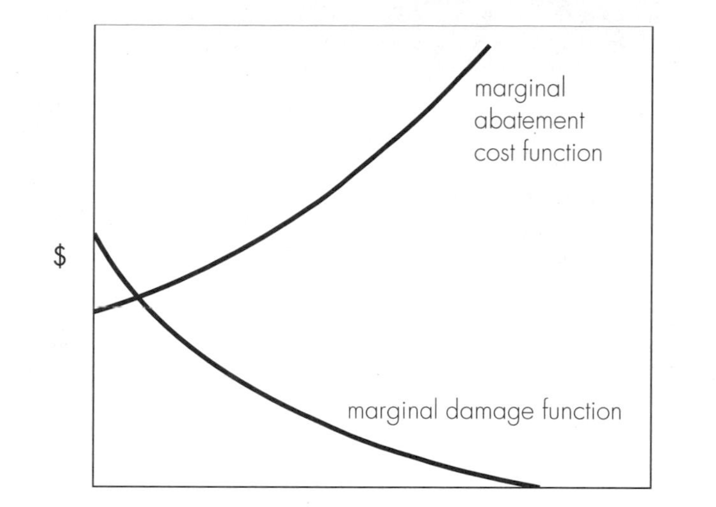

CFCs due to new evidence concerning ozone depletion to require the cessation of emissions of CFCs (and other ozone depleting chemicals) by the year 2000 in developed countries and by the year 2010 in developing countries.

More recent findings about the severity of the ozone depletion problems generated support for renegotiating the Montreal Protocol to have earlier deadlines. In fact, the Dupont Company, the world's largest manufacturer of CFCs, unilaterally decided to end sales of CFCs to developed countries by 1996 (World Resources Institute, 1992, p. 200). The Montreal Protocol was amended again in 1992 to require total elimination of CFCs in developed countries by 1996 and by 2006 in developing countries.

An important remaining issue is how to treat replacements for CFCs, which also have an ozone-depleting effect (albeit a much smaller one). The importance of developing policies to deal with these substitutes has intensified as evidence has developed which suggests that hydrochlorflourocarbons (HCFCs) and hydroflourocarbons (HFCs) may be more ozone-depleting than was originally thought (Solomon and Albritton, 1992). One possibility would be to place a tax on these substances, based on the ozone-depleting effect per kilogram of the substance. This tax would serve to decrease the use of all such substances, but discourage the most harmful substances the most. In addition, it will encourage technological development of nonozone-depleting alternatives. As discussed in Chapter 3, the same result could be achieved with

Deposit–Refund Systems for CFCs

When the ozone depletion problem first was recognized, Swedish economist Peter Bohm suggested that certain uses of CFCs would be well suited to a deposit–refund system. CFCs are contained in the coils behind refrigerators and freezers. As long as the coils remain intact, the CFCs remain isolated from the environment. However, when defunct refrigerators are disposed in landfills, the earth moving equipment and shifting garbage break the coils (or they rust if not broken), and the CFCs are released into the atmosphere. If the CFCs could be removed from the coils before disposal, their release could be prevented.

Bohm suggested that a deposit–refund system would be appropriate for accomplishing this task. A fee paid when purchasing a refrigerator would only be refundable when the CFCs were removed from the coils at a licensed recycling center. As long as the fee or deposit was greater than the costs of transporting the refrigerator and removing the CFCs, appropriate incentives would exist to recycle the CFCs. Companies that sold new refrigerators would incorporate the deposit–refund systems into their prices. A price for a new refrigerator would be lower if there was a trade-in of an old refrigerator. The trade-in differential would be lower than the refund and higher than the company's cost of complying with the recycling. Both consumer and company would profit from recycling as long as the deposit/refund was greater than the cost of recycling.

However, an international treaty which was enacted in 1987 (the Montreal Protocol) incorporated the notion that CFCs are so destructive that their use should be banned. Consequently, the practical need for a deposit–refund system for CFCs has been eliminated, since the substance is scheduled for a ban on production. Chemical companies are developing substitutes for CFCs as their use is phased out. However, since the substitutes for CFCs also have an ozone depleting impact (albeit a much smaller one), deposit–refund systems, taxes or marketable pollution permits might be necessary to generate the optimal level of use of these chemicals at the lowest social cost, as well as to provide incentives to develop substitutes which do not deplete the ozone layer.

a marketable pollution permit system, provided that the permits took into account these differences among chemicals. In either case, more research must be conducted to better understand the potential ozone depletion impact of each type of chemical.

It is interesting to note that the current illegality of CFCs has made them extremely valuable in a fashion analogous to other illegal substances such as cocaine, ivory or alcohol during the Prohibition. New CFCs are not being made in developed countries, so people occasionally try to move CFCs across national borders to earn large profits from users who are willing to violate the law. For example, the U.S. Customs Service regularly confiscates illegal shipments with street values of millions of dollars. This problem should diminish as existing stockpiles become depleted and as production in developing countries ends in accordance with the Montreal Protocol. Unlike cocaine or crack, the manufacture of CFCs requires a full-scale chemical plant, so monitoring of illegal production should be relatively easy.

GREENHOUSE GASES AND GLOBAL CLIMATE

Global warming is linked to the accumulation of a variety of gases in the atmosphere. These gases, which include carbon dioxide, methane, nitrous oxide, and water vapor, trap infrared radiation (heat) that would normally escape into the earth's atmosphere. The analogy to a greenhouse is made because the glass panels of a greenhouse allow light to enter (where it becomes converted to heat after striking the interior of the greenhouse), but block the escape of much of the heat. Actually, the analogy to a greenhouse is not an accurate depiction of what happens in the atmosphere, since the greenhouse gases absorb the heat rather than block its transfer. The presence of greenhouse gases in the earth's atmosphere and the absence of greenhouse gases in the moon's atmosphere explain the relative temperature differences between the earth and the moon, despite their approximately equal distances from the sun. The presence of very high concentrations of greenhouse gases on Venus and the absence of these gases on Mars also partially explain the temperature differentials between Venus, Earth, and Mars.

The earth's temperature (both on the surface and in the atmosphere) is always moving toward an equilibrium. If an equilibrium did not develop between the amount of heat entering the earth's atmosphere and the amount of heat leaving the atmosphere, the earth would be either continually heating or continually cooling. The injection of more greenhouse gases into the earth's atmosphere upsets this equilibrium, because there are now more gas molecules to absorb heat. The temperature of the atmosphere and the temperature at the earth's surface increase until a new equilibrium is established. The amount of heat entering and leaving the atmosphere (in equilibrium) has not changed, but the stock of heat stored by the earth and its atmosphere has increased. The capacity to absorb heat is known as the radiative forcing of the gas. If the amount of greenhouse gases diminishes, then radiative forcing decreases, and a new equilibrium is established at a lower temperature. It may take hundreds of years for a new equilibrium to become established because of the existence of lagged effects. For example, the temperature of the oceans responds very slowly to changes in radiative forcing.

There is virtually no debate surrounding the proposition that the greater the level of greenhouse gases, the greater the equilibrium temperature of the earth. There is also little debate on whether anthropogenic emissions of greenhouse gases cause a significant increase in global temperature in comparison with current temperature levels and in comparison with natural fluctuations in the temperature levels. The debate centers around the magnitude and timing of the change, and its significance to human welfare.

Water vapor is the most important greenhouse gas in the earth's atmosphere, constituting about 1 percent of the total gases in the atmosphere. Carbon dioxide has an atmospheric concentration of approximately 0.04 percent (Solow, 1991). Other important greenhouse gases include methane, nitrous oxides, and chlorofluorocarbons (predominantly CFC–11 and CFC–12). There are significant anthropogenic emissions of all greenhouse gases except water vapor.

SOURCES OF GREENHOUSE GASES

Anthropogenic emissions of greenhouse gases come from a variety of sources. In order to properly understand the emissions of carbon dioxide, which is the most important of the greenhouse gases, it is necessary to understand the "carbon cycle."

The carbon cycle refers to the movement of carbon from the atmosphere to the earth's surface. On the earth's surface, carbon is stored in the biomass of every organism. Carbon dioxide is also dissolved in surface water, with the oceans having the most significance in this regard. Carbon dioxide is removed from the atmosphere when a plant grows, and the carbon dioxide is converted to carbon in the plant's tissues. For example, a tree is approximately 50 percent carbon by weight. When an animal eats a plant, the carbon is transferred from the biomass of the plant to the biomass of the animal. When a plant or animal dies, it decays and the carbon combines with oxygen to form carbon dioxide, which is returned to the atmosphere. The carbon dioxide (CO_2) is reabsorbed when a new plant grows.

Anthropogenic activities, such as the burning of fossil fuels or deforestation, upset the equilibrium carbon cycle and cause an increase in atmospheric carbon dioxide. Fossil fuels, such as oil, coal, and natural gas, are the fossilized remains of prehistoric plants. These fuels represent stored carbon, and their combustion causes increases in atmospheric concentrations of CO_2. Similarly, when forests are cut down, all the carbon except that which is preserved in wood products (construction materials, furniture, paper in books, and so on) eventually breaks down, and its carbon is released as carbon dioxide into the atmosphere. Since much less than 50 percent of the biomass of a tree is converted into lumber or wood products, deforestation can be a significant source of carbon dioxide emissions. Of course, when deforestation takes the form of "slash and burn" clearing to create agricultural or cattle ranching areas (see Chapter 12), all the carbon is converted to CO_2, and the greenhouse effect is even more pronounced. When forests are replanted, carbon dioxide is drawn out of the atmosphere. This implies a significant environmental advantage to using biomass fuels.[2] For example, if cars burn ethanol (which can be made from wood or crops such as corn)

---

[2]This environmental advantage is relative to the generation of carbon dioxide emissions. There are other potential environmental problems associated with the large-scale production of biomass fuel crops. In particular, annual grain crops, such as corn or soybeans, which require annual plowing of the ground and large-scale use of agrichemicals, generate other environmental problems. The full social costs of biomass fuels must be compared with the full social costs of fossil fuels in selecting the mix of fuels that maximize social welfare. More discussion of this issue is contained in Chapter 8.

TABLE 6.1

ATMOSPHERIC CONCENTRATIONS OF GREENHOUSE GASES

	CO_2 (PPM)	CH_4 (PPM)	N_20 (PPB)	CFC−11 (PPT)	CFC−12 (PPT)
Preindustrial	280	0.8	288	0	0
Current	350	1.7	310	280	484
Current annual rate of change	1.6%	0.02%	0.8%	10%	17%

SOURCE: Solow (1991).
PPM = Parts per million, PPB = Parts per billion, PPT = Parts per trillion

instead of gasoline, there would be a significant reduction of CO_2 emissions. Although the burning of any fuel releases carbon dioxide, burning gasoline represents the release of stored carbon, while burning ethanol from plants represents the cycling of carbon. As long as the crop from which the fuel is made is replanted, there would be no net increase in atmospheric CO_2 concentrations from the burning of biomass fuels. However, since mature natural forests contain more biomass per acre than replanted forests, the conversion of mature natural forests to "energy plantations" would result in an increase in atmospheric CO_2. It would be better to plant the "energy plantations" in areas of previous deforestation.

Planting new forests would reduce atmospheric carbon dioxide concentrations. This process is known as sequestering carbon. The greatest opportunities for carbon sequestration exist in tropical areas, where growth rates are faster and where the amount of biomass per acre of forest is greatest.

Methane (CH_4) comes from a variety of anthropogenic and natural sources. Natural sources include wetlands and other areas where anaerobic decay of organic matter takes place. Anthropogenic sources include emissions from ruminants (cud-chewing animals, such as cattle and sheep), wet rice cultivation, emissions from coal mines and oil and natural gas wells (natural gas is methane), and leakage from natural gas pipelines. Increases in atmospheric concentrations of CH_4 may also be due to changes in atmospheric chemistry, although this is not well understood (Solow).

Nitrous oxide (N_20) originates from the burning of fossil fuels and biomass. Nitrous oxide also stems from agricultural fertilizers.

Sources of chlorofluorocarbons (CFCs) and other ozone-depleting chemicals were discussed previously in the context of ozone depletion. Reiterating, sources include refrigeration and air conditioning, propellants, foam manufacture, and solvents.

Table 6.1 contains the preindustrial and current levels of these greenhouse gases. It also contains the current annual rate of change.

FIGURE 6.3

HISTORICAL TEMPERATURE RECORD

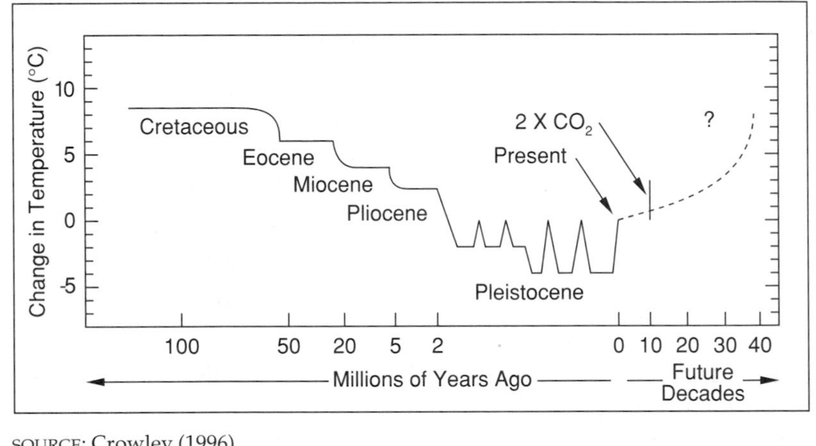

SOURCE: Crowley (1996).

IS GLOBAL TEMPERATURE INCREASING?

The question of whether global climate is increasing is a controversial question, because it cannot be answered solely in terms of an examination of the historic data. For example, using ice core samples in glaciers that can measure historic (going back over 200,000 years) temperature and carbon dioxide levels, there is a clear correlation between greenhouse gas levels (such as methane and carbon dioxide) and temperature. Concentrations of greenhouse gases can be measured from air bubbles in the ice. Temperature is estimated based on levels of hydrogen and oxygen in the ice (Crowley, 1996). However, the changes in temperature associated with carbon dioxide levels are small in relation to temperature changes generated by small changes in the earth's orbital characteristics (Solow). Figure 6.3 contains a depiction of the historical temperature record, showing how much each year was lower or higher than the temperature in 1900.

It is also difficult to reach conclusions from the temperature data that have been recorded at meteorological stations since the late-nineteenth century. The reason is that the weather stations tended to be located around cities in the Northern Hemisphere where few oceanic records were kept. Since global warming can actually lead to a variety of local effects (including a cooling in some areas, although the average effect is warming), it is important to have an appropriate distribution of temperature measurement sites. Also, as cities expand, these meteorological stations will be exposed to increased urban heat radiation. Holding the greenhouse effect and natural fluctuations constant, one would expect an increase in the temperature of urban areas. Despite the difficulty in interpreting past records, there seems to be a consensus that there has been a warming of approximately 0.5° C as a result of the carbon dioxide emissions of the last 100 years. While this is not the only evidence supporting the hypothesis of global warming, some people who are inherently skeptical of global warming have seized upon this 0.5° C change as evidence which refutes the likely development of a significant future warming.

Skeptics argue that the climatological models used to forecast the warming implications of greenhouse gas emissions predict a much stronger warming associated with cumulative carbon dioxide emissions. Skeptics also argue that the bulk of this warming was observed before 1940, while the bulk of emissions occurred after that point (Solow).

Critics of the skeptics argue that a variety of mechanisms have generated some cooling effects. In particular, there may be some carbon dioxide sinks (naturally occurring mechanisms that remove carbon dioxide from the atmosphere). Plants, which remove carbon dioxide from the atmosphere as they increase their biomass, are important sinks. Plant growth may have increased as a result of increased carbon dioxide concentration in the atmosphere (this is called the CO_2 fertilizer effect), which in turn may have removed some of the carbon dioxide emissions from the atmosphere and mitigated the greenhouse effect. Also, oceans are carbon sinks, which may be mitigating global warming. However, since scientists do not fully understand the role and extent of carbon sinks, one cannot assume that the effects of continued carbon dioxide emissions will continue to be mitigated by the functioning of carbon sinks. In particular, the capacity of sinks to absorb carbon is not known and could be fully exhausted already or fully exhausted in the near future.

Additionally, other anthropogenic activity may have had a cooling effect, which has partially offset global warming. Particulate emissions, particularly sulfate aerosols (in this context, an aerosol is a very small particle that is suspended in the atmosphere) block sunlight, creating a cooling effect. As we reduce the emissions of these particulates because of health and ecological effects (see Chapters 7 and 8), this counteracting cooling effect will disappear and, holding everything else constant, global warming may become more pronounced.

The skeptics' argument about the timing of warming and carbon emissions may also be challenged. It is quite possible that the lack of a strong warming in the period following 1940 is simply due to some naturally occurring fluctuations around a mean temperature that is increasing.

It is important to note that the controversy surrounding global warming takes place largely outside the scientific community, but rather in the popular media and in political discourse. The issue may seem controversial, but there is a relatively widespread consensus among the scientists who study global warming. This consensus is based on computer models of the atmosphere, which predict warming based on emissions of greenhouse gases.

One of the most widely cited studies of global warming is the ongoing work of the National Research Council's Board of Atmospheric Science and Climate, which predicts (based on a doubling of atmospheric carbon dioxide)

TABLE 6.2

RANGE OF ESTIMATES FROM CLIMATIC MODELS ABOUT EQUILIBRIUM IMPACT ON MAJOR VARIABLES

	PROJECTION OF PROBABLE GLOBAL AVERAGE CHANGE	DISTRIBUTION OF REGIONAL CHANGE	CONFIDENCE IN PREDICTIONS	
			GLOBAL AVERAGE	REGIONAL AVERAGE
Temperature	+2 to + 5°C	2–3 to + 10°C	high	medium
Sea level	+10 to + 100 cm	a	high	a
Precipitation	+7 to + 15%	−20 to + 20%	high	low
Soil moisture	?	−50 to + 50%	?	medium
Runoff	increase	−50 to + 50%	medium	low
Severe storms	?	?	?	?

PRIMARY SOURCE: Schneider (1991).
SECONDARY SOURCE: Nordhaus (1991).
[a]Increases in sea level are the average of the global rate. Sea level rise in particular locations will be higher or lower than this figure depending on local geological conditions.
?No basis for forecast of this variable.

a warming of 1.5 to 4.5° C (NAS, 1991). Schneider (1991) studied the scientific literature concerning predictions of global climate change and estimates the confidence of the projections. A summary of his work is provided by Nordhaus (1991), which is reproduced here as Table 6.2. As can be seen in this table, Schneider believes that the confidence in global predictions is quite strong, but confidence in how these changes will be regionally distributed is weaker.

The Intergovernmental Panel on Climate Change (IPCC), which is an international consortium designed to pool global expertise and research in climate change, recently published its "best guess" estimate of global warming. In comparison with the 1990 temperature, the IPCC forecasts that global mean surface temperature will increase by 2° C, with a confidence interval of 1 to 3.5° C. This forecast represents the consensus of the world's leading scientists who are involved in global warming.[3]

WHAT IS THE SIGNIFICANCE OF GLOBAL WARMING?

Global warming of the type associated with the consensus estimates will have a number of effects. These effects include a rise in sea level, possible increase in the intensity of storms, effects on forests and other ecosystems,

[3]The IPCC reports can be accessed on the World Wide Web at: http://www.unep.ch/ipcc/ipcc-0.html

biodiversity, agricultural effects, and effects on comfort level. Not all the effects of global warming are negative, even within the same category.

The effects on agriculture have both positive and negative components. For example, the global precipitation average should increase due to global warming.[4] Higher temperature, longer growing seasons, increased atmospheric concentration of CO_2, and increased rainfall are generally favorable for plant growth. However, this does not necessarily imply that global warming will have positive impacts on agriculture. In fact, most researchers predict that the most likely outcome is a negative effect on agriculture (although the confidence interval around this most likely outcome does include positive impacts).

How can negative impacts result from global changes that are favorable on average? The answer is that it is not just the average effect that matters, but how the variation from the average is regionally distributed. For example, if rainfall is increasing on average, but decreasing in areas of good soil and increasing in areas of poor soil, the net agricultural effect will be negative. The Great Plains of the United States, one of the world's most productive food growing areas, becomes substantially drier under some forecasts of the regional distribution of precipitation changes. The same thing may be true of rising temperature, which will shift crop zones northward. For example, areas in the northern Great Plains of the United States and the prairie provinces of Canada that currently are wheat areas may become corn areas, with the appropriate temperature zones for wheat moving even further northward. However, the soils in the former wheat zones may not be capable of supporting corn, and the more northern areas have barren soil that may not be capable of supporting any crops.

Although the effects of CO_2 enrichment are positive for plants, this positive effect does not necessarily translate into increased crop yields. The reason is that not only is the growth of valuable crops increased by CO_2 enrichment, but so is the growth of competing plants (weeds). The controlled experiments that have measured the growth response to CO_2 enrichment do not provide the data with which to model the relative effects on crops versus weeds. Also, rising temperatures and precipitation may increase pest populations, which would have a negative effect on yields.

Another example of the coincidence of both positive and negative effects is in the comfort of people. People will be less comfortable in the summer, not just because of the average rise in temperature, but also because the number and duration of heat waves is expected to increase. However, people will be more comfortable in the winter. Similarly, global warming increases the need for air conditioning but decreases the need for home heating.

---

[4]Melting ice caps and warmer oceans (hotter water takes up more space) will lead to a greater surface area of oceans, and the greater surface area and increased temperature will both lead to more evaporation.

IS CLIMATE CHANGE A BENEFIT–COST PROBLEM?

Although many economists feel that the development of global warming policy should be based on cost–benefit analysis, other scholars, such as Peter Brown of the School of Public Affairs at the University of Maryland, argue otherwise. His position is that the use of cost–benefit analysis has implications for centuries to come, so it is qualitatively different than the problems that cost–benefit analysis was designed to examine.

Brown (1991) argues that it is extremely difficult to measure costs and benefits associated with global warming. This opinion is shared by many economists. Brown references Epstein and Gupta (1990), who assert that the problem can be framed in principle in cost–benefit terms, but state that "for the moment, it is not feasible because of the profound uncertainties surrounding the MB (marginal benefits) and, for different reasons, the MC (marginal costs) curves." Brown argues that the problems of deciding what to measure, the possibility of surprises, the extremely long time horizons, and the difficulties of predicting technical change make the execution of a cost–benefit analysis virtually impossible. More serious than these problems in execution, Brown argues that problems of principle render cost–benefit techniques inappropriate. According to Brown, these problems in principle revolve around discounting, differences in kind, distribution issues, the issue of rationality, and the scope of benefits.

Brown argues that there are three conceptual problems associated with discounting. First, he argues that some things cannot be discounted. Certain resources, such as cultural resources and national parks, we regard as our duty to preserve for posterity. The use of discounting would imply that it is acceptable and even desirable to eventually consume these resources. Second, even in applications where discounting may be important, there is a problem with determining the appropriate discount rate. Finally, Brown argues that discounting "imperils the future by undervaluing it."

Brown also suggests that there are substantial differences in kind among resources that are affected by global warming and typical market goods. Cost–benefit techniques require that everything be measured in a money metric and that

> *it is foolish to think that literally everything under the sun should be subject to the measure of money. . . . The Congressional Medal of Honor cannot be bought, nor a chair in the Economics Department at the University of Chicago, at any price. New ivory cannot be legally purchased nor can a car without a seatbelt.*

Additionally, Brown points out that cost–benefit techniques do not adequately deal with distributive issues, including developed country/developing

Sea level rise is purely negative, with no offsetting positive effects. Sea level rise will lead to a loss in shoreline, although the sea level rise (10 to 100 cm in Table 6.2) is considerably lower than the sea level rise of up to 3 meters that had been predicted a decade earlier (Solow). The current IPCC estimate of expected sea level rise is 50 cm by 2100. The combination of sea level rise and the increase in storm intensities that are predicted to be associated with global warming will lead to erosive loss of shoreline, which will lead to loss of structures, beaches, and wetland habitat.

country and intergenerational issues, which are fundamental to the problem of global warming.

Brown also rejects the argument that rational decision making must be founded in the type of knowledge revealed by cost–benefit analysis. Brown argues that this assertion is based on a confusion about the meaning of rationality. He suggests that rationality has two parts: an end (somehow specified) and a rational means to that end. Although cost–benefit analysis may be a rational means to the end of maximizing discounted present value, the rationality of this means does not imply that maximized present value is an appropriate end in the context of global warming.

Finally, Brown asserts that there is the question of the scope of the benefits. Cost–benefit analysis, founded in the principles of classical economics, looks at benefits and costs to humans. Brown argues that it is not clear that the analysis should only be concerned with humans and that global climate policy should be concerned with benefits and costs to other living organisms.

Brown argues that rather than viewing global climate change in a cost–benefit context, the global warming problem should be viewed in a "planetary trust" framework. Brown argues that we should adopt the framework of Edith Weiss (1991). As Brown (p. 19) describes the framework,

> *. . . each generation has a responsibility to those who follow to preserve the earth's natural and human heritage at a level at which it is received. The responsibility to preserve certain definite things, for example, intact ecosystems and national monuments. Our obligations are not discharged simply by achieving the highest present value of consumption.*

Is Brown correct? Is the global warming problem inherently unsuited to cost–benefit analysis? Obviously, people who reject the economic paradigm would answer yes to this question. Some people who accept the economic paradigm would also argue that cost–benefit analysis cannot answer every question, and issues such as discounting the distant future, measurement difficulties, and the rights of nonhuman organisms are not addressed by cost–benefit analysis. However, many of these same economists would also argue that even though cost–benefit analysis has limitations and should not be the only basis on which decisions are made, it still can contribute insight to the decision making process. In short, cost–benefit analysis can add perspective to understanding the global warming problem, but it should not be the only perspective. See Chapter 4 for a discussion of additional and alternative criteria for evaluation.

THE ECONOMIC CONSEQUENCES OF GLOBAL WARMING

Many of the economic consequences of global warming are mitigated by the ability to adapt to climate change, since the effects of climate change take place gradually over a more than 100-year time frame. For example, as sea level rises, buildings closest to the shore can be allowed to depreciate, and new buildings can be built further inland. Alternatively, dikes or seawalls can be built to protect buildings that are too valuable to depreciate. For example, if sea level rise threatened Manhattan, a seawall could be built around the entire island. Similarly, adaptation strategies exist in agriculture. More heat resistant strains of crops can be developed, and different crops can be planted. If a particular region becomes inhospitable to agriculture in general, farming can move to a more hospitable region.

Analogous strategies are available in forestry. If climate changes so that the species of a particular forest are not well suited for the new temperature and rainfall patterns, the trees can be harvested early and a more appropriate species can be planted. Although early harvest will lead to some loss of revenue (see Chapter 11 for a discussion of how the timing of harvesting affects revenue), it is likely to be relatively small in comparison with total revenue.

Several studies have attempted to directly quantify the effect of global warming on Gross Domestic Product (GDP) or national income. For example, Nordhaus (1991) estimates the annual impact of a doubling of atmospheric CO_2 on the U.S. economy. He finds that the mean estimate is approximately 6.2 billion 1981 dollars, which amounts to roughly 0.26 percent of national income. The sectoral breakdown of these impacts is listed in Table 6.3. Cline (1992) predicts a much greater impact of 2 percent of national income, but his results include a broader definition of damages, which include nonmarket impacts. If Nordhaus' estimates were adjusted to include nonmarket impacts, they would be roughly consistent with Cline's work.

Although the above discussion could be interpreted to imply that the consequences of global warming are not severe, this conclusion should not be made for a variety of reasons, including both limitations in the ability to adapt to climate change and the possibility of "surprises." As demonstrated below, the consequences of global warming could be much more severe than the above arguments suggest.

FACTORS THAT MIGHT MAKE THE ECONOMIC CONSEQUENCES OF GLOBAL WARMING MORE SEVERE

The previous discussion suggests that the consequences of global warming may be mitigated by the ability to adapt. However, because of greater population densities, less access to new technologies, greater reliance on primary production (agriculture, forestry, and so on), and other circumstances, opportunities for adaptation may be substantially less available in developing countries.

One particular aspect of global warming that is likely to be quite costly is the effect of sea level rise on low-lying Third World countries. Many island nations may be completely inundated under plausible sea level rise scenarios. Additionally, some very populated low-lying areas, such as the river deltas of Bangladesh and Egypt (inhabited by tens of millions of people) may be lying entirely under water, or at such a low elevation above sea level that

TABLE 6.3

IMPACT ESTIMATES FOR DIFFERENT SECTORS, FOR DOUBLING OF CO_2

SECTORS	COST (BILLIONS OF 1981 DOLLARS)
SEVERELY IMPACTED SECTORS	
Farms	10.6 to 9.7
Forestry, fisheries, other	small
MODERATELY IMPACTED SECTORS	
Construction	negative
Water transportation	?
Energy and utilities	
electricity demand	1.65
nonelectric space heat	21.16
water and sanitary	positive ?
Real estate	
Damage from sea level rise	
loss of land	1.5
protection of sheltered areas	0.9
protection of open coasts	2.8
Hotels, lodging, recreation	?
TOTAL CENTRAL ESTIMATE	
National income	6.2
Percent of national income	0.26

SOURCE: Nordhaus, William D., *Global Warming: Economic Policy Responses*, MIT Press. Copyright © 1991.

they become even more vulnerable to storms. The costs to these people of losing their land and homes would be quite large, particularly because there is already a shortage of land in these countries. The developed countries have not been particularly hospitable in receiving refugees from Third World countries in the past several decades, which raises the issue of where the sea level rise refugees could resettle. The suffering of these potential refugees, the political destabilizing effects on affected and neighboring countries, and the cost of relocating the refugees could be among the major potential costs associated with global warming.

Another area in which adaptation is not likely to have an important mitigative role is with natural systems. While the pace of global climate change is relatively slow by human standards, it is extremely rapid by natural standards. For example, as global temperatures changed in North America (as glaciers advanced and retreated), forests changed. As the temperature increased, southern pine forests gradually moved north, and as the temperature decreased, northern hardwood forests gradually moved south. These changes could take place because temperatures changed at a pace that was relatively slow in comparison with the rate at which forests could expand. Animals would also migrate to their preferred climate zone and habitat.

However, the climate change associated with greenhouse warming is taking place at a relatively rapid pace. Major effects are taking place within the lifetime of an individual tree. This is far too rapid a pace for a forest to adjust by natural selection. Additionally, the migration of plant and animal species is blocked by roads, farms, cities, and suburbs. Hence, species that are disadvantaged by the new climate regime may disappear as their ability to adjust is limited by barriers.

Jesse Ausabel (1991) argues that the most significant damages from global warming may lie in damages to natural systems, particularly natural systems that are already stressed by interaction with human systems. Water resources are a prime example of this. Global warming may lead to a further drying of southern California from reduced precipitation and from reduced winter snowpack in the mountains because of higher temperatures (Gore, 1992). Water systems in southern California are already stressed from ongoing drought and overexploitation. Further stress could lead to their collapse, with profound implications for millions of people.

THE IMPORTANCE OF SURPRISES

One of the reasons to avoid being optimistic about the potential consequences of global climate change is the potential of unpredicted consequences. These surprises can come about as a result of the possible existence of threshold effects.

There are two types of thresholds with respect to the emissions of any type of pollutant. The first type of threshold is when increases in emissions generate no damages until a threshold is crossed. The second type of threshold is when marginal changes in emissions lead to marginal increases in damages until a threshold is crossed; then marginal changes in emissions lead to very large increases in damages. It is this latter type of threshold that may be most relevant for global warming, although the former type may be important in some circumstances.

One example of this latter type of threshold is if global warming progresses to the point where the tundral permafrost begins to melt. If this were to occur, anaerobic decay of organic matter could lead to a massive release of methane, which would lead to an intensification of global warming.

A second possible threshold effect would occur if temperature change became severe enough to lead to substantial melting of polar ice caps. Not only would this event lead to an increase in sea level rise, but the shrinking of the ice caps would reduce the amount of light reflected by the earth. The reduction in reflection would lead to an increase in heat absorption that would intensify global warming.

Both the melting of the permafrost and the shrinking of the polar ice caps can be classified as positive feedback effects. A positive feedback means the indirect effects of a change intensify the direct effects of a change.

Another type of threshold effect, discussed by Gore (pp. 101–103), would occur if climate changes lead to alterations in ocean currents. If the Gulf Stream stopped flowing and moving warm southern water to the colder northern regions, much of western Europe would experience the colder temperatures that are more typical of the northern latitudes. For example, the climate of Great Britain would more closely resemble that of Newfoundland, with the possibility of long periods of ice-locking of Atlantic, North Sea, and Norwegian Sea ports.

Threshold effects of the first type (no damages until a certain level is exceeded) may occur in some instances in a greenhouse environment. For example, global warming may raise summer temperatures in some areas by several degrees. This in itself may not have a harmful effect on trees and other plants. However, a small increase in average temperature may be associated with a large increase in the length or frequency of severe hot spells, which could cause stress and lead to the demise of heat-sensitive plants.

GLOBAL WARMING POLICY

There are many characteristics of the global warming problem that make it substantially different from other environmental problems. These include

1. The necessity to deal with many different pollutants (all the greenhouse gases) simultaneously.
2. The temporal separation between emissions and damages.
3. The high degree of uncertainty underlying both the scientific understanding of physical impacts and the economic understanding of costs and benefits.
4. The relative importance of equity issues, both across generations and across countries.
5. The need to achieve international cooperation.

THE NECESSITY TO DEAL WITH MANY DIFFERENT POLLUTANTS

Although carbon dioxide is the predominant anthropogenic greenhouse gas, our global warming policy should not focus solely on carbon dioxide. For example, it may be cheaper to initially concentrate on reducing CFC emissions (particularly because of the ozone depletion problem) or on methane emissions. At the margin, one wants to reduce global warming by reducing the greenhouse gas that is least costly to abate. However, one cannot merely look at the cost of reducing a kilogram of carbon dioxide emissions and

TIME PATH OF RADIATIVE FORCING OF A
HYPOTHETICAL GREENHOUSE GAS
(THE INTEGRAL OF THE TIME PATH IS THE SHADED AREA.)

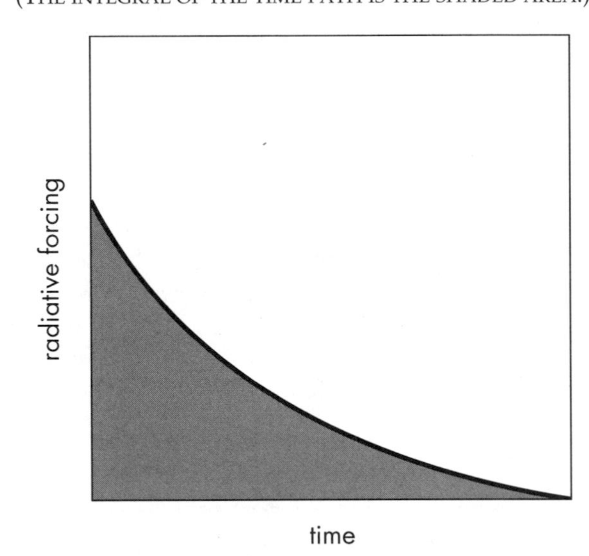

compare it to the cost of reducing a kilogram of N_20 or CFC–11 or CH_4. Each greenhouse gas has a different level of radiative forcing (heat absorbing potential), and each greenhouse gas has a different atmospheric lifetime.

In order to try to measure the equivalency of the different greenhouse gases, the Intergovernment Panel on Climate Change (IPCC) developed the concept of a global warming potential index (GWPI). The concept behind the GWPI is to compare the radiative forcing over the atmospheric lifetime of one particular gas with the radiative forcing over the atmospheric lifetime of one kilogram of carbon dioxide which serves as the benchmark. This comparison is made as a ratio, with the lifetime radiative forcing of carbon dioxide as the base (denominator) of the index. By definition, the GWPI of carbon dioxide is equal to one. The global warming potential of a greenhouse gas declines over time as the gas decays in the atmosphere. Figure 6.4 shows the time path of a hypothetical greenhouse gas. A hypothetical gas is depicted rather than an actual gas, as the choice of a time path for an actual gas would involve the discussion of atmospheric chemistry issues that are beyond the scope of this chapter. The total global warming potential of a gas is the area under the time path (the integral of the time path). The global warming potential index of the hypothetical gas is equal to the integral of the time path for a particular gas divided by the integral of the time path for carbon dioxide.

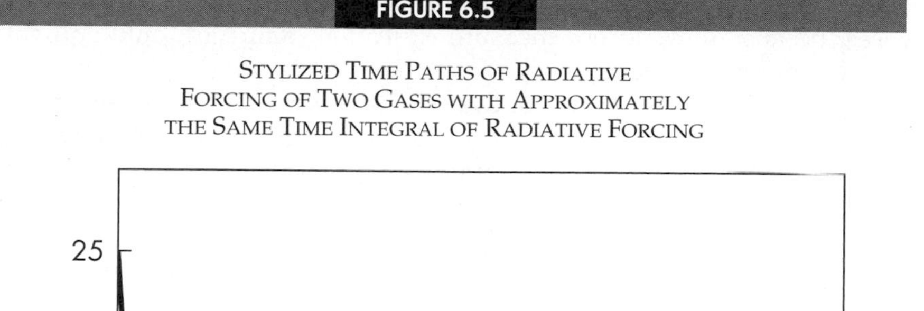

Figure 6.5 Stylized Time Paths of Radiative Forcing of Two Gases with Approximately the Same Time Integral of Radiative Forcing

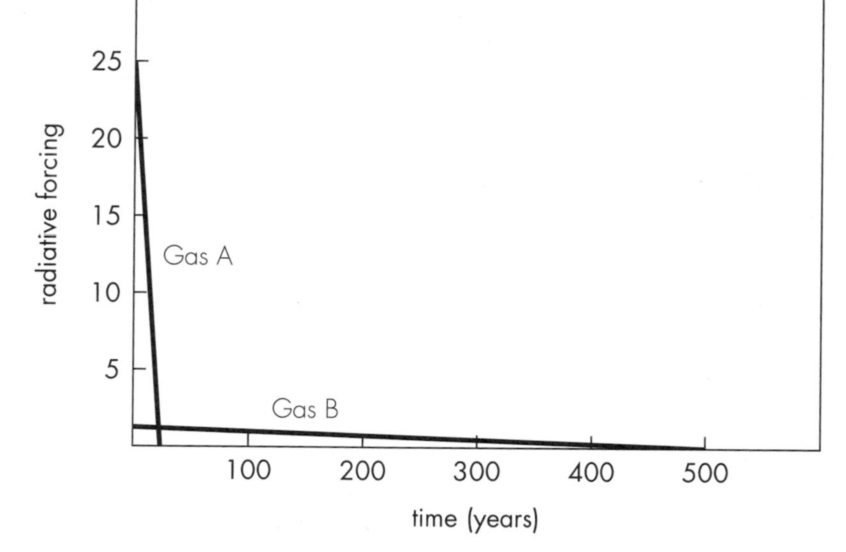

The use of the integral-based indices is controversial, as it treats a unit of radiative forcing at one point in time as exactly identical to a unit of radiative forcing at another point in time. Since different gases have differently shaped time paths, the time dimension is important. Even if one believes that the social discount rate is equal to zero (treat all costs and benefits the same, regardless of when they occur), the time dimension remains critically important.

For example, each kilogram of methane is associated with a much greater warming effect than carbon dioxide, but carbon dioxide lasts much longer in the atmosphere. Since the damages of global warming are a function of the ability to adjust to the changes, the more rapid the warming, the greater the damages that will occur. Although two gases may have the same total radiative forcing over time, they may have very different time paths. For example, in Figure 6.5, Gas A represents a short-lived gas with a powerful warming effect, such as methane. Gas B represents a long-lived gas with a weaker warming effect. Although the total warming potential (the area under the time paths) is the same for these two gases, the warming associated with Gas A occurs much earlier and will result in larger damages than the warming associated with Gas B. Thus, global warming potential indices, as defined above, do not really measure the equivalence of different greenhouse gases, because they do not measure equivalent damages, only equivalent total warming over time.

The proposition that damages associated with a given level of global warming are likely to be a function of both the time at which the warming occurs and the rapidity of the warming merits further discussion. This time dependency of damages is due to a variety of factors. The state of the world changes as time progresses: Population grows, technology improves, economies change and grow, and values and preferences change. A given level of global warming in one state of the world will be associated with a different level of damages than the same level of warming in a different state of the world. Also, damages will be a function of the speed at which global warming occurs, because the ability to adapt will be dependent upon the speed at which warming occurs. Although global warming potential indices are designed to measure the relative effects of different greenhouse gases, they do not adequately deal with the time dimension, because they do not attempt to measure the social damages associated with the emissions. At best, they should be regarded as crude approximations of the relative greenhouse effects of different gases. At this time, they are not sufficiently robust to allow for the computation of an integrated system of taxes or marketable permits for the different greenhouse gases.

However, since these GWP indices do give a coarse indication of the relative importance of the different gases, they are presented in this chapter to inform the student. Table 6.4 contains the 1992 IPCC estimates of the GWP indices for the major greenhouse gases.

Three different indices are presented in Table 6.4. They are for 50, 100, and 500 years, where the number of years is equal to the time period over which total radiative forcing is summed. For example, if one considers the total radiative forcing over 50 years, each kilogram of methane (CH_4) has 20 times the warming impact as a kilogram of carbon dioxide. If the impact is measured over 500 years, the corresponding index is equal to 4. Note the relatively high global warming indices associated with ozone-depleting gases (CFCs, HCFCs, HFCs, and CCL_4). Also, note that if the GWPI of a gas increases as the time increases, that gas has a greater atmospheric lifetime than CO_2. If the GWPI falls with increases in the time interval, then the atmospheric lifetime of the gas is lower than CO_2.

THE CONVENTION ON CLIMATE CHANGE

In May and June of 1992, a series of global environmental meetings took place in Rio de Janeiro. These meetings are usually referred to as "The Rio Summit." One of the outcomes of the Rio Summit was an international agreement on global climate change called the "Convention of Climate Change." The United States was a signatory to this convention. The main thrust of this convention was an agreement to stabilize emissions at 1990 levels, with var-

TABLE 6.4

GLOBAL WARMING POTENTIAL INDICES (GWPI) FOR VARIOUS GREENHOUSE GASES FOR 50-, 100-, AND 500-YEAR TIME INTERVALS

TIME INTERVAL	50 YEARS	100 YEARS	500 YEARS
GAS			
CO_2	1	1	1
CH_4	20	11	4
N_2O	280	270	170
CFC–11	4300	3400	1400
CFC–12	7600	6200	4100
HCFC–22	2700	1600	540
CFC–113	4900	4500	2500
CFC–114	6900	7000	5800
CFC–115	6400	7000	8500
HCFC–123	160	93	30
HCFC–124	810	460	150
HCFC–125	4600	3400	1200
HFC–134a	2000	1200	400
HCFC–141b	1100	610	210
HCFC–142b	2900	1800	630
HFC–143a	4600	3800	1600
HFC–152a	260	150	49
CCl_4	1600	1300	480

ious target dates for different countries. The United States did not place a time limit on its stabilization period, although most developed countries have committed to stabilization by 2000. There are several important facets of this agreement. First, the agreement establishes a Conference of Parties (COP) to which all signatories must report their plans for reducing greenhouse gas emissions. Second, greenhouse gas inventories will be developed for all signatories. Other important points of the treaty are that scientific uncertainty may not be used as a reason to avoid the development of an emission reduction plan and that the developed countries bear the brunt of the burden of reducing emissions.

The provision of the convention relating to scientific uncertainty merits more discussion. The convention does not stipulate that scientific uncertainty should be dismissed but that policies should be developed in the presence of scientific uncertainty. This stipulation is particularly important because the long atmospheric life of carbon dioxide and other greenhouse gases, and the fundamental changes that global warming could generate, imply that global warming is an irreversible environmental change. The existence of such an irreversibility argues for caution in the face of uncertainty and erring on the side of emissions reduction. It does not necessarily imply drastic emissions reductions, but that some current effort should be made to reduce emissions in case global warming turns out to be as severe a problem as some scientists predict. If future information indicates that the problem is less severe, we can always ease up on emissions restrictions. However, the converse is not true. If future scientific information indicates that global warming is a very severe problem, we cannot easily remove the greenhouse gases that we have added to the atmosphere.

Although this convention represents an agreement on emission levels, these levels are based less on perceptions of cost and benefits than on levels that are politically acceptable to the signatories. Although stabilization at 1990 levels represents a first step, it is possible that reductions below 1990 levels might be warranted or that new information will reveal that emissions restrictions should be eased.

As this edition of the book goes to press, a new round of discussions are taking place with respect to limiting world emissions of greenhouse gases, with a likelihood that individual countries will commit to reductions which bring them below 1990 emission levels. Many politicians and business people are concerned that these reductions will have a negative impact on the economy. This step is of particular concern in the United States. Since per capita emissions of greenhouse gases are higher in the United States than other countries, it is feared by some that the U.S. will suffer disproportionately and have increased production costs, putting the U.S. at an economic disadvantage in comparison to other countries. These consequences need not be as severe as feared, since the economic consequences of limiting greenhouse gas emissions will be a function of the policies used to achieve these reductions. If we rely on command and control strategies as we have for past environmental problems, it is likely that costs will increase significantly. However, if economic incentives were employed, these incentives would provide flexibility that would dramatically lower costs.

POLICY INSTRUMENTS FOR REDUCING GREENHOUSE GAS EMISSIONS

The ideal policy instruments reduce all greenhouse gases in a fashion that minimizes the cost of reducing the extent of global warming to the target level. At first impression it might seem relatively easy to develop a system of economic incentives (such as marketable pollution permits) for greenhouse gases, since these are global pollutants for which the damages are independent of the geographic location of the permit. In this case, there is no need to account for geographic variation in the damages caused by the pollution, so permits may be traded everywhere on a one-for-one basis, and the costs of reducing the emissions to the target level will be minimized. Similarly, a pollution tax could be formulated that could be uniform across all geographic locations.

However, rather than having characteristics which ease the creation of a marketable pollution permit system, global warming may be the most complex problem for which to develop such a system. The difficulties associated with dealing with multiple types of pollution (the various greenhouse gases) and dealing with emissions over a long period of time have already been discussed in the above section on global potential warming indices. In addition, several other serious problems must be dealt with in order to successfully implement policies to reduce the emissions of greenhouse gases. These problems are primarily related to defining the reductions which have to take place in order to achieve environmental goals. There are two types of problems. First, what should be the reductions required by each country, and second, how should the baseline level of emissions and subsequent reductions be measured?

There are significant problems associated with developing a global consensus on reductions of greenhouse gases. For example, a goal which has been frequently suggested has been to immediately freeze emissions at 1990 levels. However, should each country be required to freeze emissions at its 1990 levels, or should the total be frozen at that level with a different allocation across countries? Developing countries would argue against each country being required to freeze at 1990 levels, because they tend to have less emissions per capita since their economies are less developed. They would argue that such restrictions would make it more difficult for their economies to develop and that industrialized countries (which have significantly higher emission per capita) should be required to undertake a disproportionately high level of reductions (see boxed application on equity). Countries, such as the United States, which rely highly on fossil fuels might also argue against proportionate reduction, because they might fear that their production costs might rise relative to industrialized countries that rely more on nuclear power or hydropower. This potential relative increase in costs could erode a country's international competitiveness.

If countries can agree on the conceptual basis of a reduction plan (what levels or reduction are required of each country), many issues still must be resolved in order to implement a program. Among the foremost of these is how to measure the baseline from which reductions are measured. For example, if the U.S. exports coal to Japan, which then burns the coal to produce electricity, which country is viewed to be responsible for the emissions of carbon dioxide, the U.S. or Japan? Should a country that developed its electrical power system on a hydropower or nuclear foundation get credit for the emissions which it already has eliminated by never burning the coal or oil? It is relatively easy for the reader to develop answers to these types of questions, but different people would answer these questions in different ways, and countries will have to agree on the answers in order for the system to work. These issues and other issues will be discussed at the next round of discussions on reducing greenhouse gas emissions. Look at the following websites for information on the progress of these talks:

GLOBAL WARMING AND EQUITY

Chapter 4 discusses three types of equity issues which are important to consider when examining environmental change and environmental policy: equity across groups in a country, across different countries and across different generations. All three types of equity issues are significant when examining global climate change issues.

Intra-country equity has important implications for two reasons. First, there will be large regional differences in the impacts of global climate change. Some areas will become drier while some become wetter, some areas will become hotter while others become colder, and coastal areas have a whole set of impacts associated with sea level rise and increased storm intensity. Second, different groups within a country will have different abilities to adapt to global climate change. For example, if a country has substantially increased summer hot spells, higher income people who have access to air conditioning will suffer less than lower income people. High income coastal cities will be able to construct seawalls for protection, while low income or agricultural areas remain vulnerable to erosion, inundation and storm surge.

Equity considerations across countries are important both for the same reasons discussed with intra-country equity and for additional reasons. These additional factors have to do with the relative generation of greenhouse gas emissions across countries of different income levels. If high income countries generate most of the greenhouse gas emissions, is it equitable to ask low income countries to forego potential economic growth by limiting greenhouse gas emissions?

Finally, intergenerational equity considerations are important due to the long period of time during which today's emissions will generate damages. Addressing the problem requires the current generation to incur costs, but if the problem remains unaddressed, future generations will suffer large damages from global warming and sea level rise. Since future generations are unrepresentated in today's policy debates, it is critically important to consider the fairness of our decisions with regard to future generations.

- U.S. Global Climate Change Research Program: http://www.usgcrp.gov/usgcrp/default.html
- Intergovernmental Panel on Climate Change: http://www.unep.ch/ipcc/ipcc-0.html
- U.S. Department of State, Draft Protocol to the Framework Convention on Climate Change: http://www.state.gov/www/global//oes/protocal.htm

If the above questions can be resolved through international negotiation, then we will be in a position to develop a system of marketable pollution permits, taxes, or some other mechanism for reducing emissions. Obviously, it is critically important to minimize the cost of achieving the target level of emissions. Since emissions each year affect global warming in future years, it is important to consider both present and future emissions when looking at the cost of reducing emissions. This consideration is especially important because the cost of reducing emissions is likely to fall in the future as technological innovation enables us to use fossil fuels more efficiently and substitute alternative energy sources for fossil fuels. In addition, since damages may be a function of the time at which the gas is emitted, the timing of reductions is important for these reasons as well. In fact, there is considerable support in the U.S. Congress for postponing emissions reductions into the future.

$$Minimize\ PVSC = \sum_{t=0}^{\infty} [TD_t(\sum_{j=0}^{t} E_j) + TAC_t(E_j,t)]\,(1 + r)^{-t} \qquad (6.1)$$

Ideally, a target level of pollution for each year would be set that minimized the two costs associated with the emissions of greenhouse gases. The cost minimization problem is described in equation 6.1, where the first of these social costs is the damage from the emissions (TD_t) which are a function of the emissions in all the previous periods. The second is the cost of abatement (TAC_t) which is a function of the level of emissions in the period (E_t) and also a function of time (t). Although the mathematics of this equation are quite complicated, they suggest that if our goal was to freeze emissions at 1990 levels (or some other target) we might want to consider allowing a level temporarily higher than target level in exchange for even lower than target levels of emissions in the future. Paul Leiby and Jonathan Rubin (1997) describe the mathematics which underlies this approach and describe a rule which essentially says that it can be efficient to delay, provided that we compensate for the delay by reducing the delayed amount plus an additional amount in the future. The additional amount depends on the time path of the damages from global warming and the social discount rate (see Chapter 4 for a discussion of the social discount rate). Of course, in the process of negotiating the treaty, other countries may object to U.S. delay of emission reductions. Their willingness to accept this delay will depend on many factors, including the amount of extra reductions in the future and the extent to which they believe that the U.S. would honor its commitments.

Although the above arguments suggest that some delay *may* be a good idea (particularly for fossil fuel intensive countries such as the U.S.), this does not mean that nothing should be done in the interim. Since global warming tends to be an irreversible phenomenon, because there are long time lags between emissions and damages, and since global warming could lead to potentially large and irreversible damages, many economists argue that we should begin controlling greenhouse emissions immediately. Nordhaus (1991) lists several types of policies that could be pursued. Nordhaus and other leading economists have issued a statement stressing the need for immediate attention to the global warming problem.

Nordhaus's first suggestion is to improve our knowledge of the science that would enable us to better understand causes and impacts of global warming and improve our understanding of the economic impact of global warming. This could enable us to "fine tune" polices as more information is available.

Second, Nordhaus suggests supporting research on new technologies that would allow us to reduce the amount of greenhouse gas emissions per unit of output. He argues that economies under-invest in such technologies because of a double externality. The first externality exists because the returns to investors in such technologies are less than the returns to society, since the benefits of research and development become available to those who do not invest in it. The second externality exists because the marketplace does not reward the reduction of greenhouse gas emissions.

Third, Nordhaus suggests that we embark on a series of "no regret policies." No regret policies are defined as those that we would not regret in the future if it turns out that global warming is not the problem that we think it might be. Examples of no regret policies would be the elimination of policies that subsidize inefficient deforestation (see Chapters 11 and 12), slowing the growth of inefficient use of fossil fuels through imposition of higher taxes on gasoline and other fossil fuels (Chapter 8), and strengthening international conventions to restrict CFCs and other ozone depleting chemicals that are also greenhouse gases. Another example of a no regret policy would be an urban tree-planting program, which would sequester carbon, improve urban air quality, improve urban aesthetics, and lower energy use by lessening the need for air conditioning. All of these policies would have social benefits independent of their effect on greenhouse gas emissions.

Nordhaus suggests a modest tax on greenhouse gas emissions of $5 per ton of carbon, extended to other gases based on carbon dioxide equivalence. (Remember that expressing other greenhouse gases in carbon dioxide equivalents through the use of GWP indices is not straightforward and that any attempt is a crude approximation.) The economic costs of a $5 per ton tax are presented in Table 6.5, where they are compared to a $100 per ton tax.

In many ways, one might view a modest tax, such as the $5 per ton of carbon dioxide equivalent, as a no regret policy. Fossil fuels are currently underpriced (their marginal social cost is greater than their price) and, as can be seen in Table 6.5, the effects of a tax on the price of fuels is not large. In fact, Nordhaus finds that this low tax has a very small net benefit for the global economy (remember that nonmarket benefits are not included in this measure).

If international agreements were to require large reductions in emissions, a series of marketable permits (based on CO_2 equivalents) would probably be a preferable way to accomplish these emission reductions. There are two primary reasons. First, marketable permits could allow trades across countries to take advantage of significant cost differentials in reducing emissions. Second, marketable permits could be a mechanism for compensating developing countries for participating in an emissions reduction scheme. A disproportionate share of permits could be given to Third World countries, which they could then sell to developed countries. Also, developing countries could receive credit for carbon sequestration through reforestation schemes.

TABLE 6.5

ILLUSTRATIVE IMPACT OF DIFFERENT CARBON TAXES

SECTOR OF IMPACT	LOW TAX	HIGH TAX
TAX EFFECT		
Tax on CO_2 equivalent (per ton carbon)	$5.00	$100.00
IMPACT ON FOSSIL FUEL PRICES (BASED ON 1989 PRICES)		
Coal price		
per metric ton	$3.50	$70.00
% increase	10%	205%
Oil price		
per barrel	$0.58	$11.65
% increase	2.8%	55%
Gasoline price		
per gallon	1.4 cents	28 cents
% increase	1.2%	23.2%
Overall impacts estimated reduction of GHG emissions (CO_2 equivalent)	10%	43%
Total tax revenue, U.S. (billions of $)	$10.00	$125.00
Estimated global net	$4.00	−$114

Economic benefits (+) or costs (−), billion dollars per year, 1989 global economy

SOURCE: Nordhaus, William D., *Global Warming: Economic Policy Responses*, MIT Press. Copyright © 1991.

SUMMARY

Global warming and the depletion of the ozone layer are important environmental problems that are quite different than conventional pollution problems. The long lags between emissions and damages, the long lifetimes of the pollutants, and the complexity of the scientific relationships makes the development of policy a difficult question.

As a global society, we have moved more quickly on ozone depleting chemicals than on greenhouse gases. International agreements call for bans on the most damaging ozone depleting chemicals, and policies are being developed to address all ozone depleting chemicals. International agreements on global warming specify stabilization of 1990 emissions levels, within unspecified time periods. New treaties are being negotiated at the current time, and in all likelihood will be in place in 1998 or early 1999. The difference in the speed of movement on ozone and greenhouse gas problems is probably related to the greater certainty regarding the estimates of the damages of depleting the ozone layer, the lower cost associated with reducing CFC emissions, and the greater heterogeneity of negotiating positions on the global warming problem. Although scientific uncertainty has retarded the development of policies to reduce greenhouse gas emissions, there is now a scientific consensus that global warming is a real threat to global social welfare.

Although the benefits of slowing global warming are not known with certainty, a number of arguments suggest that it would be prudent to begin controlling greenhouse gas emissions immediately. However, the aggressiveness with which we should curtail emissions is a subject of hot debate. This chapter has attempted to present both sides of the debate, without directing the student toward a verdict. However, a short chapter like this one cannot fully inform the reader on this important question, especially since new findings occur and new policies are suggested at a fairly rapid pace. Students are strongly advised to do further reading on this controversial and important topic. The suggested readings contain a small sample of the writing on this issue.

REVIEW QUESTIONS

1. What are the major greenhouse gases, and what are their sources of emissions?
2. What is the evidence that average global warming is increasing?
3. What is the role of adaptability in mitigating the damages from global warming?
4. Why does the minimization of abatement costs require the development of policies to control all greenhouse gases and not just CO_2?
5. Assume that the time paths of the radiative forcing (RF) of a kilogram of two gases are given by the following linear equations:

   Gas A: RF = 70 − 0.25(TIME)
   Gas B: RF = 200 − 0.1(TIME)

What is the 50-year GWPI of Gas A in terms of Gas B? What is the corresponding 100-year GWPI?

QUESTIONS FOR FURTHER STUDY

1. Why is international cooperation necessary for controlling greenhouse gas emissions?
2. What is the Montreal Protocol? Do its provisions make economic sense?
3. What does Nordhaus mean by "no regret" global warming policies? Can you think of any "no regret" polices in addition to those proposed by Nordhaus?
4. Can cost–benefit analysis be used to develop global warming policy?
5. What are the equity issues surrounding global warming policy?

SUGGESTED PAPER TOPICS

1. Look at the equity issues involving global warming policy. Pay particular attention to developed versus developing country issues. Your best places to get started are the webpages of the IPCC (http://www.unep.ch/ipcc/ipcc-0.html) and the U.S. Global Climate Change Research Program (http://www.usgcrp.gov/usgcrp/default.html). Also, look at referenced works by Dornbusch and Poterba, Cline, Manne and Richels, Nordhaus, NAS, and Weiss.

2. Global warming and energy use are obviously intertwined. Look at alternative energy policies, particularly for developing countries, and see how energy policy can be used to reduce global warming. Make sure to talk about how policy can affect economic behavior and do not just concentrate on "technological fixes." Again, a good place to get started is the webpage of the IPCC. Also search the web on "global warming" and "energy" or "energy use." Also look at referenced works by Dornbusch and Poterba, Cline, Manne and Richels, NAS, and Weiss, as well as current issues of the *Energy Economist*.
3. Choose a particular sector of the economy (such as agriculture, transportation, electrical generation, or forestry) and look at how global warming will affect the industry and how the industry affects global warming. Suggest policies that can mitigate the impact of global warming on the industry. Suggest policies that can reduce the industry's impact on global warming at the least cost to the industry. Start by looking at works by Nordhaus and NAS.
4. Global warming requires international cooperation. Look at existing mechanisms for promoting international cooperation and discuss how institutions can be changed to increase cooperation. Pay particular attention to how new institutions can develop ways to increase a country's self-interest in reducing greenhouse emissions or sequestering carbon. Start by looking at referenced works by Dornbusch and Poterba, Cline, Manne and Richels, NAS, and Weiss, as well as current issues of the *Energy Economist*.
5. Evaluate the proposed global treaty on global climate change. The draft of this treaty may be found at the U.S. Department of State, Draft Protocol to the Framework Convention on Climate Change (http://www.state.gov/www/global//oes/protocal.htm).

WORKS CITED AND SELECTED READINGS

1. Ausabel, Jesse H. "A Second Look at the Impacts of Climate Change." *American Scientist* 79 (1991): 210–221.
2. Brown, Peter G. "Why Climate Change Is Not a Cost/Benefit Problem." In *Global Climate Change: The Economic Costs of Mitigation and Adaptation*, edited by J. C. White. New York: Elsevier, 1991.
3. Caldwell, Teramura and Tevini. "The Changing Solar Ultraviolet Climate and the Ecological Climate for Higher Plants." *Trends in Ecology and Evolution, 4,* 1989.
4. Cline, William R. *Global Warming: The Economic Stakes*. Washington: Institute for International Economics, 1992.
5. Crowley, T. J. "Remembrance of Things Past: Greenhouse Lessons from the Geologic Record." *Consequences,* 2:1, 1996, 2–13.
6. Dornbusch, Rudiger, and James M. Poterba, eds. *Global Warming: Economic Policy Responses*. Cambridge MA: MIT Press, 1991.
7. Epstein, Joshua M. and Raj Gupta. *Controlling the Greenhouse Effect: Five Regimes Compared*. Washington: The Brookings Institute, 1990.
8. Gore, Albert. *Earth in the Balance*. New York: Houghton–Mifflin Company, 1992.
9. Houghton, J. T., G. J. Jenkins, and J. J. Ephraums. *Climate Change: The IPCC Scientific Assessment*. Cambridge: Cambridge University Press, 1990.
10. Intergovernment Panel on Climate Change (IPCC). 1992 IPCC Supplement, 1992.
11. Krause, F., et al. *Energy Policy in the Greenhouse*. London: Earthscan Publications, 1990.
12. Leiby, Paul and Jonathon Rubin, *Bankable Permits for the Control of Stock and Flow Pollutants*. Optimal Intertemporal Greenhouse Gas Trading, unpublished paper, Oak Ridge National Laboratory, Oak Ridge, TN, 1997.
13. Manne, Alan S., and Richard G. Richels. *Buying Greenhouse Insurance: The Economic Costs of CO_2 Emission Limits*. Cambridge MA: MIT Press, 1992.
14. NAS (National Academy of Sciences). *Policy Implications of Global Warming*. Washington: National Academy Press, 1991.
15. Nordhaus, William D. "Economic Approaches to Greenhouse Warming." In *Global Warming: Economic Policy Responses*, Rudiger Dornbusch and James M. Poterba, eds. Cambridge MA: MIT Press, 1991.
16. ———. *Managing the Global Commons*, Cambridge: MIT Press, 1994.
17. ———. "To Slow or Not to Slow: The Economics of the Greenhouse Effect." *Economic Journal*, 101 (1991): 920–937.
18. Schneider, S. "Climate Change Scenarios for Greenhouse Increases." In *Technologies for a Greenhouse Constrained Society*. Oak Ridge, TN: Oak Ridge National Laboratory, 1991.

19. Solomon and Albritton. "Time dependent ozone depletion potentials for short and long-term forecasts," *Nature* (1992): 357.
20. Solow, Andrew R. "Is There a Global Warming Problem?" In *Global Warming: Economic Policy Responses,* Rudiger Dornbusch and James M. Poterba, eds. Cambridge MA: MIT Press, 1991.
21. Weiss, Edith Brown. "The Planetary Trust: Conservation and Intergenerational Equity." *Ecology Law Quarterly* 2 (1984): 445–581.
22. Yohe, G. W. "The Cost of Not Holding Back the Sea—Economic Vulnerability." *Ocean and Shoreline Management* 15 (1991): 223–255.

Acid Deposition

The only certainty is uncertainty.

Pliny the Elder, *Historia Naturalis*

INTRODUCTION

Acid deposition refers to a process by which certain types of pollutants chemically transform into acidic substances in the atmosphere and then fall to the earth. The most widely discussed vehicle for the acidity to reach the ground is through acid rain, but other forms of precipitation (including acid snow and acid fog) and dry deposition are important mechanisms in the acidification problem. Acid deposition may cause a variety of harmful effects to ecosystems, agriculture, building materials, and possibly to human health. These impacts will be discussed more fully in the next section.

Acid deposition has received considerable attention in the media and is generally regarded by the public as an important environmental problem. However, there is considerable uncertainty involving the actual damages generated by the emissions of acid deposition precursors.[1] This chapter will focus on this uncertainty and how to develop environmental policy in the presence of this uncertainty.

The concern over the acid deposition problem began to accelerate in the 1970s, culminating in the Acid Precipitation Act of 1980 which established the National Acid Precipitation Assessment Program (NAPAP). NAPAP is coordinated by an interagency task force and was established to provide information on the regions and resources affected by acidity, the extent to which acid deposition and related pollution are responsible for causing these impacts, the process by which pollutants are transformed into acids, the distribution of acid deposition, the magnitude of the effects, whether mitigation is required, and strategies to control acid deposition and related pollutants (NAPAP, 1991). NAPAP, originally established with a life of 10 years, has recently been reauthorized by the 1990 Clean Air Act Amendments.

[1]Precursor pollutants are those pollutants that are chemically transformed to generate the substances that actually cause the environmental damage, in this case, the acid deposition.

Acid rain belongs to a category of pollutants referred to as regional pollutants. Regional pollutants are those pollutants that have effects over more than just the vicinity of their emission. Their effects are felt in a broader geographic region, but they do not have global impacts in the manner of carbon dioxide or ozone-depleting chemicals. With carbon dioxide, the location of the emissions is relatively unimportant, as the gases mix completely in the atmosphere. With sulfur dioxide and nitrogen oxide emissions, however, the effects are felt primarily downwind of the emissions, so location is important.

Acid deposition problems often are manifest as transboundary (sometimes referred to as transfrontier) pollutants. Transboundary pollutants are those that are emitted in one country and are transported across a national border to another country. For example, sulfur dioxide emissions in the United States affect environmental quality in Canada, and sulfur dioxide emissions in Canada affect environmental quality in the United States. U.S. emissions of sulfur are responsible for 50–75 percent of the sulfur deposition over most of eastern Canada, except those areas northeast of the metal smelter in Sudbury, Canada. The contribution of Canadian emission to U.S. sulfur deposition is less than 5 percent, except in areas of New York, New Hampshire, Vermont, and most of Maine. In northeastern Maine, Canadian emissions are responsible for up to 25 percent of sulfur deposition (NAPAP, 1991). Acid deposition is also a transboundary pollution problem in Europe, where pollution generated in Great Britain and Germany generate acid deposition in Scandinavia.[2]

WHAT CAUSES ACID DEPOSITION?

The most important precursor pollutants in the acid deposition problem are sulfur dioxide (SO_2) and nitrogen oxides (NO_x). Sulfur dioxide is the most important pollutant in this regard. Both types of pollutants are associated with the burning of fossil fuels. Sulfur dioxide is associated with the burning of coal and oil as a boiler fuel, such as in an electric power plant. Nitrogen oxides are also associated with boiler fuels, but automobiles are also an important source of NO_x.

Acid rain and other forms of acid deposition are caused when sulfur dioxide and nitrogen oxides form sulfate and nitrate in the atmosphere, which then combine with hydrogen ions to form acids. The sulfate and nitrate molecules are formed when sulfur dioxide and nitrogen oxides combine with oxidants in the atmosphere. An important oxidant in this process is

[2]Since countries in Europe tend to be much smaller than North American countries, many pollution problems that are confined within national borders in North America constitute transboundary pollution problems in Europe. The smog problem is a good example of this.

TABLE 7.1

1985 ANTHROPOGENIC EMISSIONS OF SO_2, NO_x, AND VOCS

SOURCE	ANNUAL EMISSIONS (MILLIONS OF SHORT TONS)		
	SO_2	NO_x	VOCs
UNITED STATES			
Electric utilities	16.1	6.6	negligible
Industrial combustion	2.7	3.2	0.1
Commercial/residential/other combustion	0.6	0.8	1.9
Industrial manufacturing processes	2.9	0.9	3.7
Transportation	0.9	8.8	8.8
Other	negligible	0.1	7.6
Total United States	23.1	20.5	22.1
CANADA			
Electric utilities	0.8	0.3	negligible
Industrial combustion	0.3	0.3	negligible
Commercial/residential/other combustion	0.1	0.1	0.1
Industrial manufacturing processes	2.7	0.1	0.5
Transportation	0.1	1.2	1.1
Other	0.0	0.1	0.7
Total Canada	4.1	2.1	2.5
Total Canada and United States	27.2	22.1	24.5

SOURCE: NAPAP, 1991, *1990 Integrated Assessment*, p. 179.

tropospheric[3] ozone (O_3), which is formed when two pollutants (nitrogen oxides and volatile organic compounds [VOCs]) chemically interact in the presence of sunlight. Although VOCs are not directly responsible for acid deposition, their presence in the atmosphere leads to greater proportions of SO_2 being converted to sulfate, and NO_x being converted to nitrate. Table 7.1 lists various anthropogenic sources of sulfur dioxide, nitrogen oxides, and VOCs emissions.

The chemical relationships among pollutants that are discussed above illustrate the importance of dealing with different pollutants in a coordinated fashion. Reducing NO_x not only directly reduces acid rain, but it has an indirect effect by reducing ozone, which lessens the conversion of SO_2 to sulfate. Similarly, reducing VOCs has a direct impact of reducing tropospheric

[3]The troposphere is the lower level of the atmosphere. Ozone in the lower atmosphere causes numerous detrimental impacts. Ozone in the stratosphere (upper atmosphere) shields the earth from ultraviolet radiation. See Chapter 6 for a more detailed discussion of the relationship between the troposphere and the stratosphere.

FIGURE 7.1

MARGINAL DAMAGES OF SO_2 AS A FUNCTION OF THE LEVELS OF OTHER POLLUTANTS

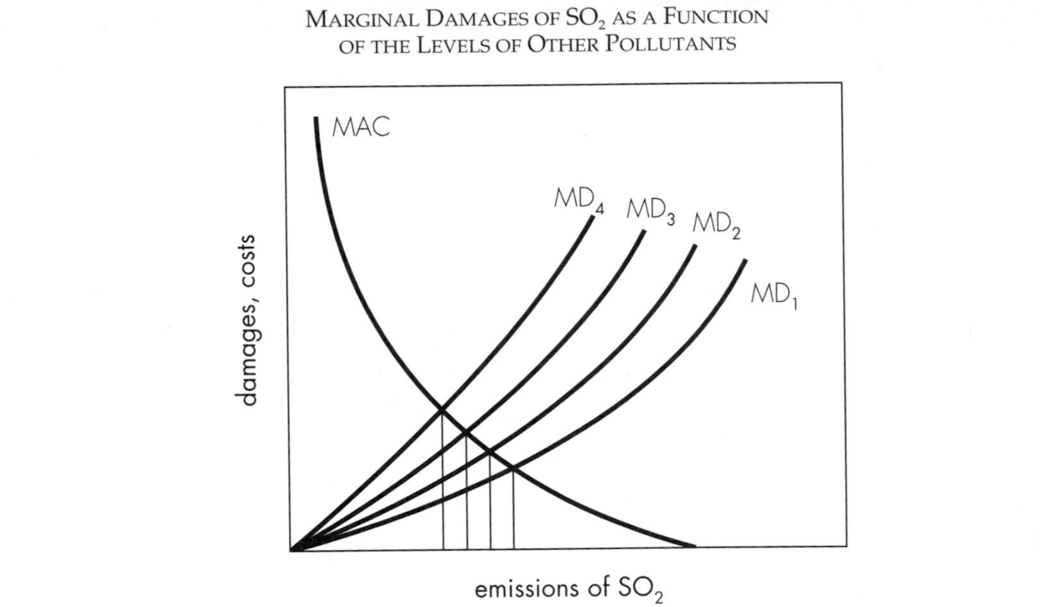

ozone, and an indirect effect on acid rain since less ozone implies that less SO_2 will be converted to sulfate. The same thing holds true for the conversion of NO_x to nitrate.

The interactions among these pollutants make the identification of the optimal level of pollution an extremely difficult problem. The reason is that because both increased NO_x and increased VOCs accelerate the conversion of SO_2 to sulfate, the marginal damages of SO_2 depend on the level of NO_x and VOCs. This relationship is illustrated in Figure 7.1, where the marginal damage function for SO_2 shifts upward as the level of NO_x and VOCs increases. Even if the marginal abatement cost function was known, one could not determine the optimal level of SO_2 emissions without also knowing the costs of reducing NO_x and VOCs, which results in shifting the marginal damage function for SO_2 downward from MD_4 to MD_3 to MD_2 and so on in Figure 7.1. In order to know the costs of shifting this damage function, one must know the marginal abatement costs for VOCs and NO_x. The problem of identifying the optimal level of SO_2 (or NO_x or VOCs) is actually even more complex than shown because the marginal abatement costs of one pollutant may be a function of the level of abatement of other pollutants. For example, changes to production processes that increase energy efficiency (increase the amount of work accomplished per unit of energy) will reduce all pollutants simultaneously. However, "end of the pipe" abatement devices,

FIGURE 7.2

ABATEMENT COSTS AND DAMAGES AS A FUNCTION OF THE LEVELS OF OTHER POLLUTANTS

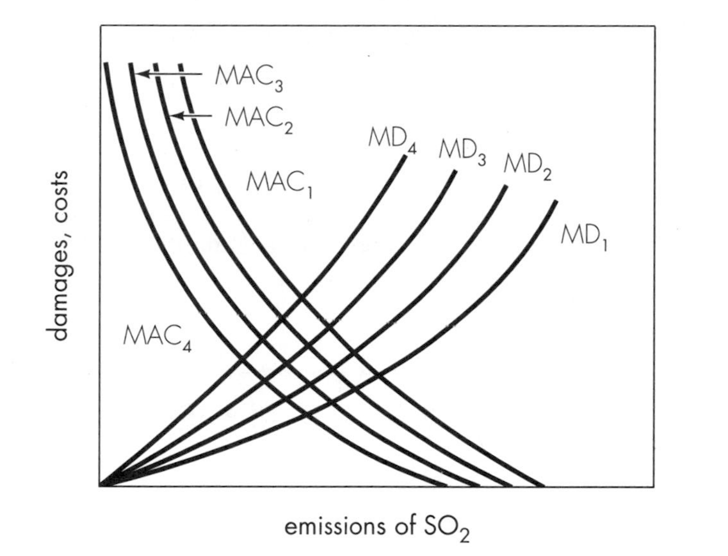

such as "scrubbers" (used in electric utilities) or catalytic combusters (used in automobiles), may reduce the emissions of one pollutant but increase the emissions of other pollutants. Figure 7.2 illustrates the case where both the marginal damage function and the marginal abatement cost functions for SO_2 are a function of the levels of emissions of the other pollutants.

The optimal level of each of the three pollutants cannot be determined independently of each other, which was the assumption which was made when the optimal level of pollution was defined in Chapter 3. In contrast, in order to achieve optimality, the level of emissions of each pollutant must be chosen to minimize the sum of the total abatement costs and total damages associated with all three pollutants. Equation 7.1 contains the total abatement costs, which are a function of the level of emissions of all three pollutants (E_1, E_2, and E_3). Similarly, total damages are a function of all three pollutants, as modeled in Equation 7.2.

$$TAC = TAC(E_1, E_2, E_3) \tag{7.1}$$

$$TD = TD(E_1, E_2, E_3) \tag{7.2}$$

The minimization of the sum of total abatement costs and total damages requires that the marginal damages of each pollutant are equal to the marginal abatement costs of each pollutant. These conditions are contained in

Equations 7.3 through 7.5. Since the marginal abatement costs and marginal damages of each pollutant are a function of all the other pollutants, these three equations must be solved simultaneously to determine the optimal level of each pollutant.

$$MAC_1(E_1,E_2,E_3) = MD_1(E_1,E_2,E_3) \tag{7.3}$$

$$MAC_2(E_1,E_2,E_3) = MD_2(E_1,E_2,E_3) \tag{7.4}$$

$$MAC_3(E_1,E_2,E_3) = MD_3(E_1,E_2,E_3) \tag{7.5}$$

As we continue to discuss the acid deposition problem, we will see that little progress has been made in identifying the optimal level of each precursor pollutant. The interactions among the pollutants that have been discussed in this section represent an important component of this problem.

THE IMPACTS OF ACID DEPOSITION

Acid deposition and related pollutants have many significant impacts on natural systems and human systems. Acidification of surface waters such as lakes and streams has detrimental effects on aquatic systems. Acid deposition is suspected to have detrimental effects on forests, particularly high elevation coniferous forests. Sulfur dioxide, sulfate particles, and acid aerosols are all suspected of having detrimental effects on human health. Ozone, caused by the emission of nitrogen oxides, has harmful effects on both vegetation (natural forests and agriculture) and humans. The particles that generate acid deposition also serve to scatter light, creating a "pollution haze" and reducing visibility. Finally, acid deposition leads to the premature weathering and degradation of materials used in buildings, monuments, fences, and other structures, particularly paints, metals, and stone. Table 7.2 illustrates the effects associated with acid precipitation, as presented in the *1990 Integrated Assessment Report* of NAPAP. Market effects indicate impacts that are felt by producers or consumers of goods that are bought and sold in markets. For example, tropospheric ozone has effects on agriculture, and acid deposition has effects on building materials. It is interesting to note the very limited measurement of economic damages occurred during the first ten years of NAPAP, as reflected in the *1990 Integrated Assessment Report*. Without this quantification of damages, it is extemely difficult to be confident that the target levels of reductions in NO_2, SO_2 and VOCs are the right level. Perhaps we should strive for more reductions or perhaps we have reduced emissions by too much. This is a question that only can be answered with more knowledge about the benefits and costs of changing the levels of emissions.

Nonmarket values can be either direct use values or indirect use values. Direct use values would include impacts on activities such as recreational

TABLE 7.2

1990 INTEGRATED ASSESSMENT DESCRIPTION OF ECONOMIC EFFECTS CATEGORIES FOR REGIONAL AIR POLLUTION[a]

	TYPE OF EFFECT		
	MARKET	NONMARKET	
EFFECTS CATEGORIES		USE	INDIRECT USE
TERRESTRIAL SYSTEMS			
Agriculture	Q	+	+
Forests	Q	+	+
Wildlife	+	+	+
Other ecosystem	NA	+	+
AQUATIC ECOSYSTEMS			
Commercial fishing	+	NA	NA
Recreational fishing	NA	Q	+
Other water-based recreation	+	+	+
Other ecosystem	NA	+	+
MATERIALS			
Building materials	+	NA	NA
Cultural materials	+	+	+
VISIBILITY	NA	+	+
HEALTH	NA	+	+

SOURCE: 1990 Integrated Assessment Report, Table 2. 7–1, page 152.

[a]A plus sign indicates that there is some potential for the economic valuation in a specific effect area to be influenced by changes in acid deposition and ozone. NA indicates that valuation is not applicable. Q indicates that NAPAP has targeted this area for valuation in quantitative terms in some part of the integrated assessment. Limited quantitative information from other studies is presented for visibility.

fishing and backpacking. Indirect use values would include existence values, option values, bequest values, and so on.[4]

Table 7.2 asserts that there are no nonmarket impacts of effects on ordinary building materials. This is not likely to be the case, for if the paint or other materials of a building deteriorate from acid deposition, it may have an effect on the aesthetic quality of the entire neighborhood. If damages were repaired instantaneously as they occurred, then there would be no "public bad" aspects to the deterioration, and the only costs would be the market

[4]See Chapter 4 for a more complete definition of the different nonmarket values associated with environmental resources.

FIGURE 7.3

THE EFFECT OF ACID DEPOSITION PRECURSORS ON THE COST OF HEALTH CARE

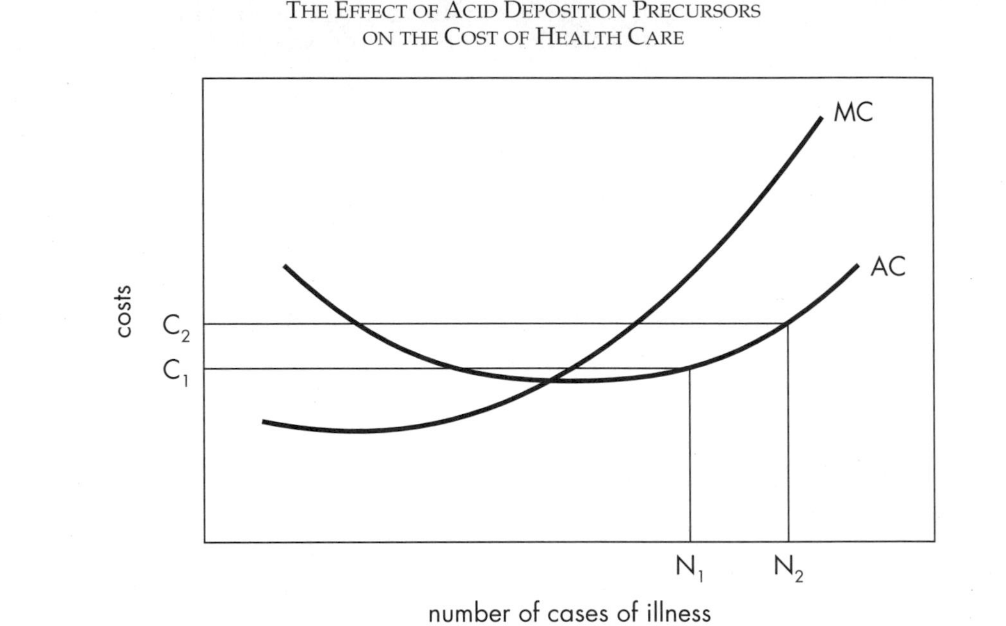

costs of repair. However, if there is any period of time in which damages remain unrepaired, there may be damage to the aesthetic quality of the entire neighborhood.

Table 7.2 also indicates that there are no nonmarket impacts associated with commercial fishing. This representation will not be true if a subset of the fishers are utility maximizers rather than profit maximizers.[5] However, since there are very few freshwater commercial fisheries in the United States, this omission is not likely to be significant (see Chapter 10).

A more serious omission is associated with the assertion that there are no market impacts associated with human health. This depiction is not likely to be the case for several reasons. First, health problems require treatment, which is costly. Second, if the marginal cost of providing health care is increasing with the number of cases of illness, then increasing acid deposition (and the pollutants associated with it, such as ozone and sulfate particulates) will increase the marginal and average cost of health care. This correlation is illustrated in Figure 7.3, where the marginal costs and average cost functions are depicted as a function of the number of cases of illness. If acid deposition

[5]See Chapter 10.

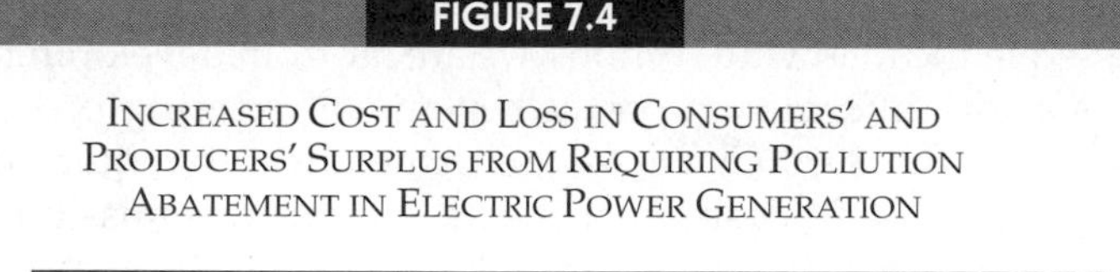

FIGURE 7.4

INCREASED COST AND LOSS IN CONSUMERS' AND PRODUCERS' SURPLUS FROM REQUIRING POLLUTION ABATEMENT IN ELECTRIC POWER GENERATION

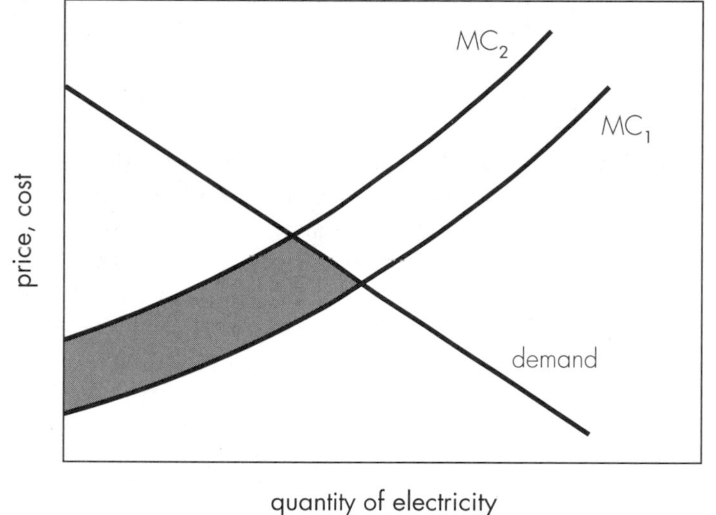

and its precursor pollutants increase the number of cases of illness from N_1 to N_2, then the average cost of treating all cases of illness increases from C_1 to C_2. Third, if the increased illness causes absenteeism and decreased productivity of labor, this change increases the cost of producing all outputs.

In order to attempt to determine a target goal for acid deposition or to conduct a cost–benefit analysis of any particular legislation, the impacts that were discussed previously must be quantified in a common metric. Economists generally look at willingness to pay in monetary terms as the common measure of the societal consequence of the impacts (see Chapter 4). It is particularly important to be able to measure the significance of these impacts in dollar terms, because the costs of reducing the impacts are dollar costs associated with market goods. For example, reducing SO_2 emissions from electric power plants requires additional inputs that increase the cost of producing electricity and lower the net social benefits derived from electricity. This increase is illustrated in Figure 7.4, where the shaded area represents the loss in consumers' and producers' surplus associated with the increase in the marginal cost function from MC_1 to MC_2.

Of course, there will be benefits associated with reducing sulfur dioxide and other precursor pollutants. Since many of these benefits are nonmarket benefits, the key to understanding the cost–benefit relationships may be valuing the impacts on these nonmarket goods and activities. Unfortunately, as discussed in the following sections, we are far from developing complete or even satisfactory information on the benefits of reducing precursor pollutants. However, the 1990 Clean Air Act Amendments require the development of estimates of the costs and benefits of reducing emissions of acid rain precursors. Although these estimates have not yet been published, they should be available in the near future.

AQUATIC IMPACTS

Acidity is measured by the pH scale, which represents the negative of the logarithm of the hydrogen ion concentration. The lower the pH level, the greater the acidity. A pH level of 7 refers to neutral acidity (no acid). Since the pH scale is logarithmic, a movement of one unit implies a tenfold increase in acidity. For example, moving from a pH of 7 to a pH of 6 means that acidity has increased tenfold and moving from 7 to 5 implies that acidity has increased by a factor of 100.[6]

As the maps in Figures 7.5 and 7.6 indicate, acid deposition is primarily an eastern problem (both in Canada and in the United States). Table 7.3 shows the number of lakes and streams that are acidified in areas of the United States that are affected by acid deposition, using different pH criteria for acidification. Lakes and streams might be acidified for reasons other than acid deposition. For example, acid drainage from coal mines and natural sources of acidity (particularly in Florida) are responsible for the acidity of some surface waters. However, the National Surface Waters Survey (NSWS) found that acid deposition was the primary source of acidification for 75 percent of the acidified lakes (natural sources: 22 percent, mine drainage: 2 percent) and 47 percent of the acidified streams (natural sources: 27 percent, mine drainage: 26 percent) (NAPAP, 1991).

Figure 7.7 illustrates the sensitivity of various aquatic organisms to different pH levels. The acidity tends to have both direct and indirect effects on aquatic organisms. First, the acidity of the water creates an environment that many organisms cannot tolerate, leading to their direct mortality. Second, acid rain can dissolve metals in the soil (such as aluminum and selenium); the metals then are transported to lakes and streams through runoff and through ground water. These metals can have toxic effects on a variety of organisms.

As of the publication of this book, the only estimates of the value of reducing the impacts on aquatic systems are for recreational fishing use values. No estimates of aquatic indirect use benefits have been made. For example, no contingent valuation studies have been conducted to determine the willingness to pay to preserve aquatic ecosystems from acidification.

---

[6]The Richter scale for measuring earthquakes is also a logarithmic scale.

FIGURE 7.5

1991 WET NITRATE ION DEPOSITION (KILOGRAMS PER HECTARE)

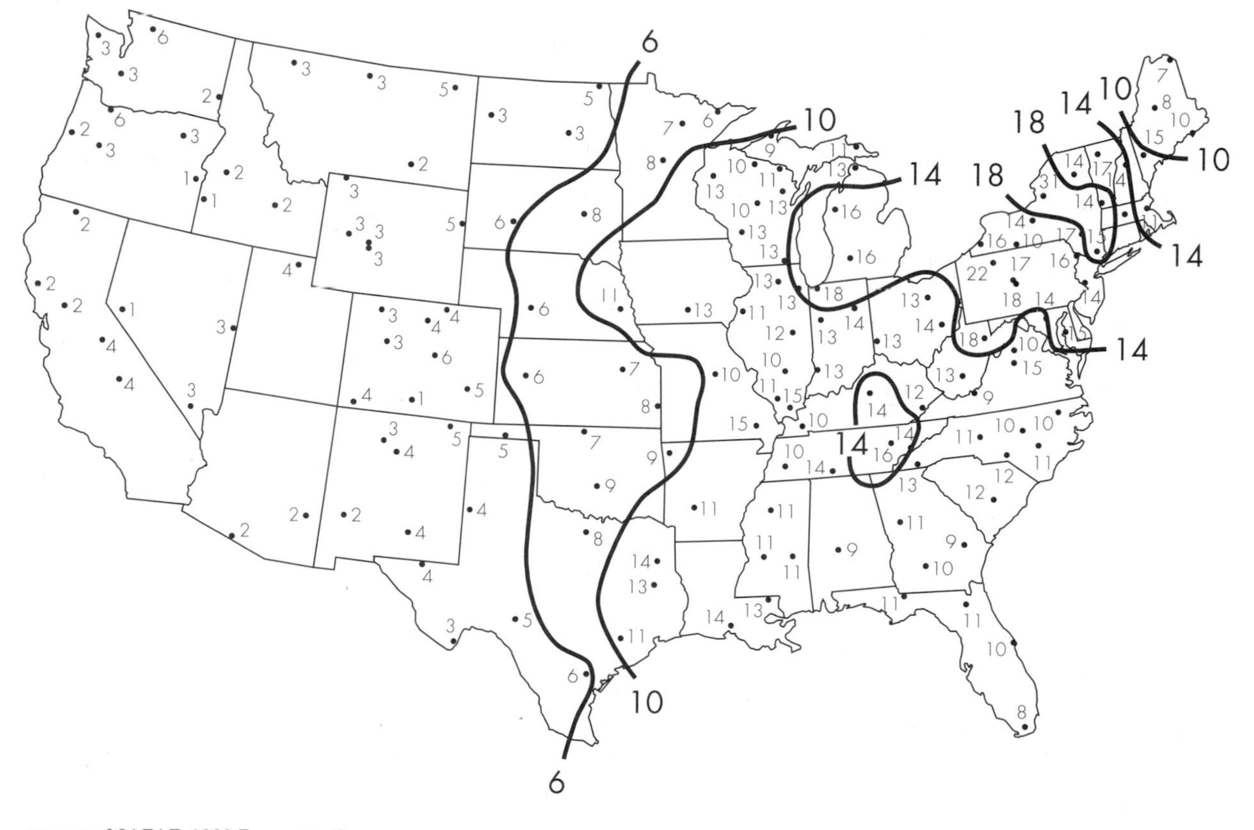

SOURCE: NAPAP, 1992 Report to Congress.

The most recent study of use values is reported in the *1990 Integrated Assessment Report*, where standard travel cost techniques (see Chapter 4) were used to derive the loss in value associated with the acidification of New England and New York lakes. Data from the Adirondack region were used to estimate values for the Adirondacks, which were then extrapolated to the New England states.

The effect of acid deposition on the net benefits from recreational fishing was estimated by determining the effect of acid deposition (measured as the acid stress index [ASI] that accounts for the direct and indirect effects of acidification) on catch per unit effort (CPUE) for each relevant species. Travel

FIGURE 7.6

1991 WET SULFATE ION DEPOSITION (KILOGRAMS PER HECTARE)

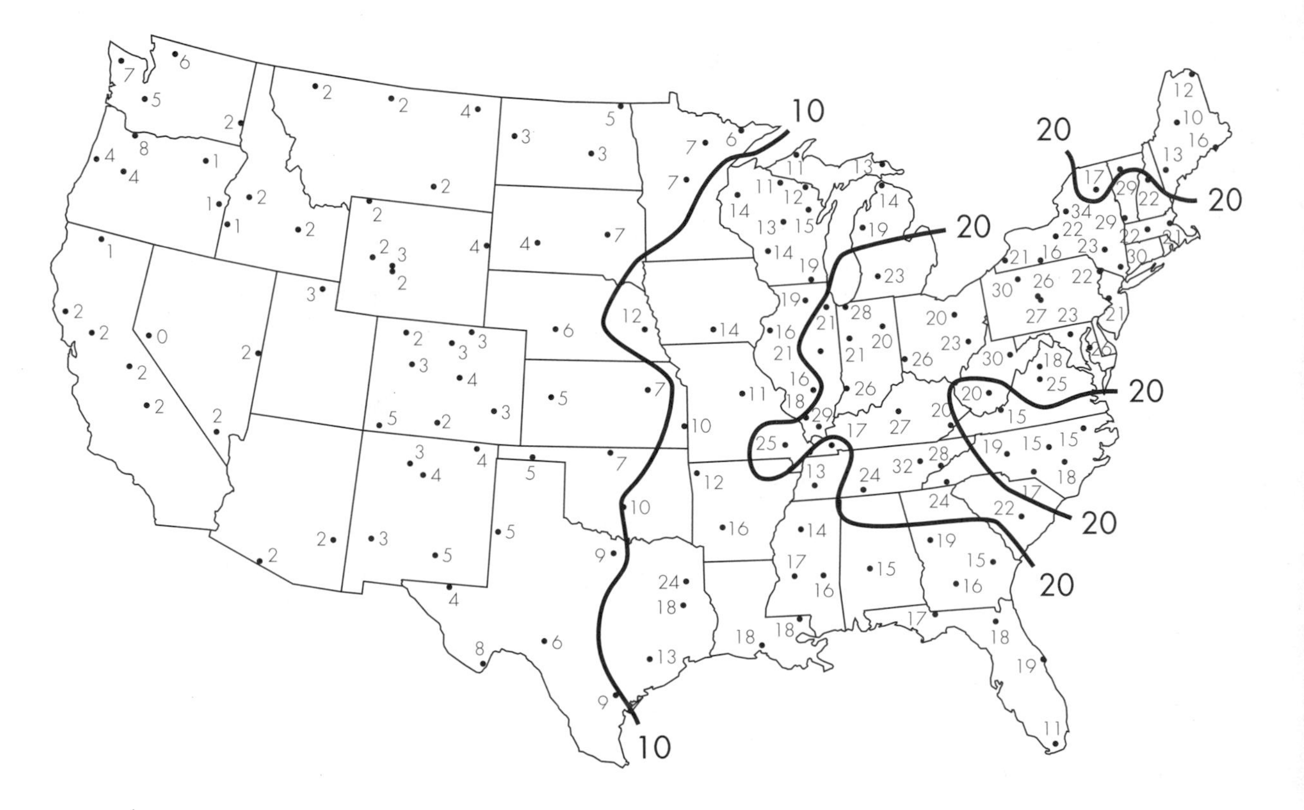

SOURCE: NAPAP, 1992 Report to Congress.

cost demand functions were then estimated with CPUE as an explanatory variable. Various scenarios concerning acid deposition were then expressed as changes in ASI, and the economic damages associated with the acid deposition were calculated using the travel cost demand functions. This is illustrated in Figure 7.8, where D_1 represents an individual's travel cost demand curve for recreational fishing with the current state of acid deposition, and D2 represents an individual's travel cost demand curve for recreational fishing with a reduced level of acid deposition. If p_1 represents the individual's actual travel cost, then the value of reducing the acidity of fishing waters would be equal to the shaded area abce.

The *1990 Integrated Assessment Report* publishes two sets of results that have fundamentally different underlying assumptions. In Table 2.7–3 of the

TABLE 7.3

ESTIMATES OF THE PERCENTAGE OF LAKES AND STREAMS WITH PH BELOW REFERENCE LEVELS

REGION	LAKE OR STREAM	TOTAL NUMBER	PERCENT OF LAKES AND STREAMS		
			pH ≤ 5.0	pH ≤ 5.5	pH ≤ 6.0
New England	L	4,330	2	6	11
Adirondacks	L	1,290	10	20	27
Mid-Atlantic Coastal Plain	S	11,300	12	24	49
Mid-Atlantic Highlands	L	1,480	1	6	8
	S	27,700	7	11	17
Southeastern Highlands	L	258	<1	<1	<1
	S	18,900	1	2	9
Florida	L	2,100	12	21	33
	S	1,730	31	50	72
Upper Midwest	L	8,500	2	4	10
West	L	10,400	<1	<1	1
All NSWS Regions	L	28,300	2	5	9
	S	59,600	7	12	22

SOURCE: NAPAP, 1991, *1990 Integrated Assessment Report*, p. 26, Table 2.2–2. See source for explanatory notes on technical measurement issues.

1990 Integrated Assessment Report, the estimated annual damages of acidic deposition on recreational trout fishing in New York, Maine, New Hampshire, and Vermont is stated to equal $1.75 million when measured with the discrete travel cost model and $0.27 million when measured by the hedonic travel cost model. These measures are estimates of losses associated with current levels of acidification compared to natural pH levels of the relevant lakes.

In Table 4.9–2 of the *1990 Integrated Assessment Report*, projections are made of potential future damages by allowing the population and number of potential anglers to increase over time. The annual damages in 2030 (measured in 1989 dollars), incurred if the current level of acid deposition continues, is estimated to be $5.3 million for the discrete model and $27.5 million for the hedonic model. Maintaining the current level of deposition will increase damages, because lakes will continue to acidify. If acid deposition is reduced by 50 percent, the annual benefits in 2030 will be $20 million (discrete model) and $31.7 million (hedonic model).

It is interesting to contrast these recreational values with the value of agricultural and forest impacts that are reported in the next section. It should also be noted that no attempt has been made to estimate the indirect

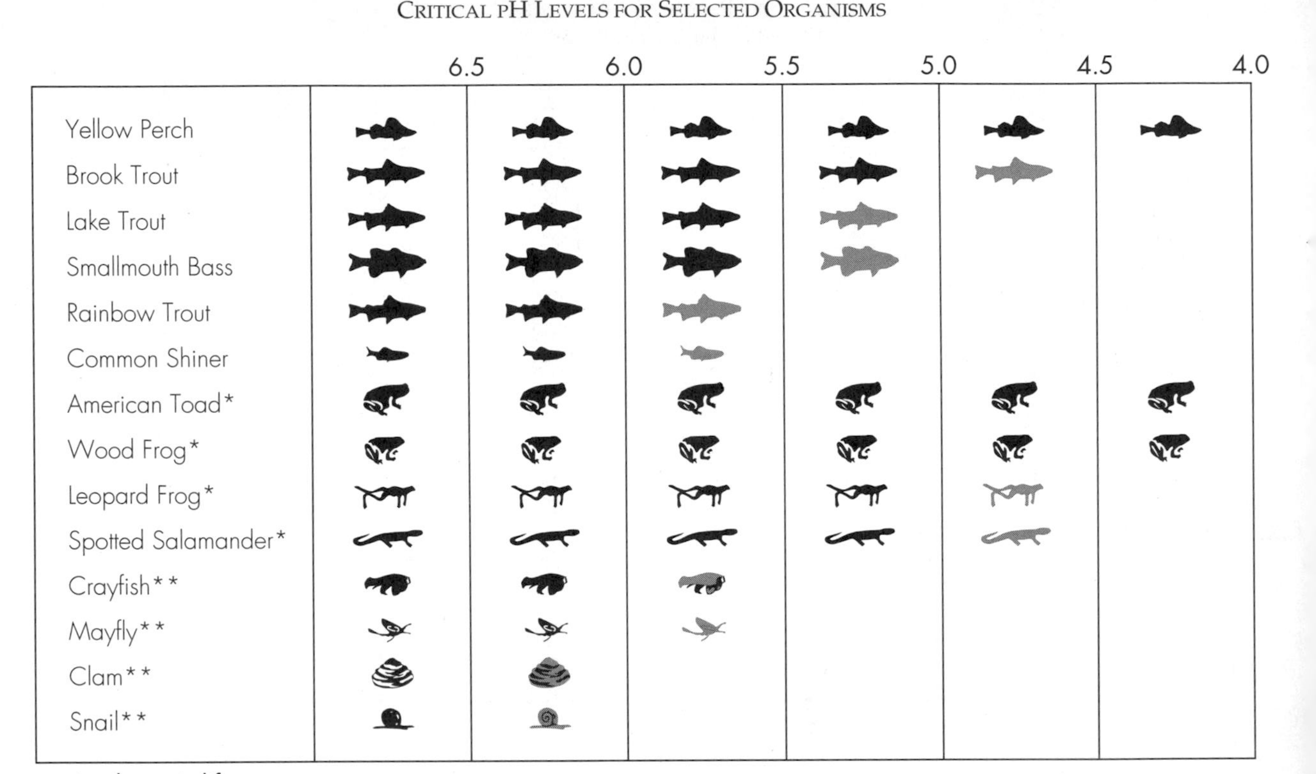

FIGURE 7.7

CRITICAL pH LEVELS FOR SELECTED ORGANISMS

*embryonic life stages

**selected species

SOURCE: NAPAP (1991), p. 227.

use values associated with reducing acid deposition impacts on aquatic ecosystems.

TERRESTRIAL IMPACTS OF ACID DEPOSITION

The two primary terrestrial impacts of acid deposition and related pollutants are on agriculture and forest ecosystems. Forest ecosystems may be adversely affected by acid deposition through a variety of mechanisms, including reducing the ability of trees to withstand severe winter cold (especially red spruce in high elevations) and through effects on forest soils. Acid deposition appears to have little negative impact on agriculture and has a positive impact by supplying sulfate and nitrate that act as fertilizer (additional sulfate

FIGURE 7.8

MEASURING THE RECREATIONAL FISHING BENEFITS OF REDUCING ACID DEPOSITION

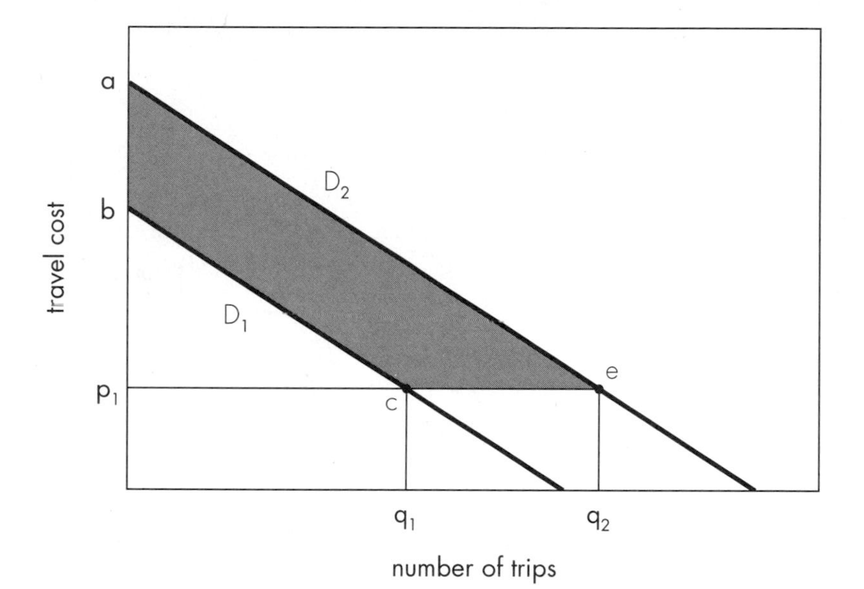

is only important in certain areas where sufficient sulfur is not present in the soil). However, tropospheric ozone, which is generated by acid deposition precursors, has a significant and negative impact on agriculture. Tropospheric ozone also has a negative impact on trees.

Table 7.4 shows the losses generated by ozone in agriculture. Relatively good estimates are available for these effects due to the existence of a large number of studies that examine the impact of ozone on agricultural yields and due to the existence of sophisticated economic models of the U.S. farming sector. These models predict how farmers and consumers react to changing economic conditions (prices, costs, and revenues) in the agriculture sector. The positive impacts of the fertilizer effects of nitrate and sulfate deposition are also reported in this table.

Results for agriculture were more readily obtained than other impact categories due to the massive research infrastructure that exists for agriculture. Many studies were conducted on the effects of ozone on agriculture in the 1980s (see Table 27B–6 in *NAPAP SOS 27*). Also, econometric models of the agricultural sector are available that compute the indirect effects of pollution by modeling how changes in the yield of one crop in one region affect the supply and demand for other crops in the same and other regions.

TABLE 7.4

EFFECT ON AGRICULTURAL SECTOR OF CHANGES IN TROPOSPHERIC OZONE AND ATMOSPHERE DEPOSITION OF NITROGEN AND SULFUR

	CHANGE IN SURPLUS (BILLIONS OF 1989 DOLLARS)		
% CHANGES IN OZONE	CONSUMERS	PRODUCERS	TOTAL
−10	0.785	−0.046	0.739
−25	1.637	0.095	1.732
+10	−1.044	0.215	−0.829
+25	−2.659	0.453	−2.206
	CHANGE IN SURPLUS (BILLIONS OF 1989 DOLLARS)		
% CHANGES IN NITROGEN	CONSUMERS	PRODUCERS	TOTAL
Yield based analysis: −50	−0.169	0.003	−0.166
Yield based analysis: +50	0.227	0.014	0.241
Cost based analysis: −50	−0.012	−0.055	−0.067
	CHANGE IN SURPLUS (BILLIONS OF 1989 DOLLARS)		
PORTION OF SULFUR-LIMITED CROPLAND (%)	CONSUMERS	PRODUCERS	TOTAL
40	−0.015	−0.030	−0.045
20	−0.008	−0.015	−0.023

SOURCE: *1990 Integrated Assessment Report*, p. 401–402.

Although the models used to estimate these losses and gains are quite complex, the concepts underlying them are relatively simple. Pollution that reduces yields for a crop increases the marginal cost function and causes a loss in producers' and consumers' surplus as in Figure 7.9.[7] Sulfur and nitrate deposition that acts as fertilizer can shift the marginal cost function downward, increasing producers' and consumers' surplus. Finally, increases in the prices of one crop will increase the demand for other crops, changing consumers' and producers' surplus for the other crops.

Corresponding measures for forestry are not available due to difficulties in scientifically quantifying the effects of acid deposition on trees. In particular, it is difficult to separate the observed overall forest decline to component causes.

[7]The problem is actually slightly more complex than this because many crops are price supported. The effects of price supports on agricultural markets is more fully discussed in Chapter 16.

FIGURE 7.9

LOSS IN CONSUMERS' AND PRODUCERS' SURPLUS ASSOCIATED WITH A NEGATIVE IMPACT OF POLLUTION ON AGRICULTURAL YIELDS

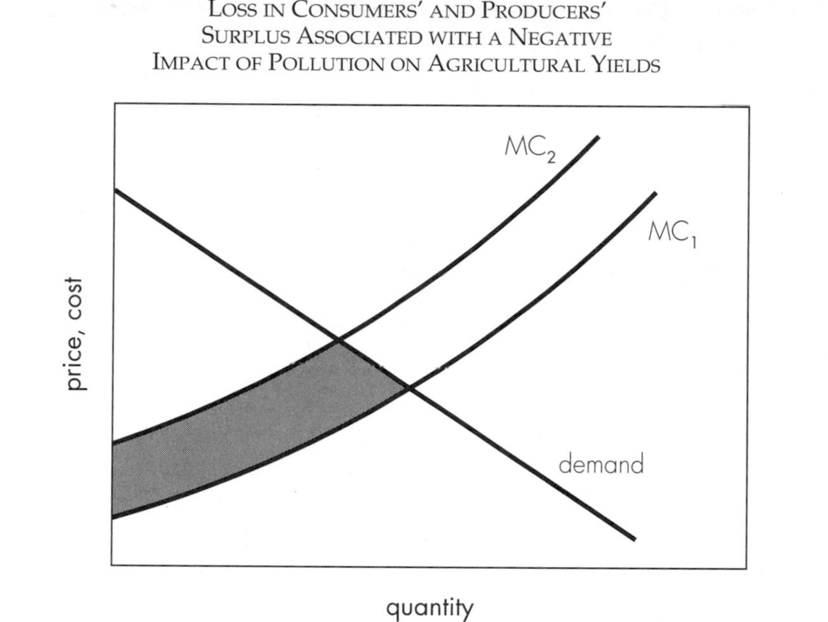

Ozone, acid deposition, other pollutants, insect pests, disease, and other factors all are sources of stress for trees. One reason it is difficult to quantify the effect is that the various sources of stress have synergistic effects. For example, acid deposition or tropospheric ozone may stress a tree, rendering it unable to withstand an attack by boring insects. It is very difficult to attribute the death of the tree to one cause or the other, or even to try to determine the relative contributions of the sources.

However, NAPAP provided estimates (contained in Table 7.5) of the economic consequences of general forest decline for southeastern pine forests as an illustrative example. As in the case of agriculture, pollution stresses shift the marginal cost function upward. It should be emphasized that these estimates are for the timber value of forests and not their overall forest ecosystem values. No studies were conducted to estimate forest recreation benefits, nor to estimate indirect use values associated with forest ecosystems.

The estimates of losses in timber value were prepared by conducting a sensitivity analysis, with the sensitivity analysis incorporating both relatively small changes of −2 percent and −5 percent (reflecting the possible range of effects due to ozone) and −10 percent (reflecting the effects due to forest decline from all causes). The estimates in Table 7.5 measure the changes in consumers' and producers' surplus associated with changes in the level of

TABLE 7.5

SENSITIVITY ANALYSIS SHOWING CHANGES IN ECONOMIC SURPLUS FOR SOUTHEAST PINES ASSOCIATED WITH CHANGES IN GROWTH RATES (MILLIONS OF 1989 DOLLARS)

	CHANGE IN GROWTH RATE			
ECONOMIC GROUP	+2.0	−2.0	−5.0	−10.0
PLANTED PINES				
Consumers	23.5	−5.6	−24.3	−61.0
Producers	−6.8	0.9	11.8	20.7
Timber owners	1.8	−4.4	−39.2	−20.2
Total	18.5	−9.1	−25.0	−60.5
NATURAL AND PLANTED PINES				
Consumers	45.2	12.8	−37.1	−117.2
Producers	−7.6	3.1	15	41.9
Timber owners	1.7	−16.2	−20.2	−33.2
Total	39.3	−0.3	−42.4	−108.5

pollution. In addition, an increase in growth rates of 2 percent was included in the analysis. The sensitivity analysis was conducted for pine trees in the Southeast. Again, it is important to note that these values are for trees as a source of wood and do not include the value of forests for recreation, production of ecological services, or other types of indirect use values.

VISIBILITY IMPACTS

The pollution particles generated by the emission of sulfur dioxide and nitrogen oxides, and the substances that are formed as the sulfur dioxide and nitrogen oxides react with other chemicals in the atmosphere, scatter sunlight and reduce visual range in polluted areas. In the eastern United States, where visibility is lower than in the west because of a variety of natural and pollution related factors, visual range is currently approximately 25 kilometers. A 20 percent reduction in sulfate aerosols would increase visibility to 29 kilometers, and a 40 percent reduction in sulfate aerosols would increase visibility to approximately 34 kilometers (Chestnut and Rowe, 1992).

Considerable evidence exists that suggests that people have a positive willingness to pay for improvements in visibility through reduction in pollution. Contingent valuation studies[8] by Brookshire, et al. (1979), Chestnut and Rowe (1992), Loehman, et al. (1981), Rae (1984) and Tolley (1986), among others, have consistently shown a positive willingness to pay for improved visibility and to prevent a significant deterioration of visibility. The question

[8]See Chapter 4 for a comprehensive discussion of the contingent valuation method.

FIGURE 7.10

STYLIZED REPRESENTATION OF
ESTIMATED CONSENSUS FUNCTION

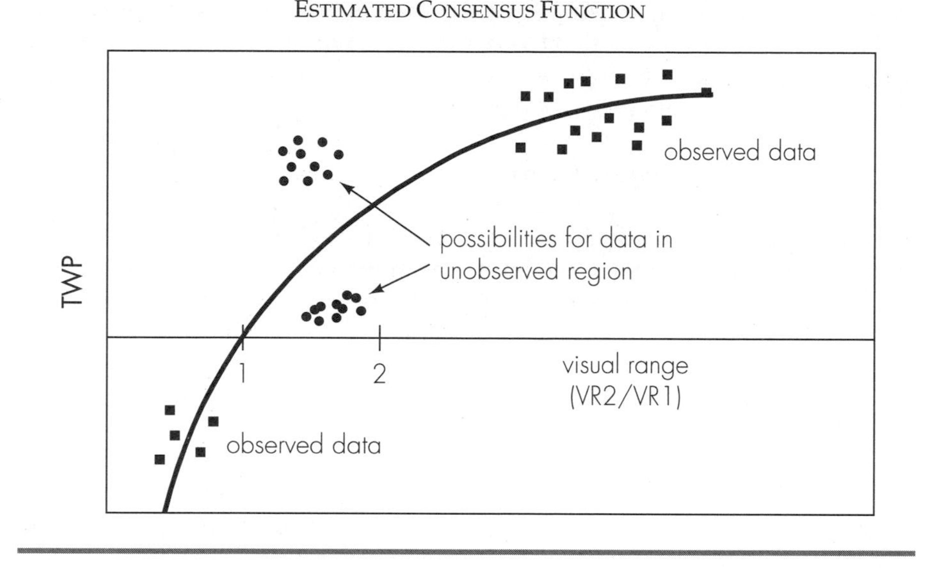

is, what is the willingness to pay for visibility improvements that could be generated by policies to reduce the emissions of sulfur dioxide and other acid rain precursors?

Despite all the research on the valuation of visibility improvements, this question is a difficult one to answer. Although Chestnut and Rowe developed a consensus function (Brown et al., 1990) that was designed to answer this question, there are information gaps that make this difficult. This consensus function was estimated by looking at the willingness to pay and the hypothesized increase in visual range associated with each estimate from each empirical study that was previously conducted. These data were then used to estimate a functional relationship between the willingness to pay for a change in visual range and the magnitude of the visual change, using a regression analysis. The problem is that no contingent valuation studies have yet been conducted that look at changes in visual range that are of a magnitude comparable to the type of changes that are likely to be brought on by new policy, such as the acid rain provisions of Title IV of the 1990 Clean Air Act Amendments. Figure 7.10 illustrates the significance of making projections into areas in which there are no observations. The vertical axis of this graph represents total willingness to pay for a change in visual range, and the horizontal axis represents a measure of changed visual range (the hypothesized new visual range [VR2] divided by the original visual range [VR1]). If the visual range is increasing, VR2/VR1 will be greater than one. If it is decreasing, VR2/VR1 will be less than one. The data from existing contingent valuation surveys are the square points labeled "observed data" in Figure 7.10. However, the region of likely change is between 1 and 1.4, and there are no studies that ask people their willingness to pay in this region. Although a regression using the observed data will provide a function with values in this region, the true willingness to pay may be above or below this function. This problem exists whenever one tries to extrapolate into a region of no data observations.

The values extrapolated throughout the use of Chestnut and Rowe's consensus function are presented in Table 7.6, although the student should keep in mind the limitations discussed previously. The last row of this table most closely corresponds to the improvements in eastern visibility in urban areas that would result from the sulfur dioxide emission reduction associated with Title IV of the 1990 Clean Air Act Amendments (Chestnut and Rowe, 1992). A 40 percent decrease in sulfate aerosols corresponds to a roughly 30 percent increase in visual range in eastern rural areas and a smaller change (approximately 20 percent) in urban areas.

There also exist several problems concerning the studies upon which the consensus function is based. The chief problem among these is that the studies generally make no attempt to ensure that the respondents' answers are driven by visibility impacts and not driven by other impacts that they felt are associated with the pollution changes. For example, people might perceive that if visibility improves, the health risks associated with air pollution should lessen, and this perception might form a basis of their willingness to pay response. The problem of other values entering the response is known as the embedding problem and is a potential problem with all contingent valuation studies. The embedding problem is discussed extensively in Chapter 4.[9]

As long as health benefits are not separately estimated, there is no double counting problem; but if a separate health estimation is conducted, then a double counting problem will arise. Also, if some people embed values and other people do not, it will be very difficult to accurately measure the sum of visibility and health benefits. Further problems may exist if survey respondents hypothesize environmental impacts that do not actually occur or if nonenvironmental values are embedded in survey responses. For example, people might incorrectly hypothesize that the pollution that reduces visibility kills animals. People also might state a positive willingness to pay for improving visibility because the act of giving to a good cause makes them feel good. These problems are discussed in more detail in Chapter 4.

HEALTH IMPACTS

Unfortunately, there has been little progress in measuring the health benefits associated with acid deposition mitigation. Human health is suspected to be

---

[9]See Brown et al. for a comprehensive discussion of the consensus function.

TABLE 7.6

CHANGE IN TOTAL WILLINGNESS TO PAY (WTP) OF URBAN RESIDENTS IN EASTERN UNITED STATES (BILLIONS OF 1989 DOLLARS)

		ROUGH BOUNDS ON Δ WTP	
% SULFATE CHANGE	VISUAL RANGE	LOWER	UPPER
+20	22 km	−\$0.256	−\$1.023
0	25 km	———	———
−20	29 km	\$0.297	\$1.187
−40	34 km	\$0.615	\$2.460

SOURCE: Chestnut and Rowe, 1992.

adversely affected by exposure to acid aerosols, SO_2, and SO_4 (Tonn, 1993). These pollutants are suspected to cause respiratory problems, such as chronic bronchitis, shortness of breath, and other ailments. Although these pollutants are suspected to be damaging to human health, there is insufficient scientific evidence with which to prove this assertion and with which to actually estimate a functional relationship between emissions and health impacts. The reason is that it is difficult to make a link between air quality and exposure to the pollutants, because exposure depends on work location, residential location, and the allocation of time between indoor activity and outdoor activity (Tonn, 1993).

MATERIALS IMPACTS

One might hypothesize that it would be relatively easy to determine the benefits of mitigating the deterioration of materials associated with acid deposition, due to the ability to conduct controlled laboratory tests of materials and the availability of markets for most of the structures and materials that are associated with this problem. However, this is not the case.

Several case studies have been conducted of galvanized steel chain-link fencing, limestone building facade, and residential painted exteriors. However, these case studies have not been extrapolated to regional or national totals. We do not have the ability to generalize from these particular studies to broad measures of damages. The first problem is that we don't know the inventory of buildings and materials that currently exists, so we have no baseline from which to measure damages. Second, we do not know the dose–response relationships for many types of materials. Third, it is difficult to know if observed damages are due to local sources of pollution (such as automobiles) or due to regional pollution (acid rain and its precursors). Finally, we do not understand how consumers respond to the damages: Do they repair immediately, take steps to prevent further damages, or do they allow damages to proceed and eventually replace the materials.

In addition to the above problems, we have not made progress toward measuring indirect use values associated with materials impacts. There are two types of these indirect use values. The first reflects people's willingness to pay to live and work in aesthetically pleasing areas. As the deterioration of materials increases, aesthetic quality will diminish unless there is an offsetting increase in preventative and remediative maintenance. If maintenance lags, aesthetic quality will decline, and the welfare of the community may be lowered in a fashion that will not be measured if the focus is entirely on the increasing costs of maintenance. The second type of indirect use value is associated with the value that people place on cultural and historical structures. Here the question should be even less focused on maintenance or repair costs, as maintenance or repair may fundamentally damage the characteristics that make these structures valuable to society. For example, while it might be possible (and cheaper) to replace a deteriorated Statue of Liberty with a fiberglass replica, that step would dramatically change the character of the statue and reduce its value to society.

ACID DEPOSITION POLICY

As the previous discussion indicates, there exists very little quantitative information concerning people's willingness to pay to prevent the impacts associated with acid deposition. Given the decade of study associated with the initial NAPAP program—and the millions of dollars spent on research—one might wonder why this is the case.

This question is difficult to answer, and it revolves around the purpose of science. The scientists who studied the acid deposition relationships were interested in testing hypotheses involving cause and effect in impacts. One of the best ways of testing these hypotheses was to focus on small geographic areas, such as individual lakes and ponds that were becoming acidified. Although the research developed important insights into cause-and-effect relationships, the knowledge gained was often specific to the location studied, and it was not easily generalizable to larger regions.

This lack of generability does not imply that the NAPAP research program did not contribute to the understanding of how to make acid deposition policy. The NAPAP research program, which spanned the decade of the 1980s, resolved many scientific questions concerning the dispersion of pollutants, the chemistry of its transformation into acid deposition, and many of the ecological effects of the acid deposition. However, many of the physical impacts of acid deposition remain incompletely understood. Effects on forests and human health are important examples. In addition, very little research money was spent looking at the willingness to pay (or other measures of the societal significance) to prevent identified impacts from occurring. The feeling was that with so much uncertainty involving the scientific relationships, it would be meaningless to try to estimate economic relationships based on the scientific relationships.

However, uncertainty will always be present in measuring the benefits or damages associated with environmental change. We have already seen the importance of uncertainty in examining the question of global warming. This fact leads to an important policy question: How do policy makers develop a set of efficient policies to deal with acid deposition when the benefits of reducing acid deposition remain largely unquantified?[10]

PRE-1990 ACID DEPOSITION POLICY

The Reagan administration chose to deal with uncertainty in costs and benefits by requiring that more information be developed before implementing any reductions in the emissions of SO_2 or NO_x. In the early 1980s, most scientists who studied the problem believed that emissions of SO_2 and NO_x led to the problem of increased acid deposition. However, at that time there was insufficient scientific evidence to prove a cause-and-effect relationship between SO_2 and NO_x emissions and acid deposition. Citing a lack of evidence of the existence of a problem, the Reagan administration did not propose any new policies for reducing acid deposition precursors (SO_2 and NO_x). This may be the reason that the research funded by NAPAP tended to focus on establishing cause-and-effect relationships between emissions of pollution and regional acid deposition.

The lack of specific acid deposition policy should not be taken to imply that SO_2 and NO_x emissions remained uncontrolled. In fact, SO_2 and NO_x are "criteria pollutants" that were regulated under the 1972 Clean Air Act and the 1977 Clean Air Act Amendments. However, these regulations (which predated the Reagan administration) focused on the local effects of emissions of SO_2 and NO_x. Unfortunately, this local focus may have exacerbated the acid deposition problem.

One way in which a local polluter can minimize the effect of pollution emissions on local air quality is to build a tall smokestack and emit the pollutants hundreds of meters above the ground. If the emissions are released at high altitude, they will be transported outside of the vicinity before they reach the ground, where they are monitored. By the time they reach the ground, they will be in another state and no longer be the concern of the state where they were emitted. The cost of reducing monitored emissions is much lower by using tall smokestacks than by using processes that remove the pollutants from the exhaust gases.

---

[10]See Chapters 3 and 4 for a discussion of marginal damage functions, marginal abatement cost functions, cost–benefit analysis, and other information requirements for developing policy.

These tall smokestacks were, to a large extent, responsible for the sulfur component of the acid deposition problem, as the smokestacks injected the pollutants into the more powerful wind currents of higher altitudes. These wind currents transported the pollutants great distances (for example, from the Midwest to New England and Quebec).

ACID DEPOSITION POLICY AND THE 1990 CLEAN AIR ACT AMENDMENTS

Another important policy question that slowed the development of acid deposition policy is, who should pay for the environmental improvements? While this question is always important when formulating environmental policy, it is particularly important given the regional nature of the acid deposition problem. For example, a large portion of SO_2 emissions are generated by electric utilities in the Midwest, and a significant portion of the damages may be associated with forest and aquatic ecosystems in the Northeast and the Southeast. Electricity consumers in the Midwest would argue that electricity is a necessity and trout fishing is a luxury, so why should their electric bills increase so that New Englanders can enjoy better recreational fishing?

This argument may at first seem to be appealing, but the appropriate comparison is not total recreation with total electricity consumption. If the price of electricity goes up, consumers will eliminate the least important uses of electricity first (for example, by turning lights off in unoccupied rooms, adjusting the thermostat by 1 degree, skipping reruns of *Gilligan's Island*). These marginal uses of electricity are not inherently more valuable than the improvements in recreation. One must compare the changes in consumers' and producers' surplus associated with the two activities. These relative values are actually an empirical issue which must be resolved through measurement and not conceptual debate.

Although one cannot justify an argument that electric consumers should have preference over recreational fishers, important political problems are associated with developing legislation that benefits one region at the expense of another. Despite the overall benefits to the nation, legislative representatives will have trouble supporting a law that reduces the standard of living of their constituents.

One way of dealing with this problem is to package several environmental policies into the same piece of legislation, so that the benefits and costs of the entire package of environmental changes are not as unequally distributed. For example, the 1990 Clean Air Act Amendments (CAAA), which occurred during the Bush administration, address not only acid rain, but local air quality problems associated with ozone and carbon monoxide, pollution from cars and trucks (VOC and N_2O), air toxics (heavy metals and other carcinogens, mutagens, and reproductive toxins), and stratospheric ozone and global climate protection.

Acid deposition is dealt with in Title IV of the 1990 CAAA, which specifies a 10 million ton reduction in annual sulfur dioxide emissions. These reductions are to be achieved by the year 2000, with electric utilities shouldering the primary burden of reduction.

An interesting aspect of Title IV of the 1990 CAAA is that it represents the first attempt by the federal government to implement a system of marketable pollution permits. Utilities that reduce pollution below the allowed levels may sell allowances to other utilities. Each allowance represents the right to emit 1 ton of SO_2. The purpose of the trading of allowances is to allow the reductions to take place among the firms who face the lowest costs of reducing pollution.[11]

Although many economists regard the incorporation of marketable pollution permits into the 1990 CAAA as an important step in improving the efficiency of environmental regulations, the system that was adopted does not have all the properties that economists believe are desirable. The primary criticism is that there is no attempt to make geographic distinctions associated with the location of emission of SO_2, as SO_2 is traded across locations on a one-for-one basis. While the acid deposition effects of sulfur dioxide emissions may be relatively independent of location, sulfur dioxide also has local pollution effects that are quite sensitive to location.

To better understand this aspect, let us look at the first trade that occurred, where a Wisconsin utility sold allowances (permits) to the Tennessee Valley Authority (a federally owned and authorized regional electric utility) in Tennessee and five other southern Appalachian states. The sale of allowances from Wisconsin to the Tennessee Valley requires the Wisconsin utility to pollute less and allows TVA to pollute more. The costs of both the Wisconsin utility and TVA must fall as a result of this trade. The costs of reducing pollution by the amount of the traded allowances must be less than the price of the allowances, or the Wisconsin utility would not have agreed to the sale. Similarly, the electricity production costs of TVA must fall, as TVA would not have entered into the trade unless their savings in abatement costs was more than the price of the allowances. Since both utilities save money, the costs of producing electricity in both regions must fall. If the only effect of the sulfur dioxide pollution is its contribution to regional acid rain, and this contribution is independent of location (within the eastern United States), then there is no change in environmental quality (positive or negative), and the trade merely reduces costs and makes everybody better off. This benefit is, in fact, the rationale behind marketable pollution permits, which are discussed more fully in Chapter 3.

However, if local pollution effects exist (the SO_2 affects local environmental quality as well as contributing to regional acid rain), then the trade

[11]Chapter 3 discusses the cost minimizing properties of marketable pollution permits.

will reduce local environmental quality in the Tennessee Valley and improve local environmental quality in Wisconsin. If the citizens of the Tennessee Valley value the reduction in electric costs more than the reduction in environmental quality, the trade is still unambiguously a good idea. Both people in Wisconsin and people in the Tennessee Valley are made better off as a result of the trade. The social desirability of the trade becomes more ambiguous, however, if the citizens of Tennessee value the reduction in environmental quality by more than the reduction in the costs of generating electricity. Then the people of the Tennessee Valley become worse off. The trade may still be a potential Pareto improvement if the citizens of Wisconsin gain by more than the citizens of Tennessee lose. In addition to efficiency implications, there may be important regional equity considerations. Because of the potential problems that may be generated by local pollution effects, many critics of this system argue that local pollution effects need to be taken into account in the trading system. See Chapter 3 for a discussion of ways to incorporate geographic variability in the damages from pollution into a system of marketable pollution permits.

Another potential problem with the acid rain provisions of the 1990 CAAA is that not all emitters of SO_2 are incorporated into the system, and NO_x is not part of any trading system. Small polluters and mobile sources are not covered by these provisions. It may be that obtaining greater reductions from small emitters and mobile sources is cheaper at the margin.

The final important criticism is that the trading system of the 1990 CAAA does not use a conventional auction system. It employs an allocation mechanism that matches the highest bidder for allowances with the lowest offerer of allowances, and then matches the next highest bidder with the next lowest offerer, and so on. This gives both the potential buyer and the potential seller an incentive to bid lower than his or her marginal abatement costs (Caisson). In theory, this system will bias price below marginal abatement costs and generate a greater than optimal amount of trading. This system was probably chosen over more traditional auction mechanisms, because the EPA wanted to ensure that a large number of trades took place. In actuality, the number of trades has been surprisingly small. Potential reasons for the small number of trades are discussed in the accompanying boxed example.

The first auction conducted under this system led to an allowance price (approximately $250) that was less than 50 percent of the expected price. This discrepancy may be because of the incentives generated by this peculiar trading system or because more abatement activity had taken place than expected. Electric utilities might have reduced pollution a greater than expected amount if they were afraid that there would not be allowances available to cover their excess pollution. Under these circumstances, it would be better to spend the money to reduce more pollution now than to pay criminal penalties if the utility cannot buy enough allowances to cover its excess pollution.

Why Have There Been So Few Trades of Allowances for Sulfur Dioxide Permits?

Economists are strong proponents of economic incentives to control the level of pollution. As discussed in Chapter 3, the trading of marketable pollution permits can lead to the attainment of the target level of pollution at the minimum abatement cost for society. If a system of permits is implemented to replace an inefficient command and control system, one would expect to see a large number of trades as polluters with high marginal abatement costs buy permits from polluters with low marginal abatement costs. As of early 1997, only a surprisingly small number of trades has taken place.

There are two potential explanations for the sparseness of trading activity. The lack of trading could be due to a lack of profit associated with trading, which would occur if most polluters had similar marginal abatement cost functions. Alternatively, polluters could have very different marginal abatement cost functions and the trades could be profitable, but other factors may diminish the ability to trade marginal abatement costs. This question is important because this type of knowledge is essential to establishing effective emissions reduction strategies for other pollutants, such as carbon dioxide.

In a report available on the internet (gopher://dewey.1.b.ncsu.edu/11/library/disciplines/gao-reports), the General Accounting Office has analyzed various factors which have caused reluctance to trade allowances for sulfur dioxide. These factors include:

- Likely buyers and sellers do not have to reduce emissions at the same time (big utilities and small utilities have different requirements for participating in the trading scheme).
- The design of EPA's auction system has generated lower than expected prices, which increases uncertainty concerning the prices at which future trades might take place.
- State regulation of electric utilities does not allow, or is not clear if it is allowable for them to recover the cost of the purchase of allowances.
- The possibility exists that EPA will change regulations on other pollutants, particularly if they are brought into an allowance trading system.

Although there are both positive and negative aspects associated with the structure of the emissions trading system that the EPA has chosen, it is difficult to make a theoretical argument about the overall social benefits of the system. Such a verdict awaits more experience with the program, which will not be fully implemented until the year 2000.

Table 7.7 contains a summary of the acid rain provisions (Title IV) of the 1990 Clean Air Act, and Figures 7.11 and 7.12 show the anticipated levels of emissions as a result of this legislation. When evaluating Title IV, it should be emphasized that the 10 million ton reduction specified by the act was not chosen based on an indication that 10 million tons was the optimal reduction. Rather, it was chosen as a reduction level that was somewhat supported by the scientific research conducted on the relationship between emissions and impacts, and it was a reduction level that was acceptable to most members

TABLE 7.7

USEPA SUMMARY OF TITLE IV OF 1990 CLEAN AIR ACT AMENDMENTS

SO_2 REDUCTION:
A 10 million ton reduction from 1980 levels, primarily from utility sources. Caps annual SO_2 emissions at approximately 8.9 million tons by 2000.

ALLOWANCES:
SO_2 reductions are met through an innovative market-based system. Affected sources are allocated allowances based on required emission reductions and past energy use. An allowance is worth one ton of SO_2 and it is fully marketable. Sources must hold allowances equal to their level of emissions or face a $2,000 excess ton penalty and a requirement to offset excess tons in future years. EPA will also hold special sales and auctions of allowances.

PHASE I:
SO_2 emission reductions are achieved in two phases. Phase I allowances are allocated to large units of 100 MW* or greater that emit more than 2.5 lb/mmbtu** in an amount equal to 2.5 lb/mmbtu multiplied by their 1985–87 energy usage (baseline). Phase I must be met by 1995, but units that install certain control technologies may postpone compliance until 1997, and may be eligible for bonus allowances. Units in Illinois, Indiana, or Ohio are allotted a pro rata share of an additional 200,000 allowances annually during Phase I.

PHASE II:
Phase II begins in 2000. All utility units greater than 25 MW that emit at a rate above 1.2 lbs/mmbtu will be allocated allowances at that rate multiplied by their baseline consumption. Fifty thousand bonus allowances are allocated to plants in midwestern states that make reductions in Phase I.

NO_x:
Utility NO_x reductions will help to achieve a 2 million ton reduction from 1980 levels. Reductions will be accomplished through required EPA standards for certain existing boilers in Phase I, and others in Phase II. EPA will develop a revised NO_x NSPS [New Source Performance Standard] for utility boilers.

REPOWERING:
Units repowering with qualifying Clean Coal Technologies receive a 4-year extension for Phase II compliance. Such units may be exempt from new source review requirements and New Source Performance Standards.

ENERGY CONSERVATION AND RENEWABLE ENERGY:
These projects may be allocated a portion of up to 300,000 incentive allowances.

MONITORING:
Requires continuous emission monitors or an equivalent for SO_2 and NO_x and also requires opacity and flow monitors.

SOURCE: This summary is taken directly from *The Clean Air Act Amendments of 1990: Summary Materials*, USEPA, U.S. Government Printing Office, Washington, D.C., 1990.

*megawatt—a measure of the electricity producing capacity of an electric power plant.

**Million British Thermal Units—a measure of the heat potential of a fuel. A BTU is the amount of heat necessary to raise one pound of water one degree Fahrenheit.

FIGURE 7.11

ANNUAL EMISSIONS OF SULFUR DIOXIDE IN THE UNITED STATES

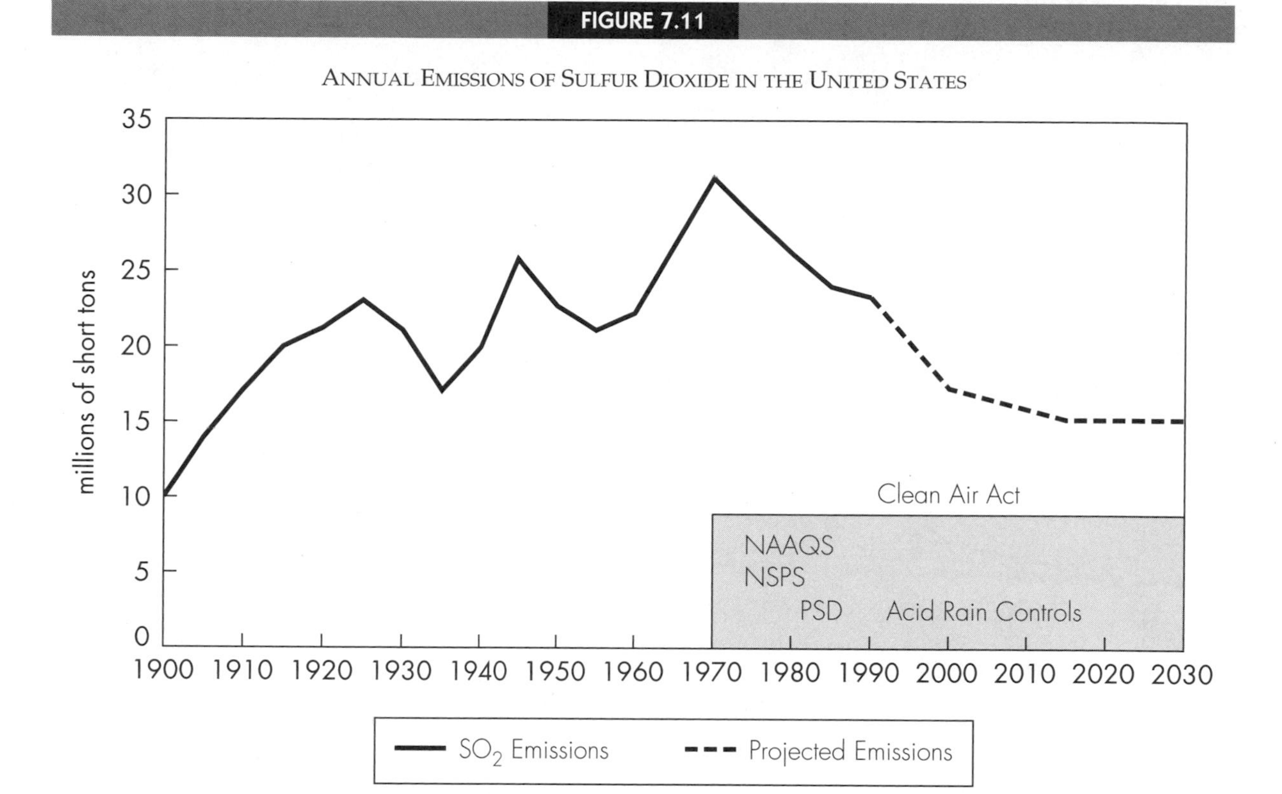

of Congress. However, as indicated earlier in the chapter, very little information has been developed on the value that people place on reducing these impacts.

The 1990 CAAA requires NAPAP to conduct an assessment of the costs and benefits associated with this reduction by 1996. NAPAP is further required to conduct assessments every four years and to identify the levels of reductions that will prevent adverse ecological impacts, although the term *adverse ecological impacts* is not defined in this legislation.

The other major point that should be noted is that Title IV relies primarily on command and control provisions. As can be seen in Table 7.7, virtually every provision is a type of command and control regulation. The only exception is the trading of allowances. Although this exception is significant, the SO_2 trading system stops far short of being a true marketable pollution permit system. Only certain classes of polluters (utilities) are required to participate in the system, and NO_x is not included in any system of market incentives.

FIGURE 7.12

ANNUAL EMISSIONS OF NITROGEN OXIDES IN THE UNITED STATES

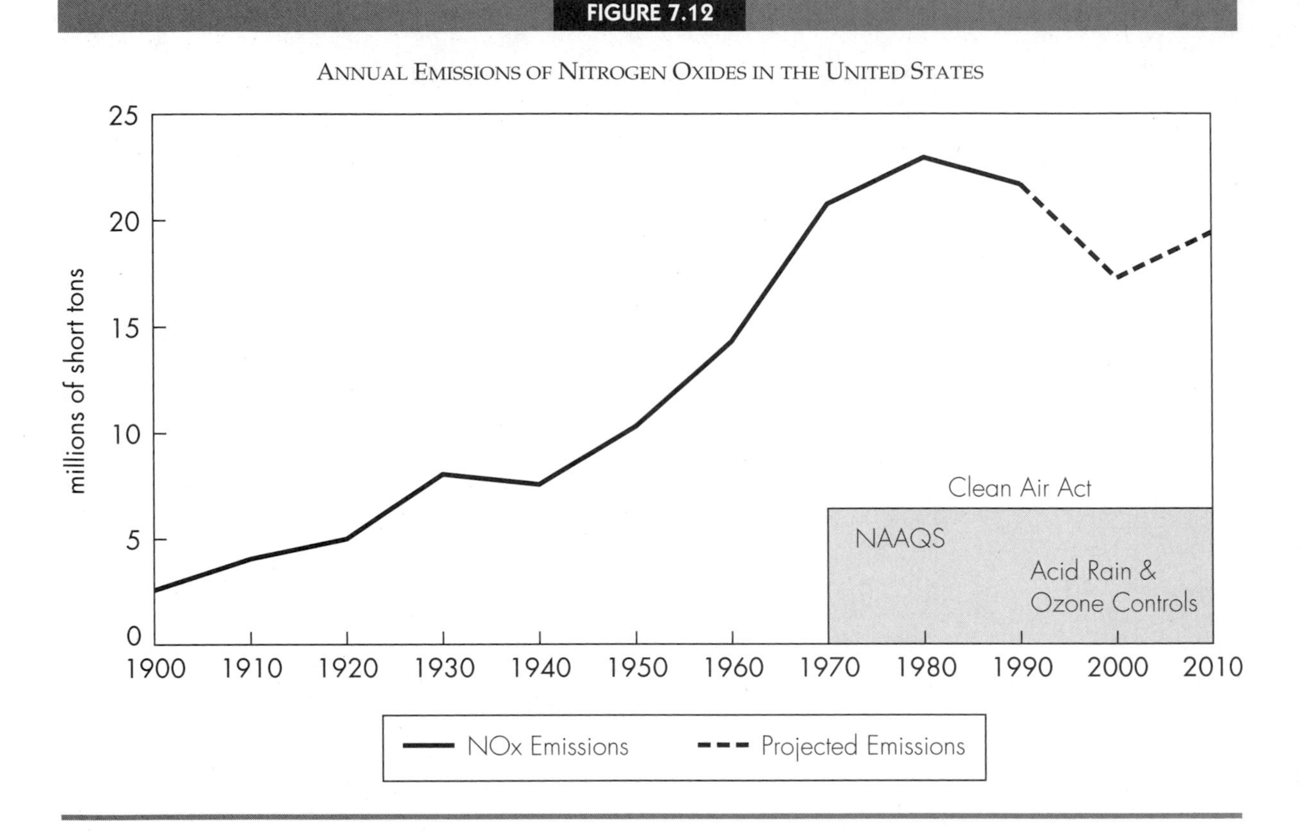

Finally, it should be noted that as the Clean Air Act was being prepared, an international agreement was being negotiated between the United States and Canada. As there was considerable disagreement between the United States and Canada on acid deposition policy in the 1980s (the United States wanted to study the situation, Canada wanted immediate action), this cooperation marked an important step. The cooperation is reflected in the Clean Air Act, whose provisions reflect the negotiated agreements between the United States and Canada. The two countries entered into a "Bilateral Agreement on Air Quality" in 1991 in order to deal with acid deposition precursor pollutants and other types of air pollution.

SUMMARY

Although there was a major research effort during the 1980s (NAPAP), there is little information about the value of changes in the level of acid deposition or changes in the level of precursor pollutants. Impacts are believed to include effects on aquatic ecosystems, forest ecosystems, agriculture, materials, visibility, and human health. Some of these effects have been demonstrated to exist, and others are hypothesized. However, even for the effects that have been demonstrated, little progress has been made quantifying the benefits of mitigating these effects. Nonetheless, the 1990 Clean Air Act Amendments (CAAA) specify a 10 million ton per year reduction in SO_2 emissions and a 2 million ton per year reduction in NO_x. However, it is not clear whether these levels are greater or less than the socially optimal levels. Research on the benefits and costs of these reductions is currently taking place. Check EPA's acid rain website (http://earth1.epa.gov/acidrain/andhome.html) for more information as it becomes available.

A novel provision of the 1990 CAAA is the trading of allowances for SO_2. Trades are allowed among electric utilities on a one-for-one basis. Although this system is designed to reduce the cost of attaining the 10 million ton reduction, there is some concern that the trading may lead to undesirable changes in local air quality in some parts of the country.

Although the 1990 CAAA specify reductions in SO_2 and NO_x, these reductions are not intended by Congress to be the final action on acid deposition. The amendments require periodic assessment of the costs and benefits of reducing acid deposition and a determination of the level of reductions that are necessary to prevent further adverse impacts.

REVIEW QUESTIONS

1. Assume that you are dealing with a very small country with one lake, one electric utility, and 1,000 people who fish in the lake. Let the inverse demand for recreational fishing trips equal

   $$Df = 200 - 10 \text{ trips} + 110 \text{ (catch per day)}.$$

   Let the inverse demand for megawatt hours (MWH) of electricty equal

   $$De = 100 - 0.001 \text{ MWH}.$$

   Assume that emissions for the power plant affect the quality of the fishing. A particular policy results in a reduction in emissions that increased catch per day from 5 to 10 while increasing the costs of producing electricity from \$7 per MWH to \$8 MWH. If each of the 1,000 people who fish have a travel cost of \$10 to gain access to the lake, does the policy increase or decrease net social benefits?
2. Look at pages 201 and 202, which discuss the SO_2 offset trading between a utility in Wisconsin and TVA in Tennessee. Assume that the trade lowers Wisconsin electricity consumers' bills by a total of \$1 million and lowers Tennessee electricity consumers' bills by \$500,000. Also assume that the trade improves local environmental quality in Wisconsin by an amount that people value at \$2 million. If the local environmental quality in Tennessee declines as a result of this trade, what is the most that Tennesseans could be willing to pay to avoid the environmental decline that would still make the trade a potential Pareto improvement?
3. What are the primary impacts of acid deposition? What valuation tools are available to measure the willingness to pay to avoid these impacts?
4. What distinguishes the acid rain problem from the global warming problem?
5. Why would a focus on local air pollution lead to increased regional transport of acid rain precursor pollutants?

6. What problems can be found in existing studies that attempt to measure the willingness to pay for increased visibility?

Questions for Further Study

1. What is wrong with Title IV of the 1990 Clean Air Act Amendments? How would you amend the amendments to make them more efficient?
2. Discuss the difficulties of projecting estimates of economic values outside the range of observed data.
3. How can the travel cost technique be used to measure the recreational fishing use values associated with reducing acid deposition?
4. Why is it easier to measure the impact of acid deposition on agriculture than on forests?

Suggested Paper Topics

1. Survey the literature on measuring the recreational fishing impacts of acid rain. If you start with Brown et al. and with NAPAP's *1990 Integrated Assessment Report*, you will find references to most of the earlier work. Ask your government documents reference librarian how to locate these reports. Also, peruse the current issues of the *Journal of Environmental Economics and Management, Land Economics, Journal of Agricultural Economics*, and the *Journal of Economic Literature* for additional references.
2. Survey the literature on measuring the value of reducing pollution-generated risks to health. Start with A. M. Freeman (1993) for a comprehensive list of references in this area. In addition to surveying this literature, you might want to discuss how the literature could be applied to the health impacts of acid rain precursors.
3. Write a paper looking at policy issues involving transfrontier pollutants. Look in the recent issues of the *Journal of Environmental Economics and Management, Land Economics*, and *Environmental and Resource Economics* for references in this area.
4. Write a paper examining the current trading system for sulfur dioxide emissions. What are the nature of the trades that have taken place? What are the faults with the trading system? How would you modify the system to make it more efficient?

Works Cited and Selected Readings

1. Brookshire, D. S., et al. *Methods Development for Assessing Air Pollution Control Benefits. Vol. 2: Experiments in Valuing Non-Market Goods: A Case Study of Alternative Benefit Measures of Air Pollution in the South Coast Air Basin of Southern California.* Prepared for USEPA, Washington, D.C., 1979.
2. Brown, G. M., et al. NAPAP SOS Report 27: *Methods for Valuing Acidic Deposition and Air Pollution Effects.* National Acid Precipitation Assessment Program, Washington, D.C., 1990.
3. Chestnut, L. G., and R. D. Rowe. "Visibility Valuation: Acid Rain Provisions of the Clean Air Act." In *Proceedings of the 1992 AERE Workshop: Benefits Transfer: Procedures, Problems, and Research Needs,* edited by Taylor Bingham. Snowbird Utah, 1992.
4. Freeman, A. M., III. *The Measurement of Environmental and Resource Values.* Washington, D.C.: Resources for the Future, 1993.
5. Loehman, E. D., et al. *Measuring the Benefits of Air Quality Improvements in the San Francisco Bay Area.* Prepared for USEPA by SRI International, Menlo Park, CA, 1981.
6. Rae, D. A. *Benefits of Visual Air Quality in Cincinnati—Results of a Contingent Ranking Survey.* RP–1742, final report prepared by Charles River Associates for EPRI, Palo Alto, CA., 1984.
7. Tolley, G. A., et al. *Establishing and Valuing the Effects of Improved Visibility in Eastern United States.* Report to USEPA, Washington, D.C., 1986.
8. Tonn, Bruce E. *The Urban Perspectives of Acid Rain: Workshop Summary.* Oak Ridge, TN: Oak Ridge National Laboratory, 1993.
9. U.S. Environmental Protection Agency. *The Clean Air Act Amendments of 1990: Summary Materials.* Washington, D.C.: U.S. Government Printing Office, 1990.
10. U.S. National Acid Precipitation Assessment Program (NAPAP). *1990 Integrated Assessment Report,* Washington, D.C., 1991.

Energy and the Environment

Energy production and use are vital to the economies and environments of all countries. Furthermore, the mix of energy sources has profound consequences for environmental quality.[1]

INTRODUCTION

The above quote, which appears in the energy chapter of *World Resources 1992–93,* illustrates the central role of energy in both economic and environmental policy. The production and consumption of energy is not only crucial to the health of economies in both developed and developing countries, but it is responsible for a large portion of the environmental problems that these countries experience. While many textbooks focus on the question of whether the market will ensure adequate supplies of energy in the future, this chapter will focus on the energy–environment interface. Questions of how the market supplies energy will also be addressed, because the nature of the supply of energy has an important impact on its interaction with the environment.

When people think of the relationship between energy use and the environment, impacts such as air pollution, global warming, and acid precipitation immediately come to mind. While these are critically important there are many additional environmental changes associated with energy production and consumption. Effects on water quality are important, because water is used and often contaminated in the production of energy through activities such as drilling for oil and gas, cooling energy facilities, coal mining, and the underground storage of oil and gasoline. Additionally, oil spills (of both the infrequent but catastrophic and small but chronic nature) pollute oceans and inland waterways. The production of energy also destroys habitat. Strip-mining removes the whole surface of the land and leads to vast changes in the landscape and acid drainage problems. Oil activities in the wetlands of Louisiana are contributing factors to the steady loss of wetlands. In addition, the mining and burning of coal creates a large amount of solid waste that

[1] *World Resources 1992–93* (Washington, D.C.: World Resource Institute, 1993), 143.

must be disposed of, which competes for scarce landfill space and may lead to environmental problems such as ground and surface water contamination, even when the waste is deposited in landfills.

Perhaps the greatest interaction between energy and the environment occurs with air pollution, where the combustion of fossil fuels is the major source of the air pollutants that were initially regulated by the 1972 Clean Air Act, with amendments in 1977 and 1990. These pollutants—which include particulates, sulfur oxides (SO_x), nitrogen oxides (NO_x), carbon monoxide (CO), volatile organic compounds (VOCs), and lead—have strong effects on human health, either directly or through the formulation of tropospheric ozone and smog. Table 8.1 lists the 1990 total emissions of these pollutants and their sources. It should be noted that in addition to its obvious role in fuel consumption, the use of energy is also responsible for virtually all the pollutants in the transportation sector and a good portion of the emissions in industrial processes (such as chemical manufacture). It is not an overstatement to conclude that the air pollution problem is primarily related to the use of energy.

Although the environmental problems associated with energy use are extremely important, they have not been the primary focus of energy policy. As will be seen in the next section of the chapter, concerns with potential scarcity, the desire to keep price low, and energy security have been the chief concerns of U.S. energy policy.

THE HISTORICAL DEVELOPMENT OF U.S. ENERGY POLICY IN THE POST-WORLD WAR II ERA

The formulation of energy policy has been substantially influenced by concerns about supply and, in turn, has influenced the way in which energy is supplied. Before examining the interactions between energy and the environment, an examination of past policy is useful, including an examination of those factors that motivated the policies, and their effects on the supply of energy.

INTELLECTUAL ANTECEDENTS

Even though they did not focus directly on energy, two authors have had a very strong influence on the way the United States and other western nations think about energy: Thomas Malthus (1798) and Harold Hotelling (1931). Malthus, whose work is discussed in more detail in Chapters 17 and 18, argued that scarcity is inevitable because population grows to exhaust its resource endowment. Hotelling (whose work forms the conceptual basis for the discussion of dynamic efficiency in Chapter 2) argued that the invisible hand of the market would optimally allocate exhaustible resources and prevent shortages since the market price of a resource such as oil reflects both

TABLE 8.1

U.S. CRITERIA AIR POLLUTANT EMISSIONS AND THEIR SOURCES, IN 1990
(MEASURED IN MILLIONS OF METRIC TONS PER YEAR)

	PARTICULATES	SO_x	CO	NO_x	VOCS	LEAD
transportation	1.5	0.9	37.6	7.5	6.4	0.0022
fuel combustion	1.7	17.1	7.5	11.2	0.9	0.0005
industrial processes	2.8	3.1	4.7	0.6	8.1	0.0022
solid waste	0.3	0.0	1.7	0.1	0.6	0.0022
miscellaneous	1.2	0.0	8.6	0.3	2.7	0.0
TOTAL	7.5	21.1	60.1	19.7	18.7	0.0071

SOURCE: U.S. Environmental Protection Agency, *Emission Levels for Six Pollutants by Source*, #cc 91-600760.

its current value and its future value. The debate about whether markets adequately address future supply continues to rage and is a motivating factor behind much of our energy policy, particularly during the so-called energy crises of the 1970s.

Although the mechanics of dynamic market efficiency are discussed in Chapter 2 (with a numerical example in Appendix 2B), it is worthwhile to summarize the intuition behind these mechanics, which are based on Hotelling's work. The fundamental proposition is that an oil producer (or producer of any other exhaustible resource) must be indifferent between selling a barrel of oil today and waiting for some future time to sell it. If the producer expects to make more money by selling later, the producer will wait. Other producers will behave similarly, and this will increase price in the present and reduce expected price in the future until all producers are indifferent between selling today and selling at some point in the future. As Hotelling demonstrates, today's price includes *user cost*, which is the opportunity cost of not having the oil available at other periods in the future. Since user cost is a component of market price, and since user cost is determined by the present demand, present supply, future demand, and future supply of oil, the market should efficiently allocate oil over the course of time. As discussed in Chapter 2, dynamic efficiency requires that the price at any point in time be equal to marginal extraction cost plus marginal user cost. As detailed both in Chapter 2 and in the following paragraph, market forces ensure that this condition is met.

If future demand is perceived to be increasing or future supply is perceived to be decreasing, this will increase present user cost. Current price will increase, which will reduce the quantity demanded, leaving more oil for the future. The higher current price will also encourage substitution of other fuels for oil, as well as increased exploration for oil, investment in increased energy efficiency, and investment in technological improvements in oil extraction. In this depiction of market processes, completely running out of the resource (absolute scarcity) never occurs because increased price causes adjustments that mitigate scarcity. Scarcity is felt through increased price, which not only reduces the quantity demanded, but causes other adjustments that increase the supply of the resource and its substitutes.

This view of the way the market works implies that there is no need to have an energy policy, as the market works efficiently and will prevent a shortage of oil or other fuels. Of course, in his theoretical work, Hotelling assumed a market that met all the usual characteristics of a perfectly functioning market, including perfect information, no externalities, perfect competition, no public goods, and no monopoly or oligopoly power. Although some "free market" advocates interpret him in this fashion, Hotelling never explicitly stated nor implied that consideration of these market failures was unnecessary.

In contrast, Malthus believed in the concept of absolute scarcity, which suggests that resources are used at an increasing rate until they are completely exhausted. Malthus argued that the population grows faster than the food supply, so the food supply not only acts as a constraint on growth but is always insufficient to allow the development of a surplus. People are always at the limit of the food supply and on the brink of starvation.

Although Malthus's original arguments were couched solely in terms of land and food resources, his arguments have been extended by the "neo-Malthusians"[2] in terms of general resources and environmental quality. Neo-Malthusians argue that the growth of the economy and population will generate a dependence on resources that will eventually exceed capacity, so that both the economy and population face inevitable collapse. Although fewer people share this apocalyptic view of scarcity, many noneconomists believe in absolute scarcity and that one day we will simply run out of oil and other resources, with no effective substitutes available.

Although these two divergent views seemed to represent dominant paradigms of intellectual thought, U.S. energy policy did not follow either theory.[3] In the period before 1970, a state agency, the Texas Railroad Commission, controlled most of U.S. oil production through a set of Texas state regulations that defined drilling rights to underground pools that lay below lands owned by multiple owners. The regulations were ostensibly to protect against a common property externality, which is generated because the rate at which oil is removed from the pool determines how much can eventually be removed. The slower the retrieval rate, the greater the total output over time. Without regulations, multiple producers operating in a common oil field will race each other to remove the oil, which will lower the total amount

[2]See Meadows, et al. (1972), for example.

[3]The following discussion of the history of U.S. energy policy draws heavily on Alfred Marcus, *Controversial Views in U.S. Energy Policy* (New York: Sage Publications, 1992).

that can be removed. Although this was the stated goal of regulation, the regulations also served to restrict present production and increase present price, creating short-term monopoly profits for oil producers.

The federal government also implemented several policies which operated mainly through favorable tax treatment. Intangible drilling expenses could be deducted as expenses, and an oil depletion allowance permitted oil deposits to be treated as depreciable assets. In addition, from 1959 to 1973, oil imports were restricted, with the goal of promoting national security. This restriction also served to increase producer profits and to reduce domestic oil reserves.

Natural gas policy was dictated by the Natural Gas Act of 1938, the purpose of which was to regulate natural gas transportation rates to keep them high enough to justify the large capital expenses of natural gas pipelines. However, price regulation was eventually extended past the transportation stage to regulate prices at the wellhead. This action seems to be part of an overall energy policy to keep prices low. As one might expect, however, this effort to keep prices low resulted in a reduction in production, a reduction in exploration activity, and a shortage of natural gas in the 1970s.

Government policy seemed to be oriented to keeping prices low, and production and consumption high, probably because of a perceived link between cheap energy and economic growth. However, the shortsightedness of these policies became apparent in the 1970s. This period saw declining U.S. oil production, an increase in oil imports, a shortage of natural gas, disruption of foreign oil supplies, and a drastic increase in energy prices. Policies designed to keep energy prices low combined with changing external factors to make prices rise through the 1970s, rather than stay low. Marcus (1992, p. 46) lists three reasons price controls received popular support:

1. The existence of a widely held belief that high energy prices lead to inflation.
2. There was greater concern with equity than with efficiency.
3. There was a perceived need to protect people from exploitation by oil companies who were believed to be earning windfall profits.

This concern with oil company profits is understandable, since user cost is a component of the price of oil and it creates a gap between marginal extraction costs and price. People who have not been exposed to the concept of user cost will perceive a disparity between price and marginal extraction cost as monopoly profits. While monopoly profits could be present, user costs alone are capable of generating such a disparity.

THE SIGNIFICANCE OF OPEC

The Organization of Petroleum Exporting Countries (OPEC) is a cartel of oil producing countries, formed in 1960 (by Iran, Iraq, Kuwait, and Saudi Arabia) as a way to counteract the economic power of the multinational oil companies. OPEC reached the zenith of its economic power in 1973 when oil prices quadrupled, and an oil embargo was imposed on the United States and other countries that supported Israel during the Yom Kippur War.

A cartel is an organization of producers who agree to act in concert as a monopolist and restrict output in order to raise prices and generate monopoly profits. All producers need not be members of the cartel; however, the cartel must be large enough so that its quantity decisions affect market price. It is interesting to note that noncartel producers will benefit from the higher market price even though they are not part of the effort to restrict output.

There is little historical experience with cartels. Cartels of firms within a country are generally illegal under antitrust law (this is certainly the case in the United States). Cartels of countries that export raw materials (such as copper or cocoa) have been successful at times in raising the price of the raw material.

One factor that weighs heavily against the long-term viability of cartels is that cartel members have powerful incentives to cheat. Remember that the cartel raises prices by restricting output among its members. Each member lowers output, and that raises market price and makes all cartel members better off. However, if an individual member were to secretly raise its output, it could take advantage of the higher prices on a greater volume of output. Thus, each individual member has an incentive to cheat and produce more, but if too many cartel members do this, it will lower price and eliminate monopoly profits. It is this incentive to cheat that has caused the collapse (and sometimes even prevented the establishment) of most cartels.

OPEC, however, was remarkably effective in raising prices in the 1970s. One of the reasons it was so effective is that there were other commonalities in addition to the goal of higher price. Geographical proximity, a common religion, and a united front against Israel served to unite the Middle Eastern OPEC members. (In the early 1970s, Venezuela was the only non-Islamic, non-Middle Eastern member of OPEC.)

Although OPEC was remarkably effective in raising prices in the mid- and late 1970s, oil prices have declined through the 1980s and 1990s. It is interesting that this decline was not foreseen in the 1970s, and policy was formulated as if energy prices would remain high and OPEC would continue to be an effective cartel.

OPEC lost effectiveness in maintaining high prices for several reasons. First, the cartel lost market power as non-OPEC sources of oil came on line in Mexico, the North Sea (Great Britain and Norway), and Alaska. In order to see how these new sources might affect oil prices, we can utilize an oligopoly model called the dominant firm model.

The dominant firm model is based on the idea that the dominant firm (in this case the oil cartel) views output of the rest of the world as beyond its control and so it attempts to set a price and then supplies the demand not met by the rest of the world. If the dominant firm is the low cost producer (and it must be in order to be the dominant firm), then it can satisfy the

FIGURE 8.1

THE DOMINANT FIRM MODEL

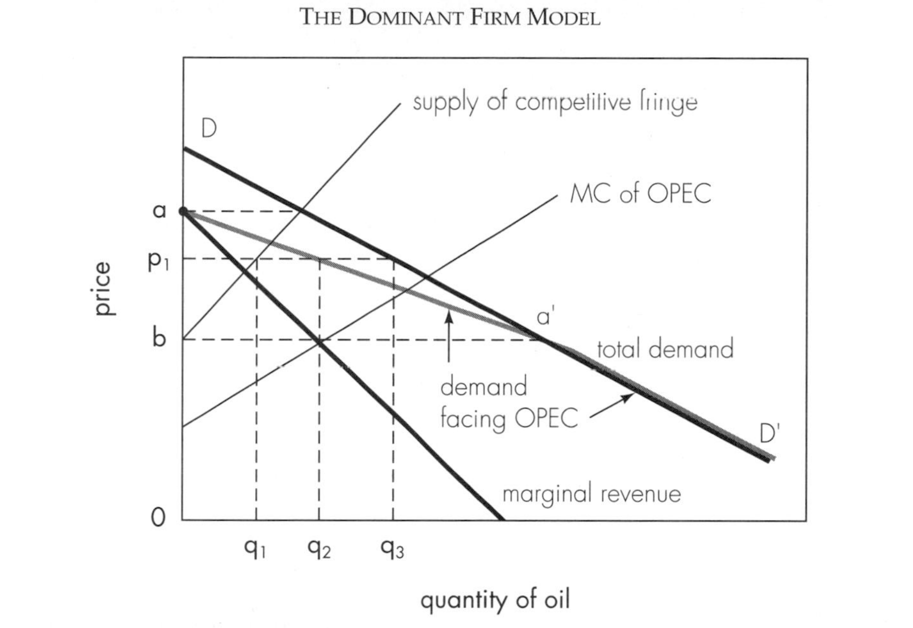

unmet demand and earn some monopoly profits on its production. However, as the dominant firm sets price higher in an attempt to earn more profits, two market reactions serve to limit its ability to earn profits. First, as price increases consumers will buy less oil, so that even though price is higher, less oil is being sold. Second, as price increases, the share of the output produced by the rest of the world (called the competitive fringe) increases so the dominant firm will sell an even smaller amount of oil. (*Note:* A rather technical microeconomic note is that economists do not talk about supply curves for monopolists; they focus on the marginal cost curve instead. In Figure 8.1, the marginal cost curve of the monopolist is labelled "MC of OPEC.")

The dominant firm model can be explained with the aid of the graph of Figure 8.1. In this graph, the supply curve of the competitive fringe is above the marginal cost (MC) curve of OPEC. The total demand curve for oil is represented by the demand curve D-D′. However, if OPEC is taking the supply of the competitive fringe as given, then OPEC views itself as facing a demand curve that is lower than the total market demand curve. The demand curve that is relevant for OPEC is one that subtracts the quantity supplied by the competitive fringe from total demand, giving the residual demand left available for OPEC (labelled "demand facing OPEC"). Notice that the vertical intercept of this residual demand curve is point a, as at prices

SOURCE: *Energy Statistics Sourcebook*, Sixth Edition, 1992. PennWell Publishing Co.

TABLE 8.2

MOTOR GASOLINE AND RESIDENTIAL HEATING OIL PRICES
(MEASURED IN CENTS PER GALLON, 1982–84 CONSTANT DOLLARS)

YEAR	GASOLINE	RESIDENTIAL HEATING OIL
1920	148.9	Not available (NA)
1925	126.6	NA
1930	119.5	NA
1935	137.3	NA
1940	131.3	NA
1945	113.9	NA
1950	111.2	NA
1955	108.6	55.9
1960	105.1	50.6
1965	98.9	50.8
1970	91.9	47.6
1975	105.3	70.0
1980	144.5	118.3
1985	103.6	97.9
1990	87.6	81.3

greater than or equal to a, the competitive fringe is capable of supplying all the demand. However, as price creeps lower than a, not all the demand can be supplied by the competitive fringe, so there is a residual demand available to OPEC. At prices lower than b, the competitive fringe cannot supply any demand, so the total demand curve becomes OPEC's demand curve at point a′. Therefore, OPEC views itself as facing the kinked demand curve, a-a′-D′.

Given this demand curve, the marginal revenue function can be defined and the profit maximizing price of p_1 can be determined. This is the price corresponding to an OPEC output level (q_2) that equates MC and MR. Notice that at this price, total quantity demanded is q_3, with q_1 units produced by the competitive fringe and q_2 units produced by OPEC.

As can be seen in this model, the greater the size of the competitive fringe (which would be reflected as a shift to the right of the competitive supply function), the lower will be the world price of oil. During the 1980s and 1990s, the competitive fringe (non-OPEC producers) has grown larger, and this has been reflected in the downward pressure on prices. Table 8.2 lists U.S. prices for fuel oil and gasoline, which shows how energy prices have fallen since 1980.

Another reason prices may have fallen is because Saudi Arabia, the dominant producer within OPEC, has vastly different incentives from many other OPEC members. For a variety of reasons, Saudi Arabia would prefer to see lower prices than other OPEC countries would like.

The first reason has to do with relative size of reserves. Saudi Arabia has enough oil reserves to continue to pump at present rates for perhaps as long as 100 years. Other OPEC countries such as Libya, Nigeria, and Iraq have reserves sufficient to last 1 or 2 decades. Consequently, Saudi Arabia has much more to lose from the development of alternative energy technologies, such as solar energy, which would reduce the future demand for oil and lower the future revenue that could be obtained by producing oil. Since higher current prices generate greater current motivation to invest in research and development of alternative energy technologies, Saudi Arabia would not necessarily advocate a price that maximized current monopoly profits. However, it would not want to forego all monopoly profits. Consequently, Saudi Arabia would advocate a price that is high enough to generate some monopoly profits, but low enough to make it difficult for alternative technologies to become established. This type of oligopoly pricing strategy is called limit pricing, since the lower price limits the ability of alternatives to develop and become established. Another reason, related to limit pricing, is that Saudi Arabia has less need for current revenue. Saudi Arabia is relatively sparsely populated, with a commensurately small economy. Under such circumstances, it is difficult for Saudi Arabia to invest all its oil revenues in its domestic economy because the economy is incapable of absorbing all the investment. Saudi Arabia can, and does, invest in other economies (such as in the United States). However, to a certain extent, it may wish to invest by protecting the future value of its oil reserves, rather than investing in foreign economies.

In contrast, countries such as Nigeria, Venezuela, and Indonesia have large populations and current pressing needs to develop their economies. They would advocate high prices now, especially because their reserves are expected to be exhausted much more quickly than those of Saudi Arabia. Of course, countries with large military expenditures such as Iran, Iraq, and Libya also would advocate higher current prices.

In summary, although OPEC has been successful in raising oil prices, it does not enjoy complete monopoly power. While monopoly profits are still being earned, many factors have served to mitigate price. Nonetheless, the existence of OPEC remains a focal point of U.S. energy policy.

OPEC and U.S. Energy Policy

In 1973, the fourth major war (the Yom Kippur War) erupted between Israel and the surrounding Arab countries. As a result of U.S. and other Western support for Israel, the Arab petroleum exporting countries imposed an oil embargo that resulted in a quadrupling of oil prices. This caused a major disruption of the U.S. economy and prompted extreme fears of an economic future dominated by high energy prices, inflation, and a shortage of oil. At the same time, the price controls on natural gas were causing shortages of natural gas. The energy future of the United States looked bleak, and a series

SOURCE: Alfred A. Marcus, *Controversial Issues in Energy Policy,* 1992, p. 40, Reprinted by permission of Sage Publications, Inc.

TABLE 8.3

THE EVOLUTION OF FEDERAL ENERGY POLICY AFTER THE 1973 ARAB OIL EMBARGO

1973	The Emergency Petroleum Allocation Act
1974	Project Independence
	Federal Non-nuclear Research and Development Act
	Energy Supply and Coordination Act
1975	Energy Policy and Conservation Act
1977	Creation of Department of Energy (DOE)
	Federal Mine Safety and Health Amendment Acts
	Clean Air Act Amendments
	Surface Mining Control and Reclamation Act (SMCR)
1978	Power Plant and Industrial Fuel Use Act
	Natural Gas Policy Act (NGPA)
	Public Utilities Regulatory Policy Act (PURPA)
	Gas Guzzler Tax
	Building Energy Performance Standards
1980	Decontrol of Petroleum Prices

of laws were passed during the Nixon, Ford, and Carter administrations to deal with the perceived problems of a world oil market dominated by the Organization of Petroleum Exporting Countries (OPEC) and the domestic shortage of natural gas. These legislative acts are summarized by Marcus (1992) and illustrated in Table 8.3.

The initial effort was to try to keep oil and energy prices at previous low levels. The Emergency Petroleum Allocation Act extended oil price controls, which had begun earlier as part of the wage and price controls that were undertaken by the Nixon administration to try to stem inflation. Of course, the attempt to keep domestic prices artificially low only served to stifle domestic production and increase U.S. dependence on oil imports. In fact, the price control efforts directly conflicted with the energy independence objectives of Project Independence. Before analyzing why so much attention was placed on keeping price at a low level, the costs of energy dependence will be analyzed.

Dependence on foreign producers of oil is not necessarily bad. In fact, if the United States and other oil importing countries were dependent upon a large number of countries for oil, foreign dependence would not be an issue. National security considerations would not be important, and prices would be lower because the large number of countries would not have monopoly power. Additionally, a high proportion of the money that flowed out of the United States to purchase oil would probably flow back to the United States in the form of increased purchases of U.S. products by these countries. However, the United States and other oil importing countries are dependent on a relatively small number of organized countries in an area of the world that has been politically unstable and militarily vulnerable. Consequently, the question of foreign dependence dominates the political and public discussion of energy policy.

Greene and Leiby (1993) analyze the question of the costs of foreign dependence in an interesting and insightful fashion by examining the cost of dependence on a foreign monopoly for oil supplies. It is the simultaneity of the monopoly problem and the foreign dependence that generates the costs for U.S. society.

Greene and Leiby separate the costs of dependence on foreign monopoly into three broad components and measure them for the period 1972 to 1991. These costs are

1. the transfer of U.S. wealth to foreign producers,
2. macroeconomic costs, and
3. political and military costs.

Greene and Leiby argue that in the absence of an oil consumers' cartel, all of the extra costs of oil imports are a loss to the U.S. economy. The reason is very simple. By raising prices through the restriction of output, the oil producers' cartel transfers consumers' surplus into monopoly profit.

This loss is illustrated in Figure 8.2, where the marginal cost function is depicted horizontally for expositional simplicity. In this figure, p_1 represents the price associated with a competitive market structure for oil, and q_1 represents the competitive output. If the market structure is monopoly, then p_2 is the price and q_2 is the corresponding output. In the competitive situation, consumers' surplus is equal to the total shaded area. In the monopoly situation, consumers' surplus is reduced to the area of triangle abp_2, with part of the loss a conversion to monopoly profits (rectangle p_2bdp_1) and with a deadweight loss equal to the area of triangle cbd.[4]

In the typical analysis of monopoly, only the deadweight loss is seen as a cost, as the monopoly profits represent a transfer from one segment of society to another.[5] However, if one is defining society from a U.S. perspective, the transformation of U.S. consumers' surplus to profits for foreign producers represents a loss.

---

[4]Deadweight loss refers to the fact that the benefits of triangle cbd simply disappear and do not become someone else's gain.

[5]The monopoly profits are not seen as a loss from an efficiency perspective. However, they are always a subject of concern from an equity perspective.

FIGURE 8.2

CONVERSION OF CONSUMER SURPLUS TO MONOPOLY PROFIT

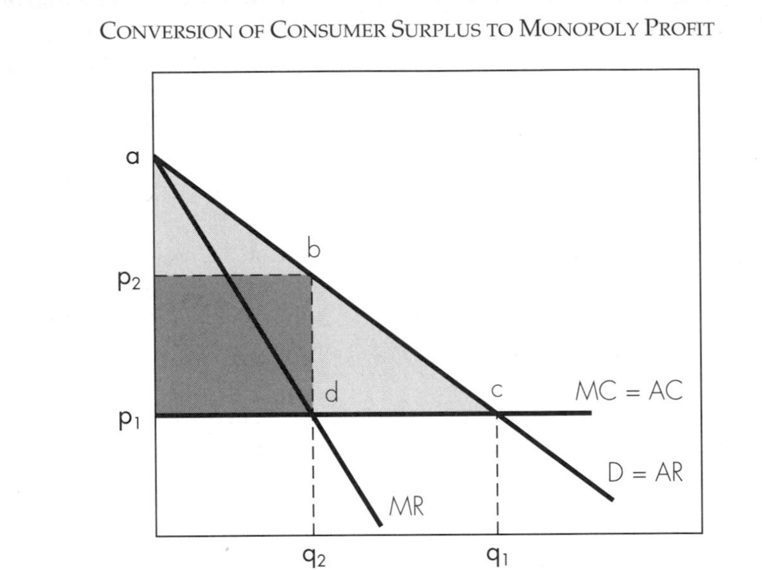

Macroeconomic losses occur when sudden price increases or shortages of oil (from an embargo, for example) shock the domestic economy and lead to inflation, losses in GDP, losses in employment, and other reductions in the health of the economy. These losses arise from the fact that a factor of production (oil) has become relatively more scarce as a result of the monopoly. With less of a factor of production, it is not possible for GDP to be as high as it was when the factor is more plentiful.

In addition to the costs of price shocks, there is a cost of mitigating price shocks. After the oil shocks of the 1970s, the United States developed the Strategic Petroleum Reserve, which buys oil and stores it for release during shortages. These releases are intended to make the shortage of oil less severe and, therefore, reduce the upward pressure on the price of oil. President Bush released reserves from the Strategic Petroleum Reserve during the 1990 Gulf crisis to help prevent potential increases in the price of oil.

Finally, there are the military and political costs of depending on oil imports from a strategically sensitive and militarily vulnerable area of the world. In the Persian Gulf region, oil shipping lanes are very narrow and easy to disrupt. Additionally, there is political instability in the area and there were concerns of possible Soviet expansion into the area in the 1970s and 1980s.

As Greene and Lieby (1993) point out, it is very difficult to ascribe military readiness costs and the actual costs of the Persian Gulf War to dependence on foreign oil. The reason is that military readiness costs are incurred to cover many possible problems, not just problems in the oil producing areas of the world. It is even more difficult to assign the costs of the Gulf War to U.S. dependence on foreign oil. Even if there was no U.S. dependence, the United States might have participated in this war to help oil-dependent allies or for other political reasons (such as to stop the proliferation of nuclear and chemical weapons). For this reason, Greene and Lieby assume that only one-third to one-half of the actual readiness and war costs can be attributed to oil dependence.

Several important patterns can be noted in Table 8.4. First, the total social costs of dependence on a foreign monopoly for oil increase through the 1970s and decline through the 1980s. This is consistent with the previous discussion that suggested a variety of market factors have weakened OPEC's monopoly power. Second, it should be observed that even though OPEC's market power has weakened, the costs of dependence on foreign oil are not trivial. The $93 billion of total costs in 1990 can be compared to Gross Domestic Product, which was $5,513 billion in 1990.

ENERGY AND THE ENVIRONMENT

As discussed earlier, the production and consumption of energy is responsible for many of our environmental problems. In Chapter 6, the association between energy use and global warming was explored. In Chapter 7, the role of energy in acid deposition production was examined. Energy use, particularly fossil fuel use, is responsible for many of our pollution problems. As mentioned at the beginning of this chapter, energy production and consumption are responsible for most of the generation of criteria air pollutants. The major impacts of these pollutants are summarized in Table 8.5.

REGULATIONS ON STATIONARY SOURCES OF POLLUTION

Stationary sources of criteria air pollutants (smokestacks from factories and buildings) have been regulated under the Clean Air Act of 1972 and its amendments (1977 and 1990). The basic thrust of these regulations has been command and control policies. The federal government established national ambient standards on the concentrations of each pollutant that would be allowed. States were then expected to develop implementation plans for meeting the ambient standards. Every state developed a set of plans based on command and control techniques, which unnecessarily increased the cost of meeting the ambient standards (see Chapter 3).

For example, the original regulations controlled air pollution in a smokestack-by-smokestack fashion. A firm that could reduce its abatement costs by

TABLE 8.4

THE SOCIAL COSTS TO THE U.S. OF MONOPOLIZATION OF THE WORLD OIL MARKET 1972–1991[a] (BILLIONS OF 1990 DOLLARS)

YEAR	WEALTH TRANSFER TO OPEC	COSTS OF STRATEGIC PETROLEUM RESERVE	TOTAL GNP LOSS	MILITARY COSTS[b]	TOTAL COSTS
1972	0	0	0	14.2	14.2
1973	3	0	17	14.2	34.2
1974	35	0	189	14.2	238.2
1975	35	0	177	14.2	226.2
1976	37	0.413	157	14.2	208.6
1977	46	0.589	167	14.2	227.8
1978	39	4.182	141	14.2	198.4
1979	60	3.954	219	14.2	297.2
1980	76	(2.63)	321	14.2	408.57
1981	64	4.382	301	14.2	383.5
1982	42	5.096	204	14.2	265.3
1983	33	3.046	147	14.2	197.3
1984	34	1.064	134	14.2	183.3
1985	27	3.299	103	14.2	147.5
1986	11	0.141	53	14.2	78.3
1987	18	0.194	61	14.2	93.4
1988	12	0.793	42	14.2	69.0
1989	19	0.546	49	14.2	82.7
1990	25	0.826	53	14.2	93.0
1991	15	NA	37	14.2	66.2[c]

SOURCE: David L. Green and Paul N. Leiby, *The Social Costs to the U.S. of Monopolization of the World Oil Market 1972–1991,* Oak Ridge National Laboratory Report number 6744, Oak Ridge, TN, 1993.

[a]Greene and Leiby report many sets of results based on different assumptions about interest rates, rate of growth of the real price of oil, macroeconomic parameters, and so on in order to show the sensitivity of these costs to other economic variables. The results that are displayed in this table have been chosen to give the student an appreciation of the magnitude of the costs involved and should not be interpreted as Greene and Leiby's best estimate.

[b]Military costs are reported as a total for the 1980–1991 period. Costs are annualized over this period by simple averaging. Average annual pre-1980 costs are assumed to equal the same value.

[c]This total does not include a 1991 estimate of SPR costs, as this was not available.

switching production from one facility to another could not do this if the pollution emanating from one facility (or smokestack) increased, even if total pollution from the firm remained constant or decreased. In other words, these regulations did not allow firms to make cost-minimizing adjustments

TABLE 8.5

MAJOR CONSEQUENCES ASSOCIATED WITH CRITERIA AIR POLLUTANTS

POLLUTANT	IMPACT
Sulfur Dioxide (SO_2)	Major contributor to acid rain problem; sulfate particles may lead to respiratory ailments and damage buildings and other materials.
Carbon Monoxide (CO)	Tropospheric ozone precursor; health impacts include interference with ability of blood to circulate oxygen within the body.
Volatile Organic Compounds (VOCs)	Contributes to formation of tropospheric ozone and smog through reactions with NO_2 and SO_2 (see Chapter 7); causes eye irritation, exacerbates respiratory problems, suspected carcinogen.
Nitrogen Oxides (NO_x)	Major contributor to acid rain problem, smog, and tropospheric ozone; causes eye irritation, exacerbates asthma (especially in small children), exacerbates other respiratory ailments.
Particulates	Leads to respiratory problems; some particulates are carcinogens.
Lead (Pb)	Lead, like other heavy metals, is extremely toxic. It damages all the body's major systems including the nervous, reproductive, and circulatory systems. Leads to permanent learning disabilities in small children.
Ozone	Formed when SO_2 or NO_2 reacts with VOCs; damages human health and the health of other organisms, especially agricultural crops and forests.

within their own facilities.[6] To rectify this situation, a modification to the regulations, called "pollution bubbles," was developed. In the bubble concept, each firm is treated as if a glass bubble encased the entire firm's operations, and all pollution from every smokestack within the bubble exited from a single imaginary smokestack. Only the imaginary bubble smokestack would be regulated, and firms could make any adjustments within the bubble as long as the pollution that left the bubble conformed to emission limitations. These command and control techniques were not particularly effective in reducing pollution either, as most urban areas do not meet the ambient standards for one or more of the criteria pollutants.

The Southern California experience with command and control techniques is particularly enlightening. Southern California, primarily because of its heavy reliance on automobiles, could not meet the federal standards, despite requiring new sources of pollution to use the best available pollution control technology. Consequently, it was declared a nonattainment area, and

[6]The minimization of abatement costs within a facility is generated by the same condition that ensures cost-minimization in general. This condition is that the level of emissions from each source of pollution should be set so that the marginal abatement costs are equal across all sources of emissions.

no new sources of pollution would be allowed in the area. This regulation meant that there could be no growth in industry in the area, which implied an economic future of stagnation. In order to deal with this problem, a modification to the Clean Air Act was implemented that allowed new sources of pollution in nonattainment areas, provided that they induced existing polluters to reduce pollution by an amount equal to 150 percent of the pollution that would be generated by the new source. This "offset" system is very similar to a marketable pollution permit system. However, it is not as efficient as a true marketable pollution permit system, because it does not allow trades among existing polluters. This allowance is important because trades among all polluters could lower the cost of meeting environmental quality standards. Southern California has had particular difficulty in meeting ambient air quality standards using command and control techniques and is currently exploring the use of marketable pollution permits for both stationary and mobile sources of pollution. Other areas of the United States (such as the Chicago area) are considering economic incentives, primarily marketable pollution permit systems.

REGULATIONS ON MOBILE SOURCES OF POLLUTION

Mobile sources of pollution (vehicles) have also been regulated in a command and control fashion. The primary instrument used is that of specifying abatement control devices for vehicles. The regulations specify a single national standard for automobile emissions, although California is allowed to employ stricter standards.

All automobiles are required to employ catalytic converters. The platinum in the converter serves as a catalyst that lowers the ignition temperature of many of the unburned hydrocarbons and other pollutants in gasoline. In other words, it facilitates the more complete burning of the gasoline. While this may seem to be an effective way of dealing with pollution, there are many problems associated with it. First, it controls all areas of the country in the same fashion, so that the same device is used in Barlow Point, Alaska, as in New York City. Obviously, an additional unit of emission has a much more severe impact in New York City than in Alaska. If Alaska and New York are treated uniformly, then either New York is being undercontrolled or Alaska is being overcontrolled, or both.

Another problem associated with this technology-based approach to pollution control is that it does not give additional incentives that would reduce pollution. It does not give people an incentive to drive less, to maintain their cars (a properly tuned and maintained car pollutes less), to take mass transit, or to purchase a car that exceeds the standards. If people have no incentive to purchase cars that pollute less, then manufacturers have no incentive to design and manufacture less-polluting cars.

Air pollution from automobiles is also indirectly controlled by the CAFE (Corporate Automobile Fuel Efficiency) standards, which specify the average miles per gallon that must be achieved by each automobile manufacturer. This restriction requires that the average fuel efficiency (miles per gallon or MPG) across all cars produced by a manufacturer achieve a minimum level, which increases over time. Higher MPG means that less gasoline is burned per mile driven, which means that there is less pollution per mile. This standard, however, does nothing to reduce the miles that are driven, and it may even increase the total number of miles driven, since more efficient cars are cheaper to drive.

Mills and White (1978) suggest a system that would give more of the appropriate incentives. Cars would be taxed based on the total amount of pollution that they generate each year. This amount would be computed by an annual diagnostic test, in which the tailpipe emissions are measured to compute the pollution per mile, which is then multiplied by the total amount of miles (read from the odometer)[7] driven. There could be a base federal tax, and then local jurisdictions would be free to establish additional taxes based on their need to further reduce pollution.

Under such a system, drivers would have the choice of buying cars that emit different levels of pollution. Drivers who drive many miles per year or live in high pollution (and high tax) areas would have an incentive to pay more money to buy a less-polluting car. Drivers who drive only occasionally or live in low pollution areas (low tax areas) would minimize their costs by purchasing less expensive but higher-polluting cars. People would have an incentive to maintain their cars, because that would reduce their pollution per mile and their pollution tax. Automobile manufacturers would have an incentive to develop less-polluting cars because, under the tax system, consumers would be willing to pay to purchase these less-polluting cars. Finally, the cost associated with driving would encourage people to live closer to work, to car pool, to use mass transit, and to make other pollution reducing adjustments.

ENERGY POLICY AND THE ENVIRONMENT

As discussed at the beginning of this chapter, energy policy has been focused on maintaining low energy prices and abundant supplies of oil. Energy policy was conducted independently of (and very often in conflict with) environmental policy throughout the 1970s and the 1980s. For example, in response to a perceived shortage of oil and natural gas, the Carter administration pursued an aggressive coal policy, substituting coal for oil and natural gas in many applications. Since coal is the dirtiest fuel, this energy policy had negative environmental consequences.

---

[7]Obviously, it would be necessary to develop odometers that were more tamper resistant and create additional penalties for tampering with odometers.

OIL SPILLS AND THE OPTIMAL LEVEL OF SAFETY

In 1989, the *Exxon Valdez* ran aground in the Prince William Sound area off the coast of Alaska, causing a spill of millions of gallons of oil which killed fish, birds, marine mammals, and other organisms, and fouled the shore. This widely publicized event sparked a debate on whether we should ship oil in this area and continue the threat to the environment or halt the shipment of Alaskan oil in order to protect the environment. As with most highly emotional political debates, this one focused on two polar extremes and did not debate the critical issue.

The critical issue in this case is not to decide a thumbs up or thumbs down on Alaskan oil, but to determine the optimal level of safety in shipping Alaskan oil. There are many actions that can be undertaken to reduce the chances of a spill taking place, to reduce the magnitude of a spill if one does take place, and to reduce the damages associated with a spill of a given size. Actions that would lessen the chances of a spill taking place include placing computerized collision avoidance systems on oil tankers (and obstacles such as shoals and islands), requiring redundancy among key personal (so that if the captain becomes incapacitated there is an equally skilled person in position to take over), and restricting other shipping when a tanker is in the shipping lanes. Actions that would lessen the amount of oil spilled if an accident did take place would include requiring double-hulled tankers and multiple oil compartments (like the cells of a car battery) in the tanker. These features reduce the amount of oil released if the hull is punctured. Actions that would lessen the damages of a spill that occurred would include requiring standby cleanup and containment equipment and crews that could be called into action the moment a spill occurs.

An important question for policy makers is how to ensure that appropriate safety measures are taken. Two options are available: economic incentives and command and control techniques. The system of economic incentives that would be most appropriate in this case is the development of liability rules. Liability rules would make oil transportation companies responsible to pay the damages for spills that they cause. Under CERCLA, the federal government, state governments, local governments, and Native American nations have the legal right to sue for damages to public resources from oil and chemical spills. In order for liability systems to be able to generate the optimal level of safety, court-awarded damages must be equal to the actual level of damages that are created by the spill. In addition, there must be a high degree of certainty that those who create the damage will be forced to pay for the damages, and that out-of-court settlements will not reduce the level of awarded damages to a level significantly below actual damages.

THE NATIONAL ENERGY STRATEGY AND THE ENVIRONMENT

The National Energy Strategy, conceived in the latter years of the Bush administration, paid more attention to environmental concerns. Table 8.6 summarizes some of the key environmental provisions of the National Energy Strategy.

Eliminating the Disparity between Marginal Social Cost and Marginal Private Cost. While these programs can help to mitigate the environmental impacts associated with energy use, they do not address the fundamental problem underlying the environmental impacts associated with energy use. This

TABLE 8.6

GOALS AND APPROACHES OF THE NATIONAL ENERGY STRATEGY—
ENERGY AND THE QUALITY OF AIR, LAND, AND WATER

GOAL	APPROACH
Improve environmental quality by National Energy Strategy actions	Increase energy efficiency and reduce demand to lower future emissions. Stimulate use of natural gas, renewable energy, nuclear power, alternative transportation, and clean coal technology to reduce air, land, and water quality impacts.
Increase flexibility in meeting environmental regulations	Improve analysis of energy impacts as a part of rule making. Expand flexibility and use of market mechanisms, such as emissions trading, to reduce compliance costs. Amend legislation to allow more flexibility in control practices while maintaining environmental protection. Provide more complete analysis of the environmental impacts of competing technologies (total fuel cycle analysis); ensure that environmental concerns with emerging energy technologies are addressed in advance.
Maintain environmental concerns in energy facility siting and licensing decisions while reducing delays and overlaps	Identify elements of the unnecessary delay and restriction; propose model programs to guide regulators. Analyze both the energy resource potential and the environmental impacts that may result from exploration and development on private lands.
Achieve and maintain the National Ambient Air Quality Standards for carbon monoxide and ozone	Implement the Clean Air Act to reduce tailpipe emissions to meet air quality standards. Reduce oil use in transportation by encouraging cleaner fuels, greater vehicle efficiency, and use of alternative vehicles.
Reduce threat of "acid rain" through cost-effective control strategies	Reduce emissions of sulfur dioxide and nitrogen oxides through implementing the Clean Air Act Amendments of 1990. Increase energy efficiency and develop new lower polluting technologies to further reduce emissions at lower cost.
Address other air quality issues	Study toxic air emissions from utilities; regulate emissions from transportation and industry. Assess radionuclide regulatory practices by state. Modify "new source review" policy for existing power plants.
Protect and improve water quality	Improve tanker safety, make better preparation for potential accidents, and provide stronger incentives for transporters to prevent accidents. Expand the use of market approaches and energy concerns in reauthorizing the Clean Water Act.
Protect human health and the environment and reduce costs through effective waste management	Propose amendments to the Resource Conservation and Recovery Act to increase land use and groundwater protection and reduce compliance costs by expanding the use of market mechanisms.
Reduce energy related waste generation	Remove regulatory disincentives to waste minimization; promote research and outreach activities.

SOURCE: *National Energy Strategy, First Edition, 1991/1992,* Department of Energy, Washington, D.C., 1991, pp. 145–146.

problem is that the price of fossil fuels does not include the social costs of the environmental impacts. There is a fundamental disparity between the private costs of energy use and the social costs. As long as this disparity exists, there will be an inefficiently high level of environmental degradation associated with energy use.

This disparity can be eliminated in a variety of fashions. As the theory developed in Chapter 3 indicates, the least costly way to do this in most cases would be through a comprehensive series of marketable pollution permits or through a system of per-unit pollution taxes. A combination of taxes and permits might be the best system, with a permit system used for large stationary sources of pollution (such as electric utilities, large manufacturing facilities, and refineries), and taxes used for mobile sources of pollution (cars and trucks) and small stationary sources of pollution (furnaces that heat homes and small commercial establishments). Taxes are suggested for small polluters since the transactions costs resulting from requiring each small polluter to participate in the permit market would be extremely high.

Since externalities are also present in the production of energy, these externalities would also need to be addressed. Liability and bonding systems would be effective in ensuring that spills, acid mine drainage, and alteration of the landscape and habitat were reduced to appropriate levels.

Fuel Taxes. If it was not politically or administratively feasible to develop a system of pollution taxes or permits, a second best solution could be to add a tax to the price of fuel based on the average amount of pollution of the fuel. For example, natural gas generates less pollution per unit of heat than oil. Oil generates less pollution per unit of heat than coal. Consequently, the per British Thermal Unit (BTU)[8] tax associated with coal would be higher than that associated with oil, and the tax for oil would be higher than the tax for natural gas. This type of fuel tax would not be quite as desirable as pollution taxes or permits, as the fuel tax would not provide as wide a range of incentives as either the pollution tax or permit system. Both the fuel tax and the pollution tax/marketable permit systems would give people an incentive to use less energy in any activity and to develop new technologies for conserving energy and using energy more efficiently (more heat produced, miles driven, or electricity generated per unit of fuel combusted). However, even though fuel taxes create incentives to burn less fuel and to be more energy efficient, they do not give energy users an incentive to reduce emissions per unit of fuel that is burned.

Many people object to more taxes on both philosophical and practical grounds. Philosophical objections center on the argument that government is too big and too intrusive. Practical objections center on the idea that government spending tends to be wasteful, so the money is better off left in

[8]A BTU is the amount of energy necessary to heat one pound of water one degree Fahrenheit.

private hands. However, people who advance these arguments about taxes and the role of government would not necessarily argue against an energy tax. The tax makes sense from an efficiency point of view, and problems with increased tax revenue and increased role of government could be handled by balancing the increased revenue collections from a fuel tax with reduced income taxes, so that total government tax revenue remained constant.[9]

The importance of increasing the price of fuel (either through systems of pollution taxes and marketable pollution permits or through the use of fuel taxes) can be seen both in the pollution problems arising from energy and in the lack of progress that has been made in developing and promoting the use of alternative sources of energy. Alternative sources of energy would include solar power, geothermal power, wind power, liquid fuels from renewable sources, and dry fuels from renewable sources. Liquid fuels from renewable sources would include ethanol and methanol from a variety of plant sources including corn, sugar cane, a variety of oil seeds (cottonseed, rape seed, etc.), grasses, and wood. They can be used as substitutes for gasoline in internal combustion engines or as boiler fuels. Dry fuels are primarily from wood and grasses and are used as boiler fuels.

These alternative sources of energy are generally less polluting than fossil fuels. For example, if wood is burned to generate steam and produce electricity, the pollution from the wood is small in comparison to the oil or coal that it would displace. Wood can be burned very efficiently, so that the combustion of the wood is very complete, leaving few unburned hydrocarbons in the emissions. Also, wood has virtually no sulfur content, so sulfur dioxide emissions would be greatly reduced. Finally, if the wood (or corn or other biomass fuel) is replanted after harvesting, then the burning of the biomass fuel does not increase global warming. Although the burning of the fuel releases carbon dioxide, if the feedstock is replanted, the growth of the new plants will pull carbon dioxide from the atmosphere. Thus, the burning of biomass fuels cycles carbon, rather than releasing stored carbon (see Chapter 6). Although biomass could fill a portion of our energy needs, it would be very difficult to grow enough energy crops to replace anything more than a fraction of fossil fuel use in the short run.

Despite government-sponsored programs of research and development into alternative energy technologies, these technologies have been slow to develop. Quite simply, alternative energy technologies are more expensive for energy users than oil or coal, so they have not become established as important sources of energy. Although fossil fuels have a lower cost to the user, they have a higher cost to society.

Alternative fuels and energy technologies would be significantly advanced if the price of fossil fuels rose to incorporate the full social cost of these fuels. This effect is illustrated in Figure 8.3. In this example, it is

---

[9]See the discussion in Chapter 5 on the "double dividend" from environmental taxation.

FIGURE 8.3

EFFECT OF POLLUTION TAXES ON ALTERNATIVE FUELS

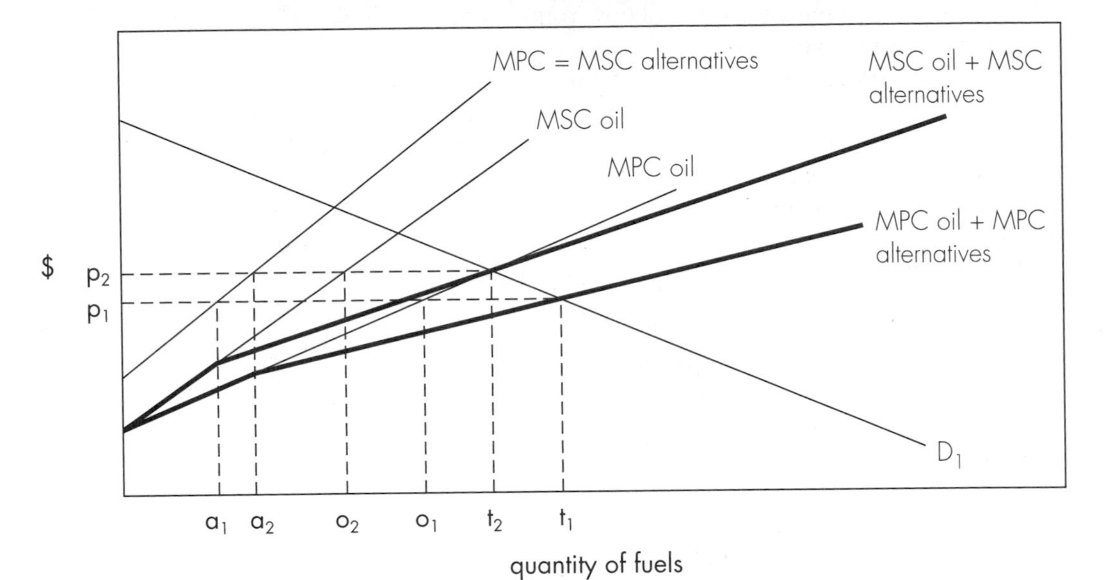

assumed that the private cost and social cost of the alternative fuels (biomass fuels, solar energy, and so on) are equal; that is, that there is no pollution associated with the use of these fuels. This assumption is obviously not true, but the full social costs of fossil fuels may frequently exceed the full social costs of alternatives. In this example, the alternative fuels and oil (which represent all fossil fuels) are measured in common units, such as BTUs (British Thermal Units).

The marginal private cost curve for all fuels is constructed by horizontally summing the MPC curve for alternative fuels and the MPC curve for oil. If D_1 represents the total demand for fuels, then the market equilibrium quantity is t_1, where the total fuel MPC curve intersects the demand function. Notice, however, that the social optimum occurs where the total MSC curve intersects the demand function, with a smaller equilibrium quantity of t_2. If the private cost of oil was increased through an externalities tax, marketable pollution permit system, or fuel tax, then the MPC curve for oil could increase to be equal to the MSC curve of oil and generate a market equilibrium equal to t_2. At this solution, which is associated with a higher price (p_2) for energy, oil usage declines from o_1 to o_2, and alternative fuel usage increases from a_1 to a_2. Notice that since oil usage declines, dependence on foreign oil will decline as well.

The increasing of the price of oil through the imposition of fuel taxes, marketable pollution permits, or pollution taxes has several desirable results. It reduces pollution, since alternative sources are generally cleaner. Also, it reduces reliance on oil, which will reduce dependence on foreign oil. At first glance, this might seem like a win-win solution, and the question might be posed, "Why haven't we raised the price of oil through pollution or fuel taxes?" The answer to that question is that there is a belief in the United States that our economic success requires low prices for energy.

THE MACROECONOMIC IMPACT OF FUEL TAXES

Let us examine the proposition that low energy prices are a requirement for U.S. economic success. This question can be answered at several levels, including looking at production processes and examining other countries.

It is clear that other economies with higher energy prices have strong and growing economies. Table 8.7 lists gasoline prices in U.S. dollars (current, not inflation adjusted) per liter for selected developed countries. As this table indicates, U.S. oil prices (as well as those of Canada and Australia) are extremely low compared to other countries. The table also lists the percentage of taxes in gasoline prices, so that it can also be seen that other countries tax much more heavily. In many of these countries, the price of gasoline is over three times higher than the price in the United States, and in virtually every country (except Canada and Australia), the price is at least twice as high as in the United States.

It might seem odd that this section of the chapter argues that the price of oil and other fossil fuels should be higher, when in a previous section of the chapter, it argued that one of the major costs of dependence on foreign oil was the loss of wealth and the macroeconomic effects associated with high prices. Clearly, the loss of wealth does not matter in the case of pollution or fuel tax because the tax component of the price stays within the United States and does not wind up in the treasuries of foreign producers. U.S. consumers lose consumers' surplus, but the money is transferred to the U.S. Treasury rather than foreign treasuries.[10]

One of the reasons that the price increases of the 1970s created undesirable macroeconomic effects is that they came so suddenly. Firms were unable to adjust quickly to the higher energy prices, so they used more energy and less labor and capital than cost minimization would dictate. However, the problem associated with rapid price increases can be eliminated by gradually increasing the tax over a long period of time. For example, if a consumer knew that the tax on gasoline was going to increase by 10 cents per gallon per year for the next 10 years, the consumer would take the future high price

[10]Of course, if the tax revenues are used unwisely, this will not be true. If they are used productively, or used to reduce income taxes, this will be true.

TABLE 8.7

GASOLINE PRICES (U.S. $/LITER) AND PERCENTAGE OF TAXES IN GASOLINE PRICE

COUNTRY	1980 PRICE	1980 % OF TAXES	1991 PRICE	1991 % OF TAXES
Australia	0.347	18.7	0.457	53.1
Austria	0.670	41.6	0.842	58.1
Belgium	0.790	53.3	0.916	66.5
Canada	0.223	24.5	0.505	42.2
Denmark	0.804	58.8	0.946	67.8
Finland	0.776	36.1	1.093	61.3
France	0.799	58.0	0.950	75.0
Germany	0.640	48.7	0.866	67.6
Greece	0.815	41.8	0.755	NA
Ireland	0.653	48.1	1.006	66.2
Italy	0.817	61.4	1.238	76.0
Japan	0.648	36.7	0.944	45.7
Luxembourg	0.610	43.8	0.661	54.9
Netherlands	0.721	52.3	0.999	70.0
New Zealand	0.478	27.6	0.577	45.7
Norway	0.752	51.7	1.118	67.4
Portugal	0.870	61.4	1.018	72.2
Spain	0.753	34.6	0.854	65.4
Sweden	0.697	49.3	1.123	67.7
Switzerland	0.688	51.1	0.771	59.5
Turkey	0.630	NA	0.735	56.1
United Kingdom	0.658	46.3	0.851	66.0
United States	0.329	11.2	0.301	32.9

SOURCE: Alfred A. Marcus, *Controversial Issues in Energy Policy*, 1992, pp. xiii–xiv.

into account when he or she decided what model of car to purchase today. Business managers would engage in the same type of advance planning, taking into account the future cost of energy when they made current decisions about production technologies, plant, and equipment.

There remains one more potential source of macroeconomic cost associated with higher prices. When analyzing higher prices in the context of OPEC and foreign dependence in a previous section, it was stated that higher prices make energy more scarce to energy users. Since energy is one factor of production, this scarcity must reduce total output or GDP. However, the same thing cannot be said to be true as a result of higher prices caused by higher taxes. The reason is that if taxes are efficiently utilized, they reduce the scarcity of other factors of production. For instance, the tax on energy can be used to finance government activities in lieu of a portion of income taxes.

The lower income tax makes labor a cheaper factor of production, which can partially offset the effects of the increase in the price of energy. Similarly, if the energy tax money is used to invest in factors of production, such as improving education or infrastructure, the economy will become more productive and partially offset the decline in productivity generated by the increase in the price of energy.

In summary, a pollution tax, system of marketable pollution permits, or fuel taxes will cause less of a negative impact on the macroeconomy than a corresponding increase in prices from OPEC monopoly power. The tax can create partially offsetting increases in productivity. In addition, the tax will serve to reduce pollution emissions that will reduce global warming, acid deposition, smog, and many other pollutants. This increase in environmental quality will also serve to increase GDP to partially mitigate the negative impacts of the tax. If the relationship between environmental quality and economic output is strong enough, the energy taxes will actually serve to enhance the performance of the macroeconomy, although not enough information is available to test this hypothesis.

NUCLEAR POWER ISSUES

Nuclear power has been the subject of considerable controversy since its emergence out of the World War II effort to develop nuclear weapons. Peaceful uses of nuclear power have been regulated heavily by the government because of the potential for disaster and the national defense implications of the use of nuclear power.

Peaceful uses of nuclear power were promoted by the government to help spur technological innovation to help support nuclear powered naval vessels. It was also felt that nuclear power had the potential for solving the nation's energy problems, even though it was apparent that the first generation of nuclear powered electric generating plants would be more expensive than conventional ways of producing energy.

One of the biggest obstacles to establishing a nuclear power industry was the liability to which an electric utility would be exposed if there was a disaster at a nuclear power plant. Consequently, Congress enacted the Price-Anderson Act, which exempted individual utilities from having to pay damages as a result of an accident. Claims would be paid by a consortium of utilities (20 percent liability) and the federal government (80 percent liability). The amount of liability from any particular accident was also limited. Critics of nuclear power argue that this limitation of liability was an inappropriate government intervention into the marketplace. If the potential for liability was too great for nuclear power to be privately optimal, it was argued, then it cannot be socially optimal.

There are a variety of other sources of disparity between the private costs and social costs of nuclear power. Table 8.8 lists some of the costs of nuclear

TABLE 8.8

COMPONENTS OF THE SOCIAL COST OF NUCLEAR POWER

COST	EXTENT OF INCORPORATION INTO PRICE
Construction and operating costs	Electricity is priced on an average cost basis so the higher costs of electricity that are generated by nuclear power is averaged across all units of electricity produced by all methods. Consequently, the price of electricity produced by nuclear power does not reflect the full private costs of producing the electricity.
Expected damages from an accident	Limited by the Price-Anderson Act
Storage of spent fuel and waste	Charged to future electricity consumers, so not incorporated into current price
Decommissioning of retired nuclear power plants	Charged to future electricity consumers, so not incorporated into current price

power and the extent to which these costs are incorporated into price. When evaluating this table, it should be noted that other sources of energy are associated with externalities that make their social costs exceed price.

Nuclear power arose out of the wartime Manhattan Project, where physicists and engineers were the project managers. This scientific management continued into the peacetime nuclear program. Consequently, it was felt that risk could be managed by engineering systems to minimize the chances of an accident occurring.

Marcus points out that this focus on technical risk issues neglected the human factor. Undertrained, bored, intoxicated, or stressed workers could make mistakes as they monitored and adjusted systems. This was the case in the Three Mile Island incident, in which worker error compounded system failure. Marcus argues that it was not until very recently that the human factor received appropriate attention from regulators.

Although it was predicted that nuclear power would be more expensive than conventionally produced electricity, it was thought that costs would fall. This has not happened, primarily due to delays in obtaining permits and in construction. One possible reason is that the design of nuclear plants varies substantially from plant to plant, so there was little opportunity to learn by experience. In contrast, France uses a standard design in all plants, so they have been able to work many flaws out of their system as time progresses.

There has been a hiatus in construction of new power plants in the late 1980s and 1990s because of high costs and concerns over the possibility of accidents. However, the recent attention to global warming has rekindled interest in nuclear power, which emits no carbon dioxide (or any of the criteria air pollutants). If new nuclear power plants are designed in the future, they will probably utilize what is referred to as "passive" safety systems. Passive safety systems rely on the laws of physics rather than on mechanical systems or human judgment to ensure that the reactor does not overheat or that the reaction does not accelerate to unsafe levels. The laws of physics, unlike mechanical systems (such as valves and pumps), do not fail. Also, there is less reliance on humans to make critical decisions.

An example of a passive system is a cooling system that moves the cooling water by relying on the property that hot water rises. Pumps can fail, valves can jam, people can make mistakes, but hot water always rises. Similarly, instead of relying on mechanical systems to lower graphite/boron rods into the reactor (to absorb neutrons and slow or stop the chain reaction if it is threatening to overheat or accelerate to the point where it could explode), passive systems have the rods held above the reactor by plastic. If the reactor begins to overheat, the plastic melts and the rods drop into the reactor. Again, there is no reliance on mechanical systems or human judgment.

An additional controversy over nuclear power is what to do with the nuclear wastes that result from the fission process. These nuclear wastes will remain dangerously radioactive for over 100,000 years. The waste must be safely contained or it will contaminate groundwater and ecosystems and threaten human health. It should be noted that the existing and future nuclear waste from the current nuclear power plants is only a small part of our total nuclear waste problem, with the vast majority of nuclear waste coming from obsolete nuclear weapons. We must develop a way to safely store nuclear waste regardless of whether we expand our nuclear power program.

The future of nuclear power has not yet been determined. No new plants have been started in recent years. Critics oppose nuclear power because of its high cost, the threat of an accident that would release radioactivity, and the problem of storage of nuclear waste. Proponents of nuclear power argue that safety can be improved by adoption of passive systems and by paying more attention to the role of people in safety. They also argue that nuclear power is environmentally preferable since it does not emit carbon dioxide or conventional air pollutants.

TRANSITION FUELS AND FUTURE FUELS

Many people who study energy believe that sometime in the future (20 or 50 or 100 years from now), there will be radically different sources of energy. Technological innovation in solar energy or hot fusion could lead to cheap and environmentally friendly sources of energy. The question that remains is, "How do we get from the present time to the time when these innovations in energy are available, or what should be our transition fuel?"

Before examining policies that try to steer the market in favor of a particular transition fuel, it would be useful to discuss how the forces of supply

FIGURE 8.4

FUEL TRANSITIONS

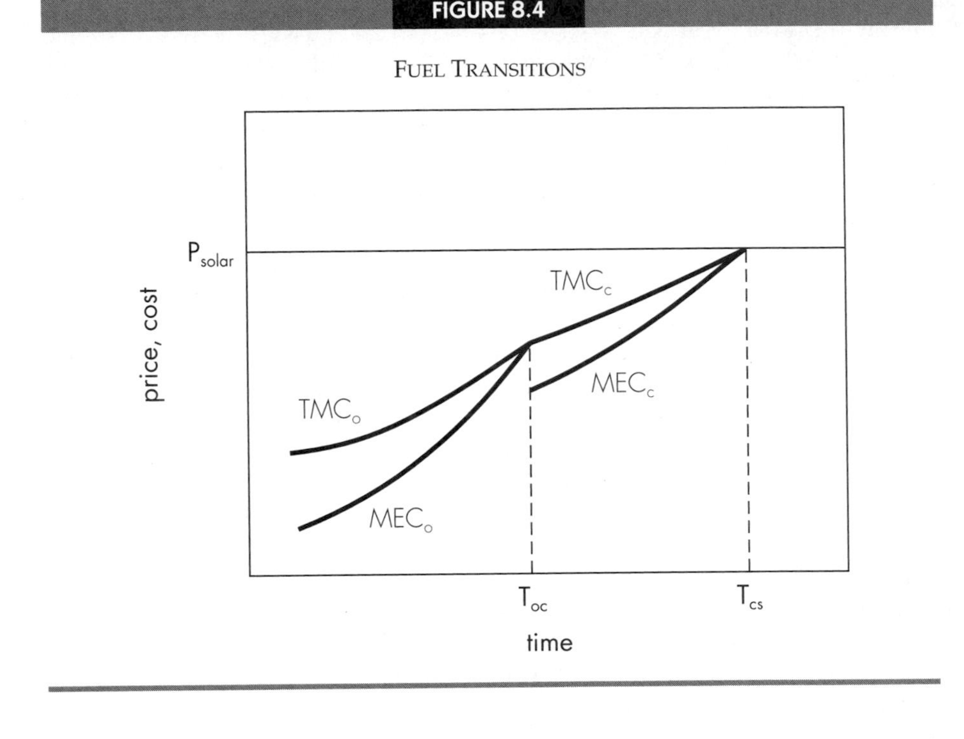

and demand would result in the switch from one fuel to another. This change is illustrated in Figure 8.4, where it is assumed that there are three possible fuels: oil, coal, and solar energy. In this example, both oil and coal have increasing marginal extraction costs, with the initial units of extraction of oil cheaper than coal. Assume that solar energy is available in unlimited quantities at a high constant marginal cost. If this is the case, then oil is used first, with price determined by total marginal cost, which in this example is equal to total marginal extraction cost plus user cost (externalities are not incorporated into price). As the marginal extraction cost of oil increases relative to the marginal extraction cost of coal, the opportunity cost of the oil (user cost) decreases towards zero. Eventually, the total marginal cost (and price) of oil is equal to the total marginal cost (and price) of coal, and the market switches from oil to coal at time T_{oc}. A similar switch from coal to solar energy occurs at time T_{cs}.

Although increasing price will serve to generate a transition from one fuel to another, policy makers have been concerned with managing these transitions. One source of concern is that the market will not adequately spur research and development into new energy technologies, especially since research and development into new energy technologies (particularly radically different technologies, such as biomass and solar energy) may be particularly risky for the investor. Another source of concern is the continual policy concern with abundant, cheap energy at a low price.

A more recent concern with energy transitions is the environmental effect of transitions. If environmental externalities are not incorporated into market price, then the transition to cleaner fuels such as solar energy will take place at a later than socially optimal date. Also, the market may choose relatively dirty fuels (such as coal) as the transition fuel, rather than cleaner ones (such as natural gas).

The Carter administration adopted a set of policies that defined coal as the transition fuel because the United States has abundant supplies of coal. This position has been largely abandoned due to high levels of pollution associated with the burning of coal. Of course, technological innovation in developing ways to burn coal cleanly could change this stance, and much progress has already been made in developing cleaner and more efficient coal-burning technologies.

Another possibility is what is referred to as deep natural gas, which is thousands of feet deeper and more expensive to produce than conventional natural gas. The deep natural gas is advocated by its proponents because natural gas is the cleanest of the fossil fuels. Biomass fuels also have their advocates, since these fuels would not exacerbate the global warming problem.

However, the debate over which fuel should constitute the transition fuel is really an unnecessary debate. If all fuels included all social costs in their prices, then the market would pick the fuel with the lowest social cost as the transition fuel, and the movement toward the future fuel would be speeded by the higher current price of energy.

ENERGY AND THE THIRD WORLD

Although U.S. citizens and citizens of other developed nations speak of an energy crisis in their respective countries, the real energy crisis is occurring in Third World nations. Third World nations, with less sophisticated economies, were much less capable of adjusting to the price shocks associated with the oil price increases of the 1970s. In addition to the types of macroeconomic effects that developed nations faced, Third World nations faced additional problems with their foreign reserve balances (holdings of internationally desirable currencies, such as U.S. dollars, German marks, Japanese yen, and so on). The need to use more foreign reserves to purchase oil stretched their already thin foreign reserves. This led many nations to borrow foreign currencies for development projects and for imports of oil and other goods that began the big debt problems of the late 1970s and the 1980s (see Chapter 17 for a discussion of this and other Third World issues).

The increase in the price of fossil fuels also forced even greater reliance on fuel wood, which is the primary source of energy in developing countries. The increased reliance on fuel wood was an important factor in the increase in deforestation in many areas of the world. The increased deforestation has led to a myriad of environmental and economic problems that are described in Chapters 12 and 17.

Environmental externalities from energy use are also responsible for the dreadful environmental quality of Eastern and Central Europe. The reliance on low quality coal and the absence of any effort to abate pollution has led to air quality that is so bad that it is one of the leading causes of sickness and death in countries such as Poland, (the former) Czechoslovakia, and Romania. These problems are also discussed further in Chapter 17.

SUMMARY

Energy policy has focused on keeping the price of energy low, probably as a result of perceived problems with oligopoly and a desire to prevent oil companies from earning monopoly profits. This process of attempting to keep prices low led to shortages and eventual increase in the price of energy.

Since energy is responsible for so many of our environmental problems, one can argue very strongly that energy is underpriced, and that proper energy and environmental policy requires significant increases in the price of energy. These increases could be accomplished by pollution taxes, marketable pollution permits, or fuel taxes.

Many people fear that raising the price of energy will lead to economic decline, but raising the price of energy through taxes is fundamentally different than when the prices are raised by OPEC. As long as the price increases are gradual, macroeconomic disruption should be relatively small.

There are many more issues surrounding energy and the environment than can possibly be addressed in a chapter of this nature. The student is encouraged to do additional reading, especially with regard to Third World issues.

REVIEW QUESTIONS

1. What important pollutants are generated by fossil fuel consumption?
2. What global and regional environmental problems are caused by fossil fuel consumption?
3. What market failures are associated with nuclear power?
4. What is a cartel and how does it function?
5. Why would Saudi Arabia advocate prices lower than those advocated by other OPEC nations?
6. How does user cost allocate an exhaustible resource over time?
7. Describe an economic incentive for dealing with pollution from mobile sources.
8. Is the price of energy too high or too low? Why?
9. What are the costs to the United States of depending on a foreign cartel for oil?

10. Under what circumstances would higher energy prices lead to economic dislocation in the United States?
11. What are the primary energy problems faced by developing nations?

QUESTIONS FOR FURTHER STUDY

1. Should solar energy be subsidized?
2. Should mass transit be subsidized?
3. Formulate an energy/environmental policy to carry the United States through the next 30 years.
4. What was wrong with the energy policies of the 1970s?
5. What was wrong with the energy policies of the 1980s?
6. Should we further develop nuclear power?
7. Why has the power of OPEC diminished?
8. Will a system of pollution controls or energy taxes adversely affect the U.S. economy?
9. How would pollution taxes affect the development of alternative clean sources of energy?
10. Why are energy price controls socially inefficient? Why would lawmakers enact an inefficient policy such as energy price controls?
11. Are biomass fuels a viable energy alternative? Why or why not?

SUGGESTED PAPER TOPICS

1. Develop an energy policy to take the United States through the next 30 years. Look at the National Energy Strategy (DOE, 1991) and current issues of Energy Economics and the Energy Journal. Also, look in general interest journals, such as *Nature* and *Science,* as well as the government documents sections of the library.
2. Examine the economic barriers to the development of solar energy. Should the government seek to reduce these barriers? What actions could it take to reduce the barriers? Search bibliographic databases on solar energy and economics, solar energy and energy policy, and solar energy and technological innovation.
3. Demand-side management is a policy that many electric utilities are undertaking in order to reduce the need to develop additional electric generating capacity. Examine this policy, how it affects utility profits, and how it affects social welfare.
4. Develop a transportation policy for U.S. urban areas. What are the objectives of your policy? (What market failures are you trying to correct?) What economic incentives would you employ? What command and control techniques would you employ? Why?
5. Investigate the relationship between per capita energy use and per capita income. What is the cause-and-effect relationship between the two? Data for an empirical investigation may be obtained from World Resources. Does your work show causation or correlation?

WORKS CITED AND SELECTED READINGS

1. Bohi, Douglas R. *Energy Security in the 1980s: Economic and Political Perspectives.* Washington, D.C.: Brookings Institute, 1984.
2. *Energy Statistics Sourcebook,* 6th ed. Oil and Gas Journal Energy Database (1992): 353–354.
3. Green, David L., and Paul N. Leiby. *The Social Costs to the U.S. of Monopolization of the World Oil Market 1972–1991.* Oak Ridge National Laboratory Report number 6744, Oak Ridge, TN, 1993.
4. Griffen, James M., and Henry B. Steele. *Energy Economics and Policy.* New York: Academic Press, 1980.
5. Hotelling, Harold. "The Economics of Exhaustible Resources." *Journal of Political Economy* 39 (1931): 137–175.
6. Malthus, Thomas Robert, 1798. "An Essay on the Principle of Population." In *An Essay on the Principle of Population: Text, Sources and Background Criticism,* edited by Phillip Appelman. New York: Norton, original 1798, reprint 1975.
7. Marcus, Alfred. *Controversial Issues in Energy Policy.* New York: Sage Publications, 1992.
8. Meadows, Donella, et al. *The Limits to Growth.* New York: Universe Books, 1972.
9. Mills, Edwin S., and Lawrence J. White. "Government Policies towards the Automobile Emissions Control." In *Approaches to Air Pollution Control,* edited by Anne Frielaender. Cambridge: MIT Press, 1978.
10. Mills, Russell. *Energy, Economics and the Environment.* Englewood Cliffs, NJ: Prentice Hall, 1985.
11. U.S. Department of Energy. *National Energy Strategy, 1st ed. 1991/1992.* Washington, D.C.: U.S. Government Printing Office, 1991.

12. U.S. Environmental Protection Agency. *Emission Levels for Six Pollutants by Source,* #cc 91-600760. Washington, D.C.: U.S. Government Printing Office, 1990.
13. Webb, Michael, and Martin J. Ricketts. *The Economics of Energy.* New York: John Wiley and Sons, 1980.
14. *World Resources 1992–93.* Washington, D.C.: World Resource Institute, 1993.
15. *World Resources 1994–95.* Washington, D.C.: World Resource Institute, 1995.
16. *World Resources 1996–97.* Washington, D.C.: World Resource Institute, 1997.

9 Material Policy: Minerals, Materials, and Solid Waste

For you are dust, and to dust you shall return. . . .

INTRODUCTION

The above passage from the Bible illustrates the cyclical nature of human life. Although it is often unrecognized by consumers, producers, and our economic system in general, our economic goods also have a cyclical life. They come from environmental resources, their form is changed, and they are returned to the environment when they leave the economic system.

In the past, we have not viewed our economic process in this circular fashion. We have viewed the process as producing goods with inputs and then consuming the goods, where the consumption of the goods eliminated them. However, it has now become increasingly apparent that the economic process does not eliminate matter, it generates wastes. As the law of mass balance implies, the mass of matter that goes into an activity must equal the mass of matter that comes out of an activity. If the mass of the good that is produced is less than the mass of the inputs used in production, then the differential has been converted into waste. After a good is consumed, all of the mass of the good becomes waste.

The circular nature of our use of materials provides a powerful argument for simultaneously examining both the input and the output side of the question of how to efficiently utilize our material resources. This circular nature is diagramed in Figure 9.1, an illustration which forms the conceptual basis of a materials balance equation, which mathematically tracks the flow of materials through a system.

Let us begin our analysis of the circular flow of materials with the environment, as that is where raw materials originate and where wastes are disposed.

Resources such as minerals are present in the environment, but must be extracted from whatever form and structure they take in the environment. For example, very few rocks contain nothing but copper. The copper is generally

The Circular Nature of Materials Use

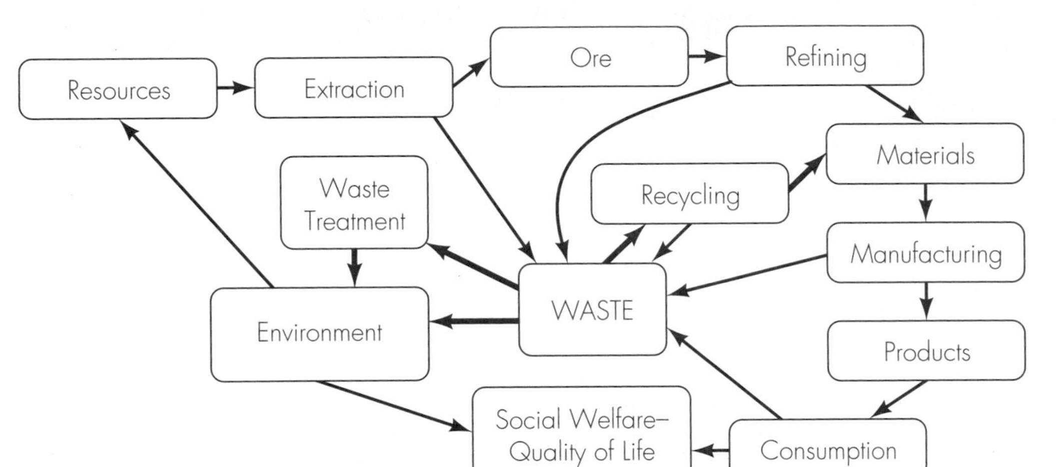

SOURCE: From *State of the World 1992,* A Worldwatch Institute Report on Progress toward a Sustainable Society by Lester R. Brown, et al. Copyright © 1992 by Worldwatch Institute. Reprinted by permission of W. W. Norton & Company, Inc.

mixed with other substances in rocks that may be buried hundreds of feet below the ground. Since the soil and other rocks must be removed before the copper-containing ore can be removed, the extraction process creates wastes.

The ore is then processed (refined) to remove the elements or compounds that are of value. Since most ores contain only a small amount of the desired substance (often a fraction of a percent), a large amount of waste is generated in this process. Since both mechanical and chemical processes are used in refining the ore, the waste can be hazardous to plants and animals and can contaminate ground and surface water.

The next step is to convert the mineral that is extracted from the ore into a material from which products can be made. For example, iron ore is converted to steel. There are two major categories of waste associated with this process. First, since these activities tend to be energy intensive, there are all the air pollutants associated with the combustion of fossil fuels (see Chapters 6, 7 and 8). Second, when the steel is made, various impurities are removed from the iron that have no economic use and must be disposed of.

The material is then used to make a product. For example, the steel could be used to make the body of a car. Again, there are wastes associated with the use of energy and with portions of the materials that do not wind up as part of the product that is manufactured.

Finally, the product is consumed, either as a final good or as an intermediate product. Both the use and the final disposal of the product will generate waste.

As we look at the outer ring of the chain of economic activities in Figure 9.1, we focus on activities that have been examined within the conventional realm of economics. The theory of the firm, the theory of the consumer, and the theory of exhaustible resources have been taught in college economics courses since World War II. However, there are market failures present in this chain of activities that have not been adequately studied. Inappropriate government interventions, imperfect competition, and externalities are present throughout this chain. In particular, we will focus on the flows within the interior of Figure 9.1 and discuss the market forces and the market failures that influence the flow of wastes.

This inner ring shows that all the activities generate wastes, which may be recycled; treated and released into the environment; or released directly into the environment without any prior treatment. Do we have the socially efficient level of total waste? Do we have the socially efficient level of recycling, and do we have the socially efficient level of treatment? If not, how do we develop policies to reduce inefficiencies in this area?

This chapter focuses on these questions. Although the flow of materials is also based on fossil fuels and activities based on renewable resources, such as agriculture and forestry, the chapter focuses on the role of minerals in this cycle, as separate chapters have already been devoted to energy (Chapter 8), forestry (Chapters 11 and 12), and agriculture (Chapter 16).

THE ECONOMICS OF MINERAL EXTRACTION

The fashion in which the invisible hand of the market allocates mineral resources is virtually identical to the way in which it allocates fossil fuels such as oil. This correlation should not be surprising, as both fossil fuel and mineral resources are exhaustible resources. The major difference between fossil fuels and mineral resources is that the use of fossil fuels tends to dissipate the resource.[1] In contrast, the use of mineral resources tends to concentrate the resource.[2] The implication of this is that it is possible to recycle minerals but not fuels.

As discussed in the analysis of Chapters 2 and 8, the market price of an exhaustible resource must be equal to the sum of marginal extraction cost and marginal user cost in order for the market to allocate the resource in a socially efficient fashion.[3] The greater the future value of the mineral, the greater the current opportunity cost, marginal user cost, and market price.

---

[1]The concentrated oil or coal is combusted and transformed into gases that go up a smokestack or out a tailpipe.

[2]The mineral is contained in a heterogenous ore (such as iron ore) and concentrated into a homogeneous material (such as steel).

[3]This equation assumes that there are no externalities in the production or use of mineral resources. Obviously, the market will not be efficient in the presence of externalites, and efficiency would require some sort of intervention (taxes, marketable pollution permits, etc.) to reduce the level of the externality to the optimal level.

It should be noted that although the potential for recycling exists with many minerals, recycling cannot eliminate the scarcity problem. The entropy law,[4] which is one of the laws of thermodynamics, says that matter (and the entire universe) is continually moving toward a less well-ordered state. Thus, there is a constant loss of the well-ordered material. For example, moving parts wear, and the material that wears away is completely disordered. Some of the material may become contaminated with toxic substances or radioactivity so that it is not safe to recycle. Some materials become mixed with other materials during the production process, rendering recycling extremely difficult. The point is that if a fraction of the material is lost every time a material is used and recycled, the material will eventually become exhausted.

As was illustrated in Chapters 2 and 8, the invisible hand will correctly allocate the mineral resource, provided that no market failures are present. However, as previously pointed out, market failures are present at every point along the circular flow of materials.

MARKET FAILURES IN THE EXTRACTION OF MINERAL RESOURCES

There are many important externalities in the extraction of mineral resources. These involve environmental externalities, inappropriate government interventions, imperfect competition and national security externalities.

Environmental Externalities. Environmental externalities are the result of the waste that is generated and the disruption of the landscape that occurs as a result of mining and refining activity. Since water is often used to help separate the usable mineral from the ore, mining tends to generate important water quality problems. Also, the disruption of the surface of the land and the exposure of impurities in the rock leads to problems from pollutants that are carried to surface waters and groundwaters through the run-off of rainwater. Acid drainage from mines is a particulary important problem.

A major component of the refining of many materials is a process known as smelting, where intensely hot furnaces are used to melt the ore and separate the mineral from the waste material. Not only does smelting release pollutants as a result of the use of the energy in the furnace, but the ores often have impurities such as sulfur that are converted to gas and released in the smelting process. Both the large copper smelter operation at Copper Basin, Tennessee, and the large nickel smelting operation in Sudbury, Ontario, are associated with significant acid rain problems. In Tennessee, the smelting was done in the late nineteenth and early twentieth centuries in pits in the mining area, and the acid rain problem was extremely local, killing all vegetation for over 145 square kilometers (Raven, Berg and Johnson, 1993,

[4]See Chapter 18 for a detailed discussion of the entropy law and its implications for economic growth.

p. 331). In the Sudbury area, the acid rain problem is more regional, and is a major contributor to acid deposition in Canada and parts of the United States (see Chapter 7).

These problems of air pollution and water pollution are dealt with in substantial detail in Chapters 7, 8, and 14. Consequently, these externalities will not be discussed in great detail in this chapter except to note that they are important, and they create a disparity between the private cost of the mineral and the social cost. It should be noted, however, that many types of mining production technologies exist which minimize these pollution problems. The policy difficulty is that these cleaner technologies may be more expensive than traditional techniques, and therefore are not often employed, particularily in small scale mining. In addition, there is an important externality associated with mining that has not yet been discussed in as much detail, and that is that the creation of a mine often completely destroys the natural environment in its vicinity. Table 9.1 lists selected environmental impacts of mining projects, including both pollution and elimination of natural environments. It should be noted that the inclusion of a mining project on this list is not an indication that this mining should not take place, but a notation of the types of environmental costs that are associated with mining projects in general.

Since most of the convenient opportunities for mining have already been exhausted, many potential new mines are located in wilderness areas or other areas that have unique environmental properties. The opportunity cost of losing these environmental resources is usually not factored into the market price of the mineral. Moreover, since the benefits of preserving the environmental resource are primarily public good benefits (see Chapter 2 for a discussion of public goods), it is difficult for the landowner to capture these benefits. Consequently, the benefits of preserving the area are seldom included in the landowner's decision making.

John Krutilla (1967)[5] was the first to argue that the focus of conservation should be less concentrated on the resources that are extracted and more directed to the unique environmental resources that the mining (and other development activities) destroys. He emphasized the public good nature of the unique natural environments and the inability of the landowner to capture the benefits of preserving the natural environment. The benefits of preservation include preserving the area for recreation, habitat, biodiversity, scientific enquiry, watershed protection, and the indirect use benefits that people derive from the existence of the area. (See Chapter 4 for a discussion of indirect use values.)

---

[5]Krutilla's article in the *American Economic Review* is a landmark article, which the author of this textbook regards as one of several articles that began the field of environmental economics. It is an extremely readable article (not a lot of graphs and equations), and any student who is serious about environmental economics, environmental policy, or environmental studies in general should read this article.

TABLE 9.1

SELECTED EXAMPLES OF ENVIRONMENTAL IMPACTS OF MINERAL EXTRACTION AND PROCESSING

LOCATION/MINERAL	OBSERVATION
Ilo-Locumba area, Peru copper mining and smelting	The Ilo smelter emits 600,000 tons of sulfur compounds each year. Nearly 40 million meters per year of tailings containing copper, zinc, lead, aluminum, and traces of cyanides are dumped into the sea each year, affecting marine life in a 20,000 hectare area. Nearly 800,000 tons of slag are also dumped each year.
Nauru, South Pacific phosphate mining	When mining is completed—in 5–15 years— four-fifths of the 2,100 hectare South Pacific island will be uninhabitable.
Pará State, Brazil Carajás iron ore project	The project's wood requirements (for smelting of iron ore) will require the cutting of enough native wood to deforest 50,000 hectares of tropical forest each year during the mine's 250-year life.
Russia Severonikel smelters	Two nickel smelters in the extreme northwest corner of the republic, near the Norwegian and Finnish borders, pump 300,000 tons of sulfur dioxide into the atmosphere each year, along with lesser amounts of heavy metals. Over 200,000 hectares of local forests are dying, and the emissions appear to be affecting the health of residents.
Sabah Province, Malaysia Mamut copper mine	Local rivers are contaminated with high levels of chromium, copper, iron, lead, manganese, and nickel. Samples of local fish have been found unfit for human consumption, and rice grown in the area is contaminated.
Amazon Basin, Brazil gold mining	Hundreds of thousands of miners have flooded the area in search of gold, clogging rivers with sediment, and releasing 100 tons of mercury into the ecosystem each year. Fish in some rivers contain high levels of mercury.

SOURCE: Young, 1992, p. 106.

Another important point about the destruction of unique natural environments as a consequence of mining is that these actions are irreversible. If we decide to carve up a wilderness because we feel that the minerals are more valuable to society, and later we discover a substitute for the minerals, it is not possible to recreate the wilderness. This potential suggests a responsibility to be cautious in embarking on a course of development. One of the major reasons for being cautious is that the benefits of preserving the natural environment will tend to increase over time, and the benefits of developing the environments (such as mining) will tend to decrease over time. The benefits of preservation increase over time because the demand for outdoor recreation is increasing over time and because the number of substitute environments is declining over time as more and more areas become developed. The benefits of development tend to decline over time as mines become

MINING OR WILDERNESS RECREATION

In *The Economics of Natural Environments,* John Krutilla and Anthony Fisher outline a process for cost–benefit analysis and apply this process to five case studies where a development project threatens to destroy a unique natural environment. One of these projects is a potential molybdenum (a strategic mineral used in metal alloys) mine in the White Cloud Peaks Mountains in Idaho.

Krutilla and Fisher discuss three potential land uses for the White Cloud Peaks: recreation, grazing, and mining. All three activities tend to be mutually exclusive. Krutilla and Fisher examine the benefits of recreation in a fashion that shows the sensitivity of recreational benefits to user congestion. They use standard techniques for valuing the benefits of range grazing. The benefits of the mining are shown to be small because there is an abundant supply of molybdenum, with prices projected to remain constant for several decades following the scheduled opening of the mine.

Although no official cost–benefit analysis was ever conducted of the conflicting uses of the White Cloud Peaks Mountains, Krutilla and Fisher's analysis shows that the decision to include this area in the Sawtooth National Recreation Area reflects sound economic judgment. Their analysis also shows that for these mountains, the benefits of wilderness recreation outweigh the benefits of alternative uses.

One potential use of the land that was not specifically examined by Krutilla and Fisher was a more intensive recreational use of the land, such as designating it as a national park. However, since this option was not proposed at the time of the study, and since the current use of wilderness recreation does not preclude the future development as a national park (mining or grazing would preclude this), there is an even more powerful argument to make the initial decision to allocate the area to wilderness recreation.

exhausted, as the capital equipment depreciates, as technology improves, and as substitutes are developed.

The significance of the decay of the benefits of development and the growth of the benefits of preservation is that development projects, for which the benefits of development are greater than the benefits of preservation in the current year, may be associated with the benefits of preservation being greater than the benefits of development in future years. Depending on the disparity between future benefits of preservation and future benefits of development, the present value of preservation may be greater than the present value of development even if current benefits and costs have a different relationship. Of course, the choice of discount rates will have an important effect on the outcome. Higher discount rates tend to minimize the impact of future values. Krutilla and Fisher (1985) and Porter (1982) discuss this issue in great detail.

Since the decision to mine or not mine a specific site is an "either/or" decision and not a marginal decision, it does not lend itself well to solutions such as incorporating marketable permits or taxes. A cost–benefit analysis (see Chapter 4) should be performed before the project is undertaken to determine if the social benefits of the mine exceed the social costs (which would include the foregone preservation benefits). Since mines require federal permitting, they are subject to the National Environmental Policy Act (NEPA) requirements for an environmental impact statement (EIS) which assesses the environmental costs of the mine, although these environmental costs are not always measured in dollar terms.

So far, we have identified two types of environmental externalities associated with mining. These are the pollution associated with the mining activity and the destruction of habitat and unique environmental areas. In addition to these two environmentally oriented market failures, there are important non-environmental market failures that tend to make mining activity greater than the socially optimal level. These market failures are created by both inappropriate government interventions and imperfect competition and serve (as do the environmental market failures) to make the marginal private costs of minerals lower than the marginal social costs.

Inappropriate Government Intervention. The most important of these government interventions is a depletion allowance. The depletion allowance is a provision in the tax code that allows firms to depreciate their mineral deposits the same way a manufacturing firm is allowed to depreciate its capital equipment. As the mining firms remove minerals from their mines, they can use the reduction in value of the mine as an expense that can be subtracted from total revenue to reduce taxable income. This deduction has the effect of reducing the private cost of production, since the taxpayers share in the cost of producing the mineral. This is shown in Figure 9.2, where MPC_1 represents the marginal private cost curve with no depletion allowance, and MPC_2 represents the marginal private cost curve with the depletion allowance. Notice that MPC is below the marginal social cost curve (MSC) due to the environmental externalities that create a disparity between marginal private cost and marginal social cost. Note that the depletion allowance exacerbates this disparity. Further, because it increases the market level of output (to q_2, which is even further from the optimal level), it increases the magnitude of the environmental costs.

Imperfect Competition. One additional market failure associated with mining is imperfect competition. Monopolistic or oligopolistic market structure in mining industries can lead to a loss in social welfare, as the monopoly or oligopoly restricts output below the optimal level in order to generate monopoly profits. (See Chapter 2 for a discussion of the social losses associated with monopoly.)

At the turn of the century and through World War II, oligopolistic market structure was an important factor in steel and many other mineral markets. However, as the economy of the United States (and other nations) became more open to international trade, foreign producers of minerals provided competition for U.S. producers. This aspect was particularly true for steel, where competition from German, Japanese, Korean and Brazilian producers

MERCURY POLLUTION AND GOLD MINING IN THE BRAZILIAN AMAZON

In the last 20 years there has been a surge of gold mining in the Amazon region, especially the Brazilian Amazon. In contrast to other types of mining in Brazil and other countries, this gold mining is not conducted by large corporations, but by small-scale entrepreneurs. Currently, there are well over 500,000 small-scale miners (known as garimpeiros) operating in the Brazilian Amazon. Although these miners are unlicensed and unregulated and have operated in an illegal fashion, the mining activity is an important source of jobs and has an important regional economic impact. Unfortunately, this unregulated mining has lead to important environmental impacts, the most important of which is mercury pollution.

Mercury is added to the gold ore, where it bonds with the gold, creating an amalgam which is easily separated from the ore. The amalgam is then heated, and the mercury boils out of the amalgam. However, since mercury is a liquid at ambient temperatures, it quickly condenses and falls back to earth, contaminating ecosystems and threatening human health (mercury has severe impacts on the central nervous system and can also lead to birth defects). Although the use of mercury in gold mining was illegal, the shear number of goldminers and their existence in the "underground" economy has made enforcement of the laws extremely difficult.

Although a technology existed to reduce the mercury pollution, it was not adopted by the garimpeiros. This technology was a retort, which is a container in which the mercury/gold amalgam is heated, but which condenses the fumes as they exit the container, capturing the mercury. This technology is similar in concept to the way a "moonshiner" distills alcohol. The advantage of this technology over existing technology is that the same mercury could be used again and again to extract the gold, lowering costs for the garimpeiros and at the same time reducing emissions of mercury into the environment.

Different people had different ideas about how to induce the garimpeiros to adopt the new technology. Some people argued for a direct control requiring the retort to be implemented, and others argued that a deposit–refund system should be placed on the mercury, which would give the garimpeiros even greater incentive to use the retort. However both solutions require extensive monitoring and enforcement among over half a million small-scale miners operating on the fringe of legality.

This problem was studied by the Centro Tecnologia Mineral, a mining research institute of the federal government of Brazil. Their conclusion was to control the problem in a way that required less reliance on monitoring and enforcement. Consequently, they implemented an extensive public information program demonstrating the proper use of the retort, emphasizing the cost saving associated with it (miners do not need to continually buy mercury if they recycle it), and educating the miners about the health (to the miners and general public) and environmental consequences of mercury emissions. Adoption of the retort technology has been impressively rapid. In the meantime, research is being conducted on developing mercury-free methods for extracting the gold from the ore.

not only eliminated the monopoly profits of U.S. producers, but took a large portion of the market as well.

National Security. For the most part, monopoly does not represent a significant problem in mineral production. The one exception is when the minerals

FIGURE 9.2

THE IMPACT OF DEPLETION ALLOWANCES

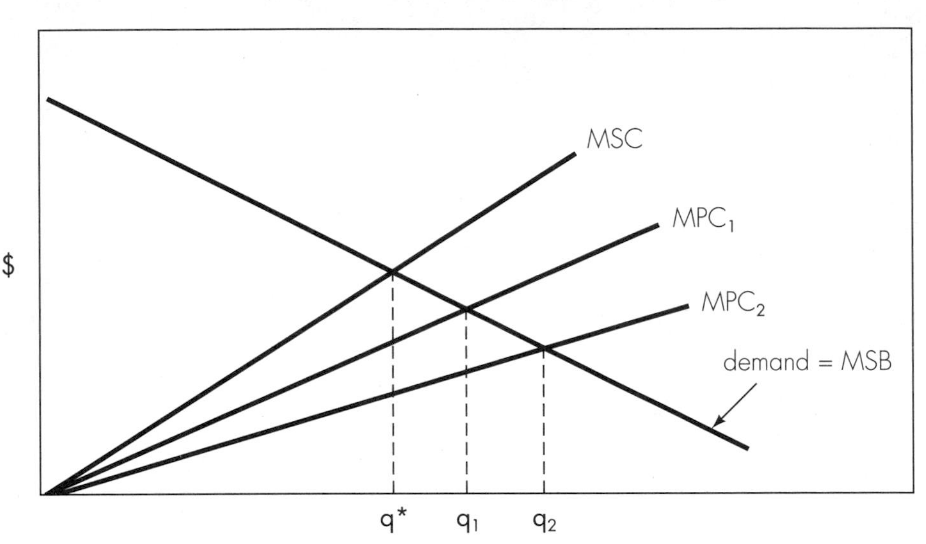

TABLE 9.2

U.S. NET IMPORT RELIANCE FOR SELECTED STRATEGIC MINERALS

MINERAL	PERCENTAGE IMPORTED IN 1990	MAJOR FOREIGN SOURCES 1986–1989
Graphite	100	Mexico, China, Brazil, Madagascar
Manganese	100	South Africa, France, Gabon
Platinum-group metals	88	South Africa, U.K., [former] USSR
Cobalt	85	Zaire, Zambia, Canada, Norway
Nickel	83	Canada, Norway, Australia, Dominican Republic
Chromium	79	South Africa, Turkey, Zimbabwe, [former] Yugoslavia
Tungsten	73	China, Bolivia, Germany, Peru

PRIMARY SOURCE: U.S. Bureau of Mines.
SECONDARY SOURCE: J.D. Morgan, "Stockpiling in the USA" in *Concise Encyclopedia of Material Economics, Policy and Management*, ed. Michael Bever, Pergamen Press: New York, 1993.

have strategic values. Strategic minerals are those that are important to aerospace and military applications, relatively rare, and generally found in countries of unpredictable political stability or potential hostility toward the United States and its military allies. They are generally rare metals that are mixed with common metals, such as steel or aluminum, to create alloys that have properties that the common metals do not have. These properties include the ability to remain stable and not become brittle at high temperatures, greater elasticity, and greater strength per unit weight.

There are several policy options available to mitigate the potential problems associated with potential politically generated shortages of strategic minerals. One option is to subsidize the development of domestic mines for these minerals, even though the extraction cost is much higher than the price on world markets (Table 9.2). In addition, these mines will generate environmental damage, as illustrated in the potential molybdenum mine discussed in the boxed example on page 263. Another option, which is probably associated with lower social costs, is to buy extra quantities of the minerals on world markets and stockpile the mineral against the possibility of future supply interruptions.

SOLID WASTE AND WASTE DISPOSAL

For most of recorded history, our civilizations have not been concerned with the solid waste that has been generated as a result of production and consumption activities. Even as we became aware of the environmental problems, such as water and air pollution, less attention was placed on the wastes that we disposed of on land.

However, as we entered the late 1970s, many areas of the United States began to experience a shortage of areas that were suitable for landfills (areas where solid wastes are buried). Urban areas sought to locate landfills at great distances from the cities, as sites in the vicinity of cities became filled. The problem of shortages of waste disposal alternatives was dramatically illustrated by the garbage barge from Islip, New York, whose unsuccessful voyages to Mexico and Caribbean countries seeking a garbage disposal site were chronicled nightly on television news shows and late-night talk shows.

Although the problem of where to put garbage is the question that receives the most public discussion, it is not the only or even the most significant problem. Developing new landfill sites or technical solutions to garbage disposal (such as incineration) do not really treat the solid waste problem, they treat the symptom of the problem. The real solid waste problem is that we are generating an inefficiently high level of waste. Solutions that do not address the problem of waste reduction are not likely to succeed, as they treat the symptom of the problem, rather than the problem itself.

FIGURE 9.3

SOCIALLY EFFICIENT LEVEL OF CONTAINERS

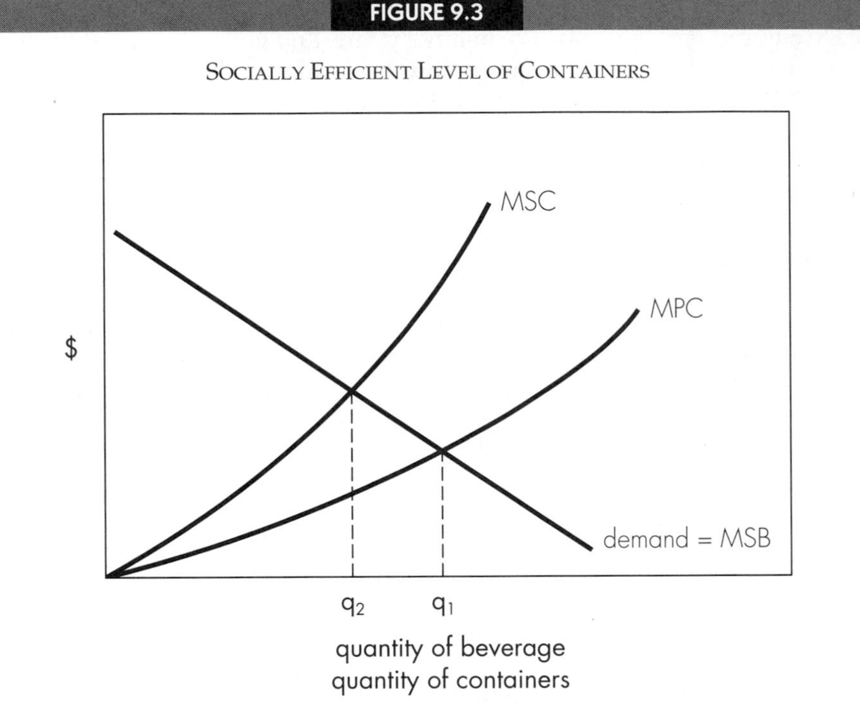

WHY DO WE GENERATE TOO MUCH WASTE?

The answer to the question of why we generate too much waste is extraordinarily simple and yet its implementation is quite complex. The simple answer is that people do not pay the full social costs of waste disposal at every level of the production process and in the consumption of the good.

This market failure is illustrated in Figure 9.3, which looks at the market for carbonated beverages. Carbonated beverages are a useful commodity to examine, because there is a one-to-one correspondence between the economic output (the beverage) and the waste output (the container).[6] The advantage of looking at a product with this type of one-to-one correspondence is that the horizontal axis can represent both the economic output and the waste output. In Figure 9.3, the marginal social cost function (MSC) is depicted as being higher than the marginal private cost function (MPC), due to the social costs from the waste associated with the product. An alternative

---

[6]This example is an oversimplification, as waste is generated in the production of the beverage and the production of the container. It is also an oversimplification because there are multiple sizes of containers, which would vary the ratio between the economic output and the waste output.

way of expressing this relationship would be to depict the marginal social benefit function to be below the marginal private benefit function. This market failure can be expressed in either fashion because one could regard the costs of the post-consumption waste as either an addition to the private costs of production or a reduction of the social benefits of consumption. As can be seen in Figure 9.3, the market level of containers (q_1) is greater than the socially efficient level of containers (q_2).

If the analysis is extended to include the amount of waste as a variable input, then the analysis must look at the demand and marginal cost of the material input. The marginal private cost of a material input (such as the packaging of a product) does not include the full cost of disposal, or the costs of the externalities associated with the disposal.

This problem is slightly more complicated than the standard externalities that have been studied, because not only must one determine the socially optimal amount of waste, one must determine the socially efficient disposition of the waste. This is very different than sulfur dioxide waste, for example, where one must only determine how much of the waste to emit at any location.

This problem of also determining how to dispose of the waste can be illustrated by assuming that there are two options for disposal, a proper option (such as disposal in a landfill) and an improper option (such as unrestricted dumping).[7] In Figure 9.4, the total amount of waste is represented by the horizontal distance between the two vertical axes. The amount of waste that is properly disposed of is represented by the distance from the left vertical axis, and the amount that is improperly disposed of is represented by the distance from the right vertical axis. The sum of the properly disposed of waste and the improperly disposed of waste must equal total waste, which is the horizontal distance between the two axes. In this example, total waste is equal to 25 units. For example, if 15 units are properly disposed of, then 10 units are improperly disposed of. Both vertical axes measure costs.

Both improper disposal and proper disposal are associated with external costs. One would expect that the external costs associated with improper disposal are greater than those associated with proper disposal, or the improper disposal would not be termed "improper." These external cost differentials are reflected in Figure 9.4, where there is a greater disparity between the marginal private cost and marginal social cost curves for improperly disposed of waste (MPC_i and MSC_i) than between the marginal private cost and marginal social cost curves for properly disposed of waste (MPC_p and MSC_p).

The market allocation of waste disposal between these two alternatives will be determined by people comparing the marginal private costs of one

---

[7]In this discussion, "proper" and "improper" refer to the relative environmental damages and do not imply anything about illegality or optimality.

FIGURE 9.4

ALLOCATION OF WASTE BETWEEN PROPER AND IMPROPER DISPOSAL

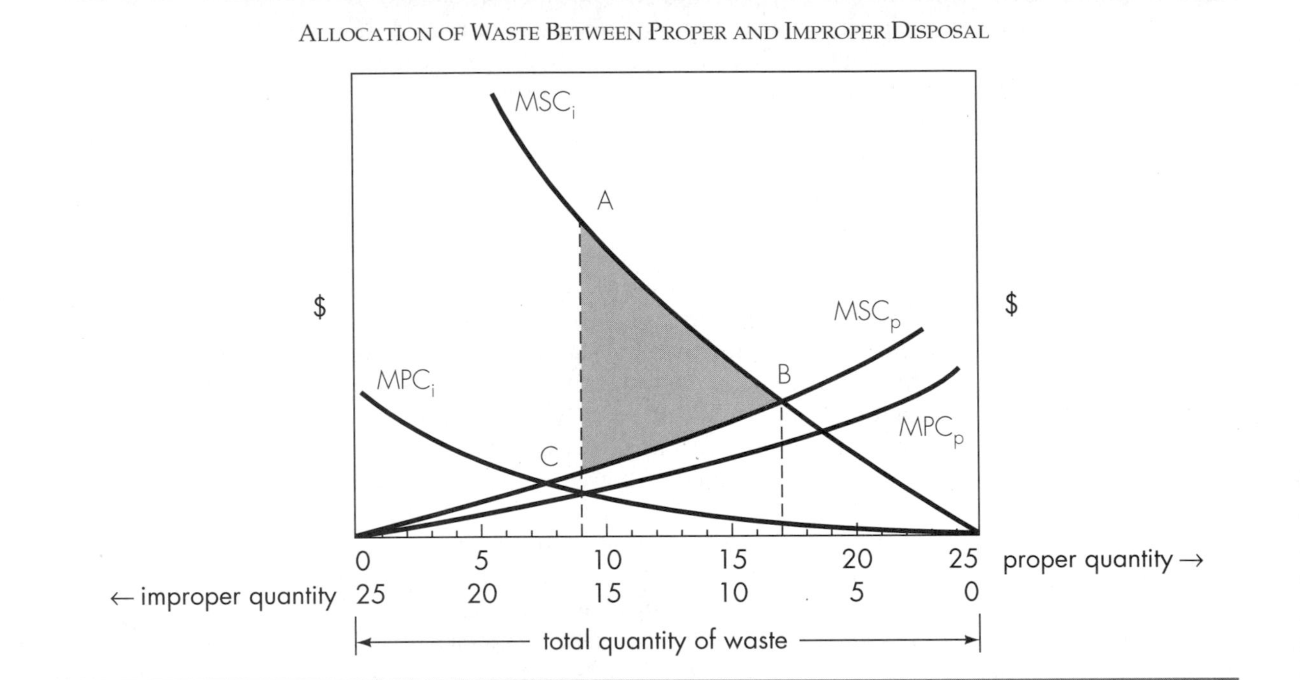

option to the marginal private costs of the other option. In the example diagramed in Figure 9.4, this equating of costs occurs when 9 units are properly disposed of and when 16 units are improperly disposed of. The optimal allocation would be 17 units properly disposed and 8 units improperly disposed, which occurs where the marginal social costs of both options are equal. The excess social cost associated with being at the market equilibrium instead of the optimal solution is equal to the area of triangle ABC.

For the purposes of expositional simplicity, the determination of the optimal amount of waste and the determination of the allocation of wastes across disposal options have been presented separately. In actuality, these decisions are mutually dependent and should be made concurrently rather than sequentially. Also, instead of the two options for waste disposal that are presented in Figure 9.4, there may be many options, each associated with different social costs.

WHY THE PRIVATE COST OF SOLID WASTE DOES NOT EQUAL ITS SOCIAL COST

Both properly and improperly disposed of solid waste generates environmental externalities. This section will focus on waste that is legally disposed of and leave the issue of illegal dumping for later discussion.

Even waste that is stored in a landfill generates external costs. The landfill itself reduces the aesthetic values of the surrounding land as odors, vermin, and unsightliness reduce the benefits that people can derive from using the surrounding land. Additionally, landfills can generate environmental harm, especially to ground and surface waters. Even if the landfill is lined with impermeable clay or plastic, which keeps water from moving downward through the waste into the groundwater, water that has been contaminated by the waste can move back upward through the garbage (this is called leachate) and be carried by rainwater run-off into surface water or groundwater under adjacent land. This leachate can carry a variety of pollutants from the mixed waste that is present in landfills. Although solid waste disposal sites are not intended to store toxic wastes, toxic wastes inevitably make their way into these disposal sites. Many household and commercial products contain toxic substances, including batteries, pesticide containers, cleaners, paint, solvents, and automotive oil.

In addition to the environmental externalities, a variety of market failures are caused by inappropriate government actions. Since trash disposal is often a government provided or government regulated activity, it is incumbent upon the government to charge a price (or set a regulated price) equal to marginal social cost. The price of trash disposal is generally not equal to marginal social cost for three reasons. First, the environmental costs associated with waste disposal are not incorporated into price. Second, the scarcity value of landfill space is not incorporated into market price. Finally, price is often based on average cost rather than marginal cost.

The lack of incorporation of environmental cost into market price has already been discussed in this book in other applications (air pollution, for example). The other two sources of the disparity between price and marginal social cost require more discussion, however.

Municipalities that operate landfill services generally charge a fee per ton of dumped garbage (called a tipping fee), which is designed to recover the costs of operating the landfill. However, since landfill sites are a scarce and exhaustible resource,[8] there is an opportunity cost or scarcity value associated with burying a ton of waste today. This opportunity cost has seldom been incorporated into the tipping fee at a landfill. As municipalities see their future trash disposal options becoming increasingly limited, they have attempted to incorporate this scarcity value or opportunity cost into tipping fees to reduce the volume of garbage and to encourage the development of alternative waste disposal options. However, this adjustment often requires raising tipping fees from the vicinity of $5 to $10 per ton to over $40 or $50

---

[8]Very few sites are physically suited for a landfill. A necessary requirement is that there be a water impermeable barrier between the area that will contain the trash and the underlying area. Of course, even if a site has the appropriate physical characteristics, it is not necessarily an acceptable landfill site. People who own land in the vicinity of a potential landfill will generally object to its nearby location. This objection is known as the NIMBY (Not In My Back Yard) syndrome, which makes it even more difficult to establish new landfill areas.

ECONOMIC INCENTIVES AND THE NIMBY SYNDROME

The NIMBY (Not in My Back Yard) Syndrome refers to the political difficulty associated with finding acceptable locations for waste disposal sites. Even though everyone recognizes the need for waste disposal sites, no one wants the sites to be located in their own community. Although the waste disposal site will create net benefits for society as a whole, the social benefits of the site are spread out among all members of society, while the social costs of the waste disposal site are borne primarily by the people living in the vicinity of the site. Everyone wants the sites to be established, only "someplace else." Of course, if every community takes this attitude, it will be very difficult to ever find a place to put the waste disposal site.

In Chapter 4, the concept of a "potential Pareto improvement" was introduced. A potential Pareto improvement occurs when an action creates more benefits for those that benefit than costs for those who are hurt by the action. The idea is that if the losers are compensated for their losses, the potential Pareto improvement could become an actual Pareto improvement, where some people are made better off, but no one is made worse off.

Clearly, the establishment of a needed waste disposal site would constitute a potential Pareto improvement, but not an actual Pareto improvement unless the community where the site is located is compensated for environmental costs which are generated by the waste disposal site. Economists have suggested a plan by which the potential Pareto improvement can be turned into an actual Pareto improvement. The plan would involve the following steps:

- Sites with the requisite physical and geographic characteristics would be identified.
- The communities in which the eligible sites are located would be given an opportunity to bid on the minimum payment they would require from the waste disposal site operators to allow the site in the community. This payment could take the form of property tax rebates for current residents, increased funding of school systems, or the provision of other public goods. Communities could make the bid as high or low as they wanted.
- The site would be located in the community which required the lowest payment to accept the waste disposal facility.

Many economists think this plan is desirable, because it creates an incentive for communities to accept the facilities, and at the same time makes the recipient community better off, since they will not accept the facility unless the compensation is greater than the social costs imposed by the plant. However, this plan is often criticized by noneconomists, who argue that some communities have less economic and political power at the beginning of the bargaining process, so the poor and less powerful communities will always wind up with lower environmental quality. The advocates of this system argue that poor communities should have the opportunity to accept the environmental costs in exchange for a more than offsetting increase in the quality of life, such as better schools. What do you think?

per ton. Needless to say, voters would not react well to such a dramatic increase in the city's price for trash disposal, which makes politicians hesitant to fully incorporate the scarcity value of landfill sites into the price which the city or county charges.

The final problem associated with the pricing of waste disposal services is that the pricing of the services is often based on average cost rather than on marginal cost. With average cost pricing, the total cost of providing services is calculated and then averaged across customers. Customers receive a monthly bill that is independent of the level of waste that they generate. Consequently, consumers do not have an incentive to conserve on the amount of waste that they purchase and dispose of into the waste stream.

WASTE AND RECYCLING

Recycling has received prominent attention as a possible solution to the solid waste problem, as well as contributing to the solution of the problems of mineral depletion and excessive use of energy. An increase in recycling would mean we would use less minerals, have less mining to disturb the environment, use less energy (it takes less energy to recycle a material than to make new material from mineral resources), and have less waste to dispose of in landfills.

Why do many people believe that too little recycling occurs? The answer to this question has two major components. First, the market failures that prevent the social cost of waste from being reflected in either the market price of a product or the market price of waste disposal make it less profitable to recycle. Second, inertia that is built into our economic system makes it difficult for recycling to become established.

The effect of failure to incorporate social costs of materials and waste disposal into market price is shown in Figure 9.5. In this graph, it is assumed that recycled materials and virgin materials (made from extracted resources) are viewed by the consumer as identical. Therefore, one can refer to the demand for materials in total, and develop a marginal private cost curve for materials by horizontally summing the marginal private cost curves for recycled materials and for virgin materials. Similarly, one can also derive a marginal social cost curve for materials in general by summing the marginal social cost curves for recycled materials and for virgin materials. In Figure 9.5, it has been assumed that there is a disparity between the marginal social cost curve and the marginal private cost curve for virgin materials, but that the marginal social cost curve for recycled material is equal to the marginal private cost curve for recycled material. Although there are externalities associated with recycled materials, they are not as severe as those associated with virgin materials. Consequently, even though it is not strictly true, the marginal private costs for recycled materials are assumed to be identical to the marginal social costs to allow us to include one less curve in an already crowded graph.

The aggregate (both recycled and marginal) private cost curve intersects the inverse demand curve at q_1 units of materials, with r_1 representing the

FIGURE 9.5

THE MARKET FOR MATERIALS AND RECYCLING

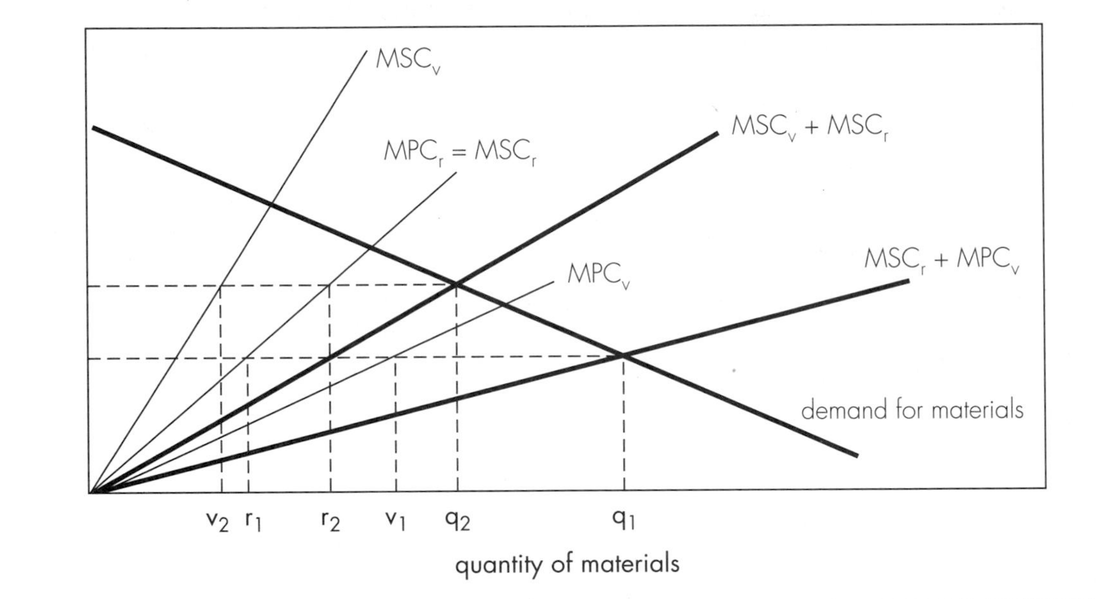

amount of recycled material, and v_1 representing the amount of virgin material. (Note that $q_1 = r_1 + v_1$.) This set of quantities represents the total amount of material (q_1), and the allocation between recycled (r_1) and virgin materials (v_1) that market forces would generate. The optimal level would occur at q_2, where marginal social costs of materials are equal to marginal social benefits. Note the optimal allocation of material between recycled and virgin is r_2 and v_2, and that $r_2 > r_1$ and $v_2 < v_1$. This allocation indicates that the market failures that generate the disparity between the MPC and MSC of virgin materials causes us to use too much virgin material and to engage in too little recycling. This problem is exacerbated because energy inputs do not reflect their true social costs (see Chapter 8). Since the production of virgin materials is more energy intensive than the production of recycled materials, the relative disparity between private cost and social costs is increased.

Another reason we may have less recycling than socially optimal is because of the "chicken and egg" problem. In order for there to be an extensive amount of recycling of materials, the cost of recycling must be relatively low. However, in order for the cost of recycling to be relatively low, an extensive amount of recycling must take place to achieve economies of scale in recycling. In other words, a lot of recycling must take place to lower costs, but a lot of recycling will not take place unless the cost is low.

FIGURE 9.6

LONG-RUN AVERAGE COST AND SHORT-RUN AVERAGE COST

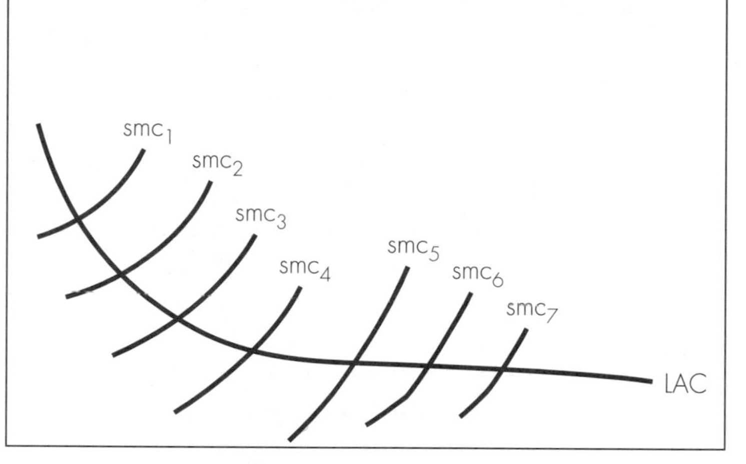

This process is illustrated in Figure 9.6. When the amount of recycling is low, the short-run marginal cost curve is relatively high, such as smc_1 in Figure 9.6. For example, in order to recycle paper, it must be collected where paper is used (urban areas) and taken to the places where paper is made (rural forested areas). As more and more paper is recycled, the infrastructure for recycling is expanded. For example, new paper factories will locate close to urban areas rather than in forested areas. Trash collection companies will develop sorting mechanisms and storage areas for recycled paper. All of these changes will lower the short-run marginal cost curve and cause a movement down the long-run average cost curve, where economies of scale will be realized.

This potential to achieve economies of scale is used as an argument for government intervention in international economics. If a particular industry is not established in the home country, the home country might provide protection from foreign competition to allow the industry to become established in the home country and to expand to the point where it achieves the low point or the flat region of the long-run average cost curve. Then the protection can be removed, because the industry will have achieved the low costs associated with economies of scale and can compete with the foreign competition without the need to continue protectionist policies.[9]

[9]The potential of reducing long-term average costs is also used as an argument for government support of research and development.

In a similar vein, one might argue the need for government intervention to allow the recycled materials industry to achieve the lower costs associated with economies of scale and to compete with the virgin material industry. In addition to incorporating the social costs of virgin materials into the price of virgin materials, the government could take other steps to help encourage recycling. It could do this in its role as a consumer rather than its role as a regulator.

Since the government is such a large consumer of materials (particularly paper to fuel the bureaucracy), it could require a certain percentage of recycled materials in the products that it purchases. This step in itself would move the recycling industry a substantial distance down its long-run average cost curve. The government could also introduce labeling requirements in which manufacturers were required to specify the amount of recycled material in their products, so that consumers with a preference for recycled materials could exercise those preferences.

A COMPREHENSIVE MATERIALS POLICY

The first step in developing a comprehensive materials policy would be to correctly price both materials and waste disposal services. There are a variety of options available to do this, at both the federal and local level.

FEDERAL MATERIALS POLICY

Although many people tend to think of materials policy as a local problem, there are many dimensions in which federal policy can reduce inefficiency. These areas include mining and minerals policy as well as solid waste policy.

The federal government can adopt direct controls or economic incentives to help reduce the environmental externalities associated with mining. Of course, many of these environmental externalities are already regulated by the Clean Air Act and Clean Water Act and a variety of mining laws and rules governing the use of federal lands. These policies tend to rely primarily on direct controls, although liability and bonding systems are often used to ensure the restoration of mined lands. Additionally, the government would need to maintain stockpiles of strategic materials to reduce the vulnerability associated with potential disruptions in supply.

An area in which the federal government could utilize economic incentives to increase social welfare is in helping to reduce the quantity of solid waste that is produced. A packaging tax, which is based on the volume (and/or weight) of the packaging of a product could be imposed at any governmental level. Similarly, a tax on the total amount of material in a product could be instituted at any governmental level. However, if this tax was sporadically instituted by a few local or state governments, it would not have much of an effect on solid waste, even in these jurisdictions. The reason is

Are the Benefits of Beverage Container Deposit–Refund Systems Greater than the Costs?

Deposit–refund systems for beverage (primarily beer and soft drinks) containers have been adopted in a number of states. The idea behind these systems is that there is an externality associated with improper disposal of beverage containers, and the social costs can be lowered by the recycling of the cans and bottles that contain the beverages.

This reduction is achieved by forcing the consumer to pay the external costs of disposal "up front." When the beverage is purchased, the consumer is required to pay a deposit that is forfeited if the container is not returned to an authorized collection center. Since the cost of noncompliance is paid when the product is purchased, no monitoring is necessary.

Such a system will generate less roadside litter and reduced exhaustion of landfill capacity. Additionally, since the manufacture of new containers consumes more energy than the recycling of old containers, the deposit–refund system will lessen the externalities associated with energy consumption.

However, such a system is not without costs. The bottles and cans must be collected, transported to authorized refund centers where the refund is issued, and transported to the factories where the containers will be cleaned for reuse, and processed into new containers or processed into other products. In other words, the system will increase the cost of providing beverages. Another cost that is often neglected in discussions is consumer inconvenience. Consumers must rinse the containers, store them, and transport them to refund centers.

Porter (1983) conducted a study of Michigan's experience with a beverage deposit–refund system and measured both the reduction in social costs associated with landfills and litter and the increase in consumer inconvenience costs. He did not reach a definite conclusion, but showed that the relationship between the benefits and costs of the deposit–refund system were sensitive to the value that was ascribed to reducing the aesthetic costs associated with roadside litter and to the proportion of the post-system price increase that was caused by the imposition of the system. (One cannot assume that the price increase was entirely due to the deposit–refund system because the beverage industry is oligopolistic [at both the production and wholesale level], and the price increases could be partially due to oligopolistic forces.)

If the costs of a deposit–refund system turn out to be larger than the benefits, there are alternative options that can be used in place of a deposit–refund system, but separate policies would have to be used to address the separate problems of littering and generating more recycling. Littering must be addressed by a system of direct controls (such as fines for littering) or by moral suasion programs ("Give A Hoot, Don't Pollute"). However, economic incentives other than deposit–refund systems may be effective in promoting recycling. For example, if households paid for trash on a marginal cost basis (per unit weight or per unit volume), and they were given recycling bins in which to place their cans and bottles, people would have a strong incentive to participate in such a voluntary recycling program. The advantage of such a program is that it reduces the inconvenience costs of households having to return the containers to a refund center. Such a program also has the advantage that other materials (such as newspapers, corrugated cardboard, and non-beverage cans and bottles) can be included in the voluntary recycling program. The key to the success of such a program is that people have to pay for each pound of garbage that they throw away, rather than paying a monthly fee for garbage collection.

that manufacturers might regard these few areas with the tax as being markets that are too small to warrant the development of special product lines or manufacturing processes. However, if the taxes were instituted at the federal level, the entire market would be affected and manufacturers would have an incentive to reduce material inputs because the cost of the inputs has increased. Differential taxes could be implemented with lower taxes levied on materials that are more easily recycled. Throughout the production process, manufacturers would have an incentive to reduce their use of material inputs and switch to materials that were more easily recycled. Similarly, consumers would have an incentive to buy products that would be associated with less waste.

For example, many products are sold with a large amount of packaging to call attention to the product and to discourage shoplifting. If the packaging became more expensive for manufacturers and retailers, they would substitute other attention grabbing and/or antishoplifting measures.[10]

While a packaging tax could reduce the quantity of waste that is generated and the associated environmental effects, it is clearly a second best policy. In the best of all worlds, both materials and waste would be priced according to marginal social cost. However, if institutional barriers or political inertia interfere with the ability to correctly set prices, a packaging tax may serve to increase social welfare.

In addition to reducing the quantity of waste, it is important to have an impact on the quality of wastes. Policy instruments are necessary to remove some of the more hazardous wastes from the waste stream, so that these wastes are not buried in landfills or incinerated in garbage burning plants. For example, flashlight batteries contain nickel, cadmium, and other heavy metals that lead to adverse effects on human health and the health of ecosystems. A deposit–refund system could be placed on batteries so that they would be removed from the waste stream and handled separately, and so that the heavy metals can be recycled into new batteries. Similarly, the unused contents of home pesticide containers, paints, solvents, drain cleaners, and other household chemicals can be extremely hazardous. A deposit–refund system on these containers (which would be allowed to be returned partially filled) could help remove these toxic chemicals from the waste stream. Of course, the chemical container deposit–refund system would be more difficult to implement than a deposit–refund system for flashlight batteries, as flashlight batteries could be safely handled by the retail stores at which the batteries were purchased. It probably would not be a good idea to bring partially used containers of pesticides and other household chemicals into the retail stores in which they were purchased as part of the deposit–

[10]It is interesting to note that music CDs and audio tapes were formerly packaged in large plastic containers to deter shoplifting. Primarily as a result of consumer demand for less waste, the CDs and audio tapes are now encased by removable (by store personnel) and reusable plastic cases.

refund system. A better idea would be to have a central toxic waste handling facility with appropriate facilities and trained personnel in order to reduce the hazards of handling and properly disposing of these wastes.

Many European countries are experimenting with producer liability for waste. In these systems, the producer is responsible for taking back the product after its useful life is over, and for paying the costs of disposal. In many ways, this liability system internalizes the cost of the externality associated with waste disposal. For example, a car manufacturer would then have an incentive to make the car in a fashion which enhances its recyclability. While this option provides many useful incentives and can reduce the volume of waste, it does not necessarily result in the minimum cost of waste reduction, as would the correct pricing of waste. However, the producer liability approach may be a good approach when correctly pricing waste is difficult, or if consumers do not perceive the price of disposal as being incorporated into the price of the product.

State and Local Solid Waste Policies

Since many of our waste problems are location specific, the solution to many of these waste problems must be developed by state and local governments. The single most important action would be to develop a system for financing the collection and disposal of solid waste that incorporated the full social cost of waste into the price of collection and disposal. Homeowners, commercial establishments, and industrial establishments must be required to pay a price for trash removal that is based on the amount of trash that has been removed.

Two obstacles can keep this marginal cost pricing from taking place. First, as mentioned earlier, incorporating the scarcity value of landfill space into the market price of trash removal will result in a drastic increase in garbage removal fees, which will generate voter dissatisfaction. Second, there is a perception that marginal cost pricing will be difficult to monitor or expensive to implement.

There is little that can be done about the first problem except educating people about the extent of the solid waste problem, the difficulty of developing options for the disposal of waste, and the importance of correctly pricing waste disposal in order to create disincentives to the generation of waste. In addition, the political opposition to increasing fees may be blunted if the scarcity value of landfill space is gradually incorporated into tipping fees and garbage collection fees.

The perception that it is expensive or administratively unfeasible to move to a marginal cost pricing system for waste removal is probably a misperception. While weighing each bag or can of trash and then charging the customer on this basis would truly be unwieldy, there are a variety of options that could be implemented to move toward marginal cost pricing. Some of these options have been adopted in several communities (see boxed example). One example of a marginal cost pricing system is an "official bag"

EXPERIENCE WITH MARGINAL COST PRICING OF GARBAGE DISPOSAL

Many economists are strong advocates of marginal cost pricing of garbage disposal. Rather than charging a monthly fee, economists argue that it is important to charge a price based on each unit of garbage generated. As the discussion in the text of this chapter emphasizes, economic theory argues that marginal cost pricing of garbage disposal will give both producers and consumers an incentive to reduce the amount of garbage that is generated.

However, there are often practical issues involved in implementing economic solutions such as marginal cost pricing of trash. Will the system which looks good in theory work when implemented in the real world?

Morris and Byrd (1990) examined this question by analyzing the impact of marginal cost pricing (which they term unit pricing) on three U.S. communities that had switched to this form of pricing. Their results are presented below:

ESTIMATED DAILY WASTE PER RESIDENT (POUNDS PER PERSON PER DAY)

	PERKASIE, PA		ILION, NY		SEATTLE, WA	
	BEFORE UNIT PRICING	AFTER UNIT PRICING	BEFORE UNIT PRICING	AFTER UNIT PRICING	BEFORE UNIT PRICING	AFTER UNIT PRICING
Waste generated	2.5	1.9	2.6	1.9	2.8	2.9
Waste recycled	0.2	0.7	0.1	0.3	0.5	0.5
Mixed waste collected	2.2	1.2	2.5	1.6	2.3	2.3

SOURCE: Glenn Morris and Denise Byrd, Unit Pricing for Solid Waste Collection, Popular Government, Fall 1990, Table 3, p. 13.

system. Garbage may only be accepted for collection if it is contained in an "official bag" that is printed with a logo or some other identifying feature. The price of the bag would be set to include the full cost of waste disposal, including the scarcity value of landfill space. Alternatively, one could adopt a sticker system, where every bag or can of trash must have an "official sticker" attached to it. The garbage collection personnel would remove the sticker from the can after it has been emptied, so a new sticker would be required for the next collection. The price of the sticker should be equal to the full social cost of the disposal of the waste. See the boxed example for a discussion of recent experience with marginal cost pricing of trash.

Recycling programs also must be local in nature, since the cost of sorting garbage (separating the recyclable items from the nonrecyclable items) is lower when the recycling is done at the household or establishment level, as the trash is generated. If the garbage is collected into large truckloads and removed to a central facility before sorting, the sorting process becomes more complicated and more expensive. Many communities have adopted mandatory or voluntary recycling programs and attempted to reduce homeowner inconvenience by allowing the homeowner to place all recyclable items in one bin. The garbage collection personnel then sort the recyclable items as they remove them and place them in a truck or trailer that has a separate compartment for each class of recyclable good. The most common types of materials that are included in these recycling programs are aluminum cans, ferrous cans, glass, some types of plastics, newspapers, and corrugated cardboard.

In addition, many communities no longer allow yard waste (leaves, lawn clippings, branches, and so on) to be mixed with garbage. These communities generally have separate pickup for these yard wastes and compost them into mulch, rather than having them take up space in the landfill. The mulch is available for gardening and landscaping at a nominal delivery charge. This program turns a waste product into an economic product.

Either of these two marginal cost pricing systems (the official bag and official sticker systems) would make the homeowner perceive a cost associated with each piece of garbage that he or she throws away. This type of pricing system can drastically improve the response to a voluntary recycling program, because the homeowner will save money by placing recyclable items in the recycle bin, saving the need to purchase as many "official bags" or "official stickers." Similarly, the homeowner would have an incentive to separate yard waste from the collected trash, either by composting the waste in a backyard compost pile, by self-delivery to the municipal compost site, or by having it collected for delivery to the municipal compost site.

PROBLEMS WITH ILLEGAL DISPOSAL

There is a potential problem with both increasing the cost of legal waste disposal and with moving to marginal cost pricing of waste. This problem arises because such policies increase the incentives for illegal disposal, as people try to avoid the increased costs of legal disposal. Since the private cost of proper disposal has increased, we would expect to see more dumping. This will increase environmental damages, as the waste is released into the environment in general, rather than at a waste disposal facility.

In order to avoid an increase in the level of illegal dumping, policies must be developed to increase the expected penalty for illegal dumping. Since the private cost of proper disposal has increased, more illegal dumping will occur unless there is an increase in the fine for illegal dumping and/or an increase in the probability of being caught or both. The former is easier to do, but the latter requires increasing the amount of resources devoted to monitoring and enforcement. Since these monitoring and enforcement costs are part of the costs of waste disposal, they should be incorporated into the price that people pay to dispose of their wastes.

SUMMARY

Although the historic focus of material policy has been on the conservation of resources, the externalities and other market failures associated with our use of materials may be of much greater concern. These market failures occur throughout the cycle of use, from the mining of minerals to the disposal of economic products when their useful life is over. A variety of policies are necessary to reduce the economic inefficiency associated with these market failures.

Mining is associated with the destruction of natural environments as well as with pollution that affects air and water quality. A variety of direct controls and economic incentives can be utilized to mitigate these externalities.

In addition to the environmental externalities, market failures associated with imperfect competition and inappropriate government intervention affect the market for minerals. A very important federal policy is the depletion allowance, which reduces the private cost of mining and leads to an inefficiently high level of mining.

Many market failures are also associated with waste disposal. The environmental externalities associated with waste disposal are made more severe by a variety of policies that reduce the private cost of waste disposal below the market price. Chief among these are prices based on the average cost of waste disposal rather than on the marginal cost.

We will continue to have a solid waste problem as long as these market failures remain uncorrected. The creation of more landfill sites or incinerators to process our increased volume of waste merely treats the symptom of the problem, rather than addressing the cause of the problem. Stated quite simply, the problem is that waste disposal is underpriced, and as a consequence we generate too much waste.

REVIEW QUESTIONS

1. What are the market failures associated with mineral extraction?
2. Assume that a particular community generates 10 million beverage containers per year. For beverage containers that are disposed in a landfill, let

   $$MSC_L = 0.05 + 0.01Q_r$$

   where Q_L is the quantity of containers disposed in the landfill and is measured in millions of cans. Let the marginal social cost of recycled cans be

   $$MSC_r = 0.03 + 0.02Q_R$$

   where Q_R is the quantity of containers that are recycled and is also measured in millions of cans. Calculate the social cost minimizing allocation of cans between landfill disposal and recycling.
3. Why might the market generate a less than socially optimal amount of recycling?
4. What is the impact of a depletion allowance on the cost of exhaustible resources?
5. What is the effect of average cost pricing of trash disposal services?

QUESTIONS FOR FURTHER STUDY

1. What should be done (if anything) to encourage more recycling?
2. What are the advantages and disadvantages of a packaging tax?

3. What is the role for deposit–refund systems in material policy?
4. Is there a role for marketable permits in material policy?
5. How should policy deal with national security problems surrounding strategic materials?

SUGGESTED PAPER TOPICS

1. Investigate the success of policies to encourage recycling. Begin with papers by Jakus and Tiller (1994); Hong, Adams, and Love (1993); and Hopper and Nielson (1991). Search recent issues of the *Journal of Environmental Management, Journal of Economic Literature, Forum, Nature,* and *Science.* Also, check for stories in major newspapers, such as *The Wall Street Journal, The New York Times,* the *Chicago Tribune,* and so on, as well as your local newspaper. Use your understanding of environmental economics to suggest policy refinements that could increase the amount of recycling.
2. Look at the impact of the mining industry on the environment and policies that have been developed to regulate this impact. How does current policy create appropriate incentives? Can policies be changed to be more efficient? Search bibliographic databases using mining and environmental policy, mining and environmental impact, and mining and environmental regulation as keywords. Check government documents of the Department of Interior (Bureau of Mines) and the EPA.
3. Investigate the issue of the impact of oil production on wetlands. How does current policy create appropriate incentives? Can policies be changed to be more efficient? Begin by examining the publications of the Louisiana Sea Grant Institute at Louisiana State University. Also, examine the wetlands references at the end of Chapter 13. Search bibliographic databases on oil and wetlands, oil and environmental impacts, and energy policy and wetlands.

WORKS CITED AND SELECTED READINGS

1. Anders, Gerhard, et al. *The Economics of Mineral Extraction.* New York: Praeger, 1980.
2. Gocht, W. R., H. Zantop, and R. R. Eggert. *International Mineral Economics.* New York: Springer-Verlag, 1988.
3. Hong, S., R. M. Adams, and H. A. Love. "An Economic Analysis of Household Recycling of Solid Wastes." *Journal of Environmental Economics and Management* 25 (1993): 136–146.
4. Hopper, J. R. and J. M. Nielson. "Recycling as Altruistic Behavior: Normative and Behavioral Strategies to Expand Participation in a Community Recycling Program." *Environment and Behavioral* 23 (1991): 195–220.
5. Jakus, Paul, and Kelly Tiller. "Rural Household Recycling: Explaining Participation and Generation." Department of Agricultural Economics, University of Tennessee, unpublished paper, 1994.
6. Jordan, A. A., and R. A. Kilmarx. *Strategic Mineral Dependence: The Stockpile Dilemma.* Beverly Hills: Sage Publications, 1979.
7. Krutilla, John. "Conservation Reconsidered." *American Economic Review,* 1967.
8. Krutilla, John, and A. C. Fisher. *The Economics of Natural Environments.* Washington, D.C.: Resources for the Future, 1985.
9. Morgan, J. D. "Stockpiling In the USA." In *Concise Encyclopedia of Material Economics, Policy and Management,* edited by Michael Bever, Oxford: Pergamen Press, 1993.
10. Morris, Glenn, and Denise Byrd. "Unit Pricing for Solid Waste Collection." *Popular Government* 56 (1990).
11. Raven, Berg, and Johnson. *Environment.* New York: Saunders College Publishing, 1993.
12. Rudawsky, Oded. *Mineral Economics: Development and Management of Natural Resources.* New York: Elsevier, 1986.
13. Porter, Richard C. "Michigan's Experience with Mandatory Deposits on Beverage Containers." *Land Economics,* 59 (1983): 177–194.
14. Porter, Richard C. "The New Approach to Wilderness Preservation through Benefit–Cost Analysis." *Journal of Environmental Economics and Management* 9 (1982): 59–80.
15. Tilton, John E. *Mineral Wealth and Economic Development.* Washington, D.C.: Resources for the Future, 1992.
16. Young, John E. "Mining the Earth." In *State of the World 1992,* edited by Lester Brown, New York: Norton, 1992, p. 106.

Renewable Resources and the Environment

Part III examines renewable resources and the interaction between renewable resources, the economy, and the environment. Renewable resources are analyzed both as harvestable outputs and as components of ecological systems.

Fisheries

And they spoke politely about the current and the depths they had drifted their lines at and the steady good weather of what they had seen.

Ernest Hemingway, The Old Man and the Sea

INTRODUCTION

The ocean currents and depths have not changed from the period of which Hemingway wrote, but his old fisherman would be very surprised by the way fisheries have changed. Modern fishing technology, coupled with increased demand and open access exploitation of fisheries, has driven many fish stocks to such low levels that they are threatened with extinction. For example, the Gulf Stream marlin, swordfish, and tuna that Hemingway's old fisherman pursued have declined precipitously in the last several decades. The Florida Bay and virtually every estuary and embayment in the world have become threatened by externalities from human economic and social activity. As rivers carry pollution into the estuaries, oil and chemical spills exact their ecological toll, and upstream water withdrawals increase the salinity of these delicate but vital ecosystems.

Although fish may not be the first environmental resource that comes to mind when the student thinks of environmental resources that deserve attention, the world's fishery resources are important for several reasons. Fish are a major source of protein for a large portion of the world's population. Even though fish are a renewable resource, they are destructible. Overexploitation and environmental change, such as pollution and loss of wetlands, threaten this important resource. Populations of many important species have declined markedly in the last several decades, causing hardships for fishing communities and lowering the quality of life of people who take pleasure from recreational fishing.

The decline in our fishery stocks has led to increased conflict among user groups. Fishing nations compete for limited stocks on the waters outside each country's 200-mile exclusion zone. Commercial fishermen and recreational anglers clash over who has access to limited near shore stocks of steelhead and salmon as they swim up the rivers to spawn. In developing nations, commercial fisheries are in conflict with those fishing families who fish solely

TABLE 10.1

MARINE FISHERIES, YIELD AND ESTIMATED POTENTIAL

AREA	AVERAGE ANNUAL CATCH (MILLION METRIC TONS)* 1977–79	1987–89	POTENTIAL (MILLION METRIC TONS)
Atlantic Ocean	22.35	21.50	29.30 to 37.21
Pacific Ocean	30.96	48.44	34.11 to 49.48
Indian Ocean	3.38	5.44	5.31 to 8.01
Mediterranean Ocean and Black Sea	1.21	1.65	1.17 to 1.53
WORLD	58.34	77.55	68.82 to 96.23

SOURCE: From *World Resources, 1992–1993* by World Resource Institute. Copyright © 1993 by World Resource Institute. Used by permission of Oxford University Press, Inc.
*This measure of world marine catch includes marine fish, cephalopods, and crustaceans.

to feed their families, and as fish stocks decline, the ability to feed their families is seriously compromised. These conflicts make the development of fisheries policy even more difficult.

In the United States, most commercially harvested species are harvested in saltwater, with the notable exception of crawfish. Some other freshwater species (catfish and trout) are important commercially, but these are mostly reared in tanks and artificial ponds and not caught in the wild. Most states in the United States have laws banning the commercial harvesting of freshwater species, although Native Americans have special fishing rights under various federal treaties.

Populations of many commercially harvested species in the United States have declined during the past several decades. The World Resource Institute references a study by the National Fish and Wildlife Foundation showing that although 17 percent of species are increasing, 30 percent are declining. Table 10.1 presents the global fisheries picture. The potential catch[1] is the maximum amount that could be harvested year after year in a sustainable fashion. It is interesting to note that for the Pacific and Indian oceans, the current catch levels are in the range of the estimates of potential catch, indicating that further exploitation of the fisheries may actually reduce catch. In the Mediterranean and Black seas, average catch in 1987–89 actually exceeded potential catch, implying that fishery stocks must be declining. More recent data which incorporates the 1990s shows that the problem has contin-

[1]The potential catch of this table is exactly the same as maximum sustainable yield, a concept that is more developed in subsequent sections of this chapter.

ued even longer. The United Nations Food and Agriculture Organization reports that 13 of 17 major ocean fisheries are in serious trouble.

Recreational fishing is also very important in the United States and other countries. According to the U.S. Fish and Wildlife Services (USFWS), approximately 35 million adult Americans (over age 16) participated in recreational fishing in 1991. These anglers engaged in approximately one-half billion days of fishing and spent approximately $35 billion on fishing related expenses.

FISHERIES BIOLOGY

Fish are like any other animal in that they require food and oxygen in an appropriate habitat. Fish reproductive strategy is generally based on the principle of large numbers. Each reproducing female generates large numbers of eggs, in some cases numbering in the millions per female. However, very few of these eggs grow and survive to reproductive maturity.

The reproductive potential of a fish population is a function of both the size of the fish population and the characteristics of its habitat. In order to better understand the relationship between growth and the size of the population, we shall initially assume that the characteristics of the habitat are held constant.

Both the growth of the population and the population itself are measured in biomass (weight) units, not by the number of individual organisms. Thus, growth can occur through the production of new organisms or the growth in the mass of existing organisms.

Figure 10.1 depicts a logistic growth function. As shown in this graph, there is no growth when the population is zero. This point is fairly obvious, but it warrants some discussion. If there are no fish to reproduce or get larger, there can be no growth. As population increases, the amount of growth increases. For example, at a population level of X_1, the annual growth will be G_1, and the rate of change of growth will be the slope of the growth function at X_1. The maximum growth rate is often referred to as the intrinsic growth rate.

Although the amount of growth will initially increase, the rate of change of growth is constantly declining. Eventually, the rate of change of growth becomes zero at X_2, and the amount of growth is at a maximum (G_2). For population greater than X_2, the rate of change of growth is negative and declining. Eventually, the amount of growth falls to zero, which occurs at the maximum population at K (population cannot get bigger if growth is zero).

Biological factors generate the constantly declining growth rate and the eventual decline in the level of growth toward zero. For example, when population is low, the resources of the ecosystem are large relative to the needs of the population, enabling rapid growth. However, as the population grows, the surplus is eliminated, and there is competition for resources, such as food, spawning areas, and nursery areas (areas where juvenile fish can

FIGURE 10.1

LOGISTIC GROWTH FUNCTION

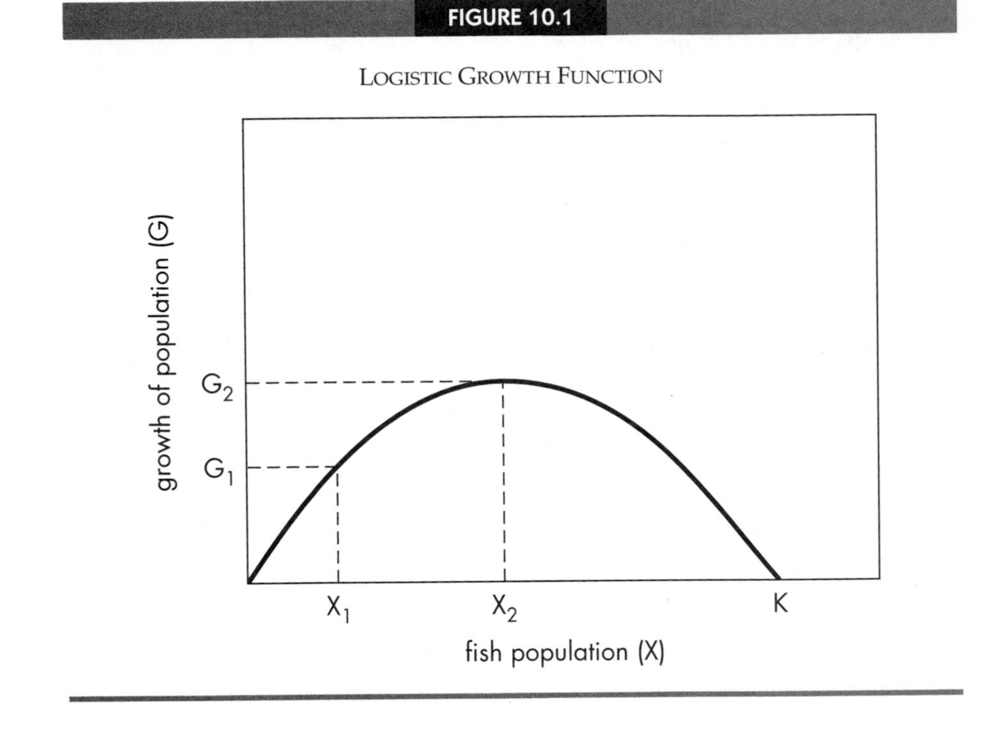

hide and escape predation). Also, as the population grows, the incidence of disease and parasites will grow, and cannibalism may increase as well.

All of these biological factors tend to slow growth as population increases. At K the environment cannot support additional growth, so population remains constant. The point represented by K is often referred to as the carrying capacity of the environment and is a biological equilibrium. By a biological equilibrium, it is meant that once the population reaches K, it will remain at K. For obvious reasons, zero population is also a biological equilibrium. Any level of population between zero and K is not in equilibrium, as positive growth will move the population toward K.[2]

The growth function of Figure 10.1 represents a particular type of growth function, where the growth rate is always declining. The technical term for this type of growth function is a compensated growth function.

---

[2]Mathematically, the logistic growth function can be represented as

$$G = rX(1 - X/K)$$

where r represents the limit of the growth rate as X approaches zero, K the carrying capacity, and G represents growth as a function of a variable population (X).

FIGURE 10.2

DEPENSATED GROWTH FUNCTION

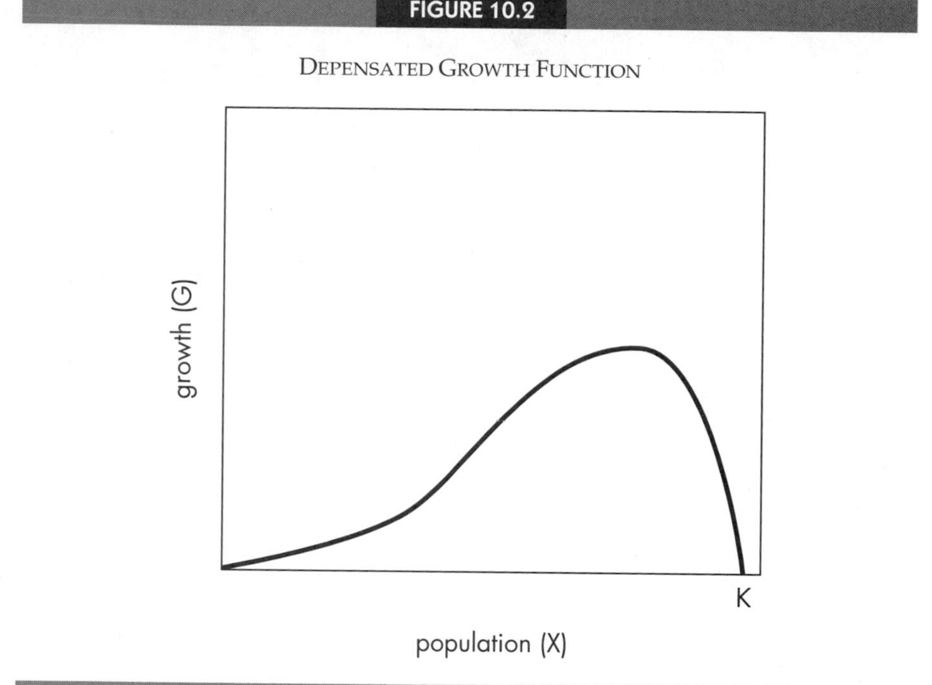

Figure 10.2 contains a depensated growth function, where the growth rate initially increases and then decreases. Figure 10.3 contains a critically depensated growth function. In this function, X_0 represents the minimum viable population. If population falls below this level, growth becomes negative and population becomes irreversibly headed toward zero.

THE OPTIMAL HARVEST

So far we have examined the behavior of fish populations independent of their interaction with humans. The interaction is important because the ability to harvest fish is influenced by the level of the fish population (also called the stock of fish), and the stock of fish is influenced by the level of the harvests.

In order to determine how harvesting affects a fish population, we will examine a growth function in Figure 10.4 and add a harvest equal to C_1. Note that since both growth and harvest are measured in biomass units, both growth and harvest can be expressed on the vertical axis of the graph.

FIGURE 10.3

GROWTH FUNCTION WITH CRITICAL DEPENSATION

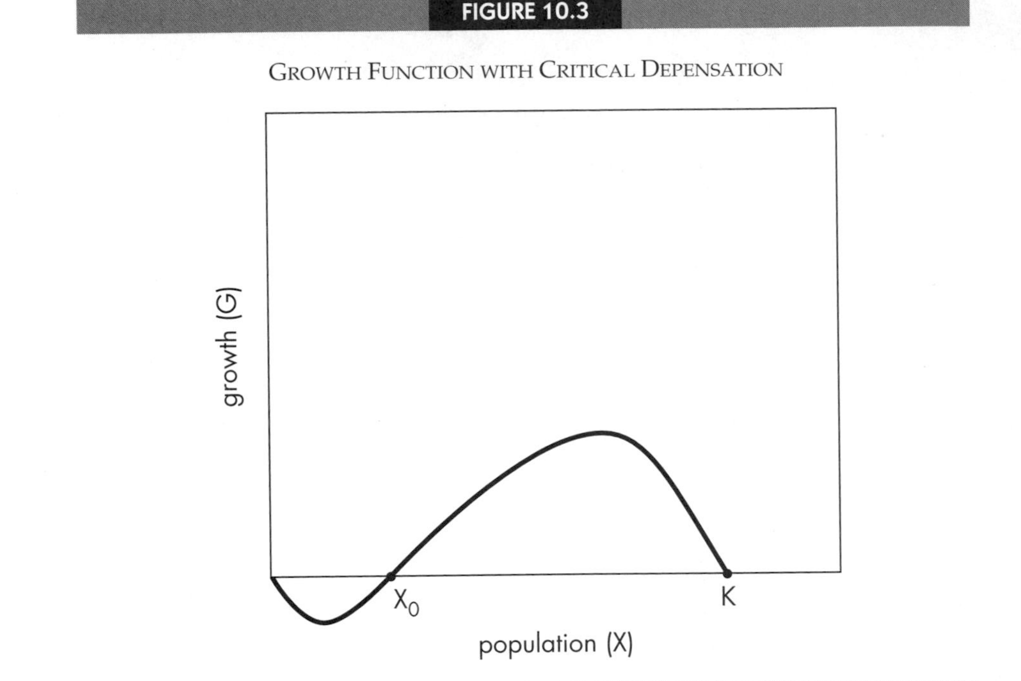

Let us assume that the fishery is initially unexploited, so the population is K. Then, a harvest of C_1 units per year is removed from the fishery. Under these circumstances, the fish population declines, because there is no natural growth, and harvesting is removing a portion of the population. Population will fall toward X_2. At X_2, the population will continue to fall, as the harvest is greater than the growth that a fish stock of X_2 can support. Simply stated, the amount of fish that humans are taking out (C_1) is greater than the amount that nature is putting back in (G_2). Therefore, the population must continue to shrink. This will happen until the natural growth is equal to the harvest. For a harvest of C_1, this equality of harvest and natural growth occurs at X_1.

In Figure 10.5, it is shown that every harvest level but C_{msy} has two equilibrium populations associated with it. For example, a harvest level of C_1 is associated with equilibrium populations of X_1' and X_1''. This means that growth is exactly equal to the harvest level of C_1, and the population will remain unchanged at either of these levels. For example, if the population is equal to X_1'', the harvest can equal C_1 year after year and leave population unchanged at X_1''. For this reason, C_1 is known as the sustainable or equilibrium catch associated with the population level X_1'', and the natural growth function can also be interpreted as an equilibrium catch or sustainable catch

FIGURE 10.4

EQUILIBRIUM CATCH

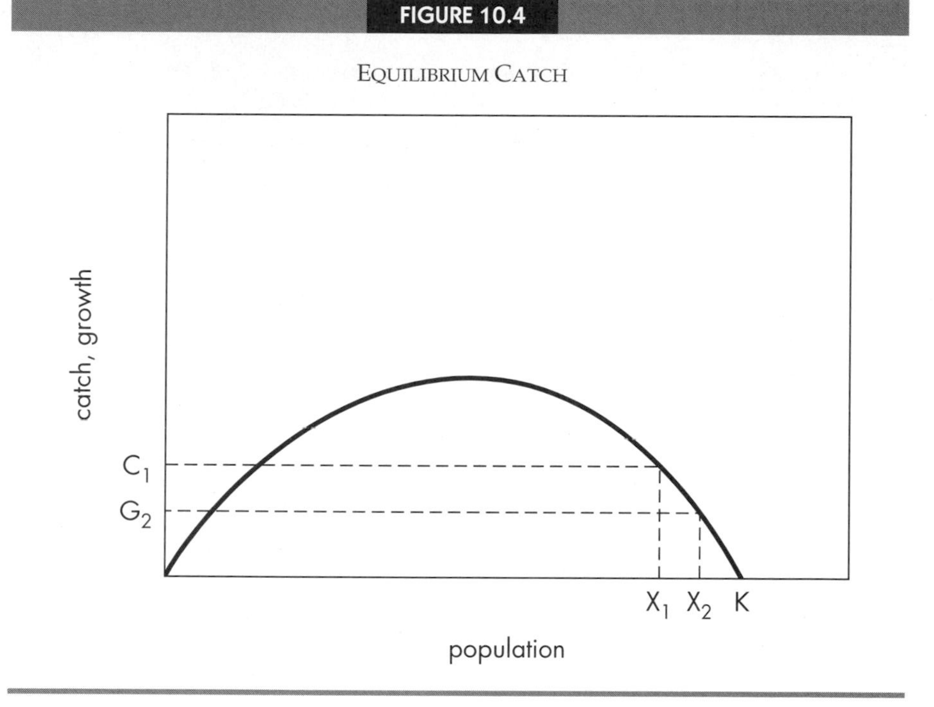

function. In the jargon of fishery management, it is known as a sustainable yield function.

C_{msy} is the maximum sustainable yield, as no level of the fish population can produce growth above this level. In the early discussions of fishery management, the achievement of maximum sustainable yield was the theoretical goal of management policies.

THE GORDON MODEL

An important general finding in economics is that the maximization of a physical quantity will not necessarily maximize the economic benefits of the activity for society. Gordon made this point in a 1955 article, in which he points out that uncontrolled access to fishery resources will result in a greater than optimal level of fishing effort.

Gordon begins his analysis by deriving a catch function that represents a "bionomic" equilibrium, which he does by looking at fishing effort and the relationship between fishing effort, catch, and fish population.

FIGURE 10.5

MAXIMUM SUSTAINABLE YIELD AND THE EQUILIBRIUM CATCH FUNCTION

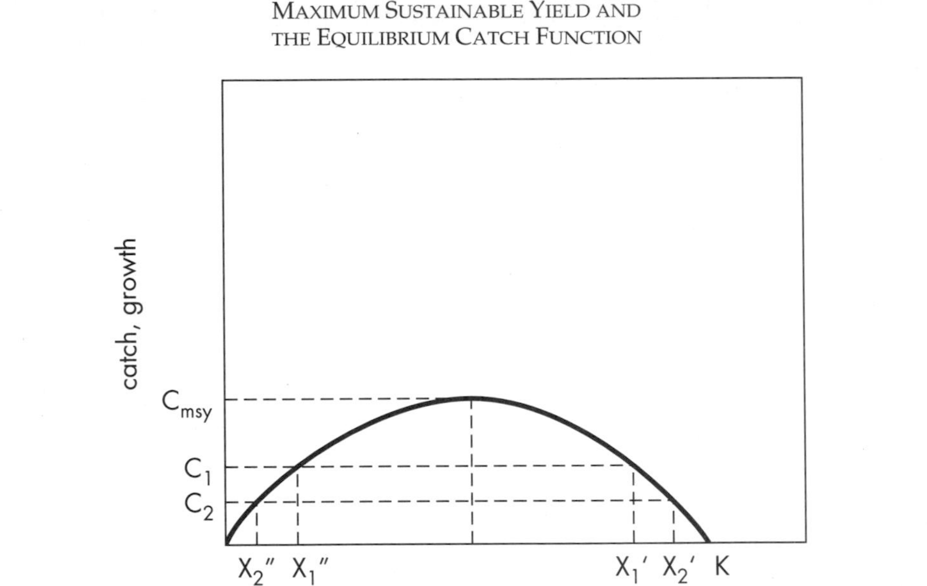

The analysis begins by assuming that, holding effort constant, catch is proportional to the fish population. For example, if effort is held constant at E_1, the catch that would result from that effort is given by the yield function Y_{E_1} in Figure 10.6. If effort is increased to E_2, the yield function shifts up to Y_{E_2}.

Of course, not every point on each yield function is a sustainable yield. The sustainable yields can be found by superimposing the equilibrium catch function on the yield functions of Figure 10.6, as shown in Figure 10.7. As can be seen, only one point on each yield function is a sustainable yield (this is not necessarily true for a depensated growth function). In Figure 10.8, the one point of sustainable catch associated with each level of effort is graphed with the corresponding level of effort, rather than the fish population, on the horizontal axis. This graph is known as the sustainable yield function, because it shows the sustainable catch associated with each level of effort. Notice that as effort increases, sustainable yield increases and then decreases.

A sustainable total revenue function can be derived from a sustainable yield function. In order to simplify matters, Gordon assumes that the price of catch is constant. In order for this assumption to hold, the particular fish population and the catch from that population must be small in relation to the total market for that fish. For example, this assumption would be appropriate for the analysis of weakfish (sea trout) in a particular embayment, such as the Pamlico-Ablemarle Sound in North Carolina. Since the price of weakfish is determined by the catch of weakfish (and substitute species) from the entire Atlantic and Gulf coasts, the catch in Pamlico-Ablemarle Sound will be too small relative to total catch to affect price.

Given the assumption that price is constant, a sustainable total revenue function can be derived by simply rescaling Figure 10.8, as done in Figure 10.9, which also contains a total cost of effort function. Gordon assumes that the marginal cost of effort is constant, so total cost is a linear function.

Gordon suggests that net economic yield, shown as TR-TC, be maximized in order to maximize social benefits. Net economic yield is more often referred to as economic rent in more modern fishery economic literature. Note that although TR-TC looks like a monopoly profit, it is very different. A monopoly profit is generated by restricting output to gain an increase in price. However, price is constant in this model, so there can be no monopoly profit. The economic rent originates in the productivity of the fish stock, as the greater the fish stock, the more fish that can be caught with a given amount of effort. The optimal amount of effort occurs at E_2, which maximizes economic rent. Note that at this level of effort, the slopes of the TR and TC functions are equal, or MR = MC.

FIGURE 10.6

YIELD FUNCTIONS

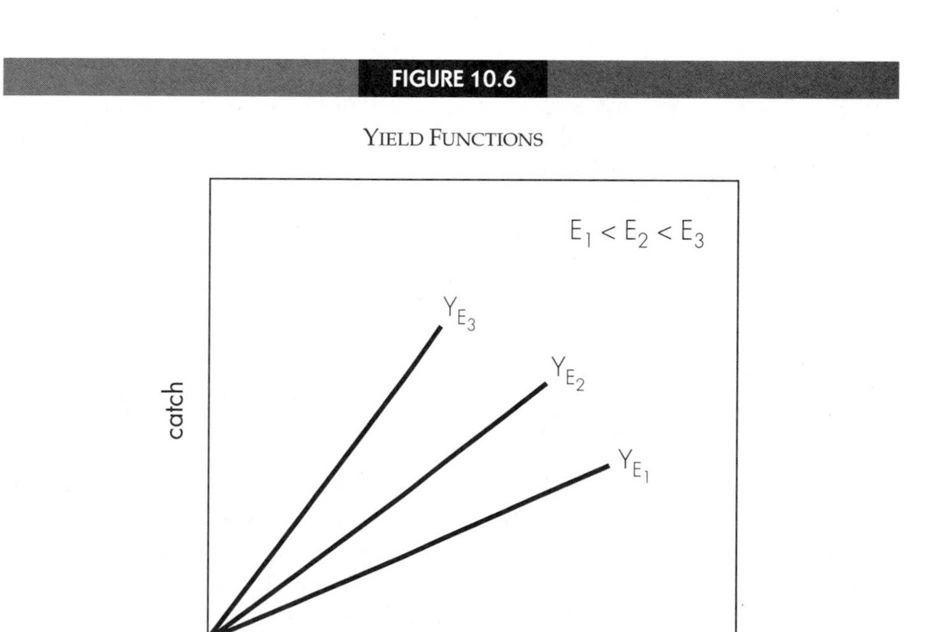

FIGURE 10.7

THE EQUILIBRIUM CATCH FUNCTION AND YIELD-EFFORT FUNCTIONS

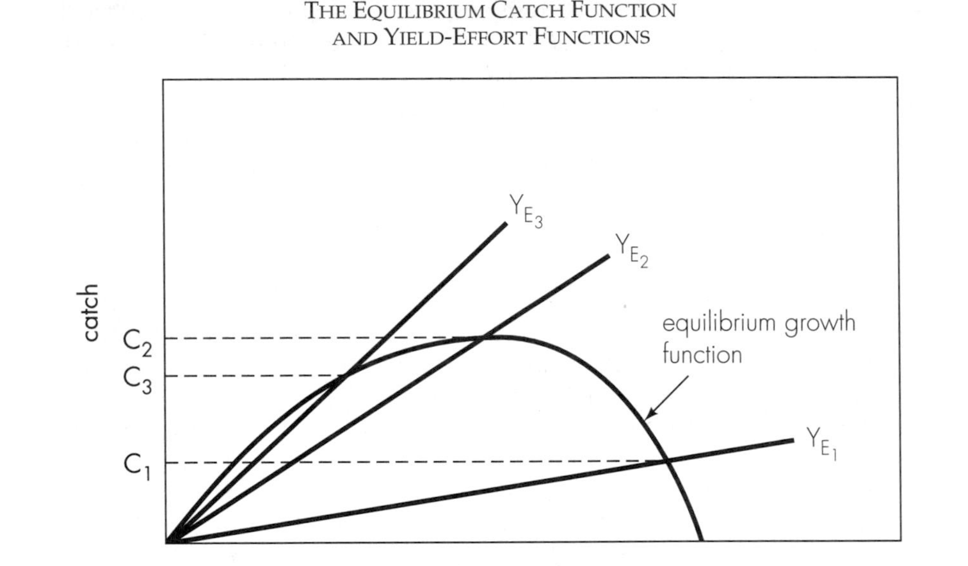

An optimal level such as E_2 is seldom realized in an actual fishery, due to the open-access nature of fisheries in general. Open-access implies that anyone can participate in the fishery. At E_2, economic rents are being earned in the fishery. In general, economic rents are not available elsewhere in the economy, so effort (labor) will enter the fishery in pursuit of these rents. Although the entrance of more effort will cause rents to fall, effort will continue to enter until opportunities in the fishery are equivalent to opportunities elsewhere in the economy, which occurs when there are no rents in the fishery. This occurs at E_1. If effort were to exceed E_1, TR would be less than TC and net losses would occur. Labor would then leave the fishery until the level of effort reached E_1, and there was no incentive to leave.

Notice that at E_1, AR = MC (also AR = AC since MC is constant). The reader may be initially confused by a market equilibrium being determined by AR = MC rather than MR = MC, but this equation is a result of the open-access externality and is based on an interaction among fishers. This interaction can best be illustrated with the aid of Table 10.2, which lists the total, marginal, and average catch associated with effort in a hypothetical fishery.

Note that initially total catch, marginal catch, and average catch increase, then marginal catch (and average catch with it) begins to fall, but total catch continues to increase because marginal catch is still positive. Eventually, mar-

FIGURE 10.8

SUSTAINABLE YIELD AS A FUNCTION OF EFFORT

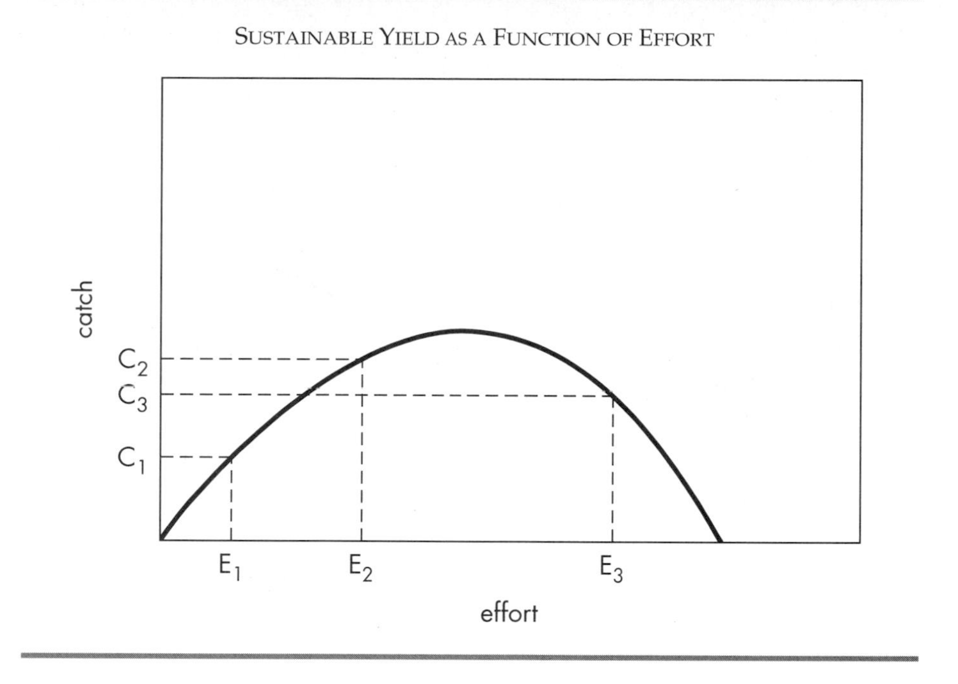

ginal catch becomes negative, and total catch begins to fall. So far, nothing is different from total, marginal, and average product functions for conventional goods.

However, there is a big difference in the allocation of total product across units of inputs (labor or effort). For example, examine the 5th fisher in the fishery. He or she adds $70 of catch to total catch. However, the 5th fisher actually catches $78 of catch. The reason is that if all fishers have the same skill levels, then they each catch the same amount of fish (of course there will be random differences across fishers). In other words, the 5th fisher catches $78 worth of fish, $70 of which represent a new addition to total catch, but $8 of the catch would have been caught by existing fishers. The important point is that when the 5th fisher decides to enter the fishery, he or she compares $78 to his or her opportunity cost, not the $70 of new catch.

This point is very important, because social efficiency requires marginal cost to be equal to marginal product. Let us assume that the opportunity cost of a unit of effort (unit of labor) is $50 per day. That means that if a worker is employed somewhere outside of the fishery, the worker would be creating $50 worth of social product. However, if the opportunity wage is $50, workers will continue to enter the fishery as long as average product is greater

FIGURE 10.9

MAXIMIZING TR-TC

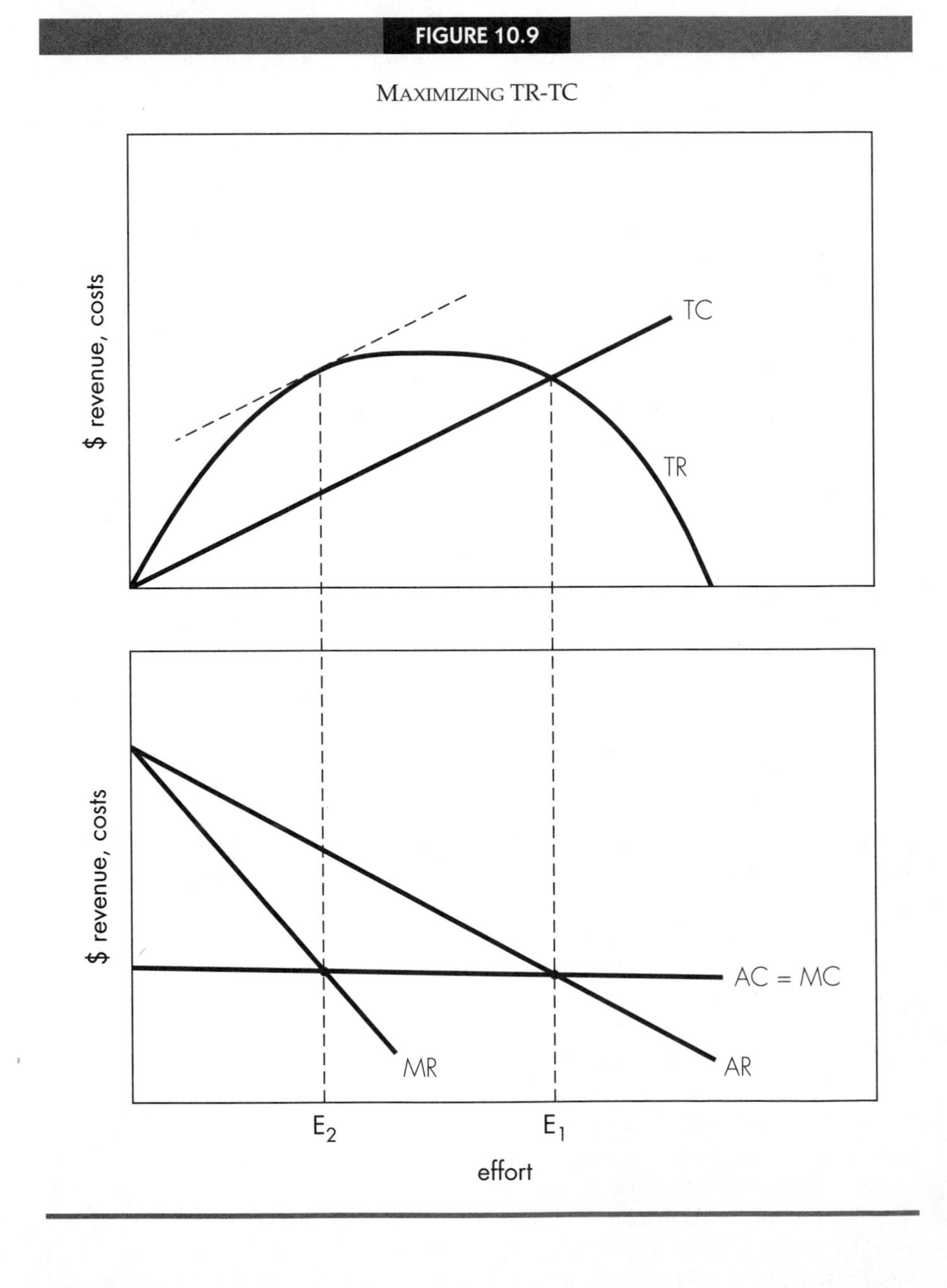

than $50. There is no further incentive to enter once average catch becomes equal to the opportunity wage of $50, which occurs at a level of effort of 12 fishers. Note that at this point, the marginal catch of the 12th fisher is actually zero. The optimal level of effort occurs at 7 fishers, where marginal catch (marginal product in the fishery) and opportunity wage (marginal product in the alternative application) are equal. The optimal number of fishers and the open-access number of fishers are shown in Figure 10.10.

Quite simply, a major inefficiency associated with the open-access resource is that too many resources are used to catch a given number of fish. Gordon suggests monopoly ownership as a way to eliminate this problem.

SHORTCOMINGS OF THE GORDON MODEL

Although Gordon's model is an excellent model for highlighting the problems associated with open-access and too much effort, it has several shortcomings. The foremost of these problems is that the model is static (one period), whereas a dynamic model would be more appropriate. The Gordon model is dynamic in the sense that all equilibrium catch levels are sustainable, but static in the sense that it does not consider future costs and benefits. If $1 of future costs or benefits are viewed to be identical to $1 of present costs or benefits, then the dynamically optimal level of effort will be the same as the statically optimal level (E_2 in Figure 10.9). Future values have the same

TABLE 10.2

THE RELATIONSHIP BETWEEN MARGINAL CATCH AND AVERAGE CATCH
(MEASURED IN DOLLARS, PRICE HELD CONSTANT)

LEVEL OF EFFORT (# OF FISHERS)	TOTAL CATCH	MARGINAL CATCH	AVERAGE CATCH
0	0	0	0
1	70	70	70
2	150	80	75
3	240	90	80
4	320	80	80
5	390	70	78
6	450	60	75
7	500	50	71
8	540	40	68
9	570	30	63
10	590	20	59
11	600	10	55
12	600	0	50
13	590	−10	45
14	570	−20	41
15	540	−30	36

FIGURE 10.10

OPTIMAL AND OPEN-ACCESS EFFORT IN A HYPOTHETICAL FISHERY (BASED ON TABLE 10.2)

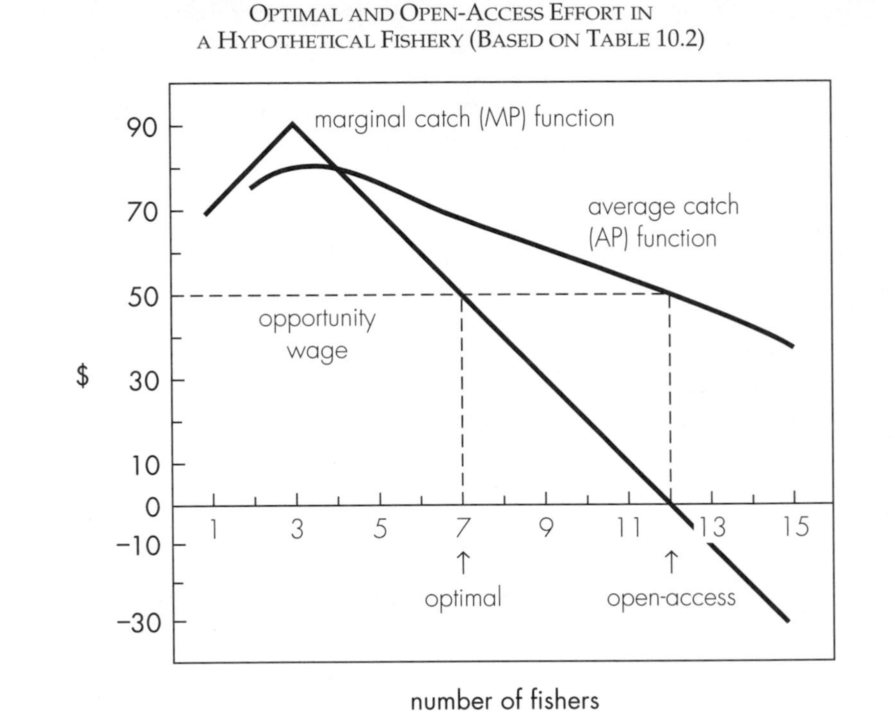

importance as present values when the discount rate is equal to zero. Clark (1985) shows that as the discount rate gets very large, the dynamically optimal level of catch approaches the open-access level of catch. As Anderson (1986) points out, this change in levels is because future income has no value with an infinite discount rate. Consequently, an owner of a fishery would use effort as long as the cost of effort is covered. In the long run, this cost will approximate the open-access level of effort. Discount rates between zero and infinity will imply a dynamically optimal level of effort between the statically optimal level (E_2) and the open-access level (E_1).

As Anderson points out, many other factors affect the optimal level of effort in a real-world fishery. In particular, pulse fishing (catching a large fraction of the population and then not fishing for several years to let the population recover) may be the best approach for certain fisheries.

INCORPORATING CONSUMERS' AND PRODUCERS' SURPLUS INTO FISHERY MODELS

Another shortcoming of the Gordon approach is that it does not consider consumers' and producers' surplus, which may exist and be important in many fisheries. The lack of consideration of consumers' and producers' surplus may be a particularly important emission for highly valued, but threatened fish such as various species of salmon, Alaskan king crab, and redfish.

Turvey (1964) was the first to highlight the possible importance of producers' and consumers' surplus. However, his model is relatively complicated and difficult to implement empirically. Rather than follow Turvey's approach, conventional demand and supply curves will be integrated into a fishery model. These types of models are developed and presented in Anderson (1986), Clark (1985), and Kahn (1985, 1987). The models discussed in this chapter will most closely parallel those of Kahn.

When switching from catch and effort models to demand and supply models, the horizontal axis is no longer measured in units of effort, but in units of catch (catch is what is demanded and supplied). While models based on effort are good for illustrating the open-access problem of inefficiently high levels of effort, they are difficult to implement empirically. One of the reasons for this difficulty is that it is not clear how effort should be defined. Should it be measured solely in labor units (that is, person hours)? This scheme is not good because different people may have different levels of capital available (that is, bigger and more powerful boats, larger nets, etc.). Effort is really an amalgam of labor, capital, and energy, and there are no clear guidelines on how to combine these separate inputs into one aggregate input. When the estimation of supply functions are discussed, it will be seen that the definition of effort is less of an issue when catch is the quantity variable, rather than effort.

Let us begin the discussion by drawing conventional demand and supply curves for catch, as in Figure 10.11. A single demand and supply curve is insufficient to describe all the important changes, as changes in catch will change the level of the fish stock, and the level of the fish stock is not measured on either axis.

As is often the case when a third variable needs to be presented in a two-dimensional framework, it can be done through a family of functions. Since greater fish populations imply that more fish can be caught with the same amount of inputs, a family of supply curves can be drawn, each defined for a different level of the fish stock, as illustrated in Figure 10.12. The greater the fish population, the closer the supply curve to the horizontal axis. Each level of the fish population has a unique supply curve associated with it.

In Figure 10.12, there are a multiplicity of potential economic equilibria. However, it must be recognized that these levels of catch, which represent

FIGURE 10.11

DEMAND AND SUPPLY CURVES

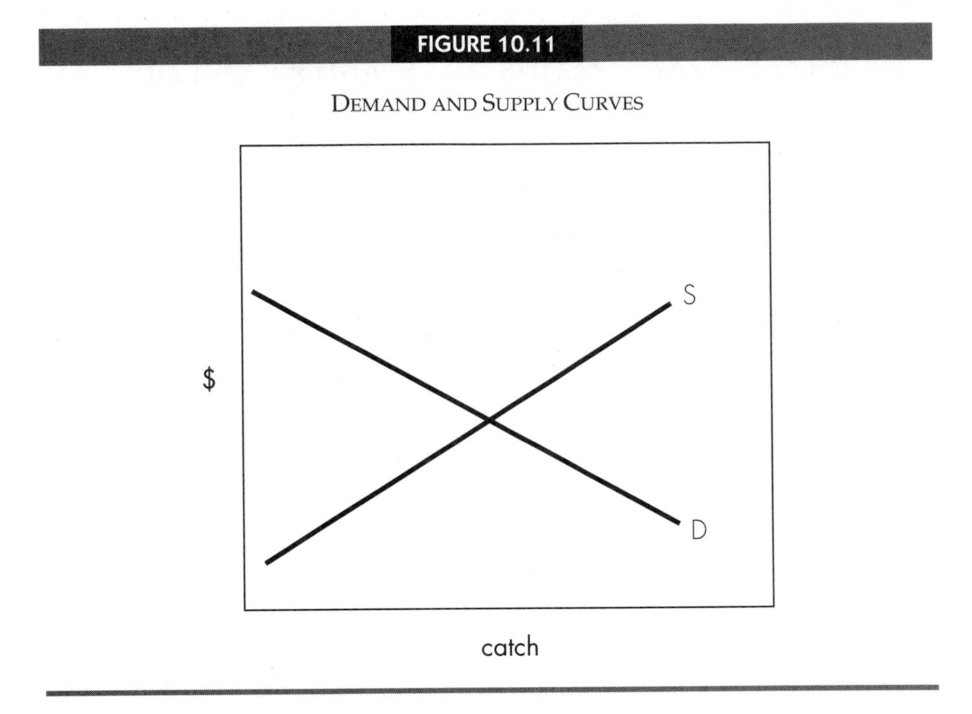

economic equilibria, are not necessarily sustainable levels of catch. As mentioned earlier, there is only one equilibrium (sustainable) catch associated with each level of population. For the six population levels depicted in Figure 10.12, the equilibrium catches are depicted in Figure 10.13.

These equilibrium catch levels are then mapped onto the family of supply curves in Figure 10.14. For example, the equilibrium catch associated with a population of F_1 (the maximum population) is zero, and point A is the only point of biological equilibrium on supply function S_{F_1}. Similarly C_2 is the equilibrium catch associated with a fish stock of F_2, so point B is the only point of biological equilibrium on supply curve S_{F_2}. This process can be repeated for each level of the fish stock, and a locus of biological equilibria can be found. This locus is presented in Figure 10.15.

Notice that a market equilibrium could occur at point E, where S_{F_6} and the demand function intersect the locus of biological equilibria. Also, notice that the sole owner of a fishery could find a better point than E. Point F in Figure 10.15 is also a point of biological equilibrium, but on a lower supply curve because the fish stock has been maintained at a higher level. The net economic benefits are shown in Figure 10.15. At point E, there would not be rent, but if the catch was produced at point F on supply curve S_{F_3}, the net economic benefits would include rent (area PEFB), consumers' surplus (area PDE), and producers' surplus (area BFA). The goal of fisheries management

FIGURE 10.12

A FAMILY OF SUPPLY CURVES FOR COMMERCIAL FISHING

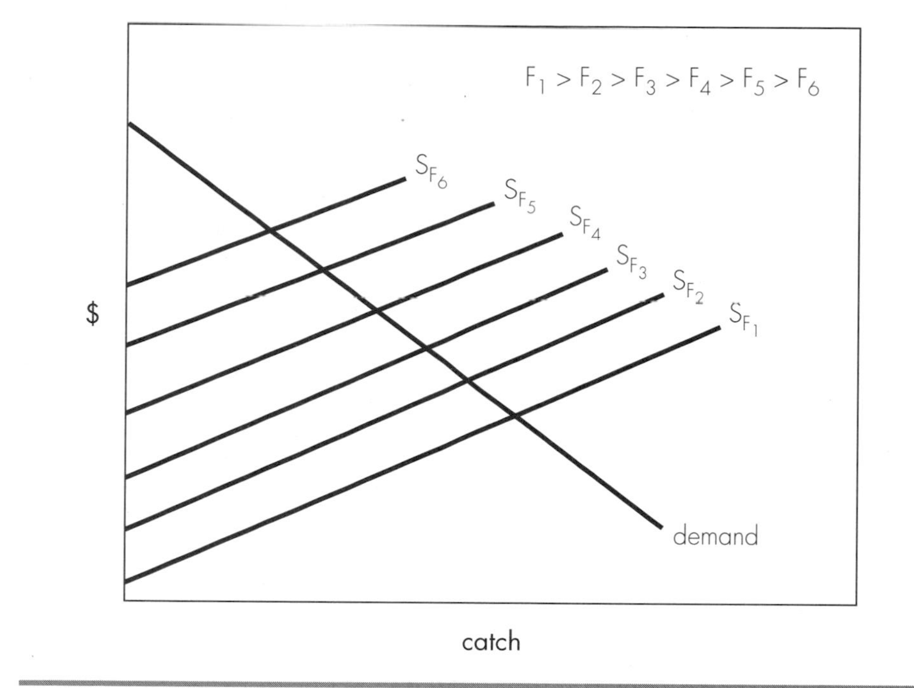

would be to choose a point along the locus of biological equilibria that maximizes the sum of these three sources of benefits. Although it is easy to find this point mathematically, there is no tangency or intersection to point to in the graph to reveal the optimal point of catch and population.

In addition to using this model to maximize the economic benefits derived from a fishery, it is possible to use the model to examine other types of fishery management problems. For example, pollution is thought to be a major factor in the decline of striped bass in the Chesapeake Bay. It is possible to use this model to compute the fishery-related damages from this pollution.

Let us assume that the pollution does not affect the edibility of the fish (that is, through toxic or bacterial contamination) but does stress the ecosystem in which the fish resides. For example, lower dissolved oxygen levels could adversely affect many organisms and diminish the productivity of the food web. The effect of pollution of this nature would be to shift the natural growth function (which is also the equilibrium catch function) downward, as in Figure 10.16. Note that the pollution has lowered the carrying capacity

FIGURE 10.13

EQUILIBRIUM CATCH FUNCTION

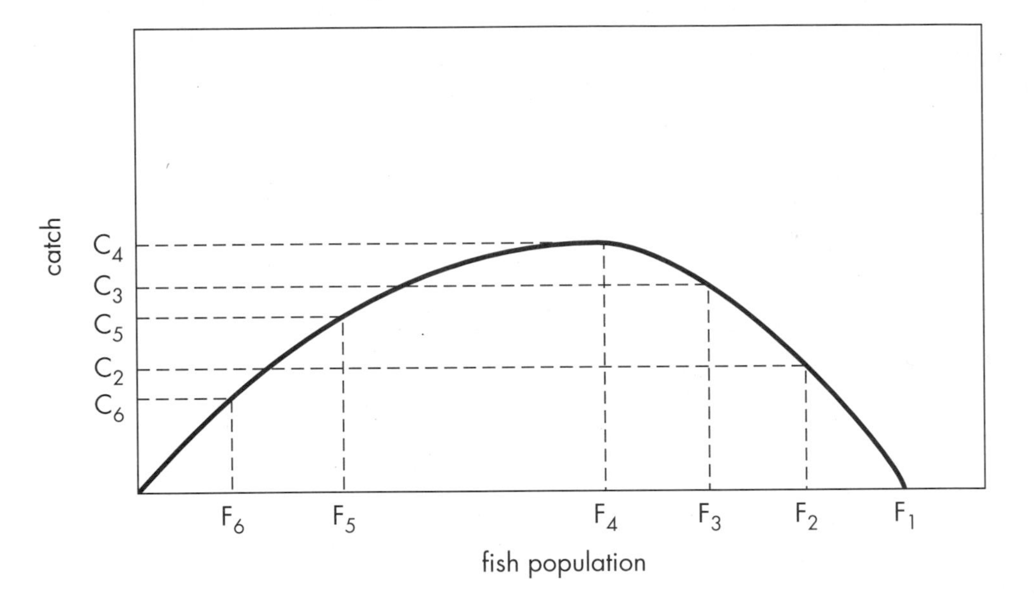

of the environment, as well as the growth that can be supported by a particular level of the fish stock.

The downward shift of the equilibrium growth function implies an inward shift of the locus of biological equilibria from B_1 to B_2 (Figure 10.17). Note that the Chesapeake Bay striped bass fishery was an open-access fishery (it is closed now). Most U.S. fisheries are open-access. Appendix 10a discusses the measurement of consumers' and producers' surplus in the context of an open-access fishery, and it also illustrates how to measure the change in consumers' and producers' surplus associated with a change in environmental quality.

CURRENT FISHERY POLICY

Past fishery policy has been focused on biological regulation, designed to protect the stock from overexploitation and potential collapse. In the past, little attention has been paid to the problem of the inefficiencies generated by open-access. Anderson terms those polices that can actually address the problem of entry as "limited-entry" techniques. Anderson calls all other reg-

FIGURE 10.14

SUPPLY FUNCTION
AND SUSTAINABLE CATCH LEVELS

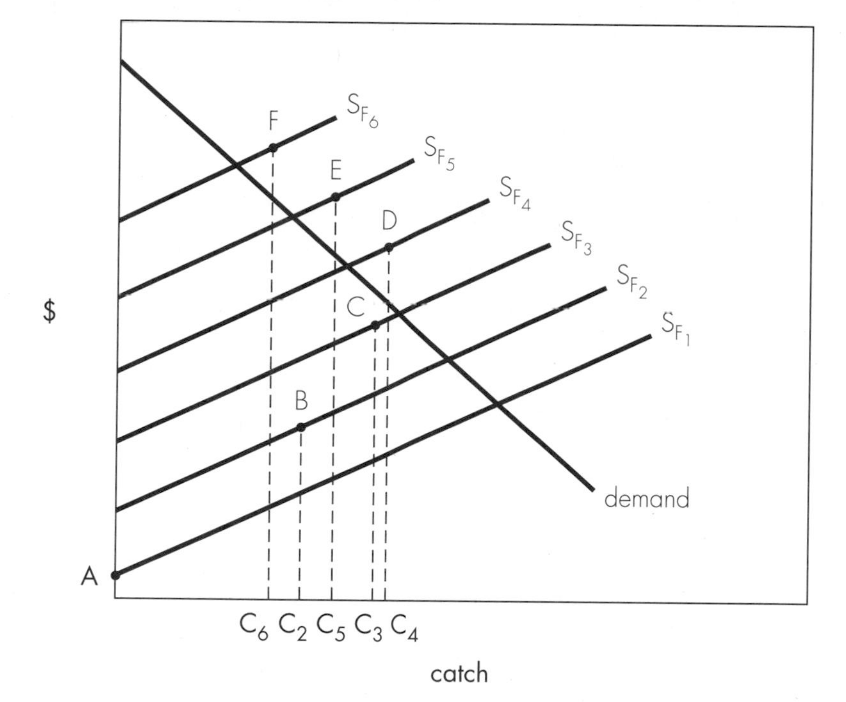

ulations or policies that do not explicitly address the problem of entry as "open-access" techniques. Both of these types of policies are discussed below. For a more in-depth discussion, see Anderson (1986).

OPEN-ACCESS REGULATIONS

Open-access techniques modify fishing behavior of those participants in the fishery without directly affecting participation in the fishery. However, since open-access techniques typically raise the cost of fishing, they may indirectly affect participation in the fishery by causing the marginal fisher to become unprofitable and leave the fishery.

These open-access techniques are designed to maintain fish stocks at some target level. The fish stocks consistent with maximum sustainable yield were often the theoretical target of fishery management, although oftentimes management schemes were not put into place until stocks had shrunk well below the level consistent with maximum sustainable yield. Open-access

FIGURE 10.15

BIOECONOMIC EQUILIBRIUM

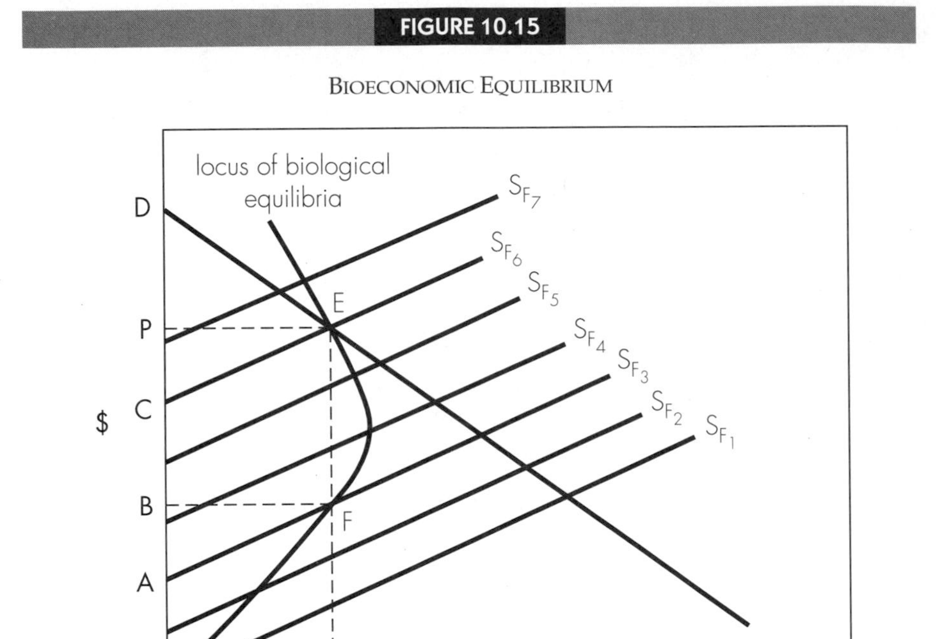

regulations generally take the form of restrictions on how fish may be caught, which fish may be caught, when fish may be caught, where fish may be caught, and how many fish may be caught.

How Fish May Be Caught. Modern fishing technology can give a fishing fleet tremendous fishing power, relative to the size of the fish populations. Sonar and spotter planes are used to locate fish. Nets and fishing lines have lengths measured in kilometers, not meters. Although such technology could generate tremendous cost savings in a properly managed fishery, its use in an open-access situation can be disastrous for a fish stock.

In open-access fisheries, it is possible to protect the fish stock by forcing inefficiency on the fishers. For example, in Maryland's portion of the Chesapeake Bay, it is illegal to dredge for oysters under motorized power. Consequently, dredging must be done under sail power, which means the boat must pull a smaller dredge and cover less bottom in a given time period.

Which Fish May Be Caught. The regulation of which fish may be caught generally revolves around restrictions on the minimum size of fish that are

FIGURE 10.16

DOWNWARD SHIFT IN EQUILIBRIUM CATCH
FUNCTION FROM INCREASED POLLUTION

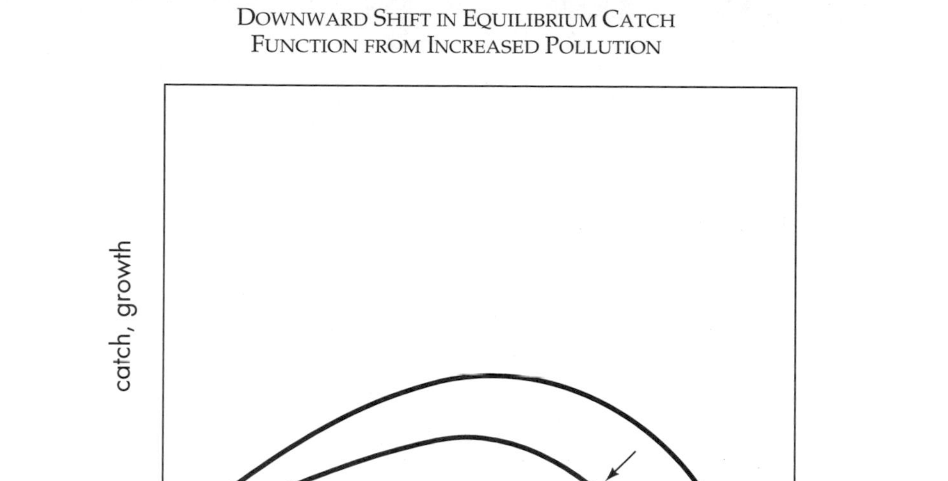

legal to harvest. The reasoning behind these regulations is to leave a portion of the fish stock in the water to provide a sufficient breeding stock to ensure future populations. Fishers generally implement this restriction by choosing a mesh size for their nets that allows the illegally small fish to pass through the net, but retains the larger legal fish.

In actuality, these regulations do not entirely conform to the biology of the situation, because it is the middle-aged fish that represent (pound for pound) the most prolific breeders in the fish stock. However, it would be impossible to design regulations that correspond to this biology, because one cannot construct a net that allows the medium-sized fish to pass through, but captures the largest and smallest fish.

When Fish May Be Caught. Regulations concerning when fish may be caught are designed to control harvests by restricting the times during which fishing is legal. Sometimes these regulations take the form of restricting certain periods on a daily basis, but more often the fishing season is closed for a certain period on an annual basis. Many times the closed season occurs during the spawning season. There are two reasons for closing fishing during the spawning season: First, the fishing activity may disrupt the spawning process. Second, some species become so extremely congregated during spawning that fishing effort could capture virtually the entire population.

FIGURE 10.17

INWARD SHIFT OF LOCUS OF BIOLOGICAL EQUILIBRIA FROM INCREASED POLLUTION

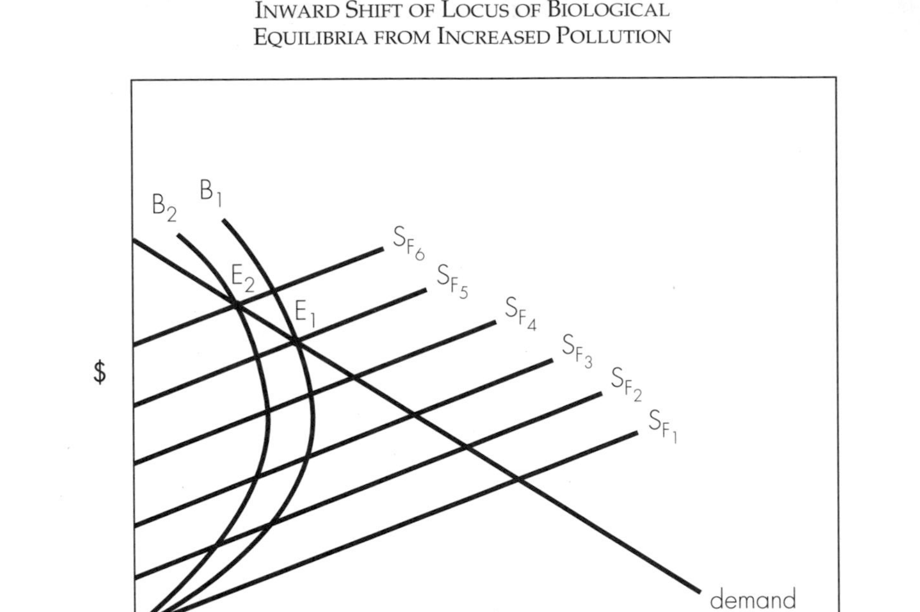

However, the congregation associated with spawning actually forms the basis of the fishery of some species, particularly anadromous species such as salmon.[3] In addition, for some species (sturgeon, shad, herring, etc.) the eggs (caviar) are the most valuable component of the fish.

Where Fish May Be Caught. Regulations on where fish may be caught are designed to protect fish stocks when they are congregated. For example, certain types of commercial fishing are banned in many embayments along the Atlantic coast, because the populations are so congregated that they are vulnerable to overharvesting. These types of regulation are also designed to protect vulnerable fishing habitats from being destroyed by the fishing process. The dragging of bottom-scooping nets and dredges through shallow

[3]Anadromous fish are those that live their adult lives in saltwater and spawn in freshwater.

areas may destroy important plant and animal communities on the floor of these areas.

How Many Fish May Be Caught. Often, open-access regulations take the form of limits on how many fish (sometimes measured in weight, sometimes measured in volume [that is bushels], and sometimes measured in number of fish) may be captured in a given time period. For example, the number of giant bluefin tuna off the Atlantic coast of the United States has declined precipitously in recent years. These fish, which usually weigh more than 500 pounds and often may weigh over 1,000 pounds, are highly sought for sushi for the Japanese market. In 1986, the dockside price for giant bluefin tuna was as high as $18 per pound. With such profitability, other forms of open-access regulation may not be effective in preventing the population from collapsing, so a limit of one fish per boat per day was established to limit exploitation. Even with this limit, the population of giant bluefin tuna is dangerously declining.

ECONOMIC ANALYSIS OF OPEN-ACCESS REGULATIONS

Economic Analysis of Open-Access Regulations. As mentioned previously, the primary effect of open-access regulations is to raise the cost of catching fish. If individual fishers are already operating in the most cost-effective manner (minimizing the private cost of catching fish), any restrictions on their activity must raise the cost of catching fish. On the other hand, since these regulations generally increase the size of the fish populations, the greater fish populations tend to lower costs. Analysis of the effects of open-access regulations must be careful to separate these two effects so that the effects of the regulations can be clearly understood.

These increases in cost generated by the regulations have the impact of eliminating the rent in the fishery at a lower level of fishing effort than would occur without the restriction. In other words, the increase in cost created by these restrictions makes it less profitable to be involved in fishing, because the restrictions increase the amount of resources required to catch a given amount of fish. At the same time, the increase in fish populations associated with the regulations and reduced level of effort serve to reduce the cost of catching a given amount of fish. However, the net effect of the regulations will be to increase costs of fishing. Finally, we must examine the impact of the regulations on the catch of fish. If the current fish population is greater than the population associated with maximum sustainable yield, then the restrictions will serve to reduce catch. However, if the fishery is highly exploited and the current fish population is lower than the level associated with maximum sustainable yield, the restrictions will increase catch. Finally, it should be noted that in the process of raising costs to protect the stock of fish, the open-access regulations actually exacerbate the problem of too many resources being devoted to the fishery by imposing additional inefficiency on the industry. Table 10.3 summarizes the impact of the open-access regulations on key variables in the fishery.

Aquaculture. Aquaculture, sometimes known as fish farming, is the cultivation of fish in artificial environments (tanks, artificial ponds, etc.) or in contained natural environments. Aquaculture is often suggested as a means of dealing with the open-access problem. However, its ability to deal with the open-access problem is limited for a variety of reasons. First, not all species can be raised in an aquaculture setting. Shellfish are particularly suitable for this purpose, because their immobility allows the use of natural environments. For example, structures upon which shellfish attach can be placed in an area, and property rights to the shellfish can be granted to the aquaculturalist who builds the structure, seeds the shellfish (distributes the juvenile shellfish into the structure area), and promotes the growth of the shellfish. This method has been practiced successfully for mussels on the North Atlantic coast of the United States and on the Atlantic coast of Canada. A number of other shellfish and finfish species have been and are being developed for aquaculture.

However, wild fish populations will still be subject to potential overexploitation. The wild stocks will be only indirectly affected through a demand effect. Producing "farmed" fish creates a substitute for "wild" fish, which will shift the demand for "wild" fish downward and, therefore, generate an open-access equilibrium associated with a higher population of wild fish. The graphical depiction of this demand effect is left as an exercise.

This demand effect may be extremely important, because if increasing human population and changing preferences continually causes the demand to increase, fisheries are threatened with collapse. Provision of a substitute can help mitigate the impacts of increasing demand.

LIMITED-ENTRY TECHNIQUES

Limited-entry techniques also raise costs for fishers, but they raise private costs in a way that lowers social costs, rather than increasing social costs. Taxes and other types of incentives may raise the cost to fishers, but they do this in a fashion in which the extra costs represent a transfer within society, rather than a loss of resources. Actually, by raising private costs in this fashion, social welfare can be increased.

A parallel argument was presented in Chapter 3 in the discussion of pollution regulations. In fact, open-access regulations can be viewed as analogous to direct controls (command and control techniques) and limited-entry techniques can be viewed as analogous to economic incentives for pollution control.

If limited-entry techniques are truly analogous to economic incentives for pollution control, then they should be available either as price policies (similar to pollution taxes) or as quantity policies (similar to marketable pollution permits). Actually, slightly more options exist for limited-entry tech-

TABLE 10.3

IMPACT OF OPEN-ACCESS REGULATIONS ON KEY FISHERY VARIABLES

VARIABLE	IMPACT
costs to fishers	increase
resources used in fishing	increase
population of fish	increase
catch of fish	increase or decrease
consumers' and producers' surplus	increase or decrease

SOURCE: Reprinted with permission of the *Southern Economic Journal.*

niques than for pollution, because either effort or catch can be taxed, and marketable permits can be established for either effort or catch.

The fishery economics literature tends to focus on permit based systems, and where limited entry techniques have actually been used in managing fisheries (New Zealand is a leader in this regard) they have also focused on permit based systems. The name that has been adopted for these systems is the individual transferable quota (ITQ).

ITQs are completely analogous to marketable pollution permits. A limit is placed on total catch, and each fisher in the fishery is allocated a portion of this total catch. This initial allocation can be done by auction, lottery, or in proportion to past catch. This initial allocation then becomes the fisher's individual transferable quota, and he or she can sell all or part of the quota. The level of effort is limited because the cost of effort increases, as people must now buy ITQs in order to fish. Note that this increase in cost has occurred without increasing the amount of resources needed to catch the fish. This increase in cost serves to eliminate the disparity between the social and private cost of fishing associated with the open-access externality.

Limited-entry techniques can also be structured relative to effort instead of catch. For example, the fishery management agency could decide that only a fixed number of boats (say N) would be allowed in the fishery. These N permits could be also allocated by auction or lottery, or based on historical participation or some other mechanism. The issues associated with the initial allocation of ITQs (either catch or permit based) are completely analogous to those discussed for marketable pollution permits in Chapter 3.

The disadvantage of using effort-based techniques is that they only indirectly influence catch. For example, a boat that has a permit could catch fish with differing levels of intensity. In particular, if people respond by fishing longer hours or using more powerful fishing technologies, the limitation on effort may not have its full intended impact. The advantage of using effort-based techniques is that it is easier and less costly to enforce. All catch-based techniques require the measurement of catch, which is costly. A

TABLE 10.4

IMPACT OF LIMITED ENTRY REGULATIONS ON KEY FISHERY VARIABLES

VARIABLE	IMPACT
costs to fishers	increase
resources used in fishing	decrease
population of fish	increase
catch of fish	increase or decrease
consumers' and producers' surplus	increase or decrease

marketable effort quota could be enforced by requiring boats to display a poster-sized certificate, which could easily be checked with binoculars from a patrolling boat of the fishing authority. No measurement or weighing need take place.

Catch-based ITQs are also subject to several potential problems. First, people might cheat on the quota, selling their catch to foreign fishing vessels outside the 200-mile limit, or surreptitiously selling the catch on shore. In addition, catch-based ITQs do not address other fishery management issues, such as the accidental capture of other species, which are discussed later in the chapter.

The main differences in the effects on the fishery between open-access regulations and limited-entry regulations can be seen by comparing Table 10.3 and Table 10.4. They prove to be symmetrical, except for the effect on the amount of resources used in fishing, which increases under open-access regulations but decreases under limited entry regulations.

Limited-Entry Techniques in Current Practice. Although most fishery regulation relies on open-access techniques, some management authorities have used limited-entry techniques. An important example of the use of limited-entry techniques is the Virginia oyster fishery, where oyster beds are treated as private property rather than as open-access resources. Not only does the creation of private property ownership of the oyster beds eliminate open-access exploitation, it gives oyster bed owners an incentive to invest in their property (by seeding with larval oysters and by creating more structure upon which the oysters may attach).

In addition, the 200-mile economic exclusion zone, which the United States (and all other coastal countries) established under the authority of the United Nations Convention of the Law of the Sea, functions as a partial limited-entry technique. A coastal country is allowed to manage the ocean and the ocean bottom within 200 miles of its coastline for its own economic benefit. Most countries, the United States included, exclude foreign fishers from operating within the 200-mile economic zone. This diminishes (but does not eliminate) the open-access problem by limiting the number of fishers (excluding all foreign fishers from access to the fishery).

WHY WE DO NOT SEE MORE LIMITS TO ENTRY

The question often arises why more fisheries do not have limited access, given that limited-access is so beneficial. There are two answers to this question. First, there is more limited access than is immediately apparent, as many limits to access are informal. Second, in most fisheries the problem of completely open access remains a serious problem, and it is probably caused by political barriers to limited entry that are generated by the fishers. It may seem strange that fishers would block limited entry, as they are the ones who would benefit most by it. This seeming paradox will be examined after the issue of informal limits to entry is addressed.

Informal Limits to Entry. Many U.S. fisheries are contained in relatively small communities in which many generations of families have participated in the fishery. Fishing families feel that their historic participation in the fishery grants them property rights to fishing areas (either individually or jointly with other families). Newcomers to the fishery are likely to have their fishing efforts or fishing equipment sabotaged by existing fishers.

For example, if a person tried to place lobster pots in an area that was not historically lobstered by his or her family, the lobster pots would quickly disappear. The movie *Alamo Bay* provides an excellent depiction of the tensions and conflicts between historic fishing/shrimping families from the Texas Gulf coast and an immigrant group of fishers from Vietnam.

Profit Maximization and Limited Entry. It seems clear that pure profit maximizers would favor limited entry. Pure profit maximizers would see the potential economic rents associated with limited entry, and most would probably support limits to entry in order to obtain these potential rents. Conversely, fishers that are not profit maximizers may not see the same gains associated with limits to entry.

A possible explanation for the opposition to limited entry among current fishers is that these fishers may be utility maximizers rather than profit maximizers. This may be particularly true for fishers from communities and families that have been fishing for many generations. Also, the less capital intensive the fishery, the more likely it is to be populated by utility maximizing fishers. Inshore fisheries that require smaller boats and less equipment are more likely to be utility maximizing than offshore fisheries that require large ships with onboard fish processing factories.

If a fisher is a utility maximizer, income will have an important positive effect on utility, but not the only effect. In particular, the ability of traditional participants (and their children) to have access to the fishery is likely to be critically important. Some of the people whose access to the fishery is limited by proposed regulation will be from these traditional groups. In small isolated fisheries, all those who are forced from the fishery will be from these close-knit communities. Consequently, in order to protect the participation of themselves, their children, and their friends and neighbors, they lobby against the imposition of limited-entry access techniques. Of course, they often oppose many types of open-access techniques because, as explained above, they raise the cost of fishing.

While the prior discussion presents a reasonable argument that utility maximizing fishers might oppose limited-entry policies, no formal evidence has yet been provided that would indicate that some fishers might be utility maximizers and not profit maximizers. The discussion presented suggests that some fishers have been in fishing families for many generations and may derive pleasure from their choice of lifestyle. They work outdoors, on the water, enjoying considerable independence as their own bosses. While these may be job characteristics that many people desire, what evidence exists to suggest that utility motives and not profit motives dominate their decision making?

One piece of evidence that would support the utility maximizing hypothesis would be if fishers turned down more profitable employment in order to participate in the fisheries. This type of evidence is not likely to be found in geographically and economically isolated fishing areas, where fishing may be the only source of employment.[4] In contrast, such evidence can be found in traditional fisheries that are located on the periphery of expanding metropolitan areas such as eastern Long Island (New York metropolitan area), the Chesapeake Bay (Baltimore, Washington, and Norfolk metropolitan areas), South Florida, and in most port cities of the U.S. Pacific coast. In these areas, fishers could use some of the skills they have developed in fishing (welding, carpentry, and engine mechanics) to take jobs in construction, manufacturing, or other areas. However, these fishers have chosen to remain in the lower paying occupation of fishing, indicating that the utility they derive from their lifestyle is an important factor in their economic decision making.

OTHER ISSUES IN FISHERY MANAGEMENT[5]

Although most of the literature in fishery economics focuses on the issue of the optimal level of effort and catch, there are many other issues in fishery management that need attention. These include problems associated with

[4]Of course, when fishing is the only source of employment, fishers may be even more concerned about limited-entry policies forcing some of the community out of the fishery.

[5]Important Internet sites which discuss fishery management issues include the U.S. National Marine Fisheries Service (http://kingfish.ssp.nmfs.gov), the Canadian Department of Fisheries and Oceans (http://www.ncr.dfo.ca), and the American Fisheries Society (http://www.esd.ornl.gov/societies/AFS/afshot.html).

FIGURE 10.18

GILL NET

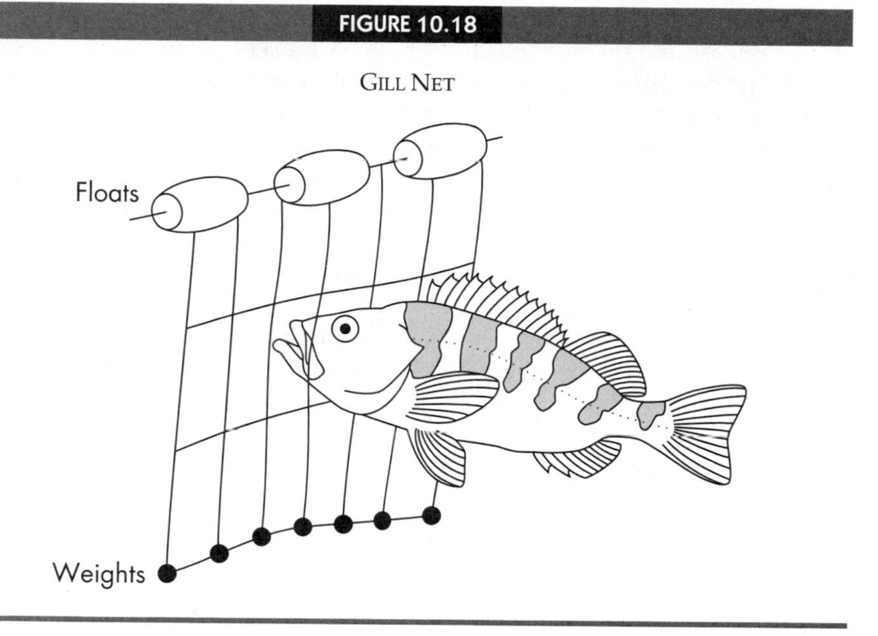

1. the incidental catch (unintended catch) of other fish species and of marine mammals and turtles
2. the pollution of fishery habitats
3. the conflicts between user groups (commercial fishers versus recreational fishers, for example)
4. international cooperation concerning the harvesting of migratory species

INCIDENTAL CATCH

Many types of fishing gear do not discriminate among fish species, and both the desired species and a spectrum of untargeted species are caught by this gear. Among the most notorious of these is the gill net, which is pictured in Figure 10.18.

These gill nets, whose lengths are often measured in kilometers, are vertically suspended in the water, like underwater fences. The fish do not see these nets because they are generally made of clear nylon or black line. The fish attempt to swim through the net, but the mesh is of a size large enough to allow their heads to poke through, but too small to allow their bodies to pass through. When the fish attempt to back out, their gill covers become ensnared in the net, and they remain trapped and tangled in the net in a fashion that renders them unable to breathe, killing them in a relatively short period of time.

The problem is that since the net kills everything it captures, the fishers cannot release the fish for which there is no market. Also, dolphins, seals, and sea turtles often become entangled in these nets (as well as other types of fishing nets) and are unable to reach the surface of the water so that they can breathe (these animals all breathe with lungs, not gills). Unable to breath, these marine mammals and turtles drown.

Sometimes these nets or pieces of these nets break away (they are often cut by boat propellers) and drift off. Since these nets are generally made of nylon, they do not biodegrade but continue to "ghost fish" for decades.

Another indiscriminate fishing method is "long-lining." A longline consists of line with baited hooks every several meters, that may be 10 kilometers long, or longer. These lines are employed off the Atlantic coast in pursuit of highly profitable swordfish, but for every swordfish caught, hundreds of nonmarketable fish—especially nonmarketable species of shark—are caught. This factor has been important in declining shark populations. The death of these sharks is especially serious because sharks take decades to reach reproductive maturity and are not prolific breeders. While viewers of movies such as *Jaws* may view this decline as beneficial, shark attacks on humans are extremely rare. As a top predator, sharks play a crucial role in the marine ecosystem, and they are also very valuable in recreational fishing, a large portion of which is catch and release.

Attempts at dealing with the problem of incidental catch have been met with stiff resistance by commercial fishers, who argue that costs will increase if indiscriminate fishing methods are banned to protect the untargeted species. This situation is, of course, a classic externality problem. The fisher who uses indiscriminate fishing methods imposes costs on society through the killing and stock depletion of untargeted species.

The question for policy makers is whether to enact policies to deal with the situation. Due to the difficulties of monitoring, restrictions on fishing methods may be preferential to policies based on economic incentives. For example, rather than tax fishers for every sea turtle that gets caught in their shrimp nets, require them to install "Turtle Excluder Devices" (TEDDIES) that allow the shrimp through the entrance to the net, but bump the turtles out of the net. The TEDDIES also keep large fish out of the shrimpers' nets. These fish are generally discarded by the shrimpers if they wind up in the nets. Another example of a restriction on fishing methods would be to ban gill nets and long-lining.

Should policy makers go ahead and implement these restrictions? The economist would answer this question by asking whether the policies provide a potential Pareto improvement: Do the people who benefit from the policy benefit by more than the losers lose? This question needs to be answered on a case-by-case basis for each potential restriction, although many analysts believe that the elimination of the indiscriminate fishing methods generates a net benefit for society.

These policies may increase costs for commercial fishers, reducing their incomes, hurting their families, and perhaps driving some of them out of business. This possibility represents an interesting equity issue because the benefits of protecting untargeted species are spread out over a large number of people, but the costs are concentrated upon very few. Should fishers be compensated for adopting more conserving fishing techniques? As in most cases, economic theory does not provide answers for these important equity issues.

Pollution of Fishery Habitat

Overexploitation is not the only anthropogenic factor responsible for the decline of fisheries; pollution also has an important impact. In the United States, many fisheries are in decline because pollution has diminished habitat. This pollution and loss of habitat has affected virtually every freshwater species and many saltwater species as well. Saltwater species that reside in estuaries and other near shore areas are vulnerable to pollution that is carried to the marine environment by rivers. Anadromous species, such as salmon, steelhead, shad, and striped bass, are particularly vulnerable to riverine pollution and other alterations of riverine areas, such as dams and reservoirs. Many species of fish are contaminated with toxic substances that make the fish unsafe to eat.

In the development of the model of demand and supply for fish presented earlier in the chapter, it was shown how the damages associated with these types of pollution could be measured. Chapter 14 discusses water pollution in more detail and presents various policy instruments that can be used to move the level of pollution toward a more optimal level.

In developing countries, soil erosion from deforestation and intensive cultivation of hillside lands has had severe impacts on fisheries. The soil erosion affects water quality not only in the rivers, but in reservoirs, estuaries, lagoons, and coral reefs into which the rivers carry the soil. Often, high levels of fecal coliform bacteria and other infectious organisms associated with human wastes are found in these water bodies and contaminate the fish, particularly shellfish. Chapter 17 discusses the market failures that lead to soil erosion and environmental degradation in developing countries and the policies that can be developed to mitigate these problems.

Management of Recreational Fishery Resources

The economic analysis of recreational fishing resources is similar to commercial fishing in many ways. In most North American recreational fisheries, there is unrestricted access to the resource, leading to open-access exploitation. Policies such as creel limits (restrictions of how many fish may be kept), restricted seasons, and size limits may help protect the fish stock from destructive exploitation. However, the theoretical goal of recreational fishing management should not be the limitation of effort, because it is the effort or

Blue Crabs and the Loss of Salt Marshes

Wetlands are under constant assault from pollution and development. People like to have waterfront homes, and much of the remaining undeveloped waterfront is covered by wetlands. This leads to an interesting question: Should wetlands be converted into marinas or vacation condominiums?

In order to answer this question, information must be developed on the value of wetlands. Gary Lynne, Patricia Conroy, and Frederick Prochaska (1981) conducted a study to measure the value of salt marshes in terms of their beneficial effect on fisheries. They used a catch per unit effort fishery model to determine the value of Florida Gulf coast salt marshes in producing blue crabs. Shellfish such as crabs and lobsters are well suited to examination with a catch per unit effort model, because it is generally easier to define effort for this type of shellfish than for finfish. The reason is that an unambiguous effort variable can be defined as the total number of traps used to commercially harvest the crabs.

Lynne et al. postulated that catch is related to the carrying capacity and to effort. The carrying capacity was posited to be related to the acreage of salt marshes. In their empirical model, Lynne et al. assumed that the carrying capacity was proportional to the natural logarithm of the total acreage of salt marshes, which implies a relationship that is increasing at a decreasing rate as in the graph below.

Lynne et al. used a regression with data on the total number of traps, dockside value of crab landings, and acreage of marshes. This function can be interpreted as a total product function, and the value of the marginal product of marshes can be determined.

Lynne et al. found a relatively low marginal product for marshes of 0.25 to 0.30 in 1974 dollars per acre (0.69 to 0.82 in 1992 dollars). One must be careful, however, in saying that the current value of marsh is less than a dollar per acre. First, this estimate is a marginal product and not an average product, so it can only be used to measure the value of small changes in total acreage of salt marshes. Given the flatness of the logarithmic function in the graph, one would expect the value of a marginal acre to be lower than the value of an average acre. Second, contribution to the viability of the blue crab fishery is only one of the many services that the salt marshes provides. Nonetheless, this study represents an excellent example of how the catch per unit effort model can be used to help understand the value of environmental resources.

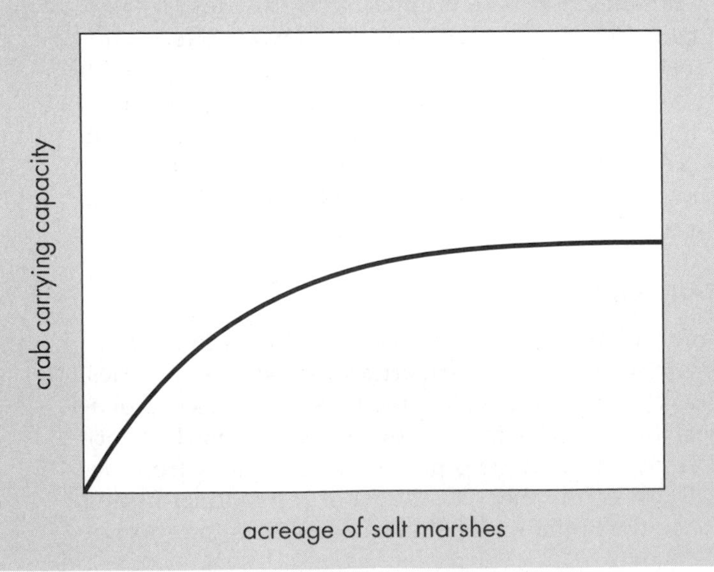

act of fishing that is the source of social benefits. Effort is not considered a cost in this case, but a source of benefits.

Each recreational fisher who enters is receiving positive net benefits from participation in the fishery. These benefits include the utility derived from the process of fishing, from catching fish (which may be released or kept and eaten), and from a variety of other sources, such as being in the outdoors, enjoying the scenery, viewing wildlife, and being with (or in some cases, being away from) family and friends. However, the goal of recreational fishing management should not be to maximize participation, as each participant imposes costs on other participants by increasing congestion and reducing the fish stock. On the other hand, recreational fishery managers seldom take steps to limit participation (in the United States, recreational fishery management is generally done on a state-by-state basis, except for special management of national parks and forests), as their budgets are to a large extent determined by the sale of fishing licenses to participants.

Nonetheless, recreational fishing management does attempt to address some of the external costs associated with the open-access nature of this resource. These policies include:

1. **Stocking Fish**—The enormous number of eggs produced by a single female fish makes the rearing of fish in hatcheries relatively inexpensive. In the controlled conditions of a hatchery (no predation, abundant food, and disease prevention) a very large number of eggs hatch into fish and grow to a size where they can be released into the wild with a high probability of survival until they are caught by recreational anglers. Obviously, this policy is designed to reduce the problem of open-access depletion of recreational fish stocks.
2. **Closed Seasons**—Recreational fisheries often have closed seasons to protect the fish stocks. As with commercial fisheries, these closures often coincide with spawning periods to protect the vulnerable spawning adults and to ensure that spawning activity is not disturbed.
3. **Access Improvements**—The creation of additional access points, such as boat launching ramps, fishing piers, parking areas, and artificial reefs tends to spread out recreational fishing activity, which reduces congestion and spreads the impact on the fish stocks over a wider area. These improvements may also reduce the cost of access to the individual angler and may improve the aesthetic quality of the trip. Of course, both the reduced cost and the increased quality may encourage the participation of more anglers, which would increase congestion somewhat.
4. **Catch and Release**—Catch and release programs are based on the idea that a recreational angler does not have to kill his or her catch to produce utility from fishing. Catch and release means that after

TABLE 10.5

A COMPARISON OF ALTERNATIVE ESTIMATES OF THE BENEFITS OF WATER QUALITY IMPROVEMENTS FROM BOATABLE TO FISHABLE CONDITIONS IN 1996 DOLLARS

STUDY	ORIGINAL ESTIMATE	1996 DOLLARS
Vaughn-Russell	$4.00 to $8.00 per person per day was the range over the models used (1980 dollars)	$1.07 to $14.80
Loomis-Sorg	$1.00 to $3.00 per person per day over the regions considered based on increment to value of recreation day for coldwater fishing (1982 dollars)	$1.58 to $4.74
Smith, Desvouges, McGivney	$0.98 to $2.30 per trip using first generation generalized travel cost model with Monongahela sites, boatable to fishable quality (1981 dollars)	$1.64 to $3.40
First Generation—Generalized Travel Cost Model	$5.87 to $54.20[a] per trip ($2.24 to $122.00 per visitor day[b])	$14.70 to $136.42 ($5.53 to $306.42)
Second Generation—Generalized Travel Cost Model	$0.08 to $5.43 per trip ($0.04 to $18.78 per visitor day) for Corps sites, change from boatable to fishable water quality (1977 dollars)	$0.47 to $13.67 ($0.09 to $47.27[c])

SOURCE: Taken directly from Smith and Desvouges (1985) and converted to 1996 dollars.

[a]These estimates relate only to the Marshallian consumer surplus.

[b]The reason for the increase in the range of benefits per day is that some trips were reported as less than a day. The appropriate fractions were used in developing these estimates.

[c]The numbers in parentheses are the per-day consumer surplus, while those above are the per-trip estimates.

capture, the fish is released back into the water to grow larger, to reproduce, and to be caught and enjoyed by future anglers. Catch and release programs are based on both moral suasion and command and control. As a moral suasion policy, catch and release is promoted by fishery managers and recreational fishing organizations (Bass Anglers' Sportsman's Society, Trout Unlimited, etc.) as a general philosophy toward fishing and conservation. Catch and release programs can also be of the command and control variety. In these command and control programs, high quality and sensitive fishing areas are declared as catch and release areas, and no fish may be kept from these areas. These areas also have special gear restrictions that are designed to maximize the chances of survival for a released fish. Live bait (which fish swallow more deeply than artificial lures) is generally prohibited in catch and release areas, as are multiple hooks. Hooks often are required to be barbless to minimize the damage from the hook.

5. **Size and Creel Limits**—Size limits place restrictions on the minimum (and sometimes maximum) size of fish that are legal to keep. Creel limits place restrictions on the maximum number of fish per day that may be kept. Both restrictions are designed to protect the reproductive viability of the fish stocks by restricting the harvesting of reproductive fish.

All of the aforementioned policies are focused on managing fishery resources relative to fishing activity. Of course, another area of policy is determining how to regulate activities when other activities impinge on recreational fishing activity. Both types of policies, but particularly the latter, require information on value on which to base cost–benefit analysis in order to use resources optimally.

In order to find the benefits associated with a particular recreational fishing activity, a valuation study must be done. Such studies generally take the form of contingent valuation or travel cost studies. (See Chapter 4 for a discussion of these valuation methodologies.) Table 4.1 in Chapter 4 presents a summary of a subset of valuation studies of recreational fishing.

Freeman and many others note that the major benefits of improving water quality can be attributed to recreational uses of water resources, including boating, swimming, and recreational fishing. Smith and Desvouges (1985) conducted a travel cost study to estimate the benefits of improving water quality in Army Corps of Engineers reservoirs from a quality level that is suitable for boating but not fishing, to a water quality level that is suitable for fishing. Smith and Desvouges summarize their results and the results from similar studies, which is reproduced as Table 10.5. In a study based on hedonic wages (compensating wage differentials), Clark and Kahn (1989) show that the annual value in recreational fishing of 10 percent improvements in dissolved oxygen, pH, and total suspended sediments are 1.34, 1.255, and 0.257 billion in 1992 dollars, respectively.

SUMMARY

Fishery resources are renewable but destructible. The destructibility problem is amplified by the open-access nature of many of the world's fishery resources. Many of the world's important fishery resources are currently collapsing.

For commercial fishing, optimal management requires the limitation of effort to a level that maximizes the sum of consumers' surplus, producers' surplus, and fishery rent. Actual fishery management seldom achieves this goal and is based on developing restrictions on how, when, where, and how

GULF COAST REDFISH

Historically, the redfish or red drum (*Sciaenops ocellata*) was pursued in the Gulf coast area only by recreational fishers and small-scale commercial fishers who responded to local demand. However, in the early 1980s, "Cajun cuisine" became a national fad, and many restaurants added a traditional Cajun dish, "blackened redfish," to their menus. This increase in demand raised the price of redfish and led to increased entry into the fishery and a drastic decline in redfish populations throughout the Gulf coast states. It also led to a highly charged political fight between recreational and commercial anglers over how the fish stocks should be allocated through regulation. This conflict in the Gulf coast redfish fishery is a prime example of the conflict between recreational and commercial fishers that occurs in many coastal fisheries.

Professor Trellis Green (1989), of the University of Southern Mississippi, conducted a study for the National Marine Fishery Service (NMFS) to determine the value of redfish in recreational fishing. He used a travel cost model with the expected success of the angler as an explanatory variable. The model was estimated using data collected by the NMFS in intercept surveys, which are surveys that are executed at randomly chosen fishing sites.

Professor Green found that increasing expected catch by 10 percent would increase net benefits (consumers' surplus) by between $5 and $14 per trip. In 1986, the typical red drum angler took approximately 30 trips per year and caught approximately 1 3/4 redfish per year. Extending this per-angler measure to the entire Gulf coast redfish fishery means that a 10 percent increase in recreational catch (which is equal to 670,000 pounds of redfish) increases social benefits by between $10 and $17 million per year.

Although Professor Green does not report value estimates for commercial fishing, he states that evidence from other studies suggests that "the marginal welfare loss incurred by the commercial fishery from red drum reallocation is not nearly so great as the gains to the recreational sector, at least in today's market." He also suggests that the "Cajun cuisine" demand for redfish could be met by the development of redfish aquaculture.

much fish can be caught. In addition to the problems associated with maximizing the value of a commercial fishery in isolation, there are other important policy questions. These questions revolve around the issue of resource utilization conflicts. Of particular importance are the incidental catch of untargeted species (fish, marine mammals, and sea turtles), conflicts between recreational and commercial fishing, conflicts between commercial fishing and subsistence fishing and the pollution of fishery habitat.

REVIEW QUESTIONS

1. Under what conditions would the imposition of open-access regulations raise the level of average cost? Illustrate this rise with a graphical model of an open-access fishery.
2. Draw a graph that shows how the expansion of aquaculture can increase the population of the "wild" stocks of fish.
3. What is meant by "carrying capacity"?

4. Why is the maximum sustainable yield not necessarily the optimal sustainable yield?
5. What is the difference between open-access and limited-entry fishery management policies?

QUESTIONS FOR FURTHER STUDY

1. What is the relevance of a depensated growth function for fisheries management?
2. Would the sole owner of a fishery ever harvest the fishery to extinction?
3. What are the major unresolved management issues in marine fisheries?
4. How would you optimally regulate a purely recreational fishery?
5. Why, in an open-access fishery, is the inverse of the supply function the average cost function and not the marginal cost function?
6. What is the difference between the bionomic equilibrium of the Gordon model, and the bioeconomic equilibrium of the supply and demand model presented in the textbook?
7. Compare ITQs with marketable pollution permits.

SUGGESTED PAPER TOPICS

1. Estimate demand and supply functions for a regional open-access fishery of your choice. The basic data on price and quantity is available from the National Oceanic and Atmospheric Administration, which publishes weight and value data for each important species in each state. Price must be obtained by dividing value by quantity. Ask your university's reference librarian for help in locating the "fishery statistics" section in the government documents section of the library. Some citations are included in the reference section of this chapter. Data on socioeconomic variables, such as CPI, GNP, population, and so on, are available from a variety of sources, including the U.S. Statistical Abstract, The Economic Report of the President, and The Survey of Current Business. Look at Buerger and Kahn (1989) and Lynne et al. (1981) for examples of empirical studies. Also, in Anderson's (1986) reference list, he notes many studies that are empirical in nature.
2. Investigate the issue of international conflict over fishery resources. Discuss the efficiency and equity implications of current attempts to deal with the conflict, such as the 200-mile economic zone and international agreements. How would you develop better policy for fishery resources that are shared by several nations? Begin with Anderson's *The Economics of Fishery Management* (1986) for discussion of this issue and also examine journals such as *The Journal of Environmental Economics Management*, *Marine Resource Economics*, *Resource Economics* and *Fisheries Bulletin*. Search bibliographic databases using international fisheries policy, treaties and fisheries, and fisheries and international law as keywords. Also, search on specific species or types of fish, such as shrimp, salmon, tuna, and groundfish.
3. If you live in a coastal state (including the Great Lakes states) you might want to look at an issue of local concern. Talk to staff members at your state's Sea Grant Institute, which is housed in your state's land grant university, as well as your resource economics professors. Also, if your university has a marine resources department or fish and wildlife department, you might look to professors there for help in locating an interesting issue. Make sure your paper does more than just state the physical dimensions of the problem. Look at the market failure and policy attempts to deal with the market failure. What are the efficiency and equity implications of the policy? Can you develop any means to improve on the policies?

WORKS CITED AND SELECTED READINGS

1. Anderson, Lee G. *The Economics of Fishery Management*. Baltimore: Johns Hopkins University Press, 1986.
2. Buerger, R., and J. R. Kahn. "The New York Value of Chesapeake Striped Bass." *Marine Resource Economics* 6 (1989): 19–25.
3. Clark, C. W. *Bioeconomic Modelling and Fishery Management*. New York: Wiley, 1985.
4. Clark, D. E., and J. R. Kahn. "The Two Stage Hedonic Wage Approach: A Methodology for the Valuation of Environmental Amenities." *Journal of Environmental Economics and Management* 16 (1989): 106–120.
5. Freeman, A. M., III. *The Benefits of Environmental Improvement: Theory and Practice*. Baltimore: Johns Hopkins University Press, 1979.
6. Gordon, H. "The Economic Theory of the Common Property Resource: The Fishery." *The Journal of Political Economy* 62 (1954): 124–142.

7. Green, Trellis. *The Economic Value and Policy Implications of Recreational Red Drum Success Rate in the Gulf of Mexico*. Report to the National Marine Fisheries Service, University of Southern Mississippi, July 31, 1989.
8. Kahn, J. R. "Measuring the Economic Damages Associated with the Terrestrial Pollution of Marine Ecosystems." *Marine Resource Economics* 4 (1987): 193–209.
9. Kahn, J. R., and W. M. Kemp. "Economic Losses Associated with the Degradation of an Ecosystem: The Case of Submerged Aquatic Vegetation in Chesapeake Bay." *Journal of Environmental Economics and Management* 12 (1985): 246–263.
10. Loomis, John, and Cindy Sorg. "A Critical Summary of Empirical Estimates of the Values of Wildlife, Wilderness and General Recreation Related to National Forest Regions," unpublished paper, U.S. Forest Service, 1982.
11. Lynne, G. D., P. Conroy, and F. J. Prochaska. "Economic Valuation of Marsh Areas for Marine Production Processes." *Journal of Environmental Economics and Management* 8 (1981): 175–186.
12. Scott, A. "The Fishery: The Objectives of Sole Ownership." *The Journal of Political Economy* 62 (1955): 112–124.
13. Smith, V. K., and W. H. Desvouges. "The Generalized Travel Cost Model and Water Quality Benefits: A Reconsideration." *Southern Economic Journal* 52 (1985): 371–381.
14. Turvey, A. "Optimization and Suboptimization in Fishery Regulation." *American Economic Review* 54 (1964): 64–76.
15. U.S. Department of Commerce. *Status of Fishery Resources off the Northeastern United States for 1986.* NOAA Technical Memorandum NMFS-F/NEC-43, 1986.
16. ———. *Current Fishery Statistics* (state landings data), various years.
17. Vaughn, William J., and Clifford S. Russell. "Valuing a Fishing Day." *Land Economics* (1982): 450–463.

APPENDIX 10.A

This appendix further amplifies the discussion in the textbook, beginning on page 303, which examines the impact of pollution on an open-access fishery. In the discussion, it was shown how one arrived at a new equilibrium associated with a new level of environmental quality. However, that discussion did not present how to measure the damages associated with a negative change in environmental quality, or the benefits associated with a positive change.

The first step is to compute the consumers' and producers' surplus associated with the pre-change equilibrium (there will be no rent since this is an open-access fishery), which is a little complicated. The reason is that in an open-access fishery, the market supply function is based on average cost, not marginal cost. Since this is the case, consumers' and producers' surplus cannot be computed as the area between the demand and supply functions. Rather, the marginal cost function must be derived from the average cost function, and then benefits can be measured as the area between the marginal cost and demand functions. This is done for the marginal and average cost curves associated with E_1 (from Figure 10.17) in Figure 10A.1. The locus of

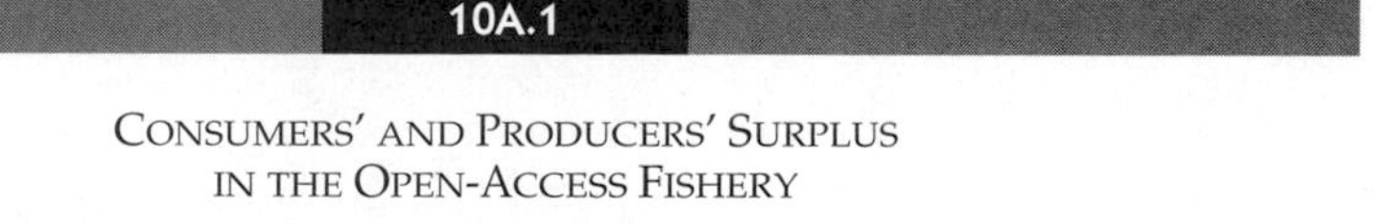

CONSUMERS' AND PRODUCERS' SURPLUS
IN THE OPEN-ACCESS FISHERY

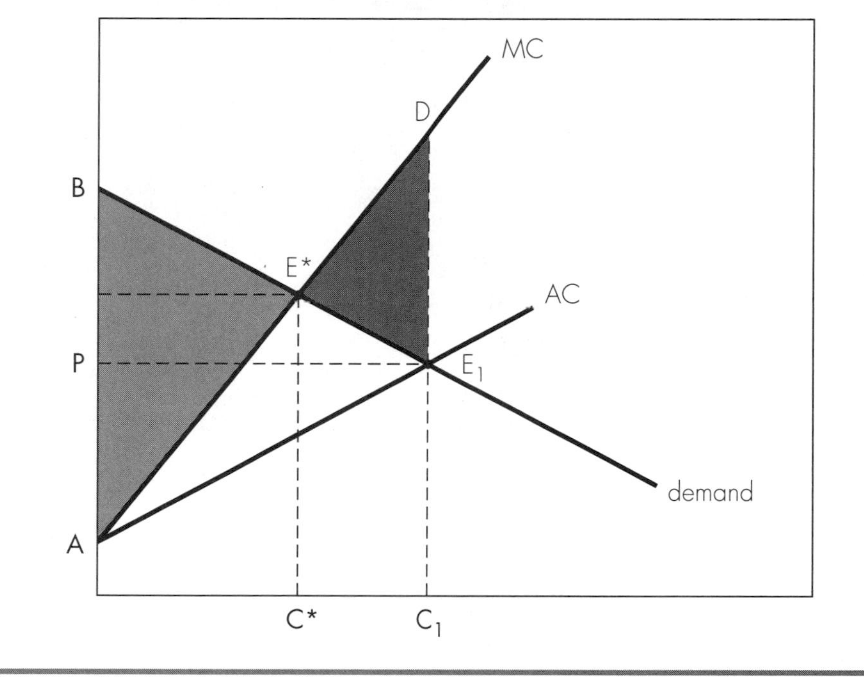

biological equilibriums is left out of Figure 10A.1, so the cost curves can be more clearly examined.

Although E_1 is the equilibrium level of catch, resources are wasted achieving E_1. In particular, the open-access leads to a catch level that is in excess of the level where marginal cost equals marginal benefits (C*). Therefore, the net economic benefits must be computed as ABE* minus area E^*DE_1. Not only is there a loss of all rent from being on the wrong point on the locus of biological equilibriums, but there is the social loss of area E^*DE_1, which must be subtracted from the positive benefits.

A similar measure could be computed for the post-pollution equilibrium of E_2 (from Figure 10.17), and the difference between the two would represent the losses that pollution generated to the producers and the consumers of the striped bass fishery.

Temperate Forests

. . . the bears don't write letters and the owls don't vote.

Lou Gold, forest advocate

INTRODUCTION

The historical emphasis of forest economics has been on how to maximize the wealth to be derived from the harvesting of the forest's wood. This emphasis is not surprising given that the first foresters were employed by large landowners who were concerned with maximizing the income to be derived from their land. This attitude toward forests was shared by many early conservationist thinkers (prior to 1900), whose primary concern was that the material wealth provided by forests not be squandered. In contrast, some of the later conservationists, such as John Muir (1838–1914), believed that the forest was a synergistic system and that the benefits derived from a forest were derived primarily through the preservation of the forest, although timbering of some forest land was certainly appropriate.

Now, as we approach the turn of the next century, we see that this debate still rages. Do we want spotted owls or lumberjacks? Is the primary purpose of the forest to provide economic benefits through harvesting the wood? Is the primary purpose of the forest the provision of environmental services? Should some forests be devoted to harvesting and some preserved for their ecological and recreational benefits? How do we make these decisions? The debate over these questions even reached the polling places in 1990, as initiatives to prevent the harvesting of pristine ("old growth") forests were placed on the ballot (and defeated) in California and other Pacific states.

This chapter looks at the traditional problem of how to maximize the income derived from timbering, as well as the more general problem of maximizing the total social benefits arising from the forest. These total social benefits would include the benefits from timbering, as well as the benefits from preservation. In order to more fully understand the benefits that can be derived from the forest, it is necessary to understand the basics of forest ecology. As the title suggests, this chapter focuses upon temperate forests. Temperate forests are found north of the Tropic of Capricorn and south of the Tropic of Cancer and include the many different types of forest found in North America.

FOREST ECOLOGY

The major point associated with forest ecology is that a forest is more than a collection of trees. It is a system of plant, animal, bacterial, and fungal organisms that interact with the physical environment and with one another. Removing one type of organism from the forest may have far reaching effects. An example of this interdependency will be examined later in the chapter in the discussion of "old growth" or "ancient" forests.

Another important point is that a forest is an example of the type of ecosystem known as a climax community. A climax community is an ecosystem that has arisen out of competition with other communities of organisms. For example, an area of land may be initially colonized by pioneer species such as grasses. Very soon, small woody plants (bushes and shrubs) become established, followed by fast growing tree species such as poplar. Eventually, the dominant plant form becomes a slower growing tree, such as an oak, maple, spruce, or fir. These taller trees are most successful in competing for sunlight, water, and nutrients.

Although it may be apparent that the different organisms interact with one another, it is less obvious how the organisms interact with the physical environment. The process of soil formation and nutrient cycling is a good example of this type of interaction.

Nutrient cycling refers to the process by which the basic life nutrients (phosphorus, potassium, and nitrogen) are absorbed from the physical environment by the various organisms in the ecosystem, transferred from organism to organism, and eventually returned to the physical environment. In temperate forests, the soil plays a crucial role in this process because it is the place where nutrients accumulate and are slowly released to the other components of the ecosystem.

Figure 11.1 illustrates nutrient cycling in a typical temperate forest. The nutrients in the soil are absorbed by the roots of trees and other plants. When these trees and plants drop their leaves or die and decay, their nutrients are returned to the soil. In addition, nutrients are cycled through the ecosystem when animals eat the plants. When these animals eliminate their wastes, the nutrients are returned to the soil. Some animals are eaten by other animals, which further extends the food web based on the trees. Finally, dead animals are eaten by a host of organisms, which eventually return the nutrients to the soil.

In addition to playing an essential role in nutrient cycling, forests play an essential role in carbon cycling. The process of photosynthesis converts

FIGURE 11.1

NUTRIENT CYCLING IN THE TEMPERATE FOREST

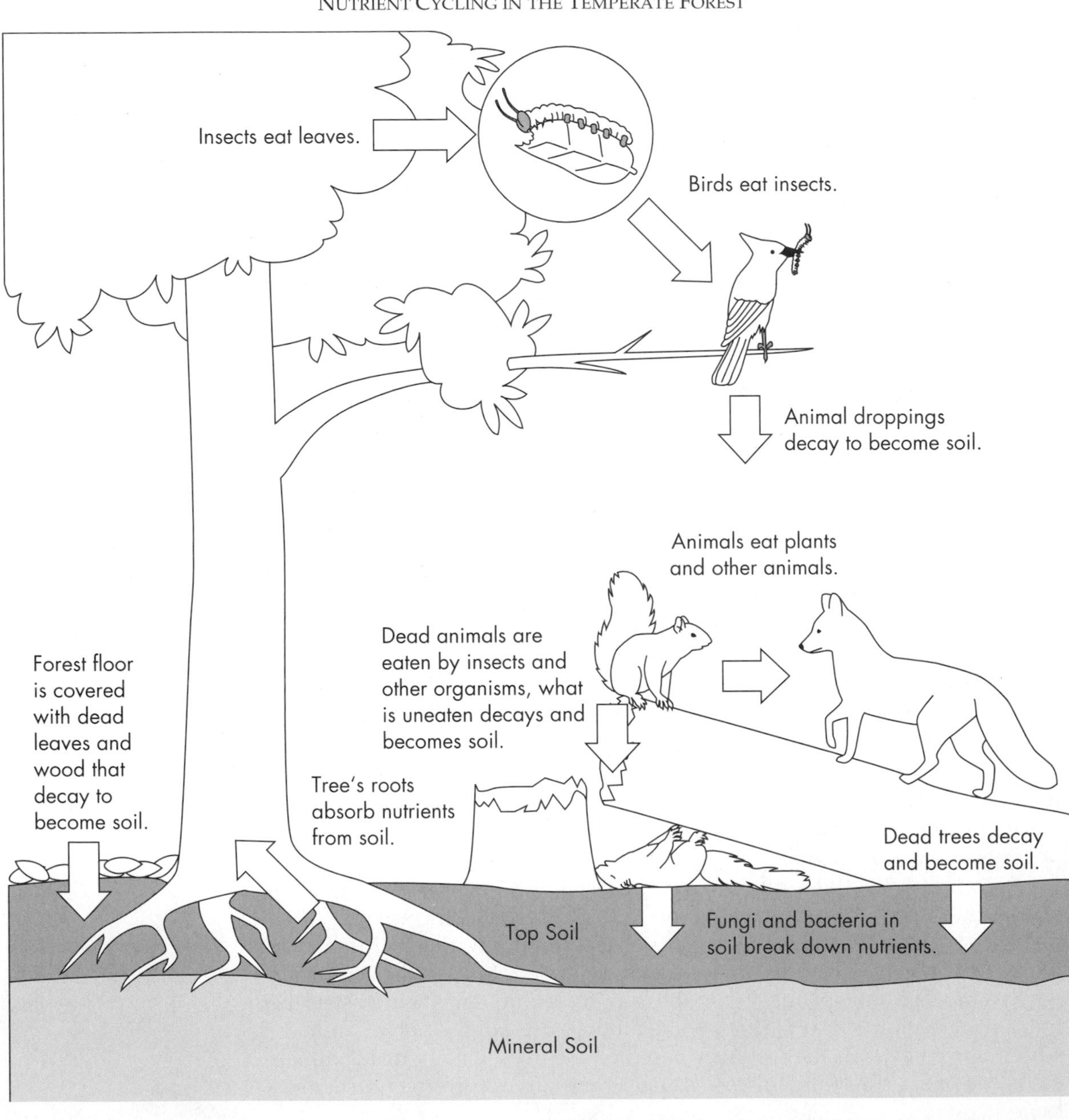

carbon from the carbon dioxide in the atmosphere to carbon in sugar in the tree's leaf cells. This carbon, the basic building block of life on earth, is then available to other organisms as it is transferred along the food chain. As discussed in Chapter 6, carbon dioxide buildup in the atmosphere is the leading contributor to potential global warming. Forests remove carbon dioxide from the atmosphere and sequester it in their woody tissue, which is approximately 50 percent carbon by weight.

Forests also play an important part in the hydrological cycle. When rain falls on the leaves of trees, the leaves slow the velocity of the rain, as the rain slowly trickles off the leaves of the trees to the ground. The organic matter in the soil acts as a sponge and absorbs the water, as the water slowly makes its way into the aquifers and surface waters. In the absence of forest cover, the rain impacts directly on the soil, which has less organic matter and less absorptive capability. Not only does the water immediately run off into surface waters (without as much reaching the underground aquifers), but the runoff leads to soil erosion that diminishes the productivity of the soil and leads to turbidity in surface waters. The reduction in the capacity to store water also means that forest plants will have less water to draw on during the drier summer months and that the surface waters will have less flow and volume during the drier months. This result has important implications for not only the plant and animal communities, but also the human communities that depend on the flow of water from the forests.

BENEFITS OF TEMPERATE FORESTS

A frequent place to begin the discussion of the benefits of forests is with a discussion of the benefits of harvesting wood. Although this practice is common in books dealing with forest management, it is important to realize that this order of discussion does not imply a ranking of benefits.

Harvested wood is used for a variety of purposes, including construction material (both boards and manufactured products such as plywood and particle board), furniture, paper, fiber (for disposable diapers and other products), and chemicals. Harvested wood is also used as a fuel for heating homes, as a boiler fuel, and can be converted into methanol (wood alcohol) that can be used to fuel internal combustion engines in cars and other vehicles.

One does not need to cut down a forest to receive benefits from it. The forest is an important habitat for plants and animals. It serves to protect surface water by slowing its runoff. The forest sequesters large amounts of carbon, which would otherwise take the form of carbon dioxide in the atmosphere, contributing to global warming. The forest is also an important area for outdoor recreation. In addition, other productive activities can take place in the forest, including harvesting animals, mushrooms, and berries; mining; and grazing of livestock.

THE PRIVATELY OPTIMAL AND SOCIALLY OPTIMAL MANAGEMENT OF FORESTS

Before discussing the privately and socially optimal management of forests, it is necessary to spend some time talking about the ownership of forests. Certainly, the privately optimal management of forests is going to depend on whether the forest owner is a utility maximizer or a profit maximizer. If the owner is a utility maximizer, preferences will play an important role in the outcome. Forest ownership can be divided into three primary categories.

The first of these are forests that are owned by households. Since households are utility maximizers rather than profit maximizers, it is difficult to talk about a single management strategy for the household forest owners. For example, some households may be concerned with maximizing the income they receive from the forest, while others may rather enjoy the aesthetic benefits of keeping their forests intact or feel that they have a stewardship responsibility to leave the forest intact. Another type of individual may engage in selective harvesting, while still keeping the forest relatively intact to enjoy the nonpecuniary benefits of having forested property. Others may be concerned with passing the forest on to their children, who can then choose how to maximize it.

The second major class of forest ownership is ownership by firms of the forest products industry. These firms are profit maximizers, who seek to maximize the present value of earnings derived from the forest. These firms own a substantial amount of forest land and would include well-known firms such as Boise-Cascade, Weyerhauser, and Georgia-Pacific, as well as a host of smaller firms. In addition to harvesting timber from their own land, forest product firms lease harvesting rights from both household forest owners and from the third category of forest ownership, public ownership.

Publicly owned forests include national parks, national forests, parks and forests operated by state and local governments, and other publicly owned tracts of forests, wildlife refuges, game management areas, and nature preserves. Different types of public forests are operated with different goals in mind, but generally these forests are managed for multiple uses and not just the generation of a stream of income from timber harvesting.

MAXIMIZING THE PHYSICAL QUANTITIES OF HARVESTED WOOD

Many forestry scientists have advocated a management strategy designed to *maximize the physical amount of wood* to be derived from the forest. There are two basic methods for doing this. One option is to let the forest grow until it reaches its peak volume of wood and then cut it. The forest is replanted, and

TABLE 11.1

TOTAL VOLUME OF WOOD, ANNUAL INCREMENT, AND MEAN ANNUAL INCREMENT (CUBIC FEET OF WOOD) ON A ONE ACRE STAND OF A HYPOTHETICAL TREE SPECIES

AGE	TOTAL VOLUME	ANNUAL INCREMENT	MEAN ANNUAL INCREMENT
10	0	0	0
20	75	7.5	3.75
30	200	12.5	6.66
40	750	55.0	18.75
50	1400	65.0	28.0
60	2100	70.0	32.5
70	2550	45.0	34.3
80	2900	35.0	36.25
90	3200	30.0	35.55
100	3450	25.0	34.5
110	3650	20.0	33.18
120	3800	15.0	31.67
130	3900	10.0	30.0
140	3950	5.0	28.2
150	3975	2.5	26.6
160	3975	0	24.84

the process is allowed to repeat itself. The other method chooses the length of the harvest-replant-harvest cycle to maximize the total harvests of wood that can be achieved over time. This harvest-replant-harvest cycle is called the rotation of the forest. In this case, the length of the rotation is chosen to maximize the flow of wood from the forest. The length of time in the rotation for either of these two strategies is critically dependent upon the way in which the trees grow. In examining this, one must keep in mind that generalizations about growth are difficult to make because growth conditions depend on the density of trees, the soil conditions, weather and rainfall conditions, and the incidence of disease and pests. It is important to look at the growth of a stand of trees, rather than an individual tree, since a tree in isolation grows more quickly than a tree that is competing with other trees for sunlight, nutrients, and water. After replanting, the trees initially grow at a rapid percentage rate, but since their size is small, the growth in the mass of wood is relatively small. As the trees grow larger, their capacity to grow increases as well. Eventually, the growth slows as the trees mature, and growth can become negative as the disease and death associated with aging has a greater impact than the production of new biomass. Table 11.1 and Figures 11.2A and 11.2B describe the growth of a one acre stand of a

FIGURE 11.2A

TOTAL VOLUME OF MARKETABLE WOOD OF
A HYPOTHETICAL TREE SPECIES—ONE ACRE STAND

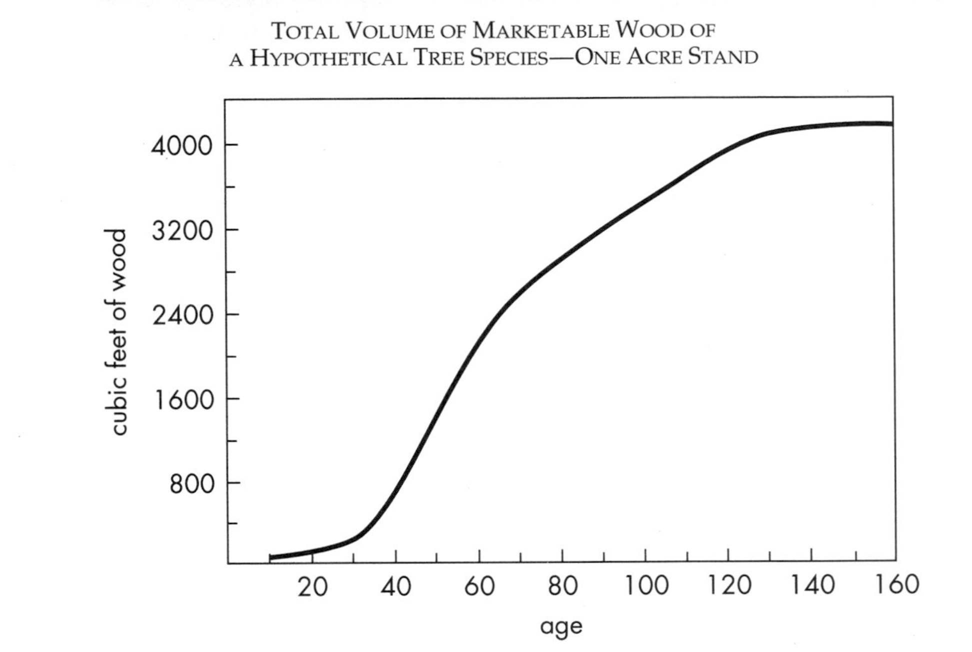

hypothetical tree species as a function of the age of the trees in the stand (all the trees in the stand are the same age). As can be seen in Figure 11.2A, the volume of wood first increases at an increasing rate and then at a decreasing rate. The volume of wood can actually decline as the trees become very old, but this decline is not shown in either the graph or the table.

It is relatively easy to see when the forest grows to its peak size: when the growth (annual increment) declines to zero.[1] The peak size for this hypothetical forest can be seen (Table 11.1, Figure 11.2A, or Figure 11.2B) to occur when the stand of trees is 160 years old.

It is not as easy to see when the total amount of wood that is harvested over time is maximized, although it is relatively easy to conceptualize the trade-offs that are involved. One way to increase the flow of wood is to harvest more frequently. However, the more frequently the forest is har-

[1]Every field of study has its own technical jargon, and forestry management is no exception. Annual increment can be thought of as the annual growth, or the marginal product of the forest (marginal with respect to time). Mean annual increment can be thought of as the annual amount of growth, or the average product of the forest (the total product divided by the number of years of the rotation).

FIGURE 11.2B

ANNUAL INCREMENT AND MEAN ANNUAL INCREMENT FOR A HYPOTHETICAL TREE SPECIES—ONE ACRE STAND

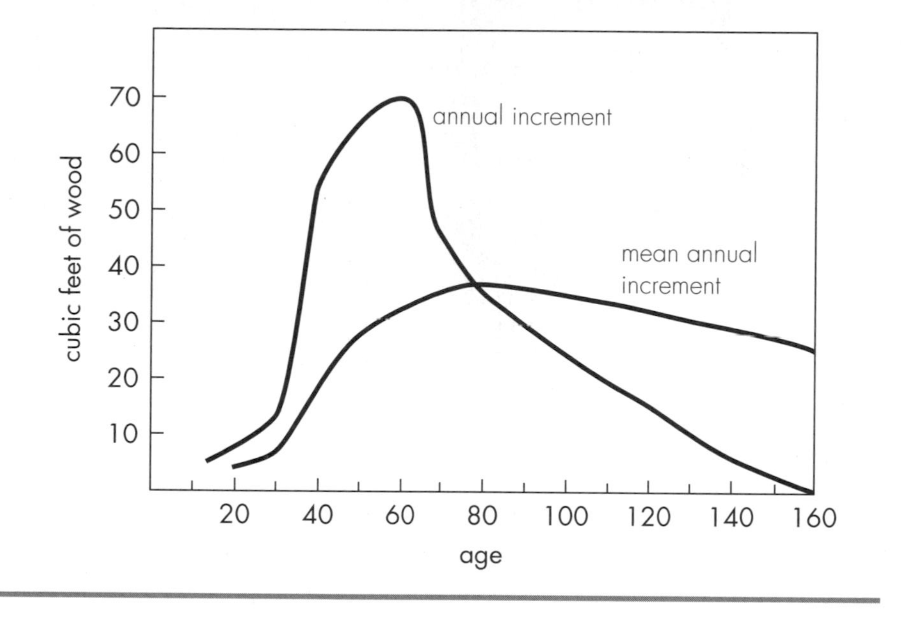

vested and replanted, the younger and smaller the trees. The other way to increase the flow of wood is to harvest less frequently and have bigger harvests. The optimal compromise between these two conflicting strategies is to harvest at the forest age that maximizes the average growth of the tree over its lifetime. If average growth is maximized over a sequence of multiple rotations, then total growth (and the total flow of wood) will be maximized as well. The average growth rate of the hypothetical tree species is contained in the fourth column of Table 11.1 and is also drawn in Figure 11.2B. Notice that the average growth rate (often called mean annual increment) is maximized when it is equal to the marginal growth rate (annual increment), which occurs at approximately 80 years for this hypothetical tree species, a time that is substantially shorter than the rotation that maximizes individual harvest (160 years). Table 11.2 shows the sensitivity of the total wood harvest to the length of the rotation, by showing the total amount of wood that can be harvested over a 500-year period, for different rotation lengths. One interesting insight that can be gained is that the penalties (in terms of lower than maximum volume of wood) for having too short a rotation are substantially higher than the penalties for too long a rotation. For example, a rotation that is approximately 40 years shorter than the flow maximizing level gives a

TABLE 11.2

LENGTH OF ROTATION, NUMBER OF ROTATIONS PER 500 YEARS, TOTAL HARVEST OVER 500 YEARS

LENGTH OF ROTATION	NUMBER OF ROTATIONS PER 500 YEARS	TOTAL VOLUME OF WOOD (CUBIC FEET) HARVESTED OVER 500 YEARS
10	50.00	0
20	25.00	1,875
30	16.66	3,330
40	12.50	9,375
50	10.00	14,000
60	8.33	16,250
70	7.14	17,150
80	6.25	18,115
90	5.55	17,775
100	5.00	17,250
110	4.54	16,590
120	4.16	15,835
130	3.84	15,000
140	3.57	14,100
150	3.33	13,250
160	3.13	12,420

500-year harvest of 9,375 cubic feet, which is approximately 52 percent of the maximum level (18,115 cubic feet). However, a rotation length that is approximately 40 years longer than the optimal level yields a 500-year harvest of 15,835 cubic feet, which is 87 percent of the maximum level. The reason for this difference can be seen in the asymmetry of the growth functions in Figure 11.2B, which is due to the small amount of growth that takes place in the first several decades of a tree's life.

Although maximizing these physical quantities of wood has been used as a management strategy in the past, and is still advocated by some natural scientists concerned with forestry management, it is an inefficient policy, even when managers are only concerned about the benefits arising from harvesting wood. The reason it is inefficient is the same reason that maximizing the sustainable yield of the fishery (see Chapter 10) is inefficient. The reason for the inefficiency in both cases is that the costs and benefits associated with different quantity levels have not been incorporated into the management strategy. In the case of the forests, one must consider the costs and benefits associated with making the rotation longer or shorter.

Although Faustmann (1849) developed a method for including these costs and benefits, and a derivation of an optimal rotation in terms of these costs

FIGURE 11.3

TIME PATHS OF BENEFITS AND COSTS FROM TIMBERING

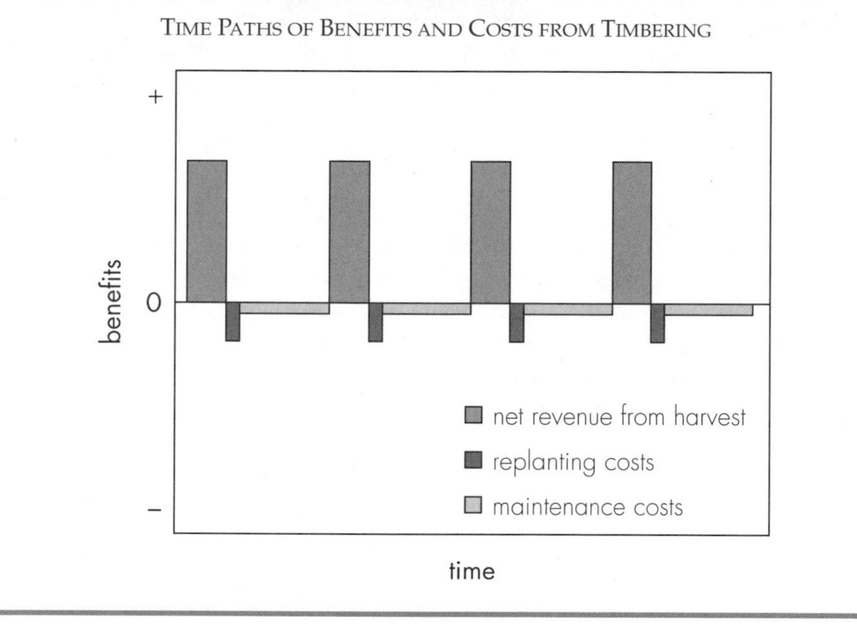

and benefits, Bowes and Krutilla (1989) indicate that his work was not widely known for approximately a century. Bowes and Krutilla credit Gaffney (1957) and Pearse (1967) with bringing Faustmann's work to the attention of American forestry researchers and forest managers.

THE OPTIMAL ROTATION

The choice of the optimal length of rotation is a conceptually simple problem. The forest manager simply asks the question, "Are the benefits of making the rotation a year longer (or a year shorter) greater than the costs?" The complexity involves determination of the costs and benefits and evaluating them over an infinite or extremely long time horizon. A good way to examine the problem is to start with a forest stand of newly planted trees and then each year thereafter evaluate the costs and benefits of letting the trees continue to grow for another year.

Figure 11.3 contains a schematic drawing of the fashion in which benefits and costs accrue over time. First, there is the revenue from the initial harvest (or the costs of planting if the area is not currently forested), then the costs of planting, then maintenance costs such as disease control, fire prevention, thinning, pruning and removal of dead wood, and pest control. Benefits come at a set of intervals, and costs come at a different set of intervals. The revenue from the harvest is often called the stumpage value and represents the proceeds from the sale of the wood, less the costs of harvesting.

The forest manager's job is to maximize the present value of this stream of costs and benefits (see Appendix 2A for a review of present value and discounting) by deciding the optimal rotation length. As mentioned above, the optimal rotation length is determined by comparing the costs and benefits of increasing or decreasing the rotation length.

The costs of letting the trees grow for another year (or a shorter period of time) include both out-of-pocket costs and opportunity costs. Out-of-pocket costs include expenses for disease prevention, thinning, pruning and removal of dead wood, fire prevention, and control of pests. Opportunity costs are based on foregone income from two sources. The first is the income that could be derived from earning interest on the revenues that would have been obtained had the trees been harvested instead of allowed to grow for another year. The second is the rent that could have been obtained had the forest been cut and the land rented. In most forestry models, it is assumed that the only use of the land is forestry, so this opportunity cost would be the rent that would be obtained from leasing out a newly replanted forest stand.

In order to simplify the analysis, the out-of-pocket costs are usually assumed to be zero, and only the two opportunity costs are considered when determining the optimal rotation. The reader can implicitly include these other costs by realizing that under most circumstances, the introduction of additional costs of waiting will shorten the rotation. An additional assumption made is that the real price of a cubic foot of wood does not change over time. If the price were to change, then there would be additional benefits (or costs) associated with increasing or decreasing the rotation.

The assumption that out-of-pocket costs are zero implies that a periodic cutting of the forest stand is efficient, unless the costs of harvesting are so high (that is, because the forest is remote or inaccessible) that stumpage value is negative for any length rotation. However, if out-of-pocket costs are sufficiently high, then it is possible that the forest should never be cut, or cut once and abandoned.

Before determining the optimal rotation, a short discussion of the benefits of allowing the trees to grow is in order. The benefits of allowing the trees to grow come from the fact that, if you wait, you have more wood to sell in the future. As one might expect, the benefits will be critically dependent upon the shape of the marginal growth (annual increment) function. This additional revenue or change in stumpage value is represented by the function $\Delta V/\Delta t$ in Figure 11.4 (based on Howe, 1979). $\Delta V/\Delta t$ is the rate of change of stumpage value [V(t)] and represents the change in stumpage value as rotation length is changed. The foregone interest opportunity cost is represented by $rV(t)$, where r is the interest rate and V(t) is the stumpage value. This function will be based on the total growth function (Figure 11.2A), and it reaches its maximum at the time when lengthening the rotation has no impact on stumpage value ($\Delta V/\Delta t = 0$). This point is labeled R_m in Figure 11.4.

FIGURE 11.4

DETERMINATION OF THE LENGTH OF THE OPTIMAL ROTATION

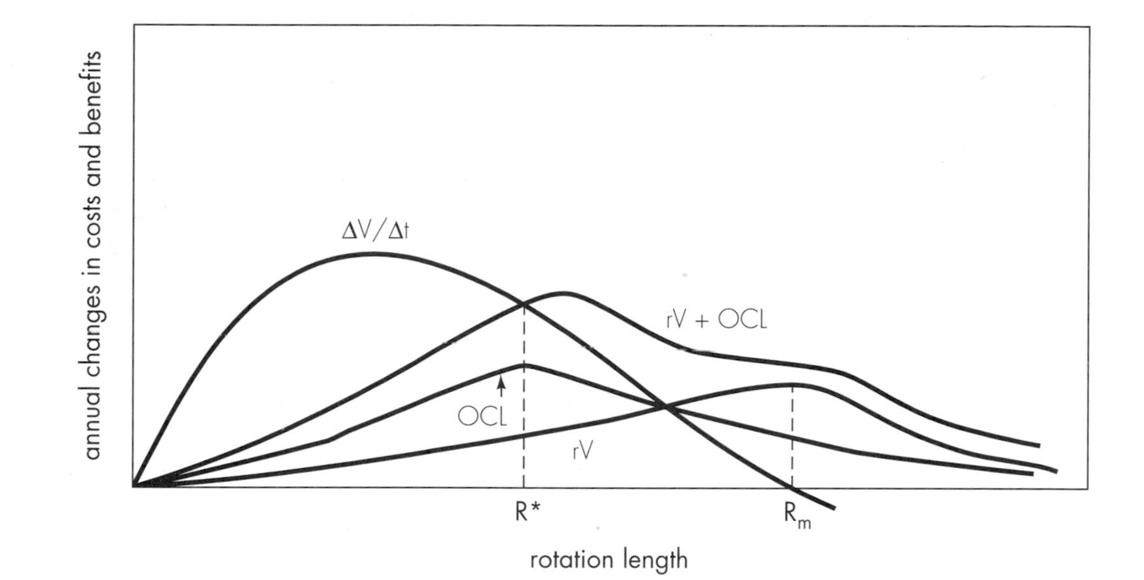

The last cost to be considered is the opportunity cost of the land (OCL). Since we have limited, by assumption, the use of the land to forestry, it may at first seem as if there is no opportunity cost associated with letting the trees grow longer. However, by letting the trees grow longer, not only is the interest income that could be realized from the sale of the harvest deferred, but the interest that could be earned on the proceeds to be derived from the sale of land is deferred as well. This opportunity cost also equals the annual rental value of the land, since equilibrium in land markets implies that the rent is equal to the product of the market value of the land and the rate of interest. Notice that the maximum of this opportunity cost of the land (OCL in Figure 11.4) will occur when the rotation is at its optimal length. This is because the forest will be most valuable (have the highest rent) when it is optimally managed.

Reiterating, the sum of the two opportunity costs is represented by the curve labeled rV+OCL. When these marginal opportunity costs are equal to the marginal benefits of changing rotation length ($\Delta V/\Delta t$), the present value of the whole future stream of harvests is maximized, and the rotation can be said to be optimal. This optimal rotation length is equal to R.

The preceding discussion has attempted to present an intuitive and graphical analysis of the determination of the optimal rotation. Students who are comfortable with calculus and would like to see a more sophisticated derivation are referred to Pearse (1967) or Howe (1979). The Howe discussion is particularly useful in deriving the effect of external changes on the optimal rotation. For example, Howe discusses the effect of changes in timber prices, severance taxes, planting and management costs, and property taxes.

These external changes discussed by Howe can also be evaluated in terms of their effect on the cost and benefit functions in Figure 11.4. For example, any external change that shifts $\Delta V/\Delta t$ upward, will *ceteris paribus,* lengthen the optimal rotation. Any external change that shifts either rV or OCL upward, will *ceteris paribus,* shorten the optimal rotation. The problem in conducting a sensitivity analysis of this nature is that many external changes affect more than one function and can therefore lead to ambiguous results.

This ambiguity can be illustrated by looking at the changes generated by an increase in the price of timber. This change would shift $\Delta V/\Delta t$ upward, since the revenue to be gained by further growth will increase as the price of wood increases. In addition, both opportunity cost functions would shift upward as well. The foregone interest opportunity cost (rV) would shift upward, because an increase in the price of timber will increase stumpage value (V). The opportunity cost of the land (OCL) will shift upward, as the rental value (or the lost interest from the proceeds of the sale of land) will increase if the revenues to be derived from the land increase.

The effect of an increase in the price of timber on $\Delta V/\Delta t$ will be to lengthen the rotation. The effect of an increase in the price of timber on the two opportunity costs (rV and OCL) will be to shorten the rotation. Which effect will dominate? The answer depends on the rate of interest. The smaller the rate of interest, the smaller the effect on rV and OCL, so the lengthening effect will tend to dominate. In other words, with low rates of interest, an increase in the price of timber will tend to lengthen the optimal rotation, while with high rates of interest; an increase in the price of timber will tend to shorten the rotation.

One obvious shortcoming of the optimal rotation model is that it assumes that the only benefits that arise from the forest are the benefits of harvesting it. In actuality, there are many benefits associated with a standing forest, such as watershed protection, biodiversity, wildlife habitat, recreation, and so on. Bowes and Krutilla (1989) point out that the relationship between the length of the rotation and the nonharvest benefits is likely to be irregular, due to the many different types of nonharvest benefits, which will have differing functional relationships with the rotation length. For instance, some benefits might be an increasing function of rotation length, while others might be decreasing. Some could exhibit a u-shaped relationship (decreasing and then increasing), and some could have an inverted u-shape. When all these different relationships are aggregated, then a function with multiple peaks and valleys is likely to result, as is the case in Figure 11.5. This function shows the relationship of the present value of the whole time stream of nonharvest benefits as a function of the rotation length. The function in Figure

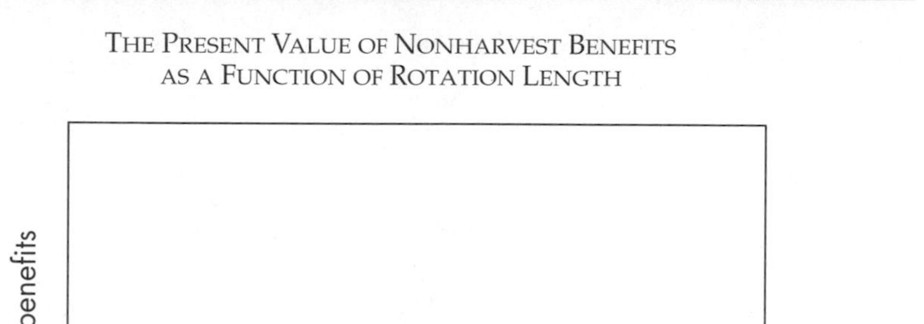

FIGURE 11.5

THE PRESENT VALUE OF NONHARVEST BENEFITS AS A FUNCTION OF ROTATION LENGTH

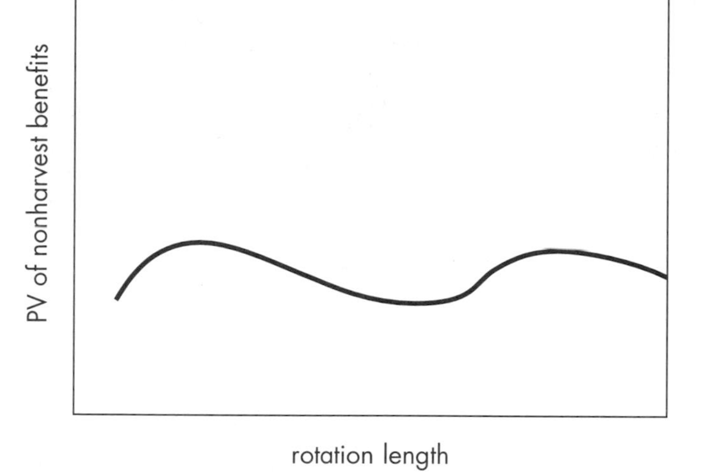

11.5 has been drawn arbitrarily for the sake of illustration. The actual shape of this function would vary from forest to forest.

In Figure 11.6, the relationship between the length of the rotation and the timber benefits (harvested benefits) is shown. The length of the rotation that maximizes the benefits from the timber harvest can be seen to be R_m.

The total benefits of the forest are the sum of the harvested benefits and the nonharvested benefits. This sum is represented by the highest function in Figure 11.6. Notice that the maximum of this function is to the right of the maximum of the timber harvest function, implying that considering nonharvest benefits will lengthen the optimal rotation. This conclusion is clearly dependent upon the shape of the nonharvested benefit function. Bowes and Krutilla point out that this optimal rotation (often called the Hartman rotation after the economist who first discussed it) will be somewhere between the rotation length that maximizes the harvested benefits and that which maximizes the nonharvested benefits. They point out that if the nonharvested benefits increase with rotation length and are large enough, it may be optimal to leave the stand unharvested forever. This is likely to be the case for our ancient growth forests.

FIGURE 11.6

OPTIMAL ROTATION WHEN NONHARVESTED BENEFITS ARE CONSIDERED

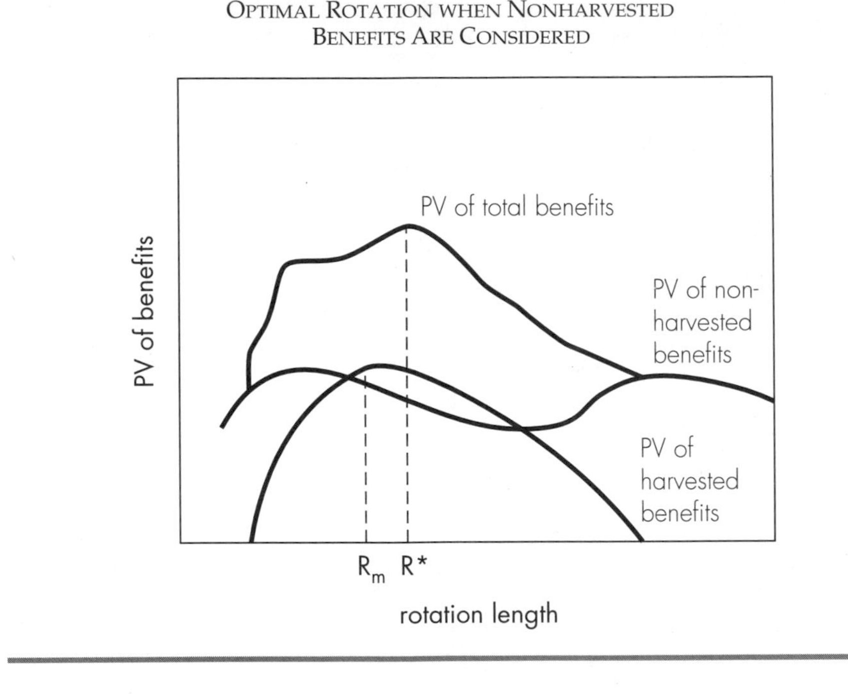

MULTIPLE USE MANAGEMENT

One shortcoming of the preceding discussion about harvested and nonharvested benefits is that it is looking at one forest stand in isolation. Both the harvested benefits and the nonharvested benefits from a particular stand of forest are dependent upon the quantity and quality of other forest stands. For harvested benefits, the price of timber is an indication of the quantity and quality of other forest stands. A similar indicator does not exist for non-forested benefits. Compounding the problem is the fact that harvesting a forest stand and eliminating its nonharvested benefits may actually reduce the nonharvested benefits from other nearby forest stands. For example, if a strategically located forest stand was clear-cut, this action could scar the whole landscape and reduce the aesthetic and recreational benefits to be derived from the surrounding forests. Although the Multiple Use Sustained Yield (MUSY) Act of 1960 specifically charges the U.S. Forest Service (USFS) with managing to promote benefits from both timber and nonharvested benefits, this task is not an easy one. One problem is that different outputs of the forest are not necessarily compatible. In order to more thoroughly understand this issue, the different uses specified by the MUSY Act are specified below.

One set of uses generates revenue for the USFS. These include timber, grazing, mineral and energy mining, and fee recreation. It should be pointed out that fee recreation is limited to special sites, where the fee is charged to recover the costs of special facilities. Most recreation on national forest land is not charged an entrance or user fee (Bowes and Krutilla, 1989, p. 20).

Readers may find it strange to see the grazing of livestock as a forest use, since forests and livestock are not generally associated. However, a forest need not be completely covered by the canopies of trees. In fact, a forest is generally defined as an area in which at least 10 percent of the land area is covered by the canopy of trees. This means that in some areas that are defined as forests, up to 90 percent of the area can be exposed to sunlight and contain the grasses and other forage plants that livestock require. About 100 million acres of national forest land is currently available for leasing to ranchers, with about half of it viewed as suitable for grazing (Bowes and Krutilla, p. 18). Bowes and Krutilla indicate there is considerable evidence to suggest that the fee the Forest Service charges for this use is considerably below the fair market value of the forage. Another set of uses does not generate revenues and are often called nonmarket uses, to differentiate them from the revenue generating or market uses. These nonmarket uses include open-access (unpriced) recreation, watershed maintenance, wilderness, and fish and wildlife.

Not surprisingly, the pursuit of market benefits often conflicts with the pursuit of nonmarket benefits. For example, a scarred clear-cut area filled with stumps and slash (the branches and other parts of the tree that are not suitable for lumber and that are left on the forest floor) is not suitable for recreation and has greatly diminished watershed attributes. It is not difficult to see how the pursuit of other market benefits, such as mining and grazing, conflict with nonmarket uses.

What is more surprising is that many nonmarket uses conflict with each other. For example, too many recreationists can lead to environmental degradation that could destroy the wilderness or reduce the vegetative cover and diminish the watershed attributes of the forest. Even different recreational uses can be incompatible with each other. A backpacker who is in the forest to enjoy its serenity is not going to appreciate a motorcyclist screaming past on a trail bike.

It is easy to say that national forests should be managed so that forest lands are allocated to the mix of competing uses that maximize the net benefits to society. However, the conflict among these uses and the difficulty of measuring many of the costs and benefits makes this a difficult task. Although the MUSY Act of 1960 states that the multiple uses should be promoted, many critics of U.S. forest policy felt that the management was slanted toward timber production.

BELOW-COST TIMBER SALES

One of the primary pieces of evidence cited by critics of USFS policy is the existence of below-cost timber sales. In the late 1970s, the National Resources Defense Council focused on the existence of below-cost timber sales and the inefficiencies that they create, including depressing the profitability of privately owned forests (Bowes and Krutilla, p. 293). Below-market timber sales are those sales of timbering rights on public land, where the revenues generated by the harvest do not cover the timber-related forest management expenses. This comparison is usually made over an intermediate length of time, such as a 5-year period.

While many complex issues surround the proper use of public forest land, a general guideline is that a forest should be used for timbering if the present value of the net benefits (net of management costs) of all the multiple uses is greater than it would be without timbering. This value comparison is difficult for a particular forest manager to evaluate, since it is difficult to allocate particular costs to specific uses. For example, roads must be built if an area is to be timbered, but the roads may also serve to help recreationists gain access to the area.

For this reason, roads are usually not included as timbering costs when the costs of timbering public forest are computed. This exclusion may lead to a flawed process for the following reasons:

1. The quantity of roads required for timbering a given forest is far in excess of the quantity of roads necessary for recreationists.
2. Roads that are unpaved may lead to erosion and other environmental damage.
3. Many types of recreational and other nonmarket benefits may be reduced by the presence of logging activity.
4. The existence of roads precludes the forest area from future designation as a wilderness area.

Timbering roads are generally built by the U.S. Forest Service when the forest land is opened for leasing to private timber concerns. The timbering concerns bid on the land (more than one company may bid on a site, but often the bids are uncontested). The USFS will decide whether or not to accept a bid based on its forest management plans for the tract of forest, but the cost of the roads is usually not a factor in its decision to accept or reject the bids.[2] In fact, since the roads are often built before the bids are let, they are treated as sunk costs. If the fees paid by the timbering companies do not

---

[2]For a complete discussion of past USFS policy and its planning systems, see Bowes and Krutilla (1989). For more recent information on National Forest Service policy, check the USFS website at http://www.fs.fed.gov.

cover all USFS timber management costs, including the roads, there is potential for inefficiencies to arise. This process is not necessarily inefficient if the forest roads serve other uses. Then, one can argue that when the USFS builds these roads, they correct a market failure by creating a public good. However, *if* these roads do not serve other uses, or if the benefits in the other uses are small relative to the costs, then the USFS may create inefficiencies by building the roads, as illustrated in Figure 11.7.

In Figure 11.7, MPC represents the costs to the timbering companies of harvesting the wood, exclusive of the cost of the roads and fees that the company pays to the USFS for harvesting rights. If neither of these costs are borne by the firm, then m_1 square miles of forest will be harvested, because that is the point at which the marginal private costs of harvesting another unit of forest is equal to the marginal private benefits (marginal revenue). However, the optimal level is either at m_3 or m_4. The optimal amount of harvesting would be m_3 if the social costs of harvesting only included the private costs and the road costs, while the optimal level would be m_4 if social costs also included the benefits from recreation and other uses of the forest that might be reduced by harvesting.

If the fees charged for harvesting rights are less than the costs that the USFS pays for road building or other services, then a greater than optimal amount of public forest will be harvested. The amount harvested will be equal to m_2, and the optimal level would be m_3 or m_4.

The optimal level will be m_3 if the social costs of harvesting consist of the sum of private costs and costs incurred by the forest service. If this case occurs, then msc_1 is the relevant marginal social cost curve, and the intersection with the marginal revenue curve occurs at m_3. Several aspects of the marginal revenue curve are worthy of discussion. First, the curve is presented as horizontal, because any individual forest is small (and therefore its harvest will not have an effect on market price). Second, marginal revenue will be equal to marginal social benefit, unless the harvesting of the wood increased the benefits derived from other multiple uses. If this increase was the case, then the marginal social benefit curve would be above the marginal revenue curve. However, for the purposes of this analysis, this situation will not be considered, as it is a straightforward extension of the arguments that are being presented.

This market failure will be even more pronounced when the harvesting of the trees reduces benefits from other uses of the forest such as recreation, wildlife, wilderness, and watershed protection. If this is the case, then the relevant marginal social cost function will be MSC_2, and the optimal level of public forest to be leased for cutting will be m_4.

As just demonstrated, below-cost timber sales may lead to an inefficiently high level of forest land being timbered. A special case is when the optimal level of harvesting is zero, but the below-cost sales lead to positive levels of harvesting. This situation is illustrated in Figure 11.8a and 11.8b. In Figure 11.8a, the marginal social harvesting cost function (MSC_1) lies entirely

EXCESSIVE HARVESTING OF PUBLIC FORESTS AS A RESULT OF SUBSIDIZED ROAD BUILDING

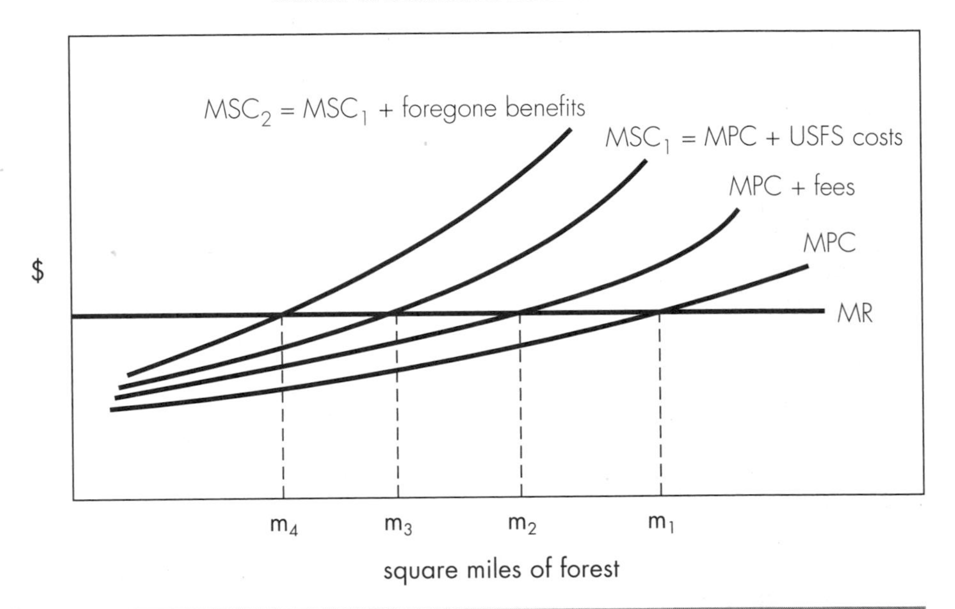

above the marginal revenue function. In Figure 11.8b, MSC_1 lies partially below the marginal revenue function, but if foregone benefits from other uses are considered, then the marginal social cost function (MSC_2 in this case) lies entirely above the marginal revenue function.

Note that there will also be a rent associated with the forest that arises from the fact that the trees have value. Many people argue that the public sector should capture this rent by charging high enough fees for harvesting rights so that the harvesters only earn normal profits and do not capture any rent. They argue that since the forests are publicly owned, the public and not private firms should earn the rents associated with harvests. If this type of pricing was implemented, fees would be likely to increase substantially. It should be noted, however, that this argument is an equity-based argument, centering on who should earn rents, not on how much should be cut. As long as property rights to the forests are adequately defined, the capture of the rents has no effect on the size of social benefits, only on who receives them. The preceding discussion of below-cost timber sales is an efficiency-oriented discussion, revealing the possibility of social losses as a result of pricing policy. It should be noted that the same arguments would apply to the leasing of grazing or mineral rights on public forest land.

FIGURE 11.8

WHEN THE OPTIMAL AREA OF HARVEST EQUALS ZERO

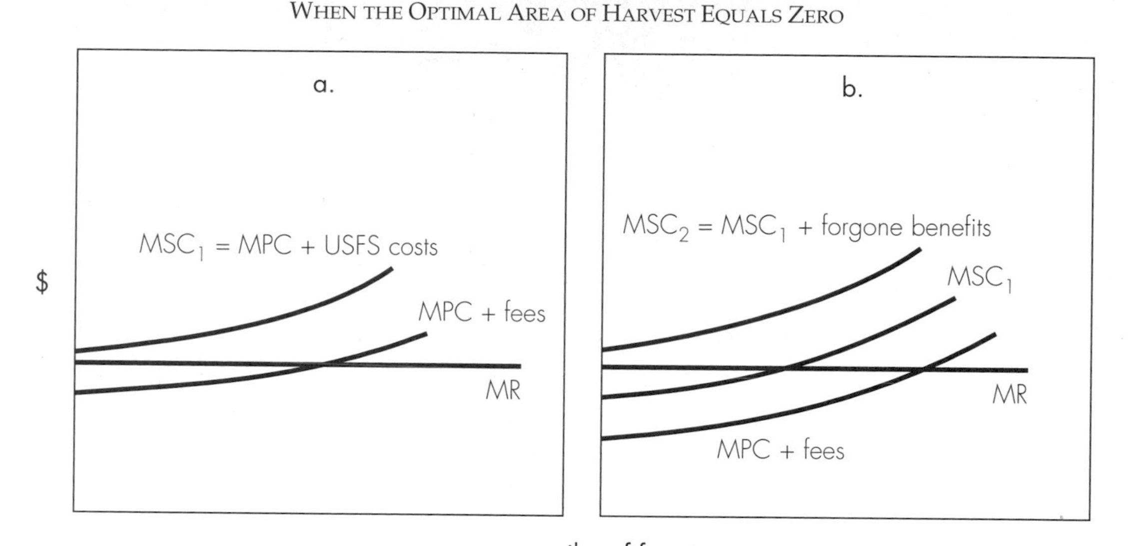

ANCIENT GROWTH FORESTS

At the beginning of the chapter, it was noted that the question of the proper use of North American "old growth" or "ancient" forests was raised and deferred until background information on forestry economics was developed. Old growth or ancient forests are forests that have never been logged and, therefore, are in their original state. Old growth forests are substantially different than forests that have been logged and replanted or logged and naturally reforested.

There are several reasons for the large differences between the ecosystems of the old growth and previously logged forests. These differences arise from the continuity of the ancient forests, the sheer size of the ancient trees (which may be 300 to 2,000 years old, depending on the species), and the diversity of the age of the trees of old growth forests (in comparison with previously logged forests that tend to have trees of uniform age).

With the exception of central prairies, western deserts, and Alaskan tundra, the United States was once almost entirely forested. Virtually the entire area east of the Mississippi River was covered by forest. Although the amount of forest land is still large, the old growth forests have dwindled to a small fraction of their original distribution. What is left is primarily in the

Measuring the Value of Spotted Owls

The bitter controversy over the use of ancient growth forests was conducted in the absence of any quantitative measures of the benefits of preservation until Jonathan Rubin, Gloria Helfand, and John Loomis (1991) conducted a study to measure the benefits of spotted owl preservation.

They used a contingent valuation method (see Chapter 4) to measure people's willingness to pay to preserve the spotted owl. Their first step was to construct a survey that discussed the spotted owl, its habitat, and the competing (commercial) uses of its habitat. People were asked to select an amount (checkoff values ranged from zero to $500) that they would be willing to pay to be completely sure that the spotted owl continued to exist in the future. This survey was conducted using randomly selected residents from the state of Washington (where a substantial amount of spotted owl habitat exists). The average willingness to pay was approximately $35.

The results were then extrapolated to the rest of the nation by using a formula from another study (Stoll and Johnson, 1984) that suggests that the willingness to pay for an environmental resource declines by about 10 percent for every 1,000 miles distance between the resource and the residence.

Then, Rubin et al. compared the preservation benefits that they calculated to the preservation costs (loss of timbering jobs and increase in the price of wood) that were estimated by the U.S. Department of Agriculture. They found that short-run preservation benefits are approximately $1.4881 billion, with short-run preservation costs equal to $1.335 billion. Long-run benefits are computed to equal $1.481 billion, as compared with long-run costs of approximately $0.497 billion, for a net benefit of preservation of almost $1 billion. The authors suggest that this represents such a large potential Pareto improvement that appropriate policy might be to compensate the people in the timber industry who lose from preservation.

SOURCE: Jonathan Rubin, Gloria Helfand, and John Loomis, "A Benefit–Cost Analysis of the Northern Spotted Owl," *Journal of Forestry Research* 89 (1991): 25–30.

temperate rain forests of the Pacific Northwest and Alaska and is shrinking rapidly due to clear-cutting of forests.

The trees in the Pacific Northwest ancient forest are composed primarily of coniferous species, among which the Douglas fir, coastal redwood, and sequoia are prominent. These trees will grow to hundreds of feet in height and 10 to 30 feet in diameter.

The huge old trees occasionally die, and their death shapes the ecosystem. Standing dead trees provide homes for many species of animals, but when they fall they provide their most significant services to the ecosystem.

When one of these huge trees falls, it knocks down everything in its path, creating a swath of sunlight that promotes the growth of plants on the forest floor. These plants provide food for a wide variety of animals. In contrast, in a previously logged forest, the trees tend to be of uniform age and height (especially if forests were artificially replanted), which allows little penetration of light to the forest floors. Since all the trees are roughly the same mass, falling trees tend to be supported by (or roll off of) their neighbors, rather than taking them down with them.

The fallen tree provides nutrients for new generations of trees. In fact, seeds of new trees germinate and become established in the rotting wood of the fallen trunk, and the new trees grow right out of the dead trunk. Since this trunk is elevated (by its own diameter) above the forest floor, the seedling trees do not face as vigorous competition from other plants as they would if they were growing on the forest floor.

In addition, fallen trees provide homes for voles, a mouse-like species of mammal, which are critical to the old growth ecosystem. The voles' primary food is a truffle-like fungus. In eating this fungus they ingest but do not digest another type of fungus, which they spread throughout the forest soil. This fungus attaches itself to the roots of coniferous trees and makes the soil nutrients more available to the trees, facilitating their tremendous growth and enabling them to deal more effectively with the summer dry season (although these forests receive a tremendous amount of rain over the year, summers tend to be very dry).

The huge fallen trees provide a haven for voles during forest fires, because the trees are so large and moist that their interiors remain cool during fires. The fires kill most of the floor plants and also the fungi that are in the top layers of soil. After the fire, the voles venture out of the logs and defecate, reestablishing the fungus in the soil. The large ancient trees are protected from the fire by their thick bark, and the fire actually helps perpetuate the forest as the heat triggers (after some delay) the dropping of seeds from the cones of these large coniferous trees.

Ecologically, these large forests are very different from logged and replanted forests.[3] One would think that the comparative advantage of ancient growth forests, particularly the spectacular coastal redwoods and sequoias, would be to leave them standing. The reason is that replanted forests are good substitutes for ancient forests as producers of timber, but poor substitutes for ecosystems. However, the U.S. Forest Service continues to allow this harvesting (as do the corresponding state agencies on state forests), even subsidizing the process through road building and other management programs.

In the beginning of the chapter, it was mentioned that this controversy revolved around what is more important: spotted owls or timber-related jobs. The jobs issue is often used by the USFS and its advocates as a reason for subsidizing the harvesting of forests (Bowes and Krutilla). Since the forest is the only source of economic activity in many remote rural regions, many

---

[3]Not all types of forest are associated with such disparities between the ecological values associated with replanted forests and the ecological values associated with old growth forests. A good example would be the spruce and fir forests of northern Maine. These forests are primarily owned by the forest product industry and are often clear-cut. Areas which were clear-cut 60 years ago are now very similar in ecological and recreational values to the original forest.

FIGURE 11.9

LOSS IN VALUE FROM CUTTING DOWN A STAND OF ANCIENT FOREST

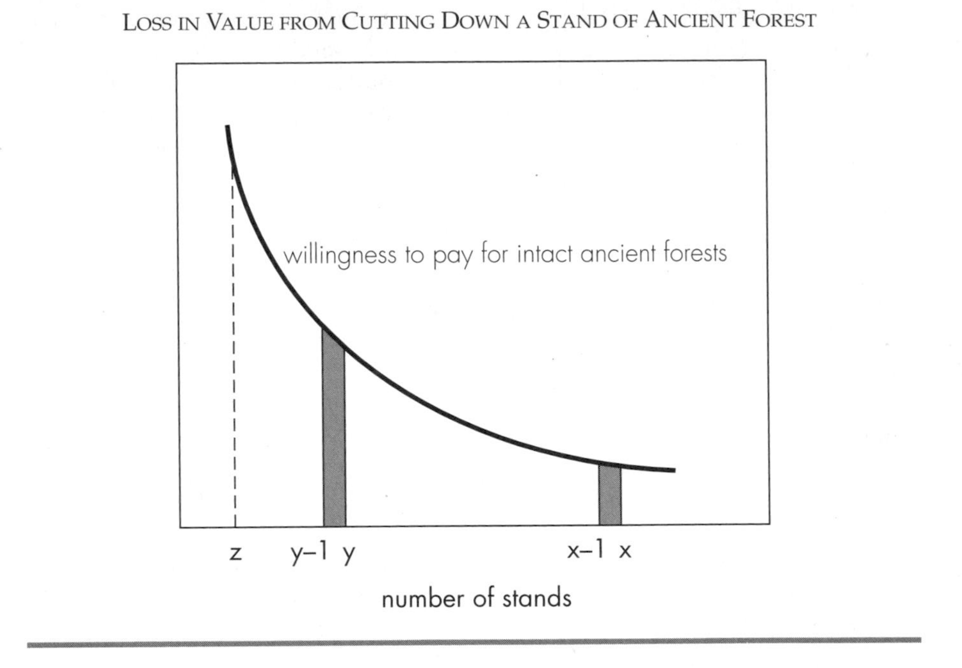

often feel that the forest must be harvested to provide jobs to support the region's population.

In actuality, this process involves a net loss for society as a whole, because it costs as much as $3 of government expenditure for every timbering job wage dollar that is created. In some cases, the timbering may eventually cost more jobs than it creates because of the ecological damages. For example, the cutting of the Tongass National Forest in southwestern Alaska may cost many salmon fishing-related jobs, because the soil erosion associated with clear-cutting makes the rivers inhospitable to salmon reproduction. In addition to these monetary costs, the subsidization of the harvest of old growth forests lowers the market price of wood, adversely affecting employment in areas of the country where wood is primarily harvested from privately owned forests.

In addition to these direct monetary costs, there are the costs to society of the loss of the ancient forests. These costs are likely to be high, since the amount of ancient forests has shrunk so drastically in recent years. If there were abundant ancient forests, then the loss in value associated with cutting a particular ancient forest would be relatively small, since a great deal of forests exist. This association is shown in Figure 11.9, where in the vicinity of x stands of ancient forests, the willingness to pay to prevent the cutting of one stand is relatively low. However, as one moves along the demand curve to a point where ancient forests are less abundant, such as in the vicinity of y, the willingness to pay becomes much larger. Many forest advocates would argue that the United States is currently to the left of a point like z, where the old growth forests are so scarce that their value becomes asymptomatically large. Note that if replanted forests were good substitutes for ancient forests, the demand curve in Figure 11.9 would not have the very inelastic portion to the left of z. Instead, the intercept would be much lower, implying that the cost of cutting the last stand of ancient forests is low, since replanted forests are good substitutes. However, it should be reiterated that for many types of forests, replanted forests are not good substitutes from an ecological and recreational viewpoint, and the opportunity costs of cutting the remaining ancient forests are likely to be quite high.

Current policy towards old growth forests is in a state of flux, with many new laws being introduced before Congress, the Endangered Species Act up for reauthorization, and many policy decisions are actually being made in the federal courts. For current information on the status of old growth forests, check the U.S. Forest Service Web site at http://www.fs.fed.gov.

SUMMARY

Forests are a resource that provide many benefits to society. In addition to providing wood, forests provide a habitat for wildlife, a reserve for biological diversity, sites for recreation, wilderness, watershed protection, and many other benefits.

If a forest is managed solely for the income to be derived from harvesting wood, which is how a profit maximizing firm would manage forests that it owns, then the present value of the stream of harvests would be maximized. This maximization is accomplished when the trees are cut at an age at which the income gains to be derived from waiting an additional year before harvest are exactly offset by the income losses generated by waiting. This time period is called the optimal rotation.

Public forests should be managed with more than this in mind, as the Multiple Use Sustainable Yield Act requires all the benefits of forests to be recognized in developing management strategies. Although the U.S. Forest Service is genuinely concerned about promoting nonharvest benefits, many critics of USFS policy argue that there still exist some biases toward harvesting and away from activities associated with leaving the forest stands intact. In particular, below-cost timber sales and subsidized road building are criticized as leading to overharvesting of our nation's public forests. This criticism may be particularly true for ancient forests for which previously logged and replanted forests are not a good substitute.

REVIEW QUESTIONS

1. Discuss the framework for maximizing the stream of income arising from timberland. What are the conditions that define the optimal rotation? What economic and biological factors influence these conditions?
2. Discuss the concept of multiple use/sustained yield management. What benefits should be considered when pursuing this type of management? How should these benefits be compared?
3. How does the inclusion of nonharvested benefits affect the optimal rotation?
4. Draw a growth function for a typical stand of trees. What factors determine its characteristic shape?
5. Use the data in Table 11.1 to calculate the optimal rotation of this hypothetical forest stand (to the nearest 10 years). Assume that the price of wood is $1.00 per cubic foot; that all harvesting, planting, and maintenance costs are equal to zero; and that the interest rate is equal to 5 percent.

QUESTIONS FOR FURTHER STUDY

1. Discuss the jobs/environment trade-off in managing ancient growth forests. How would you resolve this issue? (Make sure to consider both efficiency and equity considerations.)
2. Should forest fires be prevented? Why or why not?
3. What does the principle of comparative advantage imply for the multiple use/sustained yield management of forests?
4. How would you measure the use and indirect use benefits associated with old growth forests?
5. How would the following events affect the length of the optimal rotation of a forest?
   a. higher price of wood
   b. increased probability of fire or disease
   c. higher interest rates
   d. lower transportation costs

SUGGESTED PAPER TOPICS

1. Use data from the *World Resources 1992–93*, the *FAO Yearbooks* or the USDA sources listed in the references, and socioeconomic data from the *U.S. Statistical Abstract* or *Economic Report of the President* to estimate a demand curve for timber. Use forecasts of future economic growth in the United States and the rest of the world to forecast the future demand for timber.
2. Discuss the open-access externalities associated with rangeland and past and present federal policy to mitigate these externalities. Do you think past policy was efficient? Do you think present policy is efficient? How would you modify policy to maximize the social benefits of rangeland?
3. Examine the writings of an American conservationist such as John Muir. How do the values expressed in his writing correspond to concepts of value discussed in Chapter 4 of this book? What are the economic implications of the management concepts he discusses?
4. Compare the writings of John Muir and Gifford Pinchot. How do differences in their philosophies imply differences in emphasis on the benefits of alternative uses of the forests? How do their theories correspond to different philosophies that are articulated today?
5. Discuss the implications of acid precipitation for forest health. Are the damages associated with acid precipitation likely to be significant? How should this affect the development of air pollution and forest policy?
6. How has U.S. policy toward federal forests evolved over time? In what directions do you think it should move in the future and why?

WORKS CITED AND SELECTED READINGS

1. Bowes, M. D., and J. V. Krutilla. *Multiple Use Management: The Economics of Public Forest Land.* Washington, D.C.: Resources for the Future, 1989.
2. Clawson, Marion. *Forests for Whom and for What?* Washington, D.C.: Resources for the Future, 1975.
3. Ellefson, Paul V. *Forest Resource Economics and Policy Research: Strategic Directions of the Future.* Boulder: Westview Press, 1989.
4. Faustmann, Martin. "On the Determination of the Value which Forest Land and Immature Stands Poses for Forestry." In "Martin Faustmann and the Evolution of Discounted Cash Flow," edited by M. Gane. Institute Paper no. 42, Commonwealth

Forestry Institute, University of Oxford, 1968, pp. 27–55. Reprinted from and originally published in German in *Allegemeine Forest und Jagd Zeitung*, vol. 25, 1849.

5. Howe, Charles. *Natural Resource Economics.* New York: Wiley, 1979.
6. Gaffney, M. "Concepts of Financial Maturity of Timber and Other Assets." *Agricultural Economics Information,* series 62. North Carolina State College, Raleigh, September 1957.
7. Organization for Economic Cooperation and Development (OECD). *Market and Government Failures in Environmental Management: Wetlands and Forests.* Paris: OECD, 1992.
8. Pearse, P. H. "The Optimum Forest Rotation." *Forestry Chronicle* 43 (1967): 2.
9. Rubin, Jonathan, Gloria Helfand, and John Loomis. "A Benefit–Cost Analysis of the Northern Spotted Owl." *Journal of Forestry Research* 89 (1991): 25–30.
10. Sedjo, Roger. *Government Interventions, Social Needs and the Management of U.S. Forests.* Washington, D.C.: Resources for the Future, 1983.
11. USDA Forest Service. "An Analysis of the Timber Situation in the United States, 1952–2030." Forest Report no. 23, Washington, D.C., 1982.
12. ———. *"An Assessment of the Forest and Range Land Situation in the United States."* FS-345, Washington, D.C., 1980.
13. Stoll, John, and Lee Ann Johnson. "Concepts of Value, nonmarket Valuation and the Case of the Whooping Crane." *Transactions of 49th North American Wildlife and Natural Resources Conference.* Washington, DC: Wildlife Management Institute, 1984.

12 Tropical Forests

The unsolved mysteries of the rain forest are formless and seductive. They are like unnamed islands hidden in the blank spaces of old maps, like dark shapes glimpsed descending the far wall of a reef into the abyss. They draw us forward and stir strange apprehensions. The unknown and the prodigious are drugs to the scientific imagination, stirring insatiable hunger with a single taste. In our hearts, we hope we will never discover everything. We pray there will always be a world like this one at whose edge we sat in darkness. The rain forest in its richness is one of the last repositories on earth of that timeless dream.

Edward O. Wilson, The Diversity of Life

INTRODUCTION

Tropical deforestation is an area of environmental degradation that has captured media attention and is perceived as a metaphor for and indicator of the decline of the biosphere. Although tropical deforestation is not a new phenomenon (it has been taking place for hundreds of years), the pace of deforestation has accelerated in recent years, particularly since the late 1970s. This acceleration has raised global concern because the rain forests are important ecological resources, providing both regional and global ecological and economic benefits.

Is complete deforestation going to take place, or can policies be developed to reduce the incentives for people to destroy tropical forests? Is any deforestation a good idea? If so, what is the appropriate rate of deforestation? This chapter attempts to answer these questions by looking at the behavioral (economic and social) forces that lead to deforestation, and how policies can be developed to alter these behavioral forces.

DEFINITION OF A TROPICAL FOREST

Tropical forests are defined by the Food and Agricultural Organization as areas located between the tropics of Capricorn and Cancer, and where at least 10 percent of the area is covered by woody vegetation. Although one tends to think of rain forests when one thinks of tropical forests (in rain forests virtually all the area is covered by the canopies of trees), there are also tropical dry forest areas where there is a mixture of grassland and forests. Many of these dry forest areas are semiarid, whereas tropical rain forests will receive over 100 inches of rain per year, with the wettest areas well over twice

FIGURE 12.1

DISTRIBUTION OF TROPICAL FORESTS

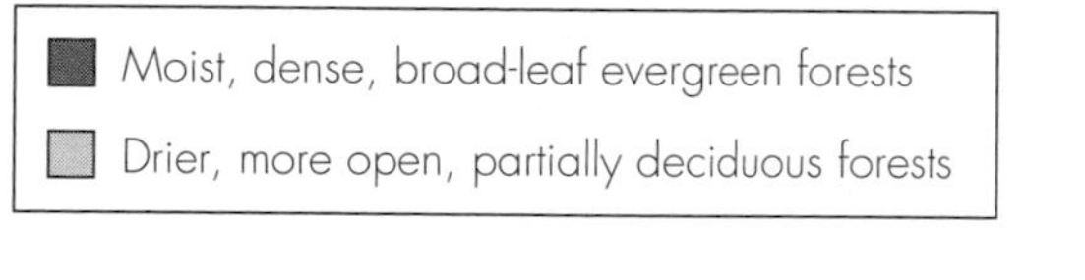

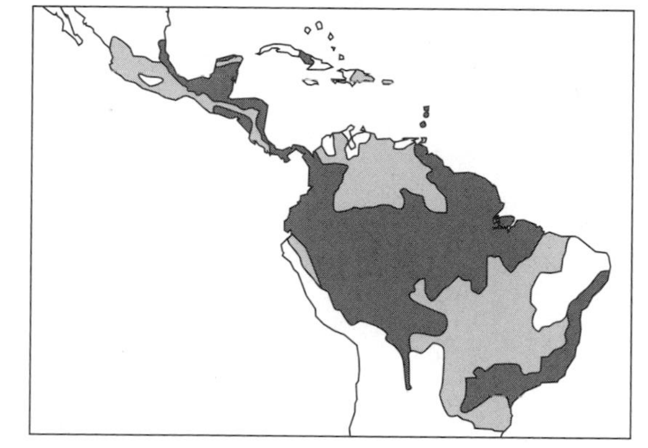

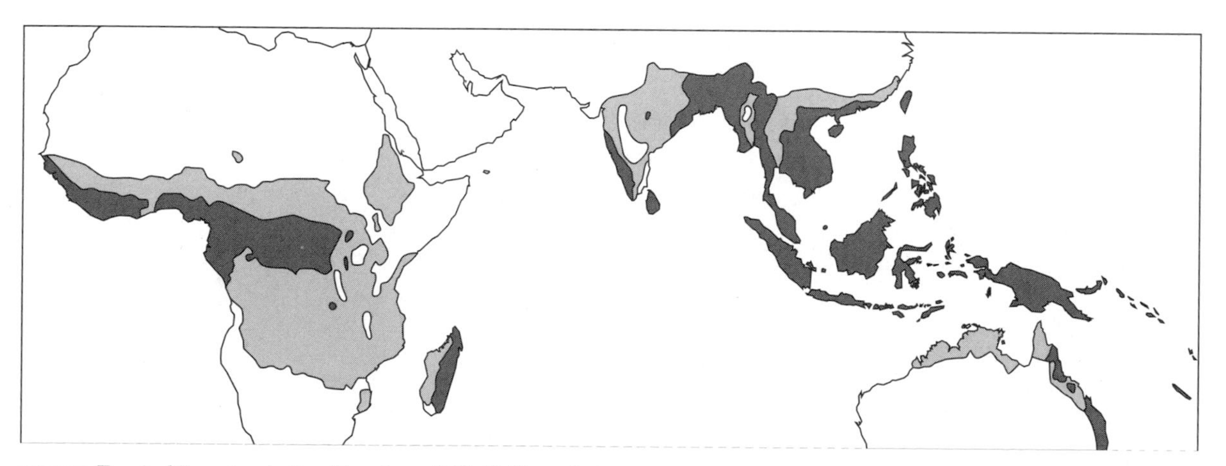

SOURCE: Tropical Forestry Action Plan, June 1987, FAO, pp. 2–3.

that amount of rainfall. Figures 12.1 and 12.2 show the distribution of tropical wet and dry forests and the share of the earth's surface that they occupy.

TROPICAL RAIN FORESTS

Tropical rain forests tend to be dominated by broad-leaved evergreens with completely interlocking canopies. In many rain forests, there are actually multiple canopies, with some giant trees rising above the first canopy. The trees that rise above the primary canopy are frequently called emergent trees. In many areas, these trees support the growth of vines that grow up the trees

FIGURE 12.2

EXTENT OF TROPICAL FORESTS (IN MILLIONS OF HECTARES)

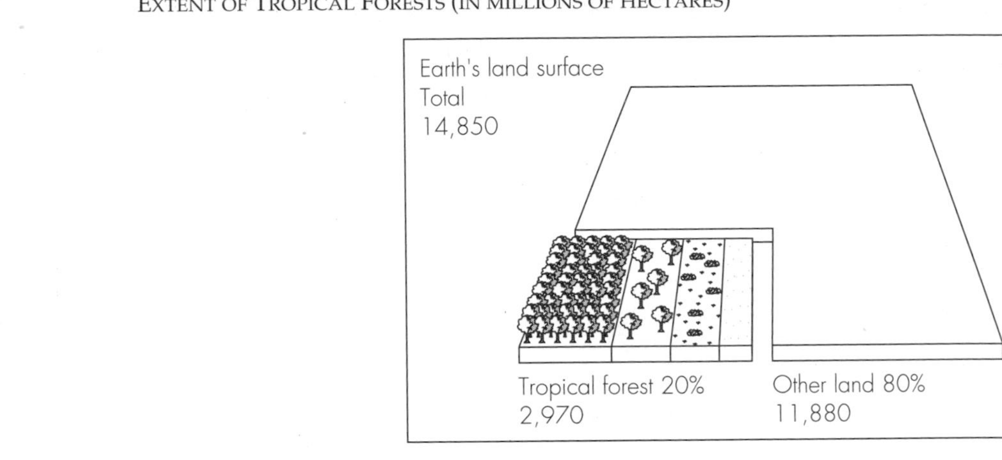

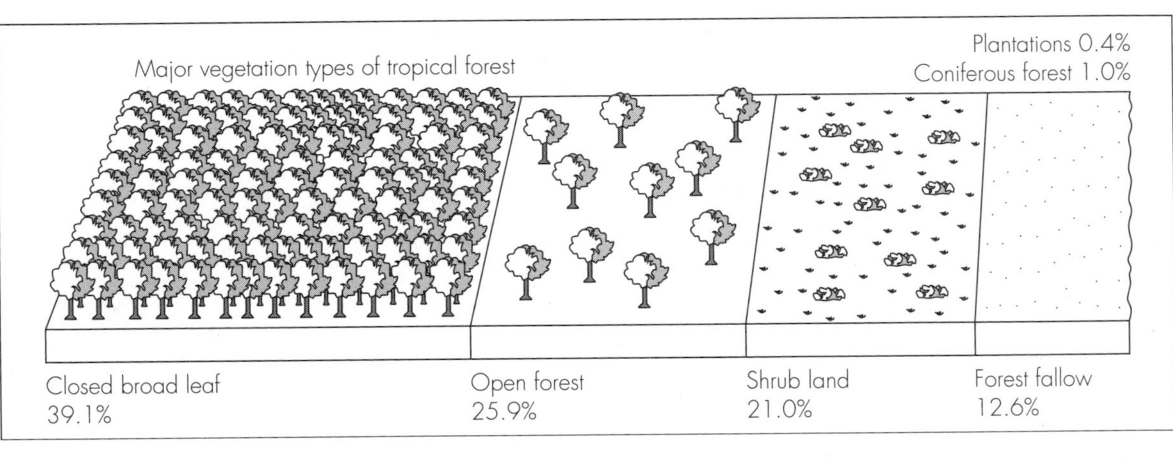

SOURCE: Tropical Forestry Action Plan, June 1987, FAO, pp. 2–3.

to try to reach the sunlight above the canopy. Epiphytes (such as orchids and bromeliads) take seed in dead organic matter (detritus) that has accumulated on the limbs of trees.

There is a constant struggle in the tropical rain forest for access to sunlight. Below the canopy, there is an understory of shrub-like plants and a series of nonwoody plants that occupy the forest floor. Typically, these understory and floor plants have less stringent requirements for sunlight. Many of the understory plants are stunted versions of the dominant trees that will experience accelerated growth when the death of giant trees allows

FIGURE 12.3

THE MULTIPLE LAYERS OF THE TROPICAL RAIN FOREST

light to penetrate to the forest floor. Figure 12.3 contains a line drawing of a crosssection of the typical forest.

The fact that little sunlight penetrates to the rain forest floor generates an environment where most of the animal species are arboreal, as it is the leaves in the canopy that form the basis for the food web. These animals, which include a multitude of insects, spiders, scorpions, birds, lizards, snakes, frogs, and mammals, live in the trees and seldom venture to the forest floor. The arboreal environment provides an ecosystem that is extremely rich in species diversity, with up to 95 percent of the world's plant and animal species found in rain forest habitat (Perry, 1990). According to Perry, as many as 10 to 30 million yet to be discovered species live in the rain forest (science has currently identified a total of about 2 million species in the entire world). It can easily be seen that tropical rain forests support many times the number of species as temperate forests, with the differential measured in terms of many orders of magnitude. Rain forest rivers are correspondingly more diverse than their temperate counterparts.[1]

Tropical rain forests have a very different nutrient cycle in comparison with temperate forests. In temperate forests, the nutrients are stored in the soil. In contrast, the soil of the rain forest is relatively barren of organic matter because it decays so rapidly in the hot, moist environment of the rain forest. Consequently, the tropical soils are mostly mineral soils with little organic matter. This lack of organic matter within the soil does not mean the nutrients are lost, however, as the first few meters of soil contain a dense lattice of the roots of the trees, vines, and other forest vegetation that immediately absorb the nutrients as they percolate through the soil and become available to the roots. Thus, rather than being stored in the soil, the nutrients are stored in the biomass of the plants, particularly in the immense trees that can approach 200 feet in height.

Tropical Dry Forests

In contrast, the tropical dry forest has decidedly less rainfall than the wet forest and, consequently, less biomass per hectare, less coverage by trees, and a greater ratio of nutrients in the soil versus nutrients in the biomass (although still a lower ratio than temperate forests). Animal life in the dry forest tends to be more terrestrial (and less arboreal) than in a wet forest, because the grass that covers the areas under the trees and between the trees also forms a significant basis for the food web. This is not to say that the leaves of the trees are unimportant, but that the grass is also important. The sparser the tree coverage, the greater the relative importance of the grass.

Benefits of Tropical Forests

One important benefit of tropical forests is that they may be harvested to supply wood, which can be used in construction, furniture, or the manufacture of paper and wood products, such as plywood and hardboard. Many valuable tree species are native to the rain forests, including mahogany, zebrawood, rosewood, padauk, bubinga, cocobolo, and teak. In addition, tropical forests, particularly tropical dry forests, are harvested for fuelwood, which is the only source of energy for the majority of people who inhabit the tropics.

Tropical forests also protect the surface waters of the areas they cover. The canopies, understories, and floor plants of rain forested areas soften the impact of the torrential rainfall and prevent it from eroding the mineral soils and generating run-off that would carry away the nutrients before they reached the roots. This protection is important even in tropical dry forests, where the annual rainfall may be low, but is often concentrated in a rainy

---

[1]See Chapter 13 for more discussion of the importance of biodiversity.

season with torrential downpours. The lessening of the impact of the rain allows it to percolate through the soil (rather than running off), where it can be absorbed by the roots of plants. What is not absorbed by the plants slowly makes its way through the soil to the rivers and streams. Since this movement is slow, it does not carry much soil into the streams and rivers. The water that is taken up by the plants is released through the leaves into the atmosphere. In large rain forests, such as the Amazon, the water that is transpired by the forest actually falls back on the forest. In other words, large rain forests create a microclimate, including their own rainfall.

Deforestation may lead to important soil erosion problems. Dams and hydroelectric facilities located downstream of deforested areas often have problems with siltation from erosion of soil. The siltation will partially or completely fill the reservoir. This siltation also causes problems in bays, lagoons, and river mouths, where aquatic life is diminished as a result of the siltation. In regions where fisheries are an important source of food and income, this impact on fisheries can have catastrophic effects on local communities.

Although tropical dry forests experience less rainfall, the role of forest vegetation is equally important to the hydrological cycle. Without the trees, shrubs, and grasses, the rainfall would roll off the ground and less would be absorbed. As a consequence, many rivers that have water throughout the year would dry up during the dry season. As will be discussed in the following section on deforestation, the removal of the trees very often causes semiarid dry forests to turn into desert.

Tropical forests also provide a habitat for many different life-forms. As mentioned previously, most of the species on earth reside in the rain forest. What is not often realized is that this reservoir of life-forms offers the key to treating many human medical problems. As much as 25 percent of the new medicines originate in a tropical rain forest life-form. Although many argue that potential medical advances are the primary benefit of maintaining species diversity, others stress that the existence of the diversity and the plants and animals themselves are important benefits of the forest.

Although there is a much smaller number of species in tropical dry forest areas, these areas also represent an important plant and animal habitat. In particular, some of our largest (and most endangered) mammals live in tropical dry forest habitats. In addition to the animals and plants, the forest represents an eco/agro system that supports populations of indigenous peoples. The forests provide fruits, edible plants, animals, and materials for construction of dwellings, clothing, and other objects. Indigenous groups of people have developed agricultural systems that are compatible with the forest ecosystems. In addition to providing food for domestic use, the rain forest provides many outputs that can be exported, including coffee, fruit, nuts, honey, cocoa, and rubber.

Finally, tropical forests, particularly tropical rain forests, play an important role in maintaining the atmosphere's chemistry. The massive amount of plant material in the rain forest is roughly half carbon by weight. This carbon is derived from atmospheric carbon dioxide through photosynthesis. This process also results in oxygen being released into the atmosphere. Thus, as the rain forests shrink through deforestation, atmospheric carbon dioxide will increase. This increase in atmospheric carbon dioxide will exacerbate global warming (see Chapter 6). Also, the rain forest is the largest terrestrial source of oxygen. An additional use of the rain forest is to convert it into something else, such as farmland or rangeland. Typically, these conversions result in agricultural systems that quickly lose their productivity and are abandoned. This will be discussed more fully in the following section, which examines deforestation.

DEFORESTATION

As can be seen in the previous discussion, tropical forests have public good benefits. These public good benefits accrue to people in both the country that contains a forest and all the other countries of the world. This potentially leads to a two-tiered market failure problem. First, since there are public good benefits, private owners of forests may cut them down at a rate that exceeds the rate that is socially optimal from the point of view of the country that contains the forest. Second, if the country seeks to develop policy to deal with this type of market failure, it has no incentive to consider preservation benefits to the citizens of the rest of the world. In other words, even if countries choose internally efficient policies to address market failure, they may not be globally efficient.

DEFORESTATION ESTIMATES

Although there is considerable controversy concerning the amount of deforestation of tropical forests that is taking place, there is a consensus among those examining the problem that deforestation is occurring at a rapid pace. The most comprehensive source of data concerning the rate of deforestation is Singh (1993). This data is summarized by Brown and Pearce (1995) and listed in Table 12.1. As can be seen in this table, the annual rate of change in forested area over the period 1981–1990 averaged a little less than 1 percent per year (0.8%), with the highest rate of deforestation in Asia and Central America/Mexico, and the greatest amount of deforestation taking place in tropical South America. The average amount of deforestation taking place in the tropics is 15.4 million hectares per year, or approximately 59,444 square miles per year. This annual amount of deforestation is a little larger than the area of Florida or the United Kingdom.

ACTIVITIES THAT LEAD TO DEFORESTATION

The three activities that are primarily responsible for tropical deforestation include cutting trees for timber, cutting trees for fuelwood, and conversion

TABLE 12.1

ESTIMATES OF FOREST COVER AREA AND DEFORESTATION BY GEOGRAPHICAL SUBREGIONS

GEOGRAPHIC REGIONS AND SUBREGIONS	NUMBER OF COUNTRIES SURVEYED	TOTAL LAND AREA (10^6 HA)	FOREST AREA (10^6 HA)		ANNUALLY DEFORESTED AREA (10^6 HA)	RATE OF CHANGE 1981–90 (% PER ANNUM)
			1980	1990		
Africa	**40**	**2236.1**	**568.6**	**527.6**	**4.1**	**−0.7**
West Sahelian Africa	9	528.0	43.7	40.8	0.3	−0.7
East Sahelian Africa	6	489.7	71.4	65.3	0.6	−0.8
West Africa	8	203.8	61.5	55.6	0.6	−0.8
Central Africa	6	398.3	215.5	204.1	1.1	−0.5
Tropical Southern Africa	10	558.1	159.3	145.9	1.3	−0.8
Insular Africa	1	58.2	17.1	15.8	0.1	−0.8
Asia	**17**	**892.1**	**349.6**	**310.6**	**3.9**	**−1.1**
South Asia	6	412.2	69.4	63.9	0.6	−0.8
Continental Southeast Asia	5	190.2	88.4	75.2	1.3	−1.5
Insular Southeast Asia	5	244.4	154.7	135.4	1.9	−1.2
Pacific Islands	1	45.3	37.1	36.0	0.1	−0.3
Latin America	**33**	**1650.1**	**992.2**	**918.1**	**7.4**	**−0.7**
Central America/Mexico	7	239.6	79.2	68.1	1.1	−1.4
Caribbean	19	69.0	48.3	47.1	0.1	−0.3
Tropical South America	7	1341.6	864.6	802.9	6.2	−0.7
Total Tropics	**90**	**4778.3**	**1910.4**	**1756.3**	**15.4**	**−0.8**

SOURCE: *Tropical Forestry Action Plan,* FAO, June 1987, pp. 2–3. Reprinted with permission of the Food and Agriculture Organization of the United Nations.
HA = hectares. 100 hectares is equal to one square kilometer or 0.386 square miles.

of land to crop or range land. While all three activities are important in all regions of the tropics, timbering activities tend to be more important in Asian countries than in other regions, fuelwood activities tend to be more important in sub-Saharan African dry forests and in mountainous regions in Asia, and conversion to agricultural land tends to be more important in Latin America and the more densely populated countries of Asia and Africa.

Although these activities directly result in forest loss, there are underlying reasons that the deforestation resulting from these activities may be excessive. Mendelsohn (1994) cites three underlying causes:

1. poor concessionaire agreements,
2. subsidies for alternative land uses, and
3. lack of adequate property rights.

In turn, there may be underlying economic conditions that generate the way in which concessionaire agreements, subsidies, and property rights are structured.

In countries that are characterized by conditions such as poor concessionaire agreements, subsidies for alternative land use, and lack of adequate property rights, the deforestation rates are likely to be inefficient from that particular society's point of view. The reason that these rates of deforestation are likely to be greater than optimal from a particular country's point of view is that the conditions prevent the realization of the maximum benefits from the forest. In addition to this internal inefficiency, there may be global inefficiency as well. This possibility is because the role of the forests as ecological treasures and storehouses of biodiversity, and their importance in global carbon cycling, make the forests valued by the rest of the countries of the world. As discussed earlier, these are global public good benefits, and a forested country is not likely to be in a position to capture the benefits that other countries derive from the forest. The country will only consider the domestic costs and benefits of deforestation and not take into account the global benefits. In other words, the tropical forests constitute a global public good, so deforestation rates that are optimal from a forested country's point of view may not be optimal from a global point of view. The significance of the forests as a global public good will be discussed more fully after the within country market failures are examined.

Timbering Activity. Given that forests regenerate, it may seem strange that timbering activity would lead to destruction of the forest. Sustainable cutting could be conducted so that each year the net growth of the forest was harvested, and the stock of wood remained constant. Rather than harvesting the stock, the flow of benefits is harvested, so the flow exists year after year after year. As discussed in Chapter 11, the forest could be harvested based on a schedule which maximizes the present value of the future stream of benefits from the forest.

This schedule of harvesting is not happening, however, and one of the major reasons is a market failure generated by the way many forested countries issue leases (concessionaire agreements) for harvesting rights. Very often, these leases are for too short a time period. This time frame generates two perverse incentives for the harvesting firms. First, they have the incentive to harvest as much as possible now, since they may not have access to the forest in the future. Second, they have no incentives to reduce environmental costs associated with the harvesting activity. As a consequence, many unharvested trees, particularly immature trees, are damaged or killed. Also, no attempts are made to leave the forest floor a suitable place for new trees to grow, as harvesting techniques may lead to massive soil erosion.

There are several solutions to this problem, and all are oriented toward giving the harvesting firms the proper incentive to protect the forest resources. First, the leases could be structured for perpetual harvesting rights, and firms could have the right to market these rights. Second, the harvesting firms could be forced to pay a penalty if they generate excessive environmental damage during harvesting.

Granting the firms long-term harvesting rights, which they can either exercise or sell, creates an incentive to maintain the productivity of the forest. If they reduce the productivity of the forest, they will bear that cost in terms of lower future harvests or a lower selling price for the harvesting rights.

Since much of the forested land is publicly owned, there is often considerable reluctance to establish a long-term lease that will give the harvesting firms perpetual control over the forest. This reluctance is particularly true when the harvesting firms are based in other countries, so that the granting of perpetual rights is viewed as relinquishing sovereignty to a foreign power. Also, since the governments of some forested countries are relatively unstable, the timbering firm may be reluctant to pay for long-term rights, which could be abrogated by a future government.

If long-term marketable timbering rights are politically infeasible, then incentives for proper care of the forest can be established by structuring penalties for excessive environmental degradation. The penalties could be structured using one, or some combination, of the following policies:

1. Harvesting companies could be required to post a large bond (deposit of money) that would not be refunded if there was excessive environmental damage. This option is discussed more fully in the following text box.
2. Harvesting companies could be required to pay a fine if they generated excessive environmental damage.
3. Harvesting companies could be banned from future harvesting activities in other forest areas if they generate excessive environmental damages in any area in which they harvest.

Fuelwood. A second activity that may lead to excessive levels of deforestation is the cutting of forests for fuelwood. In dry tropical forest areas, and in mountainous tropical forest areas, this activity is an extremely important source of deforestation, because more people rely on fuelwood as a source of energy than on fossil fuels (Food and Agricultural Organization, 1985). The reliance on fuelwood increased as the foreign exchange problems associated with the energy crises of the 1970s, falling export prices, and the debt crises of the 1980s forced countries to reduce their imports (or increase their exports and cut down on domestic consumption) of fossil fuels, such as coal and oil.

Again, the question arises of why the forests cannot be managed properly for a sustained harvest of fuelwood. The answer is, again, that there is a market failure. However, unlike the case of the timbering, it is unrelated to firm actions and concessionaire agreements. Instead, the deforestation for fuelwood consumption is largely a result of individual household action and poorly defined property rights.

Since many forest areas are open-access resources, individual households have no incentive to manage the forests for either maximum sustained yield or maximum economic benefit. Rather, there is a race to harvest the

INCORPORATING ECOLOGICAL CONSTRAINTS INTO ECONOMIC INCENTIVES

Over the past 10 or 15 years, ecologists have been conducting intensive research on the characteristics of the forest which promote its sustainability. For example, ecologists have determined that when land is cleared in timbering or for conversion to agriculture, the forest has a better chance of regenerating if several conditions are met. First, there should be a large ratio of the amount of undisturbed area to the amount of cleared area. Second, the area which is cleared should have a high ratio of the edge of the cleared area to the surface area of the cleared area. In other words, since the forest regenerates from the edge, narrow fingers of cleared area have a better opportunity to regenerate than circular or rectangular areas. These narrow fingers of cleared areas mimic natural clearings (generated by huge emergent trees being toppled by the wind), so natural processes exist to reforest the cleared area. These alternative configurations of forested and cleared area are presented in the following figure:

ALTERNATIVE PATTERNS OF TIMBER HARVEST

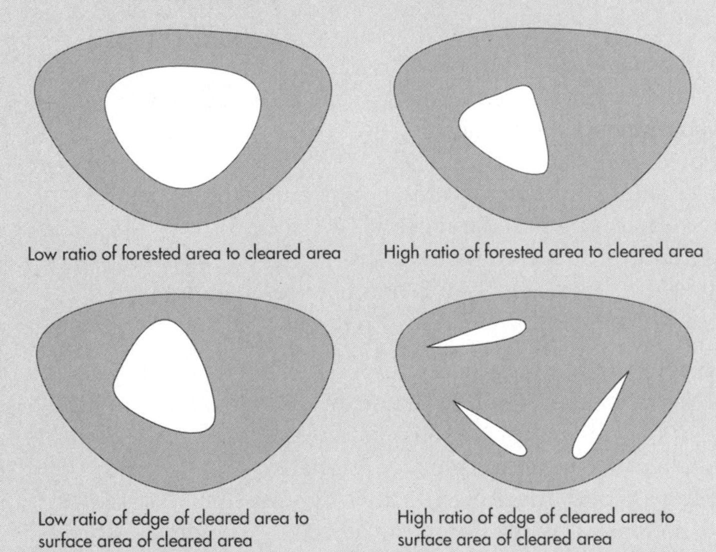

Given this knowledge of the ecology of the area, how can economic incentives (or other types of policies) be constructed to ensure that timbering activity conforms to these ecological requirements? One way to do this is to modify the bonding or deposit–refund type systems discussed in the section of this chapter on timber leases. A firm which wanted to harvest timber would have to post a large bond (pay a large deposit) which would then be returned to them if they harvest in an ecologically sustainable fashion. The greater the ratio of the amount of undisturbed area to cleared area, the larger the proportion of the deposit which would be returned. Similarly, the greater the ratio of the edge of the cleared area to the surface area of the cleared area, the greater the percentage of the deposit which is returned. This system would give harvesting firms an economic incentive to harvest in an ecologically sustainable fashion. Other important ecological variables could also be integrated into this system.

wood before someone else harvests it. This problem, combined with open-access to livestock grazing, where the livestock eat the leaves and bark from trees and shrubs as well as overgrazing the grass, may lead to rapid degradation of the forest and an inability of the forest to regenerate itself. In many areas of sub-Saharan Africa, deforestation leads to desertification.

In order to understand how to develop policies to deal with these market failures, it is necessary to examine why the forested areas lack a clear definition of property rights. In many cases, this lack of definition is due to the shift from a traditional economy to a national economy. For example, in many areas of sub-Saharan Africa, forested areas were communally owned by clans, villages, or tribes. Although these areas were common property within the clan, village, or tribe, they were not open-access and there existed a clear set of rules on the utilization of land. When these areas were colonized by European countries, the commonly owned resources became owned by the central government, and the rules of use broke down. Very often, when the countries became independent, title to the land passed from the colonial government to the new independent national governments, rather than passing back to the clans, villages, or tribes. The national government could not enforce rules governing the use of the forested land the same way a smaller unit of society, such as a clan or tribe, could. Also, the process of creating nations out of these smaller social units would sometimes generate conflict as to which smaller unit had claims to the land.

One way of reducing the overexploitation caused by open-access public forests is to give property rights to tracts of the forest to smaller units, such as a village. The village can then develop social institutions to control open-access overexploitation. The village can also prevent outsiders from exploiting the forest and has an incentive to expend some resources (such as guards) to protect its resource base.

Conversion to Other Land Uses. One of the primary causes of deforestation is the conversion of forestland to other purposes, usually agricultural. In traditional forest societies, this conversion occurred at a relatively small scale. Small patches of the forest were cut and burned for cultivation of crops in the cleared area. Since these patches were relatively small and surrounded by forest on all sides, the patches were quickly reclaimed by the forest, and the traditional farmers would then clear a new patch. This type of "slash and burn" agriculture does not result in significant permanent deforestation. It can be practiced year after year, along with sustainable agriculture and the harvesting of other forest products, such as rubber, nuts, fruit, honey, and meat from wild animals.

More recently however, the speed at which the slash and burn takes place has accelerated, and such large patches are cleared that the forest does not regenerate. The agricultural productivity of the cleared land is not great, because tropical forest soils have little organic matter and nutrients. The ash from the cleared and burned trees temporarily provides nutrients for crops,

WHY SUPERNATURAL EELS MATTER

Kenneth Taylor (1990) points out that, particularly in traditional societies, the social institutions that develop to prevent open-access exploitation need not take the form of laws or governmental bodies. He cites two examples in the Amazon which illustrate that religious or spiritual beliefs may be just as effective.

The Yanomami Indians set aside large tracts of forestlands that exist in the boundary regions between villages. These lands can only be hunted to supply infrequent inter-village festivals. The Yanomami do not think of these areas as game preserves, but rather as a means of commemorating the dead and honoring their visitors. Yet these beliefs provide a game preserve in which animals may breed in relative sanctuary, and the beliefs prevent inter-village competition for the animals living in these boundary areas.

The Kayopo believe that in certain portions of the rain forest rivers there exist supernatural electric eels that could kill any person who fishes in those waters. Consequently, the Kayopo avoid these areas. It is probably not coincidental that these waters that are supposedly protected by the supernatural eels are also the primary breeding areas of the most important species of fish. The myth serves to prevent a type of destructive exploitation (harvesting fish when they are geographically concentrated and susceptible to overharvesting), by providing a social structure to govern the use of a common property.

but, typically, after several years the land is no longer fertile enough to support crops. It is typically used for cattle ranching at this point, but the land is not productive enough to support high densities of cattle. Eventually, the soil may become too acidic to support even pasture grasses. In either the crop stage or the livestock stage of agriculture, the land is typically less productive as traditional agriculture than as forest.

An important question immediately comes to mind, and that is, if the forest is more productive in sustainable agroforestry than in traditional agriculture, why does it not remain as forest in agroforestry? There are several answers to this question, all of which revolve around the issue of market failure.

Population increases have been cited as a source of deforestation. The increasing demand for land associated with population growth leads to more deforestation. The rate of deforestation that a country regards as optimal probably will be higher in a densely populated country than in a sparsely populated country. However, deforestation, and particularly deviations between the actual level of deforestation and the optimal level, cannot be blamed solely on population, because a forest that is more productive when it remains as forest (rather than being converted to conventional agriculture) can support more people if it remains as a forest. Therefore, if one accepts the premise that the intact forest and agroforestry are more productive than cleared forest and conventional agriculture, population increase itself cannot be a significant independent source of higher than optimal deforestation. However, population increases may exacerbate certain market failures, which cause the market failures to be even more severe. Population growth may also exacerbate macroeconomic conditions that may lead to more deforestation and also result in an increase in demand for agricultural land.

In countries such as Brazil, population growth in the major urban areas, such as São Paulo and Rio de Janeiro, is very rapid. One policy that can reduce this urban population growth and somewhat mitigate the urban problems associated with this growth is to encourage population migration from the urban areas to the rural areas. In the case of Brazil, the relocation is to rain forest areas. However, the relocated settlers do not know how to exploit the rain forest as an intact forest; they only know the techniques of slash and burn, which convert the rain forest to cropland, then rangeland, then abandoned land in a cycle of about 8 years. Thus, a market failure of inadequate information is a major source of deforestation. There is a strong need to educate farmers in the techniques of sustainable agroforestry.

Education by itself, however, may not be sufficient to enable farmers to undertake sustainable agroforestry. The reason is that short-term constraints may force farmers to practice conventional unsustainable agriculture. Figure 12.4 profiles the income paths of both sustainable agroforestry and conventional agriculture. Conventional agriculture starts out with high income in the initial period immediately after clearing the forests, but tapers off quickly as the nutrients in the soil quickly become exhausted. Farmers then need to clear additional land and start the cycle over. In contrast, the income path of sustainable agroforestry starts off more slowly, as farmers plant perrenial tree crops that need time to reach maturity. In the initial time periods, income is lower but then eventually exceeds that of the first few years of conventional farming. Clearly in the long run the farmer would be better off if he or she practiced sustainable agroforestry, but the important question is how do they get through the initial years until the income from agroforestry grows to a higher level? Some potential policies to help the farmer through these initial years (until time T_s in Figure 12.4) would include low interest loans, food supplements or subsidies, or the free or low cost provision of seedlings for the perrenial crops which are being planted.

Another market failure that may be exacerbated by population growth is poorly defined property rights. If forestland is available for open-access exploitation, such as tropical dry forests in Africa that are exploited as sources of fuelwood and as grazing land, then greater population will increase the open-access exploitation and lead to faster destruction of the resource.

Open-access problems may also be exacerbated by migrants to the area. Mendelsohn (1994) develops models of squatter and pioneer behavior that show how this may be the case.

Squatters are people who occupy lands without legal ownership. Since they do not know if they will retain long-term occupancy of the land, they have little incentive to manage it for sustainable agroforestry, which will maximize its long-term benefits. Rather, they have two incentives to manage

AGRICULTURAL SYSTEMS OF INDIGENOUS AMAZONIANS

In a recent article in *The New York Times* (April 3, 1990, p. C–1), William K. Stevens reports the work of several anthropologists who have found that the ancient inhabitants of the Amazonian rain forest and their present-day descendants have developed a system of agriculture that actually shaped the character of the rain forest, so that it was capable of supporting a sophisticated civilization. This civilization, which may have lasted for several thousand years, consisted of a system of agricultural villages and chiefdoms that may have numbered into the millions of people. Some of the settlements are thought to have been inhabited by as many as 4,000 people.

Mr. Stevens reports that Dr. Posey, who studied the Kayopo tribe, found that one important technique was to clear small circular fields by felling large trees so that they fall outward from the center of the circle, which also knocks down the smaller trees. The leaves from the fallen trees provide nutrients for the most nutrient-demanding plants, which are planted along the perimeter of the circle. Tuber crops, such as yams, sweet potatoes, and manioc, are planted between the fallen trees. The tree trunks are then burned under carefully controlled conditions just before the first rains of the rainy season. The roots of the tuber plants keep the soil from washing away. The ashes from the trunks provide nutrients for sweet potatoes, yams, and manioc. Additional crops are planted, including corn, beans, rice, pineapples, papaya, mango, and bananas. The shorter crops and the tubers tend to be planted in the inner zones of the circle, and the taller crops, including the fruit trees, tend to be planted in the outer zones of the circle. Eventually, the jungle encroaches on the circle and gradually moves toward the center. The encroachment of the jungle brings other species of fruit trees, which are harvested, as well as medicinal herbs. It takes about 12 years for the jungle to completely reclaim the patch of land.

Notice that this type of agriculture does not result in the permanent destruction of the forest, nor the destruction of the soil, as do some conventional agricultural techniques. Different villages traded food crops and tree species. This trade in plants may be partially responsible for the diversity of species in the rain forest.

for the short-term only. The first incentive is the motive already mentioned—that they do not have secure long-term rights. Second, if myopic management (cutting the forests and planting crops) reduces the value of the land, they are less likely to be evicted than when the land is more valuable.

Models of pioneer behavior are based on the premise that the pioneer is moving into a relatively unoccupied and unclaimed area, and property rights to land can be established by taking actions that legitimize the pioneers' claim to the land. Such actions may include improving the land, fencing it, or posting the land and keeping others out. Of these, the action of "improvement" may be most damaging to the forest. In some forested countries, the landowner must "improve" the land to retain title to it. One legal means of improving the land is to clear and burn the forest. This action reduces the productivity of the land because it leads to the removal of the forest, but the action is taken by landowners since it establishes legal ownership rights.

FIGURE 12.4

ALTERNATIVE TIME PATHS OF INCOME FROM SUSTAINABLE AND CONVENTIONAL AGRICULTURE

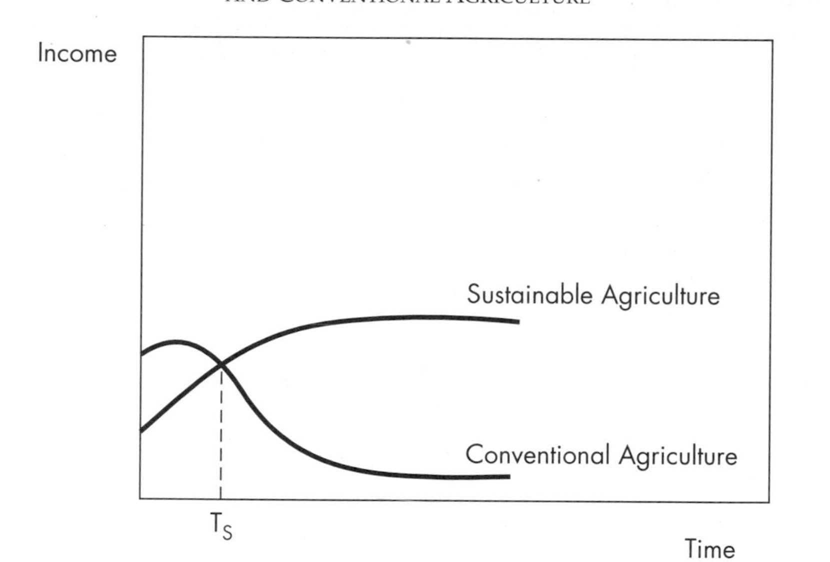

MACROECONOMIC REASONS FOR DEFORESTATION

In the previous sections, various microeconomic reasons for inefficient levels of deforestation were outlined. These reasons, which focus around market failures within the forested country, could be related to macroeconomic conditions in the country. The relationship between these macroeconomic conditions and the microeconomic problems may be illustrated by asking the following question: Why should a country establish a system of property rights or tax incentives that lead to inefficient use of the resource, when alternate systems exist?

There are three potential answers to this question:

1. The forested countries are not cognizant of the effects of these laws and policies.
2. The laws and policies are not designed with efficiency goals, but instead are a mechanism for transferring wealth from one segment of the economy to another (from indigenous forest dwellers to wealthy cattle ranchers, for example).
3. Economic conditions prevent the country from acting in its long-term interests.

The first explanation is not a particularly satisfactory one, as it means that the forested countries do not learn from experience. If outside observers are cognizant of the inefficiencies of these policies, then it would be reasonable to assume that internal policy makers must also be aware of them. While one might see unexpected detrimental impacts from policy, it is unrealistic to assume that these would continue in the long-run without some additional motivation for keeping the policies intact.

The second explanation may be important. However, it is extremely difficult to address. The use of government policy for transferring wealth leads to inefficiencies in virtually every country on earth. For example, in Chapter 11, how the U.S. Forest Service subsidized the cutting of the old growth Pacific forests to maintain the economic base of this region was discussed. If the inefficient policies leading to rapid deforestation are really a mechanism for transferring wealth, then one would not expect to see these policies reversed without substantial effort, as political and cultural traditions and barriers may sap the will of well-meaning policy makers to develop more efficient policies.

The third explanation, that economic conditions may prevent a country from acting in its long-term interests, is interesting because it is an explanation that can be addressed both by the tropical countries and by European and North American countries interested in preserving tropical rain forests. Kahn and McDonald (1995) hypothesize that macroeconomic conditions, particularly high levels of external debt, may constrain countries and force them to take actions to meet short-term cash needs that are not in their long-term interest. For example, the best long-term strategies may be to use the forests as a sustainable resource, but the need to meet interest and debt obligations may force the countries to use the forests as a depletable resource. The importance of debt on the economic conditions of the countries cannot be overstated, because many countries spend as much as $1 of every $2 or $3 dollars of GNP to meet the interest obligations on their debt.

Kahn and McDonald conducted a regression analysis that looked for a relationship between debt and deforestation. Their study found the debt elasticity of deforestation to be between 0.17 and 0.31. In other words, for every 10 percent that debt is reduced, annual deforestation may be reduced by between 1.7 and 3.1 percent. Brown and Pearce (1994) examine many empirical studies of the causes of deforestation. These studies show that other macroeconomic factors that may also contribute to this problem include falling prices for the raw materials that tropical countries export, rising energy prices, increasing population, domestic hyperinflation, high urban unemployment, and increasing urban density. More research needs to be done on the relationship between macroeconomic conditions and environmental degradation, both in developing and developed countries.

RAIN FORESTS AS A GLOBAL PUBLIC GOOD AND INTERNATIONAL POLICY

Earlier in the chapter, the public good nature of rain forests was mentioned. Rain forests provide habitats for many species, are reservoirs of biological diversity, harbor many plant and animal species that may be of medicinal or agricultural importance, and are important in terms of global carbon cycles. Consequently, one expects the tropical rain forests to be valued by people who live throughout the globe, particularly in temperate countries.

If people outside a forested country value the forests, then it is likely that the rate of deforestation taking place in the forested country is greater than socially optimal. The reason is that the costs and benefits of outsiders does not become included in the cost–benefit analysis of the forested country.

Many residents of developed countries have suggested that countries with tropical forests have an obligation to the world to preserve their tropical forests because of the global public good aspects of the forests. This argument has not been well received by the forested countries for two major reasons. First, the developed countries of Europe and North America already eliminated most of their original forests as these countries underwent economic development. The tropical countries argue that if Europe and North America fueled their industrial development with deforestation and other forms of environmental degradation, why should currently developing countries be subjected to a different standard? Second, if preserving the rain forests benefits the wealthier developed countries, why should the poorer developing countries bear the cost of this preservation?

While both of these arguments have intuitive appeal, they merit careful examination. As suggested earlier in this chapter, it is likely that the high rates of deforestation are not in a forested country's best interest. Market failures and macroeconomic considerations may generate more deforestation (and other types of environmental degradation) than is in a country's self-interest. If these problems can be addressed and deforestation slowed, the level of income should rise in these countries. One thing is clear, however: Whatever level of deforestation is optimal from a forested country's perspective, a slower rate is optimal from a global perspective, due to the global public good nature of the rain forests.

This difference in optimal deforestation rates implies that important issues surround the question of whether the poorer developing countries should be asked to bear the costs of the lower levels of deforestation. Lowering deforestation levels would make the world better off, but could make the forested countries worse off. Clearly these lower levels of deforestation will not be achieved unless the forested countries are compensated for their losses.

DEBT FOR NATURE SWAPS

One mechanism for accomplishing this compensation is the debt for nature swap, which was originally proposed by Thomas Lovejoy. A debt for nature swap is an agreement between an organization (usually a nongovernmental organization [NGO]) and a forested country. The NGO buys and retires a portion of the forested country's external debt in exchange for a promise to preserve a forest tract. External debt is the money that a country owes to foreign banks, foreign governments, and international development agencies, which must be repaid in hard currency (U.S. dollars, German marks, British pounds, Japanese yen, etc.).

As mentioned earlier, external debt is a tremendous drain on a developing country's economy, with many countries owing in the tens of billions of dollars. For some countries, as much as $1 of every $2 dollars of GNP goes to paying its debt service. Since the banks who hold the debt view their prospects for full and timely repayment as problematic, the debt can be purchased by the NGO for a fraction of its face value. The debt can be discounted to as low as 5¢ to 30¢ on the dollar.

One feature that is increasingly becoming a characteristic of debt for nature swaps is the establishment of a fund in the forested country's domestic currency, which will be used to finance the preservation of the forest tract. For example, the money can be used to hire forest rangers to enforce prohibitions against damaging uses of the forest.

Several important questions have been raised concerning debt for nature swaps. These include

1. Will the countries renege on the agreements due to the lack of enforceability of the agreements? For example, if an NGO, such as the Nature Conservancy or the Sierra Club, retires part of the debt of the forested country in exchange for protecting a forest tract, and if several years later the country decides to convert the forest to pasture, there is little the NGO can do to prevent it from happening. Of course, such an action would damage the country's international reputation and probably preclude future debt for nature swaps and other sorts of environmental cooperation, so the costs of reneging are high, but the possibility of such an action cannot be dismissed.
2. Will forested countries negotiate a debt for nature swap for a preservation action they would have taken anyway, meaning there is no net preservation gain as a result of the swap?
3. Will countries use their reduced indebtedness to qualify for more loans, so that the net debt reduction of the debt for nature swaps are small, if not zero?

The Tropical Forestry Action Plan

The Tropical Forestry Action Plan (TFAP) is a large scale effort organized by the Food and Agricultural Organization (1985), the World Resource Institute, the World Bank, and the United Nations Development Program to conserve and develop tropical forest resources on a long-term, sustainable basis. The TFAP is designed to reduce deforestation pressures by direct action to restore degraded areas and to provide methods for meeting economic needs in a sustainable fashion. There are five operational goals of the TFAP, which TFAP refers to as priority areas:

1. Integrate forestry and agriculture to conserve the resource base and make more rational use of the land.
2. Develop forest-based industry.
3. Restore fuelwood supplies and support national fuelwood and wood energy programs.
4. Conserve tropical forest ecosystems.
5. Remove institutional barriers to wise use of forests.

According to the TFAP, the achievement of these goals will lead to the following rewards:

1. restored productive capacity of the land,
2. sustainable use of forest resources,
3. improved food security through better land use,
4. increased supplies of fuelwood,
5. increased supplies of fodder,
6. increased income from locally manufactured products,
7. conservation of natural ecosystems and genetic resources, and
8. local participation in forestry and forest-based industries.

A sample of projects supported by the TFAP include:

1. restoring degraded land in Ethiopian highlands by providing "food for work" for peasants to work on conservation projects, such as terracing deforested hillsides and planting trees,
2. planting grasses and multi-purpose trees for fodder in Nepal by paying wages for tree planting for the first 2 years of the project,
3. developing forest plantations of fast growing trees in Pakistan to increase incomes of farmers, and
4. encouraging farmers in Kenya to plant trees for fruit, shade, windbreaks, and fuelwood.

The Tropical Forestry Action Plan is important in two major ways. First, it provides a framework in which forested countries can recognize the importance of their forests and develop plans to protect them. Second, it provides direct interventions to reverse degradation and mitigate some of the economic pressures that lead to deforestation.

While the Tropical Forestry Action Plan serves as an important demonstration plan and does yield important results in terms of both improving rural standards of living and reducing deforestation, it is not a panacea. It cannot, by itself, deal with all the microeconomic and macroeconomic conditions leading to greater than optimal levels of deforestation. Similarly, it does not and cannot fully address the important population pressures that contribute to deforestation.

SUMMARY

Tropical forests are diverse ecosystems that provide many benefits, including both consumable and nonconsumable benefits. Consumable benefits include wood, meat, fruit, honey, nuts, fiber, rubber, and fodder. Nonconsumable benefits include habitat, biodiversity, watershed protection, carbon sequestering, and existence values.

Tropical forests are disappearing at an alarming rate. Microeconomic conditions, such as poorly defined property rights, land tenure systems, tax incentives, and short-term concessionaire agreements, are important causes of deforestation. Macroeconomic conditions, such as the level of external debt, import and export prices, and the general standard of living, may also contribute to the rapid rate of deforestation. In addition, population growth generates additional demands for land and exacerbates the microeconomic and macroeconomic factors that lead to deforestation.

One important factor in reducing deforestation is improving the economic conditions that tend to promote it. These conditions include both market failures and macroeconomic conditions. Elimination or mitigation of these factors requires efforts at both the national and international levels. At the international level, the provision of debt relief in the form of debt for nature swaps and the provision of information are essential. Additionally, the provision of information and environmental investment provided by programs such as the Tropical Forestry Action Plan provide much promise. However, more research into causes of deforestation and more research into the development of policy remedies is necessary if the irreversible loss of tropical forests is to be avoided.

REVIEW QUESTIONS

1. What are the fundamental ecological differences between tropical rain forests and temperate forests?
2. Describe the economic activities that result in the deforestation of rain forests. What are the market

failures associated with each activity? Suggest policies to deal with these market failures.
3. What policies should developed countries pursue toward tropical deforestation?
4. Why do "supernatural eels" matter? What is the relevance of "supernatural eels" to policy?
5. Why do concessionaire agreements with timber companies tend to be short run in nature? What inefficiencies does this generate? How would you modify these agreements to eliminate these inefficiencies?
6. What are the strengths and weaknesses of debt for nature swaps as an international policy to slow deforestation?

QUESTIONS FOR FURTHER STUDY

1. Do fast food hamburgers lead to deforestation?
2. What are the strengths and weakness of the Tropical Forestry Action Plan? How would you modify the plan to make it more effective?
3. What is the relevance of the worldwide "energy crisis" for tropical deforestation?
4. Can the development of increased tourism save the tropical rain forests?
5. Is it possible for the United States to unilaterally preserve tropical rain forests? Why or why not?
6. Explain the difference between local, national, and global public goods. What is the importance of this distinction for policy?

SUGGESTED PAPER TOPICS

1. The chapter mentions several social institutions that developed among indigenous Amazonian peoples to protect common property resources. Choose an indigenous group from another region (that is North America, Australia, Africa) and discuss the role that social institutions play in maintaining the integrity of renewable natural resources. Do literature searches on the region, indigenous people, and rain forests.
2. Discuss the current experience with debt for nature swaps. What factors are likely to affect their desirability from the forested country's perspective? What factors are likely to increase their desirability from the creditor countries' perspective? What recommendations can make these programs more effective? Search bibliographic databases for rain forests and debt, and debt and the environment.
3. Formulate a hypothesis about the relationship between economic variables and deforestation. Use data from World Resources and the United Nations Human Development Report 1990 to test your hypothesis. For example, what is the relationship between deforestation and population growth, GNP, or debt? How does this relationship vary by region of the world? How does it differ between countries that contain primarily dry forest and countries that contain primarily wet forests?
4. Look at deforestation in a particular country and the activities that lead to deforestation. How do market failures affect these activities? What policies can you suggest to alleviate the market failures?
5. Development projects of the World Bank and other development agencies have been criticized as an important factor in the rapid deforestation of the 1980s. Discuss the evidence that either supports or refutes this position. If you find support for this position, what changes in policies can you suggest to reduce this type of pressure on forests? Search bibliographic databases for World Bank and the environment. Also, check World Bank reports for internal discussion on this issue.

WORKS CITED AND SELECTED READINGS

1. Bird, G. "Loan-Loss Provisions and Third-World Debt." *Essays in International Finance,* No. 176. Princeton University, Economics Department, International Finance Section, 1989.
2. Brown, Katrina, and David Pearce. *The Causes of Tropical Deforestation,* University College London Press, 1994.
3. Capistrano, A. D., and C. F. Kiker. "Global Economic Influences on Tropical Closed Broad-leaved Forest Depletion, 1967–1985." Unpublished manuscript presented at the International Society for Ecological Economics Conference, Washington, D.C., 1990.
4. Dale, V.H., et al. "Simulating Spatial Patterns of Land-Use in Rondonia, Brazil," in *Some Mathematical Questions in Biology,* edited by R. Gardner, American Mathematical Society, forthcoming.
5. Ehui, S. K., T. W. Hertel, and P.V. Preckel. "Forest Resource Depletion, Soil Dynamics and Agricultural Productivity in the Tropics." *Journal of Environmental Economics and Management* 18 (1990): 136–150.

6. Food and Agricultural Association, Tropical Forestry Action Plan, Rome, 1987.
7. Food and Agriculture Organization of the United Nations. "Forest Resources of Tropical Africa." UN32/6.1201-78-04 technical report no. 2, Rome, 1981a.
8. ———. "Forest Resources of Tropical Asia." UN32/6.1201-78-04 technical report no. 3, Rome, 1981b.
9. ———. "Los Recurses Forestales De La America Tropical." UN32/6.1201-78-04 technical report no. 1, Rome, 1981c.
10. Head, Suzanne, and Robert Heinzman. *Lessons of the Rainforest,* San Francisco: Sierra Club Books, 1990.
11. Kahn, J. R., and J. A. McDonald. "Third World Debt and Tropical Deforestation." *Ecological Economics* 12 (1995): 107–123.
12. Lanky, Jean-Paul. "An Interim Report on the State of Tropical Forest Resources in the Developing Countries." Rome: Food and Agricultural Organization of the United Nations, 1988.
13. Mendelsohn, R., "Property Rights and Tropical Deforestation." *Oxford Economic Papers* 46 (1994): 750–756.
14. Myers, Norman. *Deforestation Rates in Tropical Forests and Their Climatic Implications.* London: Friends of the Earth, Ltd., 1989.
15. Panaiotov, Todor. *Not by Timber Alone: Economics and Ecology for Sustaining Tropical Forests,* Washington, D.C.: Island Press, 1992.
16. Perry, Donald. "Tropical Biology." In *Lessons of the Rainforest,* edited by Suzanne Head, and Robert Heinzman, San Francisco: Sierra Club Books, 1990.
17. Repetto, Robert. *The Forest for the Trees? Government Policies and the Misuse of Forest Resources.* Washington, D.C.: World Resource Institute, 1988.
18. Repetto, Robert, and Malcolm Gillis. *Public Policies and the Misuse of Forest Resources.* Washington, D.C.: World Resource Institute, 1989.
19. Shilling, J. D. "Reflections on Debt and the Environment." *Finance and Development* 29 (1992): 28–30.
20. Stevens, William K. "Research in 'Virgin' Amazon Uncovers Complex Farming." *The New York Times,* April 3, 1990, p. C–1.
21. Taylor, Kenneth Iain. "Why Supernatural Eels Matter." In *Lessons of the Rainforest,* edited by Suzanne Head and Robert Heinzman. San Francisco: Sierra Club Books, 1990.
22. United Nations Development Programme. *Human Development Report* 1990. Oxford: Oxford University Press, 1990.
23. Von Moltke, Konrad. "International Economic Issues in Tropical Deforestation." Paper presented at the Workshop on Climate Change and Tropical Forests, São Paulo, 1990.
24. World Bank. *World Tables.* Baltimore: The Johns Hopkins University Press, various years.
25. ———. *Work Development Report.* Oxford: Oxford University Press, various years.
26. World Resource Institute. *World Resources 1992–93.* Oxford: Oxford University Press, 1992.

Biodiversity and Habitat Preservation

If Charles Darwin were writing today, his masterwork would probably be known as The Disappearance of Species.

INTRODUCTION

The above quotation by Edmund Wolf (1985, p. 124) is jestful, but it emphasizes an increasingly important environmental problem: the extinction of plant and animal species. Extinction is occurring at an unprecedented rate, and it results in a loss of biodiversity—which can be defined as the total variety of life on earth. Biodiversity is extremely important for ecosystem function and has many direct benefits to humans.

Although the above definition of biodiversity is general, one can specifically define the process that leads to the loss of biodiversity, which is extinction. A species becomes extinct when the last individual organisms of the species die. Viable seeds would be considered to be individual organisms under this definition, so that while all the germinated plants of a species may have disappeared, the species would not technically be extinct if viable seeds existed that could germinate into mature plants capable of producing more seed. There are many natural and anthropogenic sources of extinction, with each anthropogenic source related to a market failure.

Natural extinctions occur when the environment changes, and some species find themselves at a competitive disadvantage and are displaced by other species that are better adapted to the new conditions. Natural extinctions are always occurring, usually at a relatively slow pace although there are often rapid and massive extinctions. The disappearance of the dinosaurs is an example of a massive and rapid extinction, but it actually took place over a period of about 2 million years. Table 13.1 lists extinctions of mammals over recent history and shows the extremely rapid rate of extinction that has occurred in modern times. This rapid loss of species cannot be attributed to the types of naturally occurring environmental changes that led to the extinction of the dinosaurs. Rather, these extinctions are primarily due to anthropogenic factors.

TABLE 13.1

ESTIMATED ACCELERATION OF MAMMAL EXTINCTIONS

TIME PERIOD	EXTINCTIONS PER CENTURY	PERCENT OF PRESENT STOCK OF SPECIES LOST	PRINCIPLE CAUSE
Pleistocene (3.5 Million Years)	0.01	—	Natural Extinction
Late Pleistocene (100,000 Years)	0.08	0.002	Climate Change, Neolithic Hunters
1600–1980 A.D.	17	0.4	European Expansion, Hunting and Commerce
1980–2000 A.D.	145	3.5	Habitat Disruption

SOURCE: From *State of the World 1985: A Worldwatch Institute Report on Progress Toward a Sustainable Society,* by Lester R. Brown et al, eds. Copyright © 1985 by Worldwatch Institute. Reprinted by permission of W. W. Norton & Company, Inc.

It should be noted that the concept of marginality is different in this chapter than in many of the other chapters of this book, such as Chapter 10. In Chapter 10, the marginal unit is a fish, and the allocation question is how to choose the level of harvest that maximizes social welfare. In this chapter, the primary question revolves around providing the optimal level of biodiversity, and the marginal unit is not an organism but an entire species.

ANTHROPOGENIC CAUSES OF SPECIES EXTINCTION

There are several important anthropogenic causes of extinction. These include excessive harvesting of the species, loss of habitat, and competition from nonnative species. All of these causes stem from market failures, where people make economic decisions that do not incorporate the full social costs of their actions. Table 13.2 lists observed declines in animal species and their anthropogenic causes. Table 13.3 lists the number of species in the United States that are declining to the point that extinction is a possibility. Endangered species are those in danger of becoming extinct through all or a significant part of their natural range. Threatened species are those that are likely to become endangered in the foreseeable future.

OPEN-ACCESS HARVESTING

Excessive harvesting is often thought to be the main cause of extinction. Although this is not the case, excessive harvesting is associated with some of the most notorious extinctions in North America and some of the most visible current problems in Africa and Asia.

TABLE 13.2

Observed Declines in Selected Animal Species, Early 1990s

amphibians	Worldwide decline observed in recent years. Wetland drainage and invading species have extinguished nearly half of New Zealand's unique frog fauna. Biologists cite European demand for frog's legs as a cause of the rapid nationwide decline of India's two most common bullfrogs.
birds	Three-fourths of the world's bird species are declining in population or threatened with extinction.
fish	One-third of North America's freshwater fish stocks are rare, threatened, or endangered; one-third of U.S. coastal fish have declined in population since 1975. Introduction of the Nile perch has helped drive nearly half the 400 species of Lake Victoria, Africa's largest lake, to or near extinction.
invertebrates	On the order of 100 species are lost to deforestation each day. Western Germany reports one-fourth of its 40,000 known invertebrates to be threatened. Roughly half the freshwater snails of the southeastern United States are extinct or nearly so.
mammals	Almost half of Australia's surviving mammals are threatened with extinction. France, Western Germany, the Netherlands, and Portugal all report more than 40 percent of their mammals as threatened.
carnivores	Virtually all species of wild cats and bears are seriously declining in numbers.
primates	More than two-thirds of the world's 150 primate species are threatened with extinction.
reptiles	Of the world's 270 turtle species, 42 percent are rare or threatened with extinction.

SOURCE: From *State of the World 1992: A Worldwatch Institute Report on Progress Toward a Sustainable Society,* by Lester R. Brown et al. Copyright © 1992 by Worldwatch Institute. Reprinted by permission of W. W. Norton & Company, Inc.

In North America, the near extinction of the American bison (more commonly referred to as buffalo) was caused by excessive hunting.[1] Although the buffalo had been hunted for thousands of years by Native Americans, herds were plentiful, numbering in the millions. However, with the westward expansion of Americans of European and African descent, hunting quickly decimated the herds. Why could buffalo herds sustain hunting by Native Americans but perish under hunting by non–Native Americans?

The answer to this question is that under Native American control, buffalo herds were common property (not owned by a particular person) but not open-access resources. This crucially important distinction is often confused in the literature. Common property resources often have restrictions on their use, whereas open-access resources do not. In the case of the Native

[1]The American bison was hunted to the point that it was extinct in the wild. Today's increasing bison herds are descendant from animals that were preserved in zoos.

TABLE 13.3

ENDANGERED AND THREATENED SPECIES IN THE UNITED STATES

TYPE OF ANIMAL OR PLANT	NUMBER OF ENDANGERED SPECIES	NUMBER OF THREATENED SPECIES
mammals	55	9
birds	74	16
reptiles	14	19
amphibians	7	5
fish	65	40
snails	15	7
clams	51	6
crustaceans	14	3
insects	20	9
arachnids	5	0
plants	406	90

SOURCE: *U.S. Statistical Abstract*, 1996, p. 240.

American and the bison, cultural traditions dictated the method and magnitude of hunting, and these traditions tended to conserve the resource. Hunters who killed more than could be effectively used violated these cultural norms and received social sanctions. Also, intertribal competition for herds was prevented by the division of grazing areas into hunting areas specific to each tribe, although these boundaries were subject to fluctuation. If one tribe violated the hunting grounds of another, it could expect violent retribution. This combination of intertribal and intratribal social rules prevented the open-access exploitation and over harvesting of the buffalo herds.

In contrast, when the Native Americans lost control of their hunting grounds to non–Native Americans, the buffalo herds became open-access common property, with no restrictions on their use. Consequently, no individual hunter had an incentive to preserve the resource. Conservative hunting by one person was considered pointless, because the buffalo would be killed by someone else instead. Buffalo were killed for sport and profit. The hides were the source of profit. Meat was seldom utilized, with the exception of the tongue; most of the meat was left on the carcass to rot. A destructive race began, with each buffalo hunter seeking to shoot as many buffalo as possible before the buffalo were shot by competing hunters. Under this type of open-access harvesting pressure, the buffalo quickly disappeared from the prairie.[2]

[2]See Chapter 10 for formal economic models of open-access resources.

Modern examples of this type of open-access harvesting pressure include many of the large mammals of Africa and Asia, such as the elephant, rhinoceros, tiger, bear, and leopard. These animals are valued by hunters for body parts, such as the elephant's tusks, the rhino's horn, the bear's gallbladder, and the leopard's skin. Illegal hunting activity by poachers threatens these animals with extinction. Although hunting these animals is usually forbidden by law, it is extremely difficult to enforce these prohibitions because of the high profits associated with the illegal trade. These problems with enforcement will be further discussed in a subsequent section.

LOSS OF HABITAT

While open-access harvesting has figured prominently in the decline and demise of many species, the loss of habitat is currently a more pressing problem. Many species are found only in a limited range of habitat, and if this habitat is destroyed by conversion into another land use, or contaminated by pollution, the species will become extinct. This problem is particularly important in association with the massive deforestation occurring in tropical rain forests, as tropical rain forests contain as much as three quarters of all the earth's species (see Chapter 12). Other important losses of habitat are associated with the loss of temperate forests, destruction of coral reefs, loss of wetlands, and pollution of numerous aquatic environments.

The loss of habitat may be associated with either open-access or private property resources. For example, coral reefs tend to be open-access and are destroyed both by pollution that is carried to the coral lagoons by rivers and by fishing activity. Upstream polluters make their decisions concerning waste disposal without considering the social costs associated with the negative effects on coral reefs. Similarly, fishers make use of fishing techniques that lower their costs but are destructive of the reefs, such as fishing with explosives. Also, recreationists (particularly in Florida and the Caribbean) damage coral reefs by anchoring their boats on the reef, by removing coral from the reefs, and by harvesting shellfish that are predators to animals such as starfish (which, in turn, are predators of the animals that create the coral reefs).

On the other hand, privately owned habitat is also lost. Although the destructive interactions associated with open-access resources are not present with privately owned resources, market failures still occur. For example, an owner of a wetlands who is contemplating converting the wetlands to condominiums does not consider the social value of leaving the wetlands intact. Thus, the owner makes decisions based on equating marginal private costs and private benefits, and the ecological services of the wetlands are ignored. The wetland owner makes his or her decision concerning conversion of wetlands by comparing the private net benefits of conversion with the private net benefits of preservation. The social net benefits are generally excluded from consideration. In other words, there are negative externalities associated with the conversion of habitat to other uses.

Elephants and the Ivory Trade

The population of African elephants has been in severe decline due to illegal hunting for their ivory tusks. A single tusk will bring a poacher (person engaged in illegal hunting) several hundred dollars, which is the approximate per capita GNP in many African countries. Due to the relatively large potential income, many rural dwellers have a strong incentive to illegally hunt elephant.

There has been considerable controversy over how to deal with this problem. Several southern African countries had been relatively successful by giving property rights to elephants to neighboring villages. The local inhabitants then had an incentive to report nonlocal poachers to police authority. Also, federal governments compensated villages for crop damage by elephants. Under a plan like this, villages would manage elephants as a renewable resource, harvesting some individuals for food or promoting lucrative sport hunting by foreigners, but leaving the herd intact to produce more elephants.

Other countries that have not adopted this approach have been less successful in protecting elephant herds. In countries such as Kenya, large herds have been decimated by those engaged in ivory trade. Although this difference may be construed as evidence that creating property rights and markets works in preventing the extinction of endangered species, there may be other differences between the southern African states and the central and east African states that have more trouble maintaining their herds. First, population pressures and associated loss of habitat may be greater, especially in eastern African countries such as Kenya. Second, due to greater police and military power, borders may be more secure in southern African states (particularly South Africa). This security is critically important because, in countries such as Kenya, a large proportion of poachers cross borders to enter Kenya, particularly from Somalia and Sudan.

As a result of the decline in elephant herds, the Convention for International Trade in Endangered Species (CITES) banned the international trade in ivory in 1989. Although this ban may at first appear to be a completely positive move, some negative aspects may be associated with it. Countries that have maintained elephant herds by establishing property rights for local inhabitants lose the incentives by which local inhabitants manage and protect herds. If villagers cannot earn income from the legal sale of ivory, they have less incentive to manage and protect elephant herds. Also, the national governments of African countries lose revenue that they could earn by gathering tusks from elephants that die of natural causes.

Nonetheless, it is felt that this ban might be the only way to prevent the disappearance of elephants from most central and east African countries. Since the ban is relatively recent, it is too soon to see if it will protect the herds or if additional steps are needed. For up-to-date information on this problem search the world wide web on "elephants and CITES."

COMPETITION FROM NONNATIVE SPECIES

Competition from nonnative species can also be viewed as an externalities problem. A person introduces an exotic species (a species from another area) because he or she expects to realize benefits from this species. However, the exotic species is often associated with large ecological and social costs, because the exotic species will out-compete or prey upon native species. For example, many ground-nesting birds evolved in island ecosystems, because the islands lacked predators that would interfere with the nesting process. However, the intentional introduction of pigs and the accidental introduction of rats has led to the loss of many island bird species, including the dodo bird of Mauritius and Réunion (islands off the southeast coast of Africa). Alternatively, introduced species out-compete native species. For example, well-meaning fishery managers introduced European brown trout into eastern U.S. streams to provide more angling opportunities for recreational fishers. However, in many areas, the more aggressive brown trout has completely displaced the native brook trout. Similarly, the European House Sparrow out-competes the Eastern Bluebird for nesting sites and has displaced this bird from a large part of its natural range. Introduced insects are an important cause of the loss of ridge top forests in the United States, because Gypsy Moth caterpillar and other nonnative insects decimate forests and render them more vulnerable to other stresses such as acid rain and tropospheric ozone (see Chapter 7).

THE COSTS OF LOSSES IN BIODIVERSITY

Biodiversity is important for a variety of reasons. First, biodiversity promotes ecosystem stability. The more diverse a system, the greater its ability to withstand shocks and stresses. If biodiversity promotes ecosystem health and function, then biodiversity promotes all the services derived from ecosystems, such as protection of freshwater supplies, production of oxygen, absorption of carbon dioxide, nutrient cycling, provision of habitat, and so on. Second, plants and animal species have a value because they may be used to produce economic goods. The species may provide goods directly, such as fruits or nuts that are consumed directly, or they may be a direct source of natural chemicals and compounds (Sedjo, 1992). Third, the organisms' genes may be a source of genetic information. Such information may be used in the development of new varieties of plants with different properties than existing varieties (for example, the development of new crops with higher yields through crossbreeding).

The information from existing species may also be used to create new plants or animals through genetic engineering. Genetic engineering may transfer properties across species that are widely different (such as from bacterium to a plant), while crossbreeding transfers properties across related species or varieties within a species.

Finally, biodiversity may be important because people think that it is important. While this statement may seem tautological, many people desire biodiversity, deriving greater utility from more diverse ecosystems than less diverse ecosystems. People may also feel that it is society's ethical responsibility to maintain biodiversity.

COSTS OF LOSSES OF HABITAT

Although one of the important reasons for preserving habitat is to preserve biodiversity, habitat provides many other important services. Obviously, habitat provides an environment in which plants and animals can exist. The other services can be crucially important to ecological and social systems. For example, tropical forests and the phytoplankton layer are the principal mechanisms for producing oxygen and removing carbon dioxide from the atmosphere. A description of the services of many habitats (that is, tropical forests, temperate forests, aquatic systems) are provided in the individual chapters of this book. However, to provide an example, the costs of losing wetlands (a particularly stressed type of habitat) will be discussed.

Wetlands play a unique ecological role as a transition zone between aquatic ecosystems and terrestrial ecosystems. One critical aspect of this is nutrient cycling. The wetlands serve as a vast store of nutrients from terrestrial sources that are gradually released into aquatic ecosystems. Wetlands also serve as the beginning of the food web for many aquatic systems. The nutrients flowing from the wetlands begin the food web in the aquatic systems, and many aquatic organisms feed directly in the wetlands. Many estuarine shellfish and finfish are critically dependent upon the productivity of the wetlands. (Chapter 10 contains a boxed example that discusses the role of wetlands in blue crab production.)

One of the most important roles of wetlands is serving as a buffer against storms. Wetlands absorb storm water, preventing or lessening floods from high levels of rain. This reservoir function also reduces the impacts of droughts. Wetlands also provide a buffer to dampen storm surge and wave damage from coastal storms.

Other important roles of wetlands include serving as a habitat for many species of animals and providing a spawning and nursery area for fish and other aquatic species. In addition, many people use wetlands as an area to recreate, including canoeing and other boating activities, fishing, and hunting. Wetlands also provide a livelihood for many people through commercial fishing and trapping animals for their fur. The boxed example on page 383 examines attempts to quantify the economic value of wetlands.

POLICIES FOR MAINTAINING BIODIVERSITY

Much of the discussion of the importance of biodiversity stems from the potential of organisms to provide medicines and other important chemicals. Some of the most promising cancer treatments are associated with chemicals that occur in plants. For example, an important treatment for childhood leukemia was found in the rosy periwinkle (tropical rain forests of Madagascar), and the Pacific yew (old growth forests of the Pacific Northwest United

ECONOMIC VALUE OF WETLANDS

Robert Anderson and Mark Rockel (1991) of the American Petroleum Institute conducted a survey of various studies that have attempted to measure the value of wetlands. Some studies focused on individual functions of the wetlands, while others looked at multiple functions. Anderson and Rockel summarized the results of their survey in a table, which is reproduced below. The individual studies that are referenced in this table are listed in the Works Cited and Selected Readings section at the end of this chapter.

Although these values are subject to empirical debate, they have been included in this example to give the student an idea of what has been estimated in terms of the value of a habitat.

ANDERSON AND ROCKEL SUMMARY OF ESTIMATES OF THE FUNCTIONAL VALUES OF WETLANDS (MEASURED IN 1984 DOLLARS)

WETLAND FUNCTION	VALUE ($/ACRE/YEAR)	CAPITALIZED VALUE ($/ACRE AT 5% DISCOUNT RATE)
flood conveyance	$191[1]	$3,820
erosion, wind, and wave barriers	$0.44[2]	$9
flood storage	na	na
sediment replenishment	na	na
fish and shellfish habitat	$32,[2] $66[3]	$700–1,320
waterfowl habitat	$167[1]	$3,340
mammal and reptile habitat	$12[4]	$240
recreation	$6,[4] $25,[5] $76,[3] $70[5]	$120, $500, $1,520, $1,400
water supply	na	na
timber	na	na
historic and archaeological use	$323[6]	$6,480
educational and research use	$6	$120
water quality improvement	na	na

SOURCE: Robert Anderson and Mark Rockel, "Economic Valuation of Wetlands," Discussion Paper #065, Washington, D.C.: American Petroleum Institute, 1991. Reprinted courtesy of the American Petroleum Institute.
[1]Gupta and Foster, 1975; [2]Farber, 1987; [3]Bell, 1989; [4]Farber and Costanza, 1987; [5]Thibodeau and Ostro, 1981; [6]combines recreation, historical, nature, and cultural values, Gupta and Foster, 1975.

States) contains a chemical that is used to shrink tumors. Animals also provide important potential treatments, although usually with less frequency than plants. For example, enzymes in the skin of an African frog may provide important treatments for burn victims.

The discussion of the importance of biodiversity also revolves around the importance of maintaining a gene pool for agriculture. New properties, such as frost or disease resistance, can be transferred from wild plants to agricultural varieties by hybridization (sexual transfer) or through genetic engineering. Genetic engineering is particularly important because the cross-breeding and hybridization of important crops, such as corn or wheat, have sacrificed some of their disease resistant properties for greater yields. It is important to preserve wild species so that these traits (as well as new traits) can be brought back into agricultural crops.

Although both the medicinal potential and agricultural potential are important reasons to preserve biological resources, it is important to remember that biodiversity is critically important to ecosystem health. We should not develop policies that are only oriented toward commercial applications and merely focus protection activities on plants and animals that are judged to have the greatest potential for medicines, other chemicals, and agriculture. We should also recognize the noncommodity benefits associated with this dimension of biodiversity and compare all the benefits to costs when developing policies that focus on preserving the many dimensions of biodiversity.

POLICIES TO REDUCE OPEN-ACCESS EXPLOITATION

Since open-access exploitation problems are caused by poorly defined property rights, it would seem that a logical solution would be to better define property rights. For example, property rights to elephants could be assigned to neighboring villages (see the boxed example on page 380). This proposal has the potential to work where the market for the animal products is primarily in the vicinity of the animal's habitat, such as when the animal is hunted for subsistence. Properly defined property rights can eliminate a destructive competition among villages. If each village is assigned a management area, each village has an incentive to manage its herd as a long-term asset. Hunting to extinction would eliminate a long-run source of food. Of course, the management area has to be large enough to correspond to the entire range of the herd. If animals migrate from one management area to another, then people might feel pressure to kill them when they are in their area, because once they move to another area they might be killed by someone else.

Many times it is not practical or politically feasible to assign property rights. For example, in the United States there is a tradition of public access to wild fish and game resources. Attempts to assign private property rights would be viewed as unfair and would be political suicide for any legislator who might propose it. However, the resources are protected from overexploitation by rules governing their use. There are season limits, limits on the number of animals harvested, and restrictions on how and where the animals may be harvested. Certain animals that are rare, endangered, or threatened are not allowed to be hunted at all. Although these rules are occasionally ignored, they are obeyed on the whole within the United

States and have served fairly well in protecting animal populations from overexploitation.[3]

Both the definition of private property rights and the development of restrictions with penalties for noncompliance are ineffectual when the profits from illegal harvesting are high relative to the opportunity wage (what the hunter could earn in an alternative occupation). As the United States has found with the illegal drug trade, if the profits are high enough, people will engage in the illegal activity. This rationale is also true with illegal trade in animal products, which can create drug trade-like profits for participants. Under these circumstances, policies aimed at the supply side alone cannot be effective, because high demand will always mean high profits. However, policies aimed at the demand side can be very effective when combined with policies aimed at the supply side. A combination of prohibitions on all sales of the animal product and publicity campaigns that make it socially unacceptable to use animal products (wear furs, ivory jewelry, etc.) can eliminate the profitability of the illegal trade.

This was the rationale behind CITES's ban on ivory (see the boxed example on page 380). As long as there was some legal trade in ivory, there would be opportunities for profitable illegal trade. To dramatize the importance of banning all sales of ivory, President Daniel Arap Moi of Kenya in 1992 ordered the burning of government stockpiles of ivory. Although some people criticized this action because the government ivory could have been legally sold for several million dollars, President Moi wanted to demonstrate the importance of stopping all trade in ivory. In the United States, the Fish and Wildlife Service of the Department of Interior has the responsibility for enforcing laws that prohibit the sale of endangered species.

POLICIES TO REDUCE COMPETITION FROM NONNATIVE SPECIES

Policies designed to reduce problems from competition with nonnative species rely primarily on command and control techniques. In theory, it would be possible to rely on liability systems, where a person who introduced a nonnative species was liable for the damages that the nonnative species generates. However, this technique is not often used because it is often difficult to establish who was responsible for the introduction of the nonnative species.

Command and control regulations take two primary forms. First, one requires a permit for certain types of introductions, such as stocking a game preserve with exotic species. Second, other types of introductions are prohibited, and border inspections are established to prevent these introductions.

---

[3]Commercial fishing in marine areas is a notable exception, where the rules have not been effective in protecting the population of many species, such as tuna, swordfish, Atlantic salmon, striped bass, and so on. See Chapter 10 for more discussion on commercial fishing.

MARKET AND OPTIMAL LEVELS OF HABITAT PRESERVATION

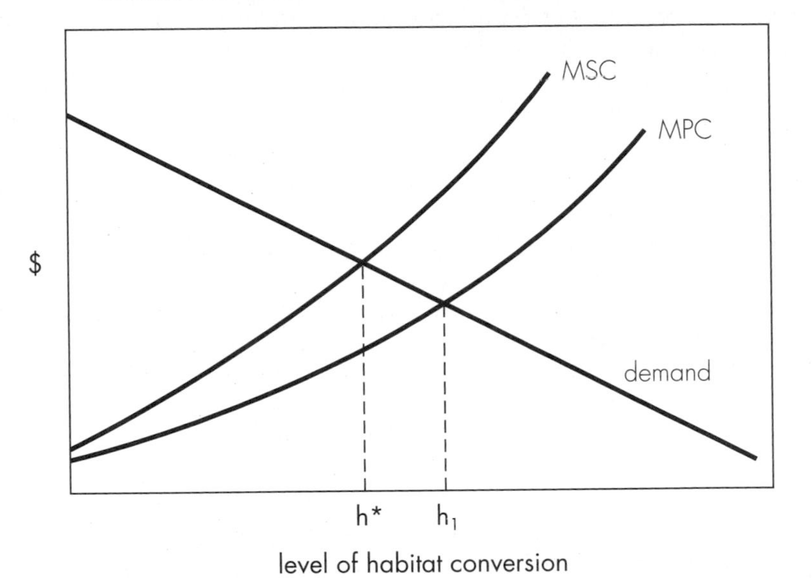

POLICIES TO REDUCE LOSS OF HABITAT

Loss of habitat occurs as human activity encroaches on natural ecosystems. Loss of habitat is an inevitable result of economic and population growth. The creation of farms, cities, roads, and so on results in a loss of habitat, as does pollution of ecosystems. However, the loss of habitat occurs at a rate in excess of what is socially optimal. This excessive loss of habitat occurs when people confront choices about how to utilize habitat, but do not have incentives to incorporate preservation values into their decision making. Consequently, the marginal private cost of converting habitat to development purposes is less than the marginal social cost. As can be seen in Figure 13.1, the market converts h_1 units of habitat, which is greater than the socially optimal level of habitat conversion, h^*.

In order for any set of policies to address the problem of loss of habitat, it must attempt to rectify this disparity between the private cost and social cost of converting habitat for other uses. Alternatively, policies could attempt to make the private benefits of preservation equal to the social benefits. Closing the disparity between the private costs and social costs of conversion and closing the disparity between the private benefits and social benefits of preservation are extremely difficult tasks. The reason is that the benefits of habitat preservation have public good characteristics (nonrivalry and nonexcludability in consumption).

Sedjo (1992) suggests a novel approach for reducing the disparity between the private benefits and social benefits of preserving tropical rain forests. As property rights are currently structured, the forested country does not generally benefit from the discovery of medicinal plants in its country. After the plant is discovered, the crucial chemical in the plant is usually identified, isolated, and then produced through chemical means or genetic engineering in pharmaceutical laboratories outside of the forested country. Consequently, the forested country does not earn much economic return from the discovery of a medicine that is based on plants within its borders. Sedjo suggests redefining property rights so that the forested country retains rights to share in the royalties associated with a medicine that is derived from a plant found within its borders, even if the medicinal chemical is eventually produced commercially outside the country with artificial means or through genetic engineering (if the plant's range spanned several countries, they would share in the rights). This type of plan would require an international treaty to be implemented.

Sedjo's plan would help internalize the costs of habitat loss with respect to the commodity value of plants and animals, but it would not deal with some of the other public good benefits of habitat preservation. For instance, it would not incorporate the carbon sequestration benefits of preserving forests. However, these benefits could be internalized by a system of marketable carbon permits, where countries received positive benefits from preserving their forests.[4]

Other values of habitat preservation may prove more difficult to incorporate into private benefits or private costs. For this reason, public policy incorporates command and control methods. Prohibitions on certain types of habitat destruction have been enacted in the United States. For example, several federal and state statutes restrict the destruction of wetlands. In addition, the creation of national parks and wildlife sanctuaries protects some of the most critical habitats. Private nongovernmental organizations, such as the Nature Conservancy, purchase important habitats and either privately manage the habitats or donate the habitats to state governments or the federal government to turn into public sanctuaries.

The Endangered Species Act of 1973 is a particularly important and controversial piece of legislation, which is designed to protect the habitat of

---

[4]The same thing could be accomplished by a marketable carbon permit system that required forested countries to purchase carbon permits for every unit of forest converted to nonforest uses. However, it is unlikely that the forested countries would agree to a system that penalized them for deforestation. They would obviously prefer a system that required the wealthier countries to partially compensate them for preserving rain forests. See Chapters 12 and 17 for more discussion of the role of wealthier countries in preserving environmental quality in developing countries and Chapter 6 for a discussion of trading greenhouse gas emission permits.

endangered species. Under this act, it is illegal to use any federal funds in a fashion that might further threaten an endangered species. Since federal funds are utilized in all sorts of infrastructure (roads, sewers, and so on), this legislation applies to a surprisingly large proportion of development projects. The Endangered Species Act also applies to projects requiring federal permits.

A major criticism of the Endangered Species Act is that it is oriented toward protecting species that are already in trouble, but does nothing to directly protect other species from becoming endangered or threatened. Critics of the Endangered Species Act argue that direct protection of habitat (both domestically and internationally) is a critical component of policy directed toward preserving biodiversity. The Endangered Species Act expired in 1993 and is still in the process of reauthorization. For current information on the status of the Endangered Species Act, check the websites of U.S. agencies such as the National Forest Service, the Fish and Wildlife Service, and the Environmental Protection Agency.

Two basic types of policies are available for protecting and preserving habitat in general. The first is to create protected areas, such as national parks and nature preserves. The second is to restrict uses of privately owned lands. For example, there are federal and state laws that restrict the destruction of wetlands.

A recent source of controversy is how one defines a wetland. Federal legislation originally defined wetlands relative to the soil type and vegetation in the area. However, in 1991, the Bush administration proposed a new manual that also stipulated a condition that the area in question must be inundated for 15 consecutive days during the growing season or saturated for 21 consecutive days during the growing season. Ecologists and environmentalists were upset with this definition because the determination of whether an area was a wetland would then be sensitive to the time at which an assessment was conducted. They also objected to these criteria because the soil and vegetation type of the area in question are direct evidence of saturation and inundation. The new definition would have eliminated some important wetland areas from legal protection and protected areas that ecologists did not consider to be wetlands. The net result of the redefinition would have been to drastically reduce the amount of protected wetlands. Landowners who object to federal restrictions on conversion of wetlands generally favored these new definitions, and environmentalists opposed them. In the end, these new definitions were not actually put into place, partially because many state governments objected to them.

The development of wildlife refuges or nature preserves is normally thought of as a government activity; however, this type of public good is increasingly provided privately. Although it is difficult for an individual citizen to protect habitat and promote biodiversity when acting alone, nongovernmental organizations (NGOs), such as the Nature Conservancy, can act as individual citizen's agents.

Endangered Species Act and the Snail Darter

The Endangered Species Act was the subject of considerable controversy shortly after enactment, when it was invoked by environmentalists to stop the construction of the Tellico Dam on the Little Tennessee River near Knoxville.[a] The project had originally been proposed before World War II to provide industrial development to this depressed corner of Appalachia, but by the time the project was reproposed by the Tennessee Valley Authority (TVA) in the mid-1960s, over 20 reservoirs had been constructed within 100 miles of the Tellico site. Also, Knoxville had developed a thriving economy based on manufacturing, the activities of the Oak Ridge National Laboratory, and the provision of services to the East Tennessee region.

Since the Knoxville area was not an area of abnormal structural unemployment, and since job programs rarely create new jobs but merely transfer their regional location, the creation of jobs in this area would not generate a net social benefit from a national perspective (see Chapter 4 for a discussion of this issue). Gramlich (1990) shows that under careful analysis, the total costs of the project would exceed the benefits.

However, TVA produced other estimates of the benefits, which included the creation of jobs. These estimates indicated that the net benefits from the dam were positive. In the meantime, a species of fish called the snail darter was discovered and believed to exist only in this portion of the Little Tennessee River. Opponents of the project seized upon this opportunity to invoke the Endangered Species Act to block the completion of the dam. However, Congress created the Endangered Species Committee, which had the authority to allow a project to cause the extinction of a species. This committee, nicknamed the God Squad because it decided on the potential extinction of species, actually ruled against the dam, but Congress still allowed the completion of the dam. The snail darter was transplanted to another stream in the region and has not become extinct.

The final result of this political process was that a dam was constructed that had more costs than benefits, and the Endangered Species Act was substantially weakened by the creation of the Endangered Species Committee. In this case, the invocation of the Endangered Species Act caused attention to be shifted from the costs and benefits of the dam and focus entirely on the snail darter.

[a]See Edward M. Gramlich, *A Guide to Benefit–Cost Analysis*, 1990, pp. 143–147 for an insightful discussion of the Tellico Project, its costs and benefits, and the Endangered Species Act.

These NGOs collect money from individual citizens and then use the funds to buy critical habitat from private landowners. The land is then usually donated to a local, state, or federal government agency that maintains and protects the biodiversity preserve.

The private provision of public goods runs counter to a particular economic theory that suggests that public goods, such as nature preserves, will seldom be privately provided. This theory offers two reasons to support this hypothesis. First, since it is not possible to exclude consumers from enjoying the benefits of a public good, it is not possible for the private landowner to extract a price and capture the public good benefits of protecting habitat and biodiversity. Second, the free rider problem will prevent a group of individuals from collectively paying for the public good. The free rider problem suggests that since people cannot be excluded from enjoying the benefit of the public good, they will not agree to pay for its provision, because they will hope that other people will pay for it and then the nonpayers can enjoy the benefits. Of course, if a large number of people have this attitude, then the public good will not be provided.

The private provision of this type of public good can take place because of the role of nongovernmental organizations, such as the Nature Conservancy. First, these groups serve to reduce the free rider problem by providing an organization through which individual citizens can commit (and seek the commitment of others) to protect the habitat. Second, they reduce the transaction costs to the individual citizen. The individual citizen is relieved of the burden of trying to decide which habitat is most important, of negotiating with landowners, and of negotiating with government agencies. The importance of these NGOs in reducing the transactions cost of habitat preservation (and other types of environmental improvement) cannot be overstated. Without these organizations, it is not likely that much private preservation activity would take place.

SUMMARY

Biodiversity, or the total variety of life on earth, is important for a multitude of reasons. Biodiversity is important to ecosystem stability and contributes both directly and indirectly to social welfare. Among the most important reasons to preserve biodiversity is the preservation of genetic information that can be a source of medicines, industrial products, and agricultural crops.

There are three primary anthropogenic sources of loss of biodiversity. These are over harvesting, competition from nonnative species, and loss of habitat. All of these sources of loss of biodiversity stem from market failures, where the benefits of preserving (or opportunity costs of destroying) biodiversity are not incorporated into private decision making. The loss of habitat is the most important contemporary source of loss of biodiversity.

Both command and control and economic incentives can be used to reduce the socially inefficient loss of habitat. Economic incentives include a variety of mechanisms to allow landowners to capture the benefits of preserving habitat. For example, countries that have tropical rain forests can be given the right to share the profits associated with medicines developed from rain forest plants. Command and control techniques include a variety of prohibitions on land use, including prohibitions on the direct destruction of the habitat (such as the draining of wetlands) or restrictions on pollution and other activities that may damage the habitat.

Review Questions

1. What are the economic forces that lead to extinction?
2. What is the difference between an open-access resource and a common property resource?
3. What are the economic benefits of preserving biodiversity?
4. What policies are available to deal with open-access exploitation?
5. What policies are available to deal with loss of habitat?

Questions for Further Study

1. How can the definition of property rights protect biodiversity on both a domestic and an international basis?
2. The Pacific yew tree has a chemical that can be used to treat cancer, but there are a limited number of trees, and the tree must grow for many decades before it produces the chemical. How would you allocate the trees between current cancer patients, future cancer patients, and leaving the trees in their natural state?
3. What are the flaws of the Endangered Species Act? How would you improve it?
4. What types of costs and benefits of preserving biodiversity can be measured with current valuation techniques? (See Chapter 4 for discussion of techniques.) What types of costs and benefits are difficult or impossible to measure?
5. Should a private landholder's options on how to use his or her land be limited by legislation to protect biodiversity?
6. Wetlands tend to have more protection than other types of habitat. Is there justification for this practice?
7. Define the elements of a national policy to protect biodiversity.
8. Define the important elements of an international treaty that would protect biodiversity.

Suggested Paper Topics

1. The measurement of biodiversity is a critical issue that spans both natural science and economics. How should one measure biodiversity in a way that is both scientifically credible and useful for policy? Start with articles by Solow et al. and Weitzman. Also, look at biodiversity related articles in *Nature, Science,* and the environmental economics journals (*Journal of Environmental Economics and Management, Land Economics, Environmental and Resource Economics,* etc.).
2. Look at the measurement of the value of biodiversity. Look at both articles related to the measurement of the value of biodiversity and existence values in general. Start with the article by Brown and Goldstein (1984). Search bibliographic databases on biodiversity and value, biodiversity and society, biodiversity and economics, and biodiversity and benefits. Also, look at articles on contingent valuation and other nonmarket valuation techniques (see Chapter 4 for references).
3. Look at the cost–benefit analyses surrounding the Tellico dam and the snail darter. Were these analyses conducted correctly? What improvements can you suggest? Develop a protocol for valuing the preservation of endangered species. Start with the reference by Gramlich (1990) and also search bibliographic databases on snail darter and economics, snail darter and environmental policy, and snail darter and cost–benefit analysis.

Works Cited and Selected Readings

1. Anderson, Robert, and Mark Rockel. "Economic Valuation of Wetlands." Discussion Paper #065, Washington, D.C.: American Petroleum Institute, 1991.
2. Bell, F. *Application of Wetland Valuation Theory to Florida Fisheries.* SGR-95. Sea Grant Publication, Florida State University, 1989.
3. Bishop, R. C. "Endangered Species and Uncertainty: The Economics of a Safe Minimum Standard." *American Journal of Agricultural Economics* 60 (1978): 10–18.
4. Brown, G. M., and J. H. Goldstein. "A Model for Valuing Endangered Species." *Journal of Environmental Economics and Management* 11 (1984): 303–309.

5. Eisworth, M. E., and J. C. Haney. "Allocating Conservation Expenditures across Habitats: Accounting for Inter-Species Distinctness." *Ecological Economics* 5 (1992): 235–250.
6. Farber, S. "The Value of Coastal Wetlands for Protection Against Hurricane Damage." *Journal of Environmental Economics and Management* 14 (1987): 143–151.
7. Farber S. and R. Constanza. "The Economic Value of Wetland Systems." *Journal of Environmental Management* 24 (1987): 41–51.
8. Gramlich, Edward M. *A Guide to Benefit-Cost Analysis*. New York: Prentice-Hall, 1990.
9. Gupta, T. R., and J. H. Foster. "Economic Criteria for Freshwater Wetland Policy in Massachusetts." *American Journal of Agricultural Economics* 57 (1975): 40–45.
10. Lynne, G. D., P. Conroy, and F. Pochasta. "Economic Valuation of Marsh Areas for Marine Production Processes." *Journal of Environmental Economics and Management* 8 (1981): 175–186.
11. McNeely, Jeffrey A. *Economics and Biological Diversity: Developing and Using Economic Incentives to Conserve Biological Resources*. Gland, Switzerland: IUCN, 1988.
12. Ryan, John C. "Conserving Biological Diversity." In *State of the World*, edited by Lester Brown. New York: Norton, 1992.
13. Sedjo, Roger A. "Property Rights, Genetic Resources, and Biotechnological Change." *Journal of Law and Economics* 35 (1992): 213.
14. Solow, Andrew, Stephan Polasky, and James Broadus. "On the Measurement of Biological Diversity." *Journal of Environmental Economics and Management* 24 (1993): 60–68.
15. Thibodeau, F. R., and B. D. Ostro. "An Economic Analysis of Wetland Protection." *Journal of Environmental Management* 12 (1981): 19–30.
16. U.S. Statistical Abstract. Washington, D.C.: U.S. Government Printing Office, 1992.
17. Weitzman, M. L. "On Diversity." *Quarterly Journal of Economics* 107 (1992): 363–406.
18. Wolf, Edmund C. "Conserving Biological Diversity." In *State of the World*, edited by Lester Brown. New York: Norton, 1985.

Water Resources

Water's seeming ubiquity has blinded society to the need to manage it sustainably and to adapt to the limits of a fixed supply.

Sandra Postel, "Managing Freshwater Supplies"

INTRODUCTION

In the history of the United States, water has played a central role, but water has always been taken for granted. Water supplies were abundant in the east, and as the eastern population expanded and moved westward, there was sufficient land with available water to support the populations that were moving westward. Although much of the west was (and is) semiarid, and the availability of water influenced choice of location, the overall availability of water did not appear likely to place constraints on the growth of population or the economy.

Since water was not viewed as a scarce resource, it was not managed to prevent or mitigate increasing scarcity. As we moved through the later half of the twentieth century, important water problems have arisen because our use of water has lessened its availability.

Three distinct problems affect the availability of water. First, in many areas of the country our use of water exceeds the rate at which it is being replenished. Second, many activities use water as an input, and when the water is returned to surface or groundwaters, its quality is diminished. Finally, many activities use surface or groundwaters as a means to dispose of waste, or waste inadvertently escapes into these waters. As these three types of problems indicate, our water problems have both quantity and quality dimensions.

Of course, these quantity and quality problems also exist outside of the United States, often on a more serious scale. Lack of clean drinking water is probably the most pressing environmental problem that the world faces. In many developing countries, intestinal disorders from contaminated drinking water are the primary cause of death of children under 5 years of age. The World Health Organization (WHO) has estimated that more than one billion people do not have access to uncontaminated drinking water (Raven, Berg, and Johnson, 1993, p. 281).

HYDROLOGICAL CYCLES

An understanding of hydrological cycles is essential to an understanding of water resource problems. The hydrological cycle refers to the movement of water from the atmosphere to the ground and then its evaporation and return to the atmosphere.

Water vapor in the atmosphere condenses and falls to the ground as some type of precipitation (rain, snow, sleet, and so on). Some of the precipitation lands directly in surface water (not surprising since three-fourths of the earth's surface is covered by water). Some of the precipitation falls on land, where most of it evaporates before it can be available for use. Water that falls on land may also move downhill along the surface of the ground until it reaches surface water (river, lake, and so on). Some water penetrates the surface of the ground and becomes groundwater. Some groundwater remains underground in aquifers, and some groundwater flows into surface waters. In the United States, 66 percent of the precipitation returns to the atmosphere directly by evaporation; 31 percent runs off to rivers, streams, and lakes; and 3 percent seeps underground. These percentages are averages and vary across the country as topography, soil type, ground cover, and land use vary.

NUTRIENT CYCLING

Another important cycle that is relevant for water issues is the nutrient cycle. The nutrient cycle refers to how the basic nutrients (nitrogen, potassium, and phosphorous) move through the ecosystem. For example, nutrients are contained in the tissues of living and dead organisms. When the organisms die and decay, the nutrients are released in the soil or water and become available to plants, which absorb them. The plants are then eaten by other organisms that absorb the nutrients, and the cycle continues. This cycle is part of a healthy ecosystem, be it aquatic or terrestrial. Problems arise when additional organic wastes from human activity are introduced into the ecosystem. This causes special problems in aquatic ecosystems, as the processes that break the wastes down into basic nutrients remove dissolved oxygen from the water. Dissolved oxygen is essential for aquatic life. Furthermore, the nutrients can contribute to algae blooms (massive growths of algae) that block light from penetrating the water. The algae also have short life spans, and when they die and decay they further remove oxygen from the water. If all the oxygen is depleted, then the decay process shifts from bacteria that operate in the presence of oxygen to bacteria that operate in the absence of oxygen. This anoxic decay leads to the "rotten eggs" smell associated with polluted waters. Agricultural runoff, suburban runoff, paper plants, food processing, stockyards, and discharge from sewage treatment plants are among the most significant anthropogenic sources of nutrients.

WATER CONSUMPTION

In Chapter 1, a resource taxonomy was developed that defined renewable resources as those in which the stock regenerates itself, resource flows as never-ending flows that come from a nondepleting stock, and exhaustible resources as those that have no regenerative capacity. Different water resources meet each of these definitions.

Water, in general, meets the definition of a renewable resource. The evaporation from the oceans and other water sources creates the precipitation that replenishes the oceans. However, smaller water bodies do not generally replenish themselves. The evaporation from a river does not provide the water source for the river. An important exception to this is the Amazon rain forest, where water vapor from evaporation from surface waters and transpiration by trees generates rain, much of which falls to earth in the confines of the rain forest. Reducing the size of the rain forest could thus reduce the amount of rain that falls in the region and is an important consequence of deforestation.

Water in riverine systems can be viewed as a resource flow. The water that arrives at any particular point along the river is independent of the amount of water that is taken out at that point. More importantly, a decision to remove water today will not affect the amount of water that can be taken out tomorrow.

Some water resources can be viewed as an exhaustible resource, because the rate of growth of the stock is small in relation to the use of the water. This exhaustibility is especially true for what is often called "fossil water," which is water that accumulated slowly in underground aquifers over millions of years. Although these aquifers may contain very large amounts of water, they are being recharged at a very slow pace. For all practical purposes, the rate of recharge can be viewed as negligible. Thus, the resource can be viewed as exhaustible—once removed it will never be replaced. A good example of this type of water resource is the Ogallala Aquifer, which underlies eight western and midwestern states (Montana, Wyoming, Nebraska, Colorado, New Mexico, Texas, Oklahoma, and Kansas).

All three classifications of water resources may be adversely affected by degrading uses of water that add contaminants of some nature to the water. Uses of water that do not adversely affect quality are called benign uses.

WATER AS AN INPUT TO PRODUCTION AND CONSUMPTION ACTIVITIES

Water is essential to all life processes and most economic processes. However, as will be demonstrated, water is not always used in a socially optimal fashion. We will examine water as both a resource flow and as an exhaustible resource and then discuss how market failures inhibit the optimal use of water in both cases.

FIGURE 14.1

SCARCITY AND THE PRICE OF WATER

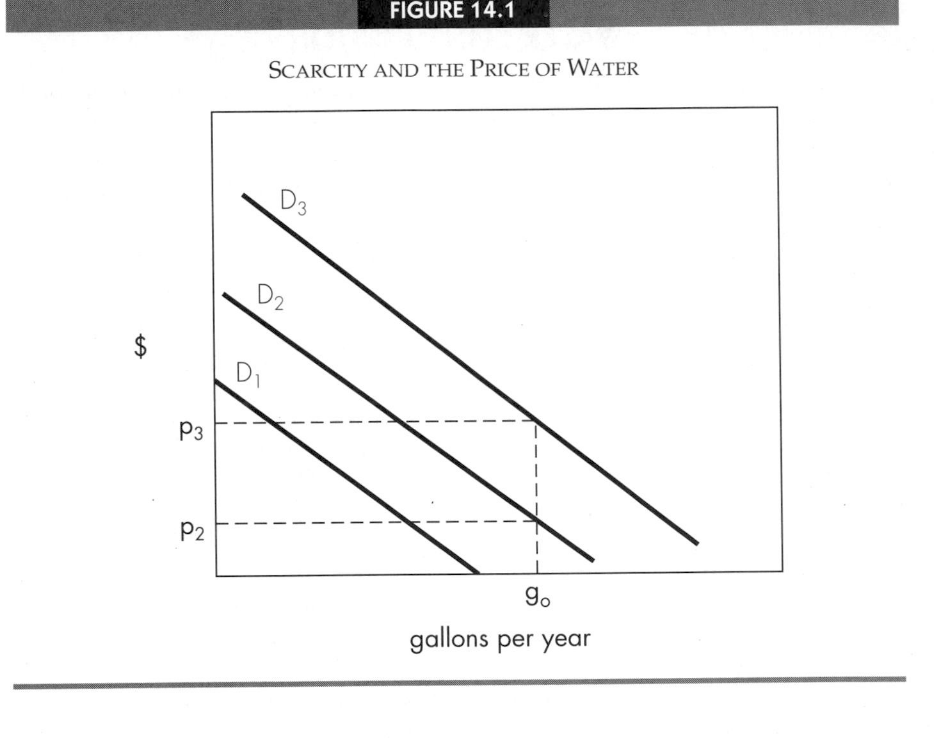

WATER AS A RESOURCE FLOW

Imagine a river flowing down from the mountains, through farms, to a city. The rain and snow in the mountains assure that there will be a continual flow of water in the river, year after year. If the flow of the river is much larger than the withdrawals of water to meet consumptive and productive needs, and if the uses of the water are benign, then there will be no resource allocation problem. Even if the water is available at zero cost and zero price, all the needs for water will be satisfied, and there will be no market failure.

Now let us change the scenario so that the flow of the river is not capable of meeting all the needs at any point in time. Even though there is a continual supply of water at any point in time, there will be a shortage of water at each point in time. At a price that is equal to zero, more water will be demanded than is available. As price is increased above zero (but the actual cost of obtaining water remains zero), the least valuable needs will be left unsatisfied, and the more valuable needs will be met. As long as the value of a gallon of water is greater than the price, people will purchase that gallon. If price is continually raised, eventually the quantity demanded will exactly equal the amount of water that is available. At this point, the willingness to pay for water will be equal to its price.

An alternative way of looking at this problem that illustrates how essential scarcity is in determining price is contained in Figure 14.1. In this figure, g_o represents the daily volume of water that may be removed from the river; the cost of extraction is equal to zero; and demand is represented by D_1. Under these circumstances, all the demand that exists at zero price will be satisfied. However, if demand shifts upward to D_2, all the demand that exists at zero price cannot be satisfied. Since there is unsatisfied demand at zero price, there will be upward pressure on price, which will increase to p_2. Notice that in this case, the price is solely determined by the opportunity cost of the water, since the cost of producing water has been assumed to be zero. As demand increases, the opportunity cost of consuming the fixed flow of water increases, which is reflected in an increased price. For example, if demand shifts further outward to D_3, price will increase to p_3.

The reflection of scarcity in the price of water will ensure that water is used in its most valuable applications, as only applications whose value equals or exceeds opportunity cost will be willing to pay the price that reflects the opportunity cost. Property rights to the water must be well defined in order for a market and a price to exist. Of course, if there is a cost of producing the water (purification, transportation, and so on), the marginal costs of production will also be reflected in price.

Even though this type of water resource is being continually replenished, it does not mean that there will be no shortage of water. If property rights are not defined appropriately, or if some other market failure keeps price from incorporating the scarcity value of water, then price will be too low. The quantity demanded will exceed the quantity supplied, and a shortage will exist. This shortage is depicted in Figure 14.2, where it is again assumed that there are no costs associated with producing a fixed supply of water of g_o. Here, the price at which the quantity demanded would equal the fixed quantity supplied is p_0. However, if some mechanism caused price to be at a lower level, such as p_1, then the quantity demanded would be g_1, and the shortage of water would be $g_1 - g_o$.

In eastern and Great Lake states, water resources can be viewed primarily as resource flows. Most cities and agricultural areas in this region depend on surface water or groundwater that is generally replenished by normal rainfall. However, where chronic water shortages exist, it is almost always a result of the price being too low. Of course, any area can have water problems caused by a prolonged or episodic drought.

One mechanism that often leads to urban water problems is the process by which water is priced and distributed to customers. In virtually every urban area, this distribution is done by either a regulated water utility or a municipal water company. In both cases, political or regulatory forces can push the price of water below its opportunity cost.

Let us first examine the case in which water is provided to consumers by a regulated monopoly. A city would award the exclusive right to distribute water to a private company. The general practice for determination of the

FIGURE 14.2

WATER SHORTAGE CAUSED BY LOW PRICE

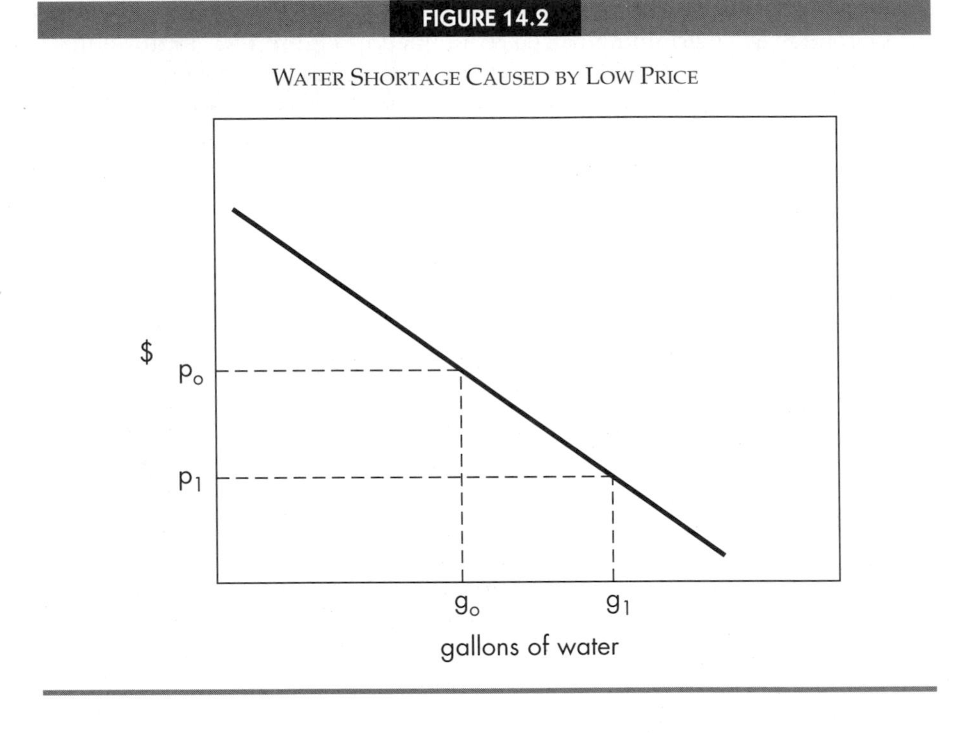

price that a regulated monopoly is allowed to charge is rate of return regulation (see a regulatory economics or industrial organization textbook for a more detailed explanation of rate of return regulation). Regulated utilities are allowed to charge a price that yields them a reasonable rate of return (after expenses) on their capital investment. Notice that such a price does nothing to incorporate the scarcity value of water, it only incorporates the scarcity value of the other inputs used in purifying and distributing the water.

If the price is below opportunity cost, more water will be demanded than is available. Then some other mechanism must be used to allocate water, and there is no guarantee that this mechanism will allocate water efficiently. For example, new uses of water may be prohibited, even though they may be more highly valued than some existing uses.

Alternatively, the city may elect to provide the distribution of water as a city service. In this case, there will be political pressures from the electorate to keep the price of water low. Again, there will be no mechanism to ensure that price includes opportunity cost.

Another potential problem exists when users are not forced to pay the marginal costs of water, but only pay average cost. This system is often the case for apartment buildings, where the building as a whole is metered, but individual apartment dwellings are not metered. Although the total cost of water will be incorporated into all tenants' rent, tenants pay no extra cost for

WATER SHORTAGES AND WATER CONFLICTS IN NEW YORK CITY

New York City obtains its water from reservoirs in the Catskill Mountains and other areas in upstate New York. This system of reservoirs and distribution pipelines began in the nineteenth century. Despite the elaborate system, New York City has chronic problems with shortages of water, especially during summer periods of less than normal rainfall in upstate New York. During the 1970s and 1980s, the shortage of water forced New York City to adopt emergency water regulations, in which many uses of water become prohibited.

New York City proposed a series of technical responses to this problem, including taking more water from currently exploited river systems, exploiting new river systems (mostly in the Adirondack region), and repairing the antiquated delivery systems so that less water leaks from the system. Upstate residents object to additional water withdrawals from upstate river systems, as the withdrawals created low flows in the river, increasing their temperatures and reducing dissolved oxygen. These effects reduce the ecological productivity of their rivers, particularly as a trout habitat.

The technical solutions that were proposed could not provide long-term relief from shortages, as they only addressed the symptoms of the water problem and not the cause of the water problem. The cause of the water problem was that many users did not pay for the water that they consumed. Water use was almost completely unmetered in New York City. People had no incentives to conserve water because they did not pay any price for water. The price should include both the opportunity cost of not having the water available for another use and the recreational and ecological opportunity cost of removing the water from the rivers. New York City's water problems and the misallocation of upstate water resources would be present as long as the scarcity value of water was not reflected in the price people paid.

New York City found itself in an interesting dilemma because it was impossible to correctly price water, as installing water meters for every single use in New York City would be prohibitively costly. In the absence of an ability to charge consumers the appropriate marginal price for water, what is the next best solution to this problem of excess demand for water?

New York City has embarked on a multifaceted program to reduce the demand for water. First, water is metered wherever practical. In particular, water is metered as it comes into big apartment buildings (although not in the individual residential units of the building), so apartment owners have a financial incentive to repair leaking pipes and faucets and install water saving appliances. Second, water-efficient appliances (shower heads, toilets, etc.) are required in new construction and when old appliances are replaced. Third, and perhaps most importantly, New York City has embarked on an innovative set of public education and public participation programs to reduce the general demand for water and to eliminate leaks and other wasteful uses of water.

extra water use. Therefore, they have limited incentives to conserve their use of water.

WATER AS AN EXHAUSTIBLE RESOURCE

When the rate of use of a stock of water is much greater than the rate of recharge of the stock, then the resources can be regarded as an exhaustible resource. This state describes the reality of water resources in many western

WATER SHORTAGE AND CONFLICT IN SOUTHERN CALIFORNIA

Southern California has been plagued by water shortages that were exacerbated by prolonged drought through the late 1980s and early 1990s. Although one expects water problems when a metropolis of millions of people develops in a desert, the problem has also been exacerbated by the way water rights are defined for the water that flows down from the mountains (which usually receive plentiful rain and snowfall).

Farmers in southern California have appropriation rights to much of the available water, paying a fraction of what the water's value would be in alternative uses (residential, commercial, and industrial uses). Farmers use this water to irrigate crops that can be cheaply grown elsewhere in the country (such as soybeans and rice). However, their rights to inexpensive water make it profitable to plant these crops in southern California. In effect, the farmers are subsidized by the availability of water at a fraction of its true value.

This definition of water rights and the resulting price of water, which is below opportunity cost, will continue to stress the water resources of southern California and reduce the amount of water available for residential uses. There is simply not enough water in southern California for both agriculture and population to continue to grow. In addition, the removal of much of the water from the rivers has led to severe ecological damage in many river systems.

states, particularly those overlying the Ogallala Aquifer and other slowly recharging aquifers. The economic analysis of water as an exhaustible resource is very similar to the analysis of water as a resource flow, only there is an additional opportunity cost to consider.

First, there is the same opportunity cost of not having the water available for another current use, which can be termed the contemporaneous opportunity cost. Second, because current use depletes the stock and makes it unavailable for future use, there is the opportunity cost of not having the water available for future use. This opportunity cost may be termed the intertemporal opportunity cost. Both the contemporaneous and the intertemporal opportunity costs are components of user cost and must be incorporated into price if water is to be efficiently allocated among all its alternative uses, both present and future.

WATER AND PROPERTY RIGHTS

In the previous discussion of urban uses of water, it was assumed that the municipality had the right to withdraw water from the river and aquifer, and distribute it to consumers (or to grant this right to a regulated utility). The only allocation question that remained was if the municipality or utility would price the water appropriately so that water would be allocated to its most valuable uses.

COMBINED SEWER OUTFLOW

One of the most serious threats to water quality in the United States comes from combined sewer outflow. Combined sewer outflow refers to the fact that in most urban areas, storm sewers are connected to the sewage treatment system. Storm sewers carry the runoff from rainfall that falls on paved urban areas and consequently does not sink into the ground. The problem is that since sewage treatment plants are generally operating at near capacity in treating the water that is returned in the sanitary sewer system, the storm runoff from heavy rainfalls overloads the sewage treatment system and causes the treatment system to be bypassed. Not only storm water, but also untreated sewage bypasses the treatment system. As a consequence, rivers and lakes in most urban areas (not just the largest cities) receive periodic doses of untreated sewage, which can substantially reduce water quality and make it unhealthy to recreate in the water body in the period following the heavy rainfall. The bypass of the system also allows the roadside trash that is swept into storm sewers to bypass the sewage treatment systems, and the trash is released into rivers and other water bodies. Combined sewage outflow is suspected to be one of the important causes of the beach debris problem that has been experienced on the Atlantic coast, especially in the area between New Jersey and Massachusetts.

Although this problem is well understood, the solution to the problem is not easy to develop. The reason is that solving the problem calls for rerouting the storm sewers and/or increasing the capacity of the sewage treatment plants, both of which are extremely expensive, especially in an era when local governments have little budget flexibility. For example, it has been estimated that solving the combined sewage outflow problem in New York City would cost over $10 billion.

Many cities are looking at innovative methods for dealing with sewage treatment needs. New development projects (subdivisions, malls, industrial parks, etc.) are generally required to construct holding ponds to store storm runoff and allow it to be absorbed into the ground or released more slowly (after sediments drop out) into surface waters. Some municipalities are also exploring the possibility of constructing large wetlands to cleanse sewage by natural processes.

Arcata, California, has developed a 63 hectare wetlands on a former garbage dump to deal with a portion of the town's sewage. The purified water that flows out of the wetlands is used to irrigate other wetlands and used to support an oyster aquaculture operation in the adjacent bay. Processed wastewater is also used to support aquaculture in many other countries, particularly in Southeast Asia.[a]

[a]World Resources Institute, *World Resources 1992–93*, 169.

In the real world, the problem is more complicated than this, as non-transferable property rights exist, which prevent the flow of water to its most beneficial uses. The critical factor is that these rights to use water, unlike most property rights, cannot be transferred, and it is the lack of transferability that prevents the water from being sold to its most beneficial uses.

Property rights to water are a big issue in the Western United States, where the natural dryness of the region exacerbates conflict over water. It is very difficult to generalize about property right structures for water, as the definition of property rights varies substantially across Western states. Perhaps the most important type of property rights to water are appropriation-based water rights.

Appropriation-based water rights make water available for use by anyone who can apply it to beneficial purposes. Priority goes to the user who established his or her appropriation-based rights first. Initially in many states, these rights tended to be nontransferable which meant that water rights could not be transferred to other uses, even if the new uses had higher values. However, the increasing demand for water associated with growing urban areas has led many of the Western states to make appropriation-based water rights transferable. Some states even allow the state or private parties to buy appropriation-based rights to leave the water in the river for beneficial ecological purposes, an important way to generate environmental improvement as a win-win situation. People who contribute money to buy water rights for ecological protection are better off, because they value the ecological improvement by an amount equal to or greater than their contribution. Farmers and other current users of water are better off, for if they voluntarily agree to sell a portion of their water rights, they must value these rights less than the revenue they receive from selling their rights. Unfortunately, the opportunities for this type of market-based environmental improvement are limited, as some Western states do not define in-stream uses such as recreation or ecological protection to be legally "beneficial uses," and thus are ineligible for transfer. Consequently, in these states, people cannot use the water rights system to generate ecological improvement.

DEGRADING USES OF WATER

So far, the only opportunity cost of water use that has been discussed is that one use of water makes it unavailable for other uses (either current or future). However, uses may have an additional cost if the use of water renders it unfit for other uses. A particular use of water may have an adverse effect on water quality, leading to negative impacts on other water uses.

There are three types of water uses that degrade water quality. The first type of use is when removal of water from surface water bodies or groundwater aquifers generates ecological damage. For example, heavy withdrawal of water in coastal areas, such as southern Florida, has led to saltwater intrusion into the aquifer. The second is when a direct consumer of water uses it and then returns it to the hydrological cycle with wastes or contaminants. For example, residential use of water adds human waste and other contaminants. Even after treatment in a sewage treatment plant, the water that is returned to the river or lake will be augmented by nutrients and other contaminants that can generate ecological damage and reduce the usefulness of the water for other purposes. The third type of use is represented by activities that generate wastes that are either directly deposited into, or make their way through, natural mechanisms (such as the runoff of rainfall) into water bodies. For example, rainfall runoff carries agrichemicals (pesticides, herbicides, and fertilizers) from agricultural fields into surface and groundwater. Table 14.1 summarizes the sources and impacts of pollutants that adversely affect water quality.

U.S. POLICY TOWARD WATER POLLUTION

U.S. policy toward water pollution has historically focused on large point sources of pollution. Point sources of pollution are those where the pollution enters the water body at a specific point, such as the end of an effluent discharge pipe. Attempts to control pollution have met with mixed success. Organic pollutants from large sources have been reduced. However, nonpoint sources of pollution and inorganic toxic pollutants continue to represent a major problem.

One of the major thrusts of attempts to reduce water pollution were programs to reduce the impact of the discharge of municipal sewage. Even as recently as the 1960s, many small and medium-sized cities had no sewage treatment facilities, and untreated waste was often discharged directly into rivers and lakes. In large cities (which often had treatment plants), treatment was inadequate. This inadequate sewage treatment led to very severe degradation of water quality in virtually every river that flowed through metropolitan areas.

Lake Erie, which borders on Ohio, Pennsylvania, New York, and Ontario, was particularly hard hit. Pollution from municipal and industrial discharge was so extensive that many parts of Lake Erie became completely anoxic and incapable of supporting aquatic life. Thriving recreational and commercial fisheries were decimated.

However, the amendments to the Clean Water Act required all municipalities to develop and upgrade their sewage treatment facilities. Both primary treatment (removal of suspended particles) and secondary treatment (breakdown of organic wastes) were required, and eventual goals were established to minimize the release of all nutrients. The federal government would pay 75 percent of the costs of the facility, and local governments would be responsible for the remainder of the construction costs and for operating costs.

The primary reason for the federal government to subsidize wastewater treatment plants has to do with the difference between the local social benefits of treatment and the regional or national benefits of treatment. The benefits of wastewater treatment are felt by all the people who live downstream of a particular community, not just the people residing in the community. Therefore, the social benefits to the nation of treating a particular community's wastewater are greater than the social benefits to the community. Consequently, the level of treatment provided by individual communities would be lower than socially optimal, necessitating federal government intervention to correct this market failure. Of course, the federal government could

TABLE 14.1

SOURCES AND IMPACTS OF SELECTED POLLUTANTS

POLLUTANT	SOURCES	IMPACT ON AQUATIC ORGANISMS	IMPACT ON HUMAN HEALTH AND WELFARE
Sediment	Agricultural fields, pastures, and livestock feed lots; logged hillsides; degraded streambanks; road construction	Reduced plant growth and diversity; reduced prey for predators; clogging of gills and filters; reduced survival of eggs and young; smothering of habitats	Increased water treatment costs; transport of toxics and nutrients; reduced availability of fish, shellfish, and associated species; shortened life span of lakes, streams, and artificial reservoirs and harbors
Nutrients	Agricultural fields, pastures, and livestock feed lots; urban areas; raw and treated sewage discharges; industrial discharges	Algal blooms resulting in depressed oxygen levels and reduced diversity and growth of large plants; release of toxins from sediments; reduced diversity in vertebrate and invertebrate communities; kills fish	Increased water treatment costs; risk of reduced oxygen carrying capacity in infant blood; possible generation of carcinogenic nitrosamines; reduced availability of fish, shellfish, and associated species; impairment of recreational uses
Organic materials	Agricultural fields and pastures; landscaped urban areas; combined sewers; logged areas; chemical manufacturing and other industrial processes	Reduced dissolved oxygen in affected waters; kills fish; reduced abundance and diversity of aquatic life	Increased water treatment costs; transport of toxics and nutrients; reduced availability of fish, shellfish, and associated species
Disease-causing agents	Raw and partially treated sewage; animal wastes; dams that reduce water flows	Reduced survival and reproduction in fish, shellfish and associated species	Increased costs of water treatment; river blindness; elephantiasis; schistosomiasis; cholera; typhoid; dysentery; reduced availability and contamination of fish, shellfish, and associated species

TABLE 14.1

CONTINUED

POLLUTANT	SOURCES	IMPACT ON AQUATIC ORGANISMS	IMPACT ON HUMAN HEALTH AND WELFARE
Heavy metals	Atmospheric deposition; road runoff; industrial discharges; sludges and discharges from sewage treatment plants; creation of reservoirs; acidic mine effluent	Declines in fish populations due to failed reproduction; lethal effects on invertebrates, leading to reduced prey for fish	Increased costs of water treatment; lead poisoning, itai-itai and minimata diseases; kidney dysfunction; reduced availability and healthfulness of fish, shellfish, and associated species
Toxic chemicals	Urban and argicultural runoff; municipal and industrial discharges; leachate from landfills	Reduced growth and survivability of fish eggs and young; fish diseases	Increased costs of water treatment; increased risk of rectal, bladder, and colon cancer; reduced availability and healthfulness of fish, shellfish, and associated species
Acids	Atmospheric deposition; mine effluent; degrading plant materials	Elimination of sensitive aquatic organisms; release of trace metals from soils, rocks, and metal surfaces, such as lead pipes	Reduced availability of fish, shellfish, and associated species
Chlorides	Roads treated for removal of ice and snow; irrigation runoff; brine produced in oil extraction; mining	At high levels, toxic to freshwater life	Reduced availability of drinking water supplies; reduced availability of fish, shellfish, and associated species
Elevated temperatures	Urban landscapes; unshaded streams; impounded waters; reduced discharges from dams; discharges from power plants and industrial facilities	Elimination of cold water species of fish and shellfish; reduced dissolved oxygen due to increased plant growth; increased vulnerability of some fishes to toxic wastes, parasites, and diseases	Reduced availability of fish, shellfish, and associated species

SOURCE: From *World Resources, 1992–1993,* by World Resource Institute.

NOTE: See source for references on specific effects.

require the treatment without the associated subsidies, but this action would not be politically expedient, and many communities simply would not have the tax base to improve their sewage treatment plants.

One interesting aspect of these subsidy programs is that they do not cover the costs of operating the plants, only building the plants. Since local governments have to pay the operating costs, they have an incentive to choose plant designs that rely more on capital equipment (which the federal government subsidizes) than on variable inputs (labor and energy) that are unsubsidized. Thus, the subsidy system leads to a nonoptimal mix of inputs in the production of sewage treatment, and in extreme cases, local governments have not had the funds to properly maintain the capital equipment.

The construction of many of these plants was completed during the 1970s and had a rapid effect on water quality. Aquatic systems are capable of processing and assimilating organic wastes such as sewage, but the problem was that the assimilative capacity of the environment had been exceeded by the massive releases from municipal sewer systems. As the inputs of organic waste were reduced, water quality improved. Drastic changes in water quality occurred in areas such as Lake Erie, which now support thriving walleye, trout, and bass populations. The Potomac River, whose recovery was speeded by several tropical storms that helped to flush existing wastes from river bottom sediments, is another example of an urban water body that has rapidly recovered and now supports fish populations that require clean water. Bass fishing guides now take clients on fishing trips in downtown Washington, D.C. A favorite fishing spot is along the river wall at the famous Watergate building.

These subsidized improvements in municipal sewage treatment plants were required by the Water Pollution Control Act of 1972, the Clean Water Act (CWA) of 1977, and 1977 and 1987 amendments to the CWA. In addition to focusing on sewage treatment, these acts also focused on other large point sources of pollution, such as paper plants, food processing facilities, and other industries. These acts were based on command and control techniques. All discharges were illegal unless authorized by the National Pollution Discharge Elimination System (NPDES). Dischargers were required to meet two technology-based standards. Polluters were required to use best practical technology (BPT) for conventional pollutants and best available technology (BAT) for toxic pollutants. The difference between BPT and BAT is that BPT allows for the consideration of the cost of the technology and BAT does not consider the cost of the technology.

Economic incentives have not been employed to deal with water quality problems. Although a marketable pollution permit system was established for paper plants on the Fox River in Wisconsin, so many restrictions have been placed on the ability to make trades that few trades have been made.

In some senses, it would be easier to develop a functioning system of marketable pollution permits (or taxes) to control water pollution because of the greater ease in measuring the dispersion of emissions (pollutants move with water and wind currents). However, the development of such a system would require more interstate cooperation, since all the major river systems span several states. For example, rivers that bring nutrients into the Chesapeake Bay flow through Maryland; Virginia; Delaware; West Virginia; Washington, D.C.; Pennsylvania; and New York. Similarly, some watersheds span our international borders with Mexico and Canada. Of course, this problem is even bigger in Europe where countries tend to be geographically smaller.

As might be expected, the system of command and control policies led to the problems associated with command and control technologies that were discussed in Chapter 3. Firms were not free to choose the cheapest way to reduce pollution, which led to much higher than normal costs. Industries had to employ the same types of control mechanisms, regardless of the effect of their emissions on environmental quality. For instance, a polluter in a region where emissions were lower than the capacity of the environment to break down and assimilate the pollutants had to install the same control devices as in an area where the environment's capacity is exceeded.

Unfortunately, the Clean Water Act and associated amendments have not been as successful as possible in meeting the goals of the legislation. The overall goal is to restore and maintain the chemical, physical, and biological integrity of the nation's waters. Operational goals included water bodies that were clean enough to support both fishing and swimming by 1983 and the elimination of all discharges (EOD) by 1985.

Of course, anyone who understands the laws of mass balance knows that the elimination of all discharges is impossible to obtain. While it is possible to recycle, to reuse, to change the form of waste and to change the media into which it is disposed, it is impossible to eliminate the waste. The law of mass balance states that the mass of the inputs going into an activity must equal the mass of the outputs coming out of the activity. The law of mass balance cannot be repealed.

We have had mixed success in meeting the other goals. Reduction in organic wastes from point source polluters and municipal waste treatment plants has improved water quality in some water bodies to the point that it sustains healthy fish populations. However, in regions where nonpoint sources of pollution are important, water quality remains poor. Combined sewer outflow is an important problem in many urban areas. The text box on page 401 discusses this problem in greater detail.

The major sources of nonpoint pollution are agricultural, urban, and suburban runoff. In particular, nutrients from agricultural pollutants adversely affect water quality in areas like the Chesapeake Bay (and its tributaries) and the San Francisco Bay (and its tributaries). In recent years, more attention has been paid to nonpoint pollution, with new regulations requiring farmers to institute "best farming practices" to control nutrient runoff and soil erosion. These issues are further discussed in Chapter 16 (Agriculture and the Environment).

Even though some progress has been made in controlling organic pollutants, the problem of toxic pollutants has not been similarly reduced (it may even be increasing), and many areas that have healthy fish populations (such

TABLE 14.2

IRRIGATED LAND DAMAGED BY SALINIZATION, TOP FIVE IRRIGATORS, AND WORLD ESTIMATE, MID-1980S

COUNTRY	AREA DAMAGED (MILLIONS OF HECTARES)	SHARE OF IRRIGATED LAND DAMAGED (%)
India	20.0	36
China	7.0	15
United States	5.2	27
Pakistan	3.2	20
(Former) Soviet Union	2.5	12
Total	39.7	24
World	60.2	24

SOURCE: From *State of the World 1990: A Worldwatch Institute Report on Progress Toward a Sustainable Society*, by Lester R. Brown et al. Copyright © 1990 by Worldwatch Institute. Reprinted by permission of W. W. Norton & Company, Inc.

as the trout and salmon fisheries of the Great Lakes) have advisories and prohibitions against eating the fish because of contamination with PCBs, mirex, dioxin, heavy metals, and other toxic substances.

INTERNATIONAL WATER ISSUES

As mentioned in the introduction, water problems in other countries (particularly developing countries) may even be more severe than in the United States. The primary water problem in developing countries, without doubt, is the contamination of water by untreated human waste. This problem is not just associated with small backwater villages. Large cities with modern industries that employ the latest industrial technologies (such as Rio de Janeiro and Mexico City) also suffer from this problem. In many of these cities, the problem is intensified because of the large slums (completely devoid of any sanitary facilities) filled with migrants from rural areas that have developed around these cities.

This is not just a Third World problem however, as problems with drinking water quality still exist in many developed countries. For example, in the Po River valley in northern Italy, many cities (including Milan) dump untreated wastes into the river. Heavy chemical runoff from agriculture further degrades Po River water quality. The Mediterranean Sea, which travel posters depict as paradise, suffers from extreme water pollution problems. These problems are likely to become even worse as increased population growth, urbanization, and industrialization in North Africa leads to more effluents and overwhelm the reductions in pollution that are being undertaken in European countries.

TRANSFRONTIER EXTERNALITIES

One of the major problems associated with water extraction and water pollution in Europe, Asia, Africa, and Latin America is transfrontier externalities. The water consumption or waste disposal activities in one country affect water availability and water quality in neighboring countries. This affect occurs particularly in areas such as the Middle East, where geographically small countries overlie common aquifers and where rivers such as the Jordan River and the Tigris-Euphrates River drain several countries. The fact that political hostility exists in the region exacerbates the problem and makes it more difficult to develop solutions to prevent the deterioration of water resources.

These transfrontier externalities cannot be internalized without international agreement. One hopeful note is that international commissions, sometimes sponsored by the United Nations and sometimes developed regionally, are being initiated to solve these problems. Although different countries have different perspectives, different needs, and different institutions, the countries are recognizing that if they do not cooperate, all will suffer.

The United States has had long-standing agreements with Canada concerning water use and water quality in boundary areas and a special commission to deal with Great Lakes issues. Agreements with Mexico are being developed at the current time.

AGRICULTURE AND WATER QUALITY IN DEVELOPING COUNTRIES

In addition to untreated sewage, agricultural pollution is extremely important in affecting water quality in developing countries. Problems with deforestation, over-tillage, tillage of hillsides, heavy use of dangerous pesticides, and runoff of fertilizer (including human and animal waste as fertilizers) has seriously affected water quality in many developing nations.

One of the most degrading uses of water is the irrigation of agricultural fields. Not only may irrigation lead to the rapid depletion of groundwater and reduced flows in rivers, but the irrigation causes major changes in the water that flows out of the agricultural fields and also degrades soils.

One of the major problems with irrigation is that the repeated soaking of the soil and the evaporation of the water in the soils draws salts from lower levels of the soil and deposits them in the top layers of the soil, where they adversely affect many of the crops that the farmers are trying to grow. In addition, the salts wash into nearby streams and rivers, causing ecological damage and reducing the quality of water for downstream water users.

The other major problem with irrigation is that it often diverts so much water out of the river systems that there is massive ecological degradation of the aquatic systems. The premier example of this is an irrigation project in

TABLE 14.3

SELECTED CASES OF EXCESSIVE WATER WITHDRAWALS

REGION	STATUS
Colorado River Basin, United States	Yearly consumption exceeds renewable supply by 5 percent, creating a water deficit, Colorado River is increasingly salty; water tables have fallen precipitously in areas of Phoenix and Tucson.
High Plains, United States	The Ogallala, a fossil aquifer that supplies most of the region's irrigation water, is diminishing; over a large area of the southern plains, the aquifer is already half depleted.
Northern China	Groundwater overdrafts are epidemic in northern provinces; annual pumping in Beijing exceeds the sustainable supply by 25 percent; water tables in some areas are dropping up to 1–4 meters per year.
Tamil Nadu, India	Heavy pumping for irrigation has caused drops in water level of 25–30 meters in a decade.
Israel; Arabian Gulf; and Coastal United States	Intrusion of seawater from heavy pumping of coastal aquifers threatens to contaminate drinking water supplies with salt.
Mexico City; Beijing, China; Central Valley, California; Houston-Galveston, Texas	Groundwater pumping has caused compaction of aquifers and subsidence of land surface, damaging buildings, streets, pipes, and wells; hundreds of homes in a Texas waterfront community have been flooded.
California, United States	Water from Owens Valley and Mono Basin has been diverted to supply southern California water users; Owens Lake has dried up, and Mono Lake's surface area has shrunk by a third.
Southwestern [former] Soviet Union	Large river withdrawals have reduced inflows to Caspian and Aral seas; the Caspian sturgeon fishery is threatened, the Aral fisheries are virtually gone, and the sea's volume may be halved by the turn of the century.

SOURCE: From *State of the World 1985: A Worldwatch Institute Report on Progress Toward a Sustainable Society,* by Lester R. Brown et al, eds.

the former Soviet Union, which has so reduced river flow that the Aral Sea (into which the rivers drain) has been reduced to a fraction of its former size, with the loss of an important commercial and subsistence fishery. This problem has been intensified by the fragmentation of the former Soviet Union, as the rivers flowing into the Aral Sea flow through several newly independent republics, increasing the difficulty of developing a solution to the problem. Table 14.2 describes the extent of irrigation and damage done in the countries that are the biggest irrigators. Table 14.3 lists selected cases of excessive withdrawals, which include withdrawals for irrigation and withdrawals for other purposes.

SUMMARY

Although the earth's surface is three-quarters covered by water, uncontaminated freshwater is a scarce resource. Many market failures, including externalities, nontransferable property rights, and poorly conceived regulatory practices, exacerbate the scarcity. These market failures adversely affect the quantity of water that is available for use, as well as diminishing its quality. This diminished water quality has important effects for social and ecological systems that are dependent upon water resources.

From a U.S. and international perspective, one critically important policy change would be to price water so that it included its full opportunity cost. This would include the opportunity cost of both current and future uses of water, as well as the costs associated with reductions in the quality of water resources.

Water scarcity problems are critically important internationally, particularly in developing countries. This chapter has identified some of these problems, but has not extensively discussed solutions, as different institutional, economic, social, and environmental factors imply differences between U.S. policies and the policies that would be most appropriate in a developing country context. These policies and the factors that determine their structure are discussed in depth in Chapter 17.

REVIEW QUESTIONS

1. What market failures are associated with the use of water for consumption?
2. How does average cost pricing of water affect social efficiency?
3. Assume that the inverse demand function for water in a town can be described

   $D = 25 - 0.04Q$, while $AC = 0.5Q$ and $MC = Q$.

   If the town prices water at average cost, consumers' and producers' surplus will be lower than if priced at marginal cost. Calculate the loss in this example.
4. What is wrong with the Clean Water Act and its amendments?
5. Why is the elimination of all discharges an ill-founded goal for water policy?

QUESTIONS FOR FURTHER STUDY

1. What extra difficulties are involved with devising water quality policy for internationally shared waters?
2. What are the equity issues involved with rapid consumption of "fossil water"?
3. Should California farmers be forced to pay full social cost for the water they use?
4. How do Third World water issues differ from those in the United States?
5. How should property rights to water be defined? Does it make any difference if the resource in question is a resource flow or an exhaustible resource?

SUGGESTED PAPER TOPICS

1. Investigate the controversy over Native American water rights. Start with the references by Dumars and Burton. Search bibliographic databases using Indian water rights and Native American water rights as key words. Also, search government documents databases, paying particular attention to the Bureau of Indian Affairs (Department of Interior) and Congressional Committees on Interior and Indian Affairs.

2. Investigate the benefits of water quality improvements. Start with Freeman's 1993 book for current references. Also, look at recent issues of the *Journal of Environmental Economics and Management, Land Economics,* and *Water Resources Research.*
3. Investigate a water quality issue in your area. Is it caused by a pollution externality or a lack of property rights to consumptive uses of the water? What is current policy toward this issue? Use your knowledge of economics to develop an improved set of policies.
4. Look at comparative property rights regimes toward water resources and the incentives that they create. Do we need a national policy or is a state-by-state policy good enough? Start with references by Anderson, Carriker, Wahl, and Waite. Search bibliographic databases on water rights.
5. Investigate the development of policy to manage internationally shared water resources. Start with the *Natural Resource Journal,* which has been active in publishing articles in this area. Search bibliographic databases on international water issues, the Great Lakes Commission, boundary waters, the Colorado River, and the Rio Grande.

WORKS CITED AND SELECTED READINGS

1. Anderson, Terry Lee. *Water Rights: A Scarce Resource Allocation, Bureaucracy, and the Environment.* Cambridge, MA: Ballinger Publishing Company, 1983.
2. Boris, Constance. *Water Rights and Energy Development in the Yellowstone River Basin: An Integrated Analysis.* Washington, D.C.: Resources for the Future, 1980.
3. Burton, Lloyd. *American Indian Water Rights and the Limits of Law.* Lawrence, KS: University of Kansas Press, 1991.
4. Carriker, Roy R. *Water Law and Rights in the South.* Mississippi State University: Southern Rural Development Center, 1985.
5. Dewsnup, R. L., and D. W. Jensen. *A Summary Digest of State Water Laws.* Arlington, VA: National Water Commission, 1973.
6. Dumars, Charles T. *Pueblo Indian Water Rights: A Struggle for a Precious Resource.* Tucson: University of Arizona Press, 1984.
7. Foster, Harold. *The Emerging Water Crises in Canada.* Toronto: J. Lorimar, 1981.
8. Freeman, A. M. III. *The Measurment of Environmental and Resource Values.* Washington, D.C.: Resources for the Future, 1993.
9. Postel, Sandra. "Managing Freshwater Supplies." In *State of the World 1985,* edited by Lester Brown. New York: Norton, 1985.
10. ———. "Saving Water for Agriculture." In *State of the World,* edited by Lester Brown. New York: Norton, 1990.
11. Raven, Peter H., Linda R. Berg, and George B. Johnson. *Environment.* Philadelphia: Saunders College Publishing, 1993.
12. Wahl, Richard W. *Markets for Federal Water: Subsidies, Property Rights and the Bureau of Reclamation.* Washington, D.C.: Resources for the Future, 1989.
13. Waite, G. Grahame. *A Four-State Comparison of Public Rights in Water.* Madison: University of Wisconsin Law Extension, 1967.
14. World Resources Institute. *World Resources 1992–93.* Washington, D.C.: World Resources Institute, 1993.

Further Topics

Part IV examines important environmental and resource issues that do not relate directly to either exhaustible or renewable resources. Toxic substances, agriculture and the environment, and Third World issues receive special treatment in Part IV. The concluding chapter looks at how the interaction between the economic system, the environment, and social policy determines the prospects for the future.

Toxins in the Ecosystem

. . . The chemicals to which life is asked to make its adjustment are no longer merely the calcium and silica and copper and all the rest of the minerals washed out of the rocks and carried in rivers to the sea; they are the synthetic creations of man's inventive mind, brewed in his laboratories, and having no counterparts in nature.

Rachel Carson

INTRODUCTION

In 1962, Rachel Carson alerted the world to the problems associated with Western society's increasing chemical dependence. Although her important book, *Silent Spring,* was couched mostly in terms of our increasing use of pesticides, the book alerted the public to the ecological and human health consequences of the increasing presence of toxic chemicals in the environment.

In subsequent years, public attention has been refocused on toxic substances. As events such as Love Canal, Times Beach, and the kepone contamination of the James River unfolded during the 1960s and 1970s, we became aware that toxic substances are a critically important environmental problem.[1]

One way of categorizing the toxic problem is to distinguish between exposure as a result of toxic substances being present in economic processes and exposure as a result of wastes that result from economic processes. The

[1]Love Canal is a site in Niagara, New York, where a variety of toxic substances were secretly buried in an area upon which homes and a public school were later built. Health problems surfaced and the extent of the toxic contamination became apparent. The Love Canal is the "textbook" example of illegal and unethical disposal of toxic substances by a chemical producing firm.

Times Beach, Missouri, is a small rural town in which dioxin-laden waste oil was dumped on dirt roads to control dust. Although many critics say the government overreacted, it bought the town and forced residents to move to other locations.

The kepone problem in the James River was due to waste products in the manufacture of pesticides in a town near Richmond, Virginia. The kepone releases led to nervous system problems of people who were exposed to the kepone, and it contaminated the fish in the river so badly that it is still prohibited to eat fish from this area.

first category of exposure would include on-the-job exposure to toxic substances, pesticide residues on fruits and vegetables, and radon exposure in people's homes. The second category of exposure would include environmental exposure to humans and ecosystems, such as through toxic contamination of lakes and rivers. There is little ecosystem exposure in the first category (which is primarily human health-oriented); the primary ecosystem exposure is from the wastes that generate the second category of exposure.

The toxic waste part of the problem can be further dichotomized into separate policy dimensions. The first dimension of the problem is to develop solutions to deal with past releases of toxic substances into the environment. The second dimension of the problem is to develop policies to control new sources of exposure to toxic substances.

Before discussing the issues involving toxic wastes, it is necessary to define exactly what is meant by toxic wastes and the implication of toxic waste for ecosystem and human health. The definition of a toxin is a poison produced by the metabolic processes of an organism (*Webster's,* 1989), but when *toxic* is used as an adjective to describe wastes or substances, it generally means a waste or substance that is poisonous and is of inorganic origin. Since almost any substance can be poisonous at some level of exposure, it is necessary to derive a more operational definition of toxic wastes and toxic substances. Toxic wastes are those wastes that have the following properties:

1. Exposure to small amounts of the waste generate adverse health effects in living organisms. These effects include interference with a variety of life processes, including reproduction and often include carcinogenic and tetragenic[2] effects.
2. The wastes are generally of synthetic or inorganic origin and persist in the environment for long periods of time (many generations of human life).
3. A variety of mechanisms exist for the waste to move through both the physical environment and the ecosystem.

The first property is one that makes toxic wastes a particularly important problem. Exposure to toxic waste can kill an organism or stress the organism to the extent that it succumbs to natural causes. From a human perspective, one of the most important health consequences of exposure to toxic waste is that this exposure can lead to an increased risk of cancer.

The second property is important because synthetic materials that do not exist in nature will persist for long periods. Since these substances do not exist in nature, nature has no mechanisms for breaking these products down into simpler compounds that are less toxic. Examples of synthetic toxic compounds include PCBs, DDT, benzene, dioxin, and radioactive substances such as plutonium. There is also a problem with naturally occurring inor-

[2]Creating mutations such as birth defects.

ganic compounds that are collected and concentrated by economic activity. Among the most important of this class of chemicals are the heavy metals (such as mercury, lead, cadmium, and arsenic).

The third property is important because it provides a mechanism for exposure. Toxic chemicals are transported through physical means, such as the movement of ground and surface waters. They are also transported by movement through the food web. For example, algae may absorb toxic substances in a body of water to which toxic waste has been transported. This algae is then eaten by zooplankton that absorb the toxic substance. The zooplankton is eaten by small fish, which are in turn eaten by large fish, which are eaten by piscivorus birds, such as eagles, herons, and ospreys. As the toxic substance passes through the food web, the concentration of the toxic substance in the organism increases. For example, a DDT concentration of 0.00005 ppm (parts per million) in water would lead to a DDT concentration in plants and algae of 0.04 ppm. This would lead to a concentration of 0.2 to 1.2 ppm in plant-eating fish and 1 to 2 ppm in larger fish. The DDT concentration in fish-eating birds may be as much as 76 ppm, which is 1 million times greater than the DDT concentration in the water (Raven, Berg and Johnson, 1993). Of course, this increase in concentration has extremely important implications for humans, who are at a terminus on the food web.

THE NATURE OF THE MARKET FAILURE

MARKET FAILURES IN EXPOSURE TO TOXIC SUBSTANCES IN CONSUMPTION AND PRODUCTION

Toxic substances are present in the production processes that create our economic goods and in the economic goods themselves. For example, in agriculture, pesticides are used to reduce the negative impact of pests on agricultural yields. People may receive exposure to these toxic substances in their role as laborers and in their role as consumers. The workers who apply the pesticides and who do other work in the fields on subsequent days may absorb pesticides through their respiratory systems or through contact with their skin. The consumer may be exposed through ingestion of pesticide residues, which may remain on fruits and vegetables even after washing.

A variety of economic goods contain toxic substances to which either workers, consumers, or both may be exposed. These include solvents, paints, inks, cleaners, fuels, plastics, electronics, pharmaceuticals, pesticides, fungicides, and herbicides. In addition, a variety of foods have additives in them that are suspected to be harmful.

There is considerable discussion of whether this type of exposure represents a market failure, because it can be argued that the people who receive the exposure are a part of the market transaction and *choose* their level of exposure. However, even if this type of exposure does not constitute an

RADON EXPOSURE IN HOMES

One of the most interesting exposure issues facing policy makers is exposure to radon. Radon is a naturally occurring radioactive gas that percolates up through the soil as small amounts of uranium decay in the soil. This exposure is not a problem that is limited to uranium-mining areas. Many areas of the country have small amounts of uranium in the soil.

The gas infiltrates building foundations through cracks, places where pipes penetrate the foundation, and through pores in the cement. As people began to seal their homes more tightly to save energy, it limited the ventilation of the house and increased indoor concentrations of radon. In houses in the most radon-prone areas, the health risks associated with radon are equivalent to those associated with smoking several packs of cigarettes per day. Radon is believed to be the second leading cause of lung cancer in the United States (although many of these cases are associated with uranium miners).

Since radon is a naturally occurring hazard, one might expect that there is no role for public policy. Yet there is a very important market failure. Since radon is odorless, colorless, and does not lead to immediate health consequences (it leads to an increased chance of cancer many years in the future), people do not adequately comprehend the nature of the risk to which their family will be exposed. Thus, information is imperfect and as a result, the level of remediation is lower than it would be if people had better information.

There are two major ways to address this market failure. First, one can try to educate the public as to the nature of the risk that they face. William Desvousges and Kerry Smith have studied alternative mechanisms for communicating this risk. Second, if it is felt that the imperfect information problem cannot be adequately addressed, then a series of direct controls can be enacted. These controls could include mandatory testing of houses, especially before their sale, and mandatory remediation of houses that exceed safe levels of radon. These remediation procedures would include sealing foundation cracks and installing basement ventilation systems.

externality because the people choose their level of exposure, there still may be important market failures.

The first potential market failure concerns imperfect information. In order for the choices of individuals to promote optimal social welfare, they must be aware of the risks that they are facing as a result of their choices. Most consumers probably do not adequately understand the risks associated with using solvents, consuming additives in their hot dogs, or consuming pesticide residues on their fruit. The government can rely on two types of policies to mitigate this problem. The first is to provide information about the hazard on the product label in a fashion that is understandable by the consumer. A good example of this policy is the rotating warning system on cigarette packages and cigarette advertisements. The second is to use direct controls, limiting the use of certain substances in certain applications. For example, certain pesticides can only be applied by licensed applicators and are not available for home use.

An important dimension of the imperfect information problem is that the general public does not adequately understand risk in general. Part of this misunderstanding is due to a lack of understanding of probability. Many people would argue that if a coin flips to "heads" three times in a row, then there is a greater than 50 percent chance that it will be "tails" on the next flip. Many people believe that low probability events such as a house fire or flood simply are events that only happen to other people and could not possibly happen to them. Then there is an additional problem of comprehending the very small probabilities (1 in 1 million versus 1 in 100 million) associated with exposure to toxic substances.

Policy makers also have to deal with the problem that many of the risks cannot be perceived by a person's senses. Radon gas, for example, is odorless and colorless and does not stem from a visible hazard (such as a waste dump or a weapons facility). Additionally, many of the risks are of future consequences. This factor is particularly true of exposure to carcinogens, which increases the probability of contracting cancer, but this increased risk may lag behind exposure by several decades.

Finally, it is not necessarily possible to look at voluntary exposure to risk to determine the way that the public values reductions or increases in risk. The reason for this is that people may not regard voluntarily accepted risk as being the same good as externally imposed risk. People regard exposing themselves to risk (that is, not wearing a seatbelt, smoking, bungee jumping, and so on) as a basic right. They also believe they have a basic right to be free of externally imposed risks (secondhand cigarette smoke, toxic waste, etc.). Consequently, they view an externally imposed risk as a much greater consequence than a voluntarily accepted risk, even though the two types of risks may have equivalent implications for risk of illness or death.

The primary piece of legislation that addresses this market failure is the Toxic Substances Control Act of 1976 (TSCA). This legislation gives the Environmental Protection Agency the authority to test existing chemicals, conduct pre-market screening of new chemicals, control unreasonable risks of chemicals, and collect and distribute information about chemical production and risks. The EPA has implemented several policies to accomplish this authority, including requiring manufacturers to test the chemicals that they produce, limiting or prohibiting the manufacture of certain chemicals, requiring record keeping and labeling, requiring notification of hazards and consumer recalls, and controlling disposal methods. It should be noted that TSCA does not apply to pesticides, drugs, and nuclear materials, which are covered by other statutes.

The second source of market failure applies to on-the-job exposures. As stated above, one could argue that on-the-job exposures do not constitute a market failure, as people voluntarily expose themselves to the risk. In fact, wages would adjust so that (holding other job characteristics constant) the riskier jobs would have the highest wages, as the workers would have to be compensated to accept the additional risk.[3] Of course, the problem of imperfect information immediately comes to mind, and it is apparent that the system will only give us the optimal level of risk if people have perfect information as to the exact nature of the risks. However, even if perfect information exists, there could be another important market failure related to the mobility of labor.

The definition of a perfectly competitive market requires that there is perfect mobility throughout the system. This assumption may be violated in many labor markets, where only one employer or type of employment exists in a region, and where a variety of barriers keep the workers from moving to another region for another type of employment. For example, migrant farm workers may not regard their wages as high enough to compensate them for exposure to pesticides, but they have little option except to keep working in the fields.

Again, the provision of information and the utilization of direct controls can help reduce this type of exposure toward a socially optimal level. For example, "right to know" laws require employers to inform their workers of actual or potential exposure to toxic substances. Additionally, both the Occupational Safety and Health Administration (OSHA) and the Environmental Protection Agency (EPA) regulate exposure through prohibiting the use of certain chemicals and by requiring safety procedures (protective clothing, respirators, ventilation, etc.) for use with toxic chemicals.

Should all toxic substances be banned? Although this step might seem to make sense at first, when the question is examined more closely, it can be seen that this resource allocation problem is the same as any resource allocation problem. The marginal social benefits of the toxic substance must be compared with the marginal social costs to determine the socially optimal level of production and consumption of the toxic substance. For example, smoke detectors contain a radioactive material, but the benefits of reduced risk from dying in a house fire far outweigh the increased risk from exposure to the low-level radioactivity contained in the smoke detector. Similarly, artificial sweeteners reduce obesity and its associated health risks (it also reduces health risks for diabetics). Even if artificial sweeteners result in an increase in other types of health risks (such as increased risk of cancer), their use may result in a net reduction in risk.

THE DELANEY CLAUSE

The "Delaney Clause" is a controversial piece of legislation that does not recognize the trade-offs between costs and benefits. The Delaney Clause is contained in amendments to the Food, Drug, and Cosmetics Act that were

---

[3]In Chapter 4, this voluntary exposure and adjustment of wages is discussed as a basis for measuring the value of reducing risk.

enacted in 1958. The purpose of the Delaney Clause is to ban the use of additives in processed food that are shown to cause cancer in animals. In addition to not recognizing cost–benefit trade-offs, other problems are associated with the Delaney Clause. First, it does not regulate exposure in unprocessed foods, so cancer-causing pesticides are prohibited in tomato sauce but not in fresh tomatoes. Second, many pesticides that have been in use are not covered by the legislation, so new pesticides associated with a smaller risk of cancer are not allowed to replace old pesticides with a higher risk of cancer. For example, fungicides are used to treat hops, which is an ingredient of beer. A new fungicide called fosetyl A1 was developed that had a 1 in 100 million additional risk of cancer. Since it was associated with this cancer risk, EPA prohibited its use. However, the older fungicide (called EBDC) that it would have replaced was estimated to have a risk of cancer that is 1,000 times as great, on the order of 1 in 100,000 (Raven, Berg and Johnson, p. 516).

MARKET FAILURES IN THE GENERATION OF TOXIC WASTE

Market failures related to the generation and illegal disposal of toxic waste are completely analogous to market failures associated with other types of pollution. Firms do not take social costs into account when deciding how much toxic substances to produce and use, in deciding the safety features of containment systems (such as underground storage tanks for gasoline and other toxic substances), in deciding the level of safety associated with transportation, and in determining the method of disposal of the waste.

Earlier in the chapter, toxic substances were defined in terms of strong health effects (human or ecosystem) that can be generated by relatively small amounts of the substance. They were also defined in terms of the persistence of the chemical in the environment. These properties, especially the first property, imply that the marginal damages from the first few units of emissions are high. These high initial damages will often imply that the optimal level of pollution is near zero, as illustrated in Figure 15.1. As discussed in Chapter 3, when this is the case, it is probably more efficient to simply ban the release of the substance than to try to achieve a level of E^* emissions through economic incentives. The reason is that the resource costs of administering the tax or permit system are probably greater than the excess social costs (shaded triangle) that are created when emissions are regulated to zero. Of course, for some toxic substances, the optimal level is zero, rather than near zero.

The problem is made even more complex because the damages associated with today's emissions are a function of past emissions, as past emissions linger in the environment and contribute to current concentrations of pollution. Even if initial levels of concentration were to be associated with low levels of damage, as long as the marginal damage function is positively sloped, the past emissions that persist in the environment imply that current emissions will have greater marginal damages than past emissions. Rather

FIGURE 15.1

THE OPTIMAL LEVEL OF EMISSIONS OF TOXIC WASTES

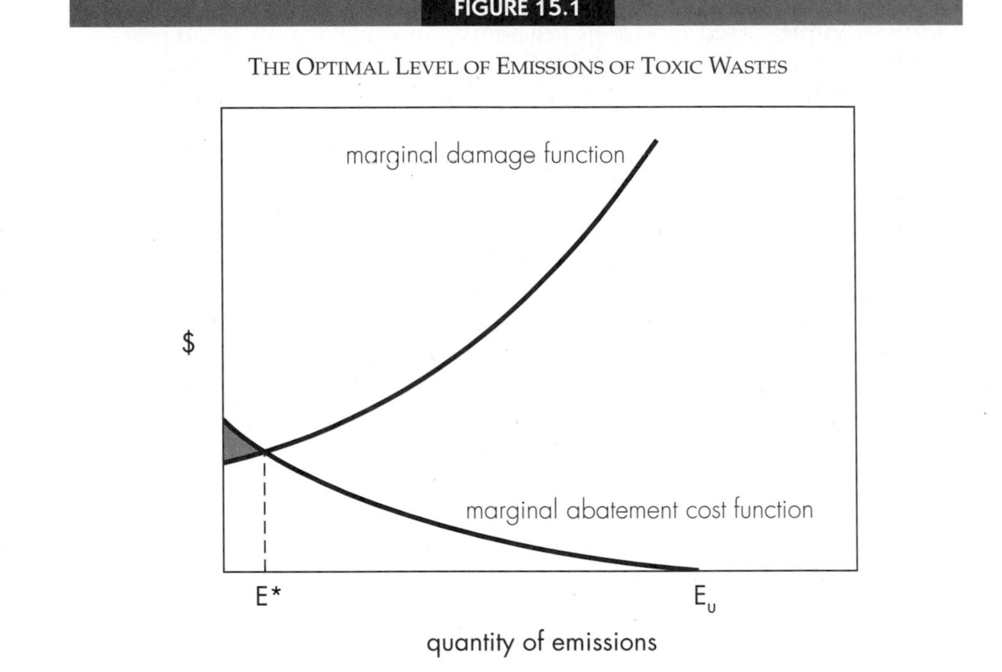

than simply comparing marginal abatement costs and marginal damages in each period, one must look at the present value of total damages and the present value of abatement costs, and try to minimize the sum of the two. This minimization would be done by choosing an optimal time path of emissions, rather than focusing on each period independently.

When the optimal level of emissions of these toxic pollutants is zero or near zero, economic incentives, such as taxes or marketable pollution permits, are not often discussed as a means for preventing the release of toxic substances into the environment. Direct controls that prohibit the release of toxic substances and that mandate required disposal techniques have been implemented for many classes of toxic substances, along with "cradle to grave" manifest systems that create a written record of the movement of the toxic substances from the time they were manufactured to the time they are disposed of in a licensed toxic waste disposal facility. Other extremely hazardous materials have been banned from some or all types of use. For example, the use of DDT, asbestos, PCB, benzene, and a host of other toxic chemicals have been banned.

Although direct controls have played and should play an important role in the control of toxic substances, there is a role for economic incentives—but not the per unit emissions taxes and marketable pollution permits that economists advocate for conventional types of pollution emissions. The development of both deposit–refund systems and liability systems can help to reduce the release of toxic substances into the environment.

DEPOSIT–REFUND SYSTEMS

Deposit–refund systems can help in reducing the release of toxic substances, particularly releases by households. Consumers use a variety of toxic substances in products such as batteries, solvents, and pesticides. The government could impose direct controls that specify exact procedures for the disposal of these products. However, since there are so many millions of households, each of which is disposing a small amount of toxic substances, the monitoring cost of administering such a system would be relatively high. One way of avoiding this monitoring problem is to make the households pay for improper disposal "up front" in the form of a deposit. When the product (or product container) is properly disposed of, then the deposit is refunded. The deposit–refund system establishes an automatic penalty for improper disposal.

A deposit–refund system would be more difficult to establish for large-scale generators of toxic waste such as chemical manufacturers. The reason for this is that with the large-scale generators there is not an economic product that has a one-to-one correspondence with the waste, such as a battery or a container of pesticides. Firms assemble a variety of chemical substances as inputs and combine them to produce an economic output and waste outputs (which may be toxic substances). However, there is a variable, rather than fixed relationship between the inputs and the economic or waste output. Because of this variable relationship, a deposit–refund system could backfire and lead to the generation of more toxic waste than without the system.

Such a deposit–refund system would likely be based on the law of mass balance.[4] Let the mass of the inputs be equal to X and the mass of the waste output be equal to W. Then the government could require a per unit deposit of δ on each unit of mass of inputs and collect a per unit refund of ρ on each unit of mass of waste outputs. The total deposit would be equal to δX and the total refund would be equal to ρw. However, nothing in this system would discourage the production of waste; it is aimed at ensuring its proper disposal. Firms might even adjust the relationship among outputs and produce more of the waste output (and less of the economic output) in order to collect a greater refund.

One possibility for avoiding this undesirable incentive is to rely on the law of mass balance to develop a deposit–refund system, but modify the system presented above. If firms pay a per unit of input deposit of δ and purchase X weight units of inputs, the firm's total deposit payment will be

[4]The law of mass balance states that the mass of the inputs is equal to the mass of the outputs.

δX. If the firm simply received a per-unit-weight refund of ρ on each unit of its total legal disposal of waste, it would have a greater incentive to produce more waste output and less economic output. However, if the per unit refund is constructed by multiplying ρ by X/W, a firm cannot increase its total refund payment by producing more waste. The per unit refund would be equal to $\rho\ (X/W)$, and the total refund would be equal to $\rho\ (X/W)W$. Although the refund is received per unit of W legally disposed of, the per unit refund is increased by reducing W. The refund is based on W, but it is mathematically equal to ρX.

For example, if the per unit deposit (δ) was equal to $2.00, and the firm utilized 2,000 pounds of inputs (X), the total deposit that it would be required to pay would be equal to $4,000. If the firm produced 1,500 pounds of economic output, 500 pounds of waste output (W), and the per unit refund (ρ) was equal to $2.00 per pound, then if the firm legally disposed of its 500 pounds of waste, it would receive a total refund of $2(2,000/500)(500)= $4,000. It is important to note that the firm cannot increase its refund by producing more waste. If it increased its waste production to 1,000 pounds (holding constant the 2,000 pounds of input), its total refund would equal $2(2,000/1,000)(1,000), which still equals $4,000.

Although such a deposit–refund system has some desirable properties in that it could reduce the amount of waste that was actually produced, it would require more information and monitoring than the types of command and control regulations that are currently in place. The high information requirements of this type of deposit–refund plan probably indicate that the role of deposit–refunds in toxic waste management may be more for household disposal of toxic waste than for large-scale waste generators.

LIABILITY SYSTEMS

Although deposit–refund systems may not be effective for large-scale generators, another type of economic incentive is being used that has important incentive effects. This incentive is a liability system that requires toxic waste generators, transporters, and disposal site operators to pay for the damages associated with any release of any waste for which they were responsible. If people who use toxic substances or generate toxic wastes are required to pay any damages that they generate, this requirement will increase their incentives to provide the appropriate amount of safety in utilizing and disposing of these substances.

This rationale was behind Comprehensive Environmental Response, Compensation and Liability Act of 1980 (CERCLA), which creates several different types of liabilities for toxic waste generators and transporters. In particular, it establishes liability for wastes that are disposed of improperly and allows the government to force the waste generators, transporters, or disposal site operators (jointly or separately) to clean up a site into which any of these parties were responsible for the release of hazardous substances.

Under this legislation, the generator of the waste is responsible for future damages from improper disposal of the waste, even if the generator of the waste contracts the disposal to a licensed transport and disposal firm. Firms that spill chemicals may also be sued for the environmental damage that they generate, with local, federal, state, and Native American government agencies acting as the trustees of these public environmental resources.

The combination of high fines and prison terms for accidental release, extremely high fines ($250,000) and up to 15 years in jail for "knowing endangerment," liability for damages, and liability for cleanup are strong incentives for firms to comply with laws governing the treatment and disposal of toxic substances. These penalties, combined with adequate record keeping and strict enforcement, may be sufficient to control the generation of new toxic sites. However, the nation is faced with an incredibly large legacy of existing toxic sites, where wastes have been accidently released or intentionally dumped in past years. Surprisingly, the federal government is one of the primary culprits in generating this toxic legacy.

THE GOVERNMENT AND IMPROPER WASTE DISPOSAL

Readers may be surprised to find that the federal government is responsible for many of the most severely contaminated toxic sites. Defense-related activity is the primary culprit in this problem, with horrendous contamination at many Department of Defense and Department of Energy weapons sites. Contamination with heavy metals, radioactive waste, carcinogenic solvents, ammunition and other ordnance, nerve gas, and other substances is a legacy of the cold war. Tables 15.1 and 15.2 show examples of the magnitude of the militarily generated toxic sites.[5]

It may seem surprising that the government is a primary source of toxic contamination problems, since one views the purpose of government as promoting social welfare. However, the people who actually make the decisions in government have objectives that include other factors in addition to social welfare. In the pursuit of these other objectives, they incur costs to their programs and costs to society. Just as with corporate managers, these program managers compare private costs and private benefits (program costs and program benefits) rather than social costs and social benefits. Since program costs and program benefits do not constitute the full spectrum of social costs and social benefits, a disparity arises between private values and social values, and an excessive level of toxic releases occurs.

[5]The military is also responsible for other toxic substance–related problems, as a result of military activity. For example, Vietnam veterans suffer a variety of medical effects from exposure to the defoliant "Agent Orange," of which the primary component is dioxin. Persian Gulf War veterans are also experiencing health problems from breathing the toxic smoke of burning oil wells in Kuwait, as well as suspected exposure to chemical weapons.

TABLE 15.1

UNITED STATES SELECTED MILITARY HAZARDOUS WASTE SITES

LOCATION	OBSERVATION
Otis Air Force Base, Massachusetts	Groundwater contaminated with trichlorethlyne (TCE), a known carcinogen, and other toxins. In adjacent towns, lung cancer and leukemia rates 80 percent above state average.
Picatinny Arsenal, New Jersey	Groundwater at the site shows TCE levels at 5,000 times Environmental Protection Agency (EPA) standards; polluted with lead, cadmium, polychlorinated biphenyls (used in radar installations and to insulate equipment), phenols, furans, chromium, toulene, and cyanide. Region's major aquifer contaminated.
Aberdeen Proving Ground, Maryland	Water pollution could threaten a national wildlife refuge and habitats critical to endangered species.
Norfolk Naval Shipyard, Willoughby, Virginia	High levels of copper, zinc, and chromium discharges. Contamination of Elizabeth River and of and Chesapeake bays.
Tinker Air Force Base, Oklahoma	Concentrations of tetrachloroethylene and methyl chloride in drinking water at levels far exceeding EPA limits. Also, TCE concentration highest ever recorded in U.S. surface waters.
Rocky Mountain Arsenal, Colorado	Approximately 125 chemicals dumped over 30 years of nerve gas and pesticide production. The largest of all seriously contaminated sites, called "the most contaminated square mile on earth" by the Army Corps of Engineers.
Hill Air Force Base, Utah	Heavy on-base groundwater contamination, including volatile organic compounds up to 27,000 parts per billion (ppb), TCE up to 1.7 million ppb, chromium up to 1,900 ppb, lead up to 3,000 ppb.
McClellan Air Force Base, California	Unacceptable levels of TCE, arsenic, barium, cadmium, chromium, and lead found in municipal well system serving 23,000 people.
McChord Air Force Base, Washington	Benzene, a carcinogen, found on-base in concentrations as high as 503 ppb, nearly 1,000 times the state's limit of 0.6 ppb.

SOURCE: From *State of the World 1991: A Worldwatch Institute Report on Progress Toward a Sustainable Society,* by Lester R. Brown et al, eds.

For example, many of our toxic contamination problems are a result of nuclear weapons manufacture at Department of Energy and Department of Defense facilities. The managers in charge of developing these weapon systems looked toward optimizing values associated with the weapons systems. For example, they might try to produce the maximum defense from a given-sized budget (of course, they will also be working to increase the size of the budget), or they might try to minimize the cost of each system so they can

TABLE 15.2

UNITED STATES RADIOACTIVE AND TOXIC CONTAMINATION AT MAJOR NUCLEAR WEAPONS PRODUCTION FACILITIES

FACILITY (TASK)	OBSERVATION
Feed Materials Production Center, Fernald, Ohio (converts uranium into metal ingots)	Since plant's opening, at least 250 tons of uranium oxide (and perhaps six times as much as that) released into the air. Off-site surface and groundwater contaminated with uranium, cesium, and thorium. High levels of radon gas emitted.
Hanford Reservation, Washington (recycles uranium and extracts plutonium)	Since 1944, 760 billion liters of contaminated water (enough to create a 12-meter-deep lake the size of Manhattan) have entered the groundwater and Columbia River; 4.5 million liters of high level radioactive waste leaked from underground tanks. Officials knowingly and sometimes deliberately exposed the public to large amounts of airborne radiation in 1943–56.
Savannah River, South Carolina (produces plutonium)	Radioactive substances and chemicals found in the Tuscaloosa aquifer at levels 400 times greater than government considers safe. Released millions of curies of tritium gas into atmosphere since 1954.
Rocky Flats, Colorado (assembles plutonium triggers)	Since 1952, 200 fires have contaminated the Denver region with unknown amount of plutonium. Strontium, cesium, and cancer-causing chemicals leaked into underground water.
Oak Ridge Reservation, Tennessee (produces lithium-deuteride and highly enriched uranium)	Since 1943, thousands of pounds of uranium emitted into atmosphere. Radioactive and hazardous wastes have severely polluted local streams flowing into the Clinch River. Watts Bar Reservoir, a recreational lake, is contaminated with at least 175,000 tons of mercury and cesium.

SECONDARY SOURCE: Michael Renner, "Assessing the Military's War on the Environment," in *State of the World*, 1991, ed. Lester Brown, taken from Table 8.7 on p. 148. PRIMARY SOURCES: "Status of Major Nuclear Weapons Production Facilities: 1990," *PSR Monitor*, September 1990; Robert Alvares and Arjum Makhijani, "Hidden Legacy of the Arms Race: Radioactive Waste," *Technology Review*, August/September 1988.

produce more systems. Since every dollar of budget that is devoted to environmental protection reduces their ability to meet other goals, they devote insufficient attention to environmental protection. This behavior is no different than how a profit maximizing firm would behave, except, until very recently, the Department of Defense and Department of Energy facilities were not subject to the oversight of the Environmental Protection Agency or to corresponding state agencies. The urgency (real or imagined) of the cold war and the perceived objective of preventing world domination by communism exacerbated the narrowness of the objectives of these activities.

It is interesting to note that this disparity between the private cost and social cost associated with government activities is the fundamental reason that Eastern Europe and the former Soviet Union are so horribly polluted. For example, a manager of a steel factory (which is a government enterprise) in a socialist country was rewarded based on meeting production quotas and containing costs. Environmental protection was not the charge of the steel factory manager, so even though the steel plant was a government enterprise, it led to socially inefficient levels of pollution.

HOW MUCH SHOULD WE PAY FOR CLEANUP?

Although the prevention of further toxic contamination is an extremely important policy question, it is not the only dimension of toxic substance policy that must be considered. There remains the immensely important policy question of how much to spend to clean up the contaminated sites that currently exist. This question is important because there are over 50,000 contaminated sites in the United States, and when potentially leaking underground storage tanks are factored into the question, the number of potential sites may be greater than 2.5 million.

There are two important questions that must be addressed in developing a cleanup or remediation policy. The first question is, which sites should be cleaned and in what order should they be cleaned? The second question is, to what extent should each site be cleaned? A third question might be, who should pay for the cleanup? However, this is more a question of equity than efficiency, although who pays for cleanups could have an effect on the level of safety that firms choose.

Russell, Colglazier, and Tonn (1992) discuss five major programs that deal with past contamination. These programs include the Superfund program of the Comprehensive Environmental Response, Compensation and Liability Act of 1980 (CERCLA), the Resource Conservation and Recovery Act (RCRA), a program established under RCRA to deal with underground storage tanks, the federal facility program, and state and private programs.

The Superfund program charges the Environmental Protection Agency with identifying, evaluating, and remediating hazardous waste sites in the United States. These sites include landfills, manufacturing facilities, mining areas, and illegal hazardous waste dumps. According to Russell, Colglazier, and Tonn, more than 1,200 sites have been placed on the national priority list of hazardous waste sites that are eligible for CERCLA funding, but the total number of sites that belong on the list may be from 2,000 to 10,000 sites.

Subtitle C of RCRA regulates facilities that treat, store, and dispose of hazardous wastes. These facilities are required to obtain permits from EPA, and corrective action to deal with past or potential releases may be required to obtain permits. According to Russell, Colglazier, and Tonn (1992), there are approximately 2,500 non-federal facilities of this sort in the United States.

TOXIC WASTE SITES AND RACE

For more than a decade, the Commission for Racial Justice of the United Church of Christ has been investigating the issue of the location of toxic waste facilities with respect to the minority characteristics of the community. Based on anecdotal evidence, they formulated a hypothesis that minority communities were disproportionately exposed to toxic and hazardous waste. In 1987, the Commission for Racial Justice released a comprehensive cross-sectional study that provided a statistical examination of toxic sites across the country.[a]

Their study found that race was the most significant of the variables that they tested to explain the location of commercial hazardous waste facilities. Socioeconomic status also mattered, but race proved to be more significant, even after controlling for urbanization and regional differences.

Their study also examined uncontrolled toxic waste sites and found that blacks were heavily overrepresented in the cities with the most uncontrolled toxic wastes sites (Memphis, TN: 173 sites; St Louis, MO: 160 sites; Houston, TX: 152 sites; Cleveland, OH: 106 sites; Chicago, IL: 103 sites; Atlanta, GA: 94 sites).

This study has served as an impetus for increased concern over environmental equity and environmental justice. Current policies call for an examination of the environmental equity and justice provisions of EPA actions.

[a]Commission for Racial Justice, United Church of Christ, *Toxic Wastes and Race in the United States*, 1987.

An additional program established under RCRA deals with underground storage tanks. There are an incredible number of underground storage tanks in the United States (approximately 1.7 to 2.7 million, according to Russell, Colglazier, and Tonn), such as tanks at gasoline stations, vehicle fleet facilities, and chemical manufacturing companies. Many of these tanks have been in place for 30 years or more and have developed or will develop leaks that allow the release of toxic substances.

The fourth program described by Russell, Colglazier, and Tonn is the federal facility program. As mentioned previously and listed in Tables 15.1 and 15.2, many federal facilities, such as Department of Defense and Department of Energy sites, have released toxic waste into the environment. In addition, a large amount of waste is currently stored at these facilities.

The final program examined by Russell, Colglazier, and Tonn is a catchall for state and private programs that do not fall under the other four categories. It is estimated that there are 40,000 sites in the United States in this category.

As the numbers of sites and complexity of cleanup would indicate, the cleanup of contaminated sites in the United States will be a tremendously expensive task. Russell, Colglazier, and Tonn estimate these costs, which are listed in Table 15.3. The current policy emphasizes destruction of the waste and a complete cleanup of the contaminated sites. A less stringent set of policies would look toward the containment and isolation of wastes at contaminated sites, to prevent their spread into groundwater or into ecosystems.

TABLE 15.3

ESTIMATED COSTS OF U.S. HAZARDOUS WASTE REMEDIATION FROM 1990 TO 2020

		COST (IN BILLIONS OF 1990 DOLLARS)		
REMEDIATION PROGRAM	RANGE POINTS	LESS STRINGENT SCENARIO	CURRENT POLICY	MORE STRINGENT SCENARIO
Superfund	Plausible upper bound	180	302	704
	Best guess	90	151	352
	Plausible lower bound	63	106	246
Resource Conservation and Recovery Act	Plausible upper bound	316	377	423
	Best guess	199	234	258
	Plausible lower bound	150	170	188
Underground storage tanks	Plausible upper bound	*	*	*
	Best guess	67	67	67
	Plausible lower bound	32	32	32
Federal facilities	Plausible upper bound	*	*	*
	Best guess	110	270	430
	Plausible lower bound	*	140	*
State and private programs	Plausible upper bound	*	*	*
	Best guess	18	30	70
	Plausible lower bound	*	*	*
Total	Plausible upper bound[a]	691	1,046	1,694
	Best guess	484	752	1,177
	Plausible lower bound[a]	373	478	966

SOURCE: *Environment* 34, 1992: 12–39. Reprinted with permission of the Helen Dwight Reid Educational Foundation. Published by Heldref Publications, 1319 Eighteenth Street, NW, Washington, DC 20036-1802.

*Substantial data limitations and program uncertainties preclude the determination of plausible cost bounds.
[a]In cases where there are no plausible upper and lower bounds for the program's cost estimate, the best guess was added to the total.

The less stringent policy would not have significantly different impacts on human or ecosystem health. However, it pushes cleanup costs to future generations if they decide that more complete remediation is desirable.

As can be seen in Table 15.3, these cleanup costs are not trivial. For example, Russell, Colglazier, and Tonn point out that the $752 billion best guess estimate for current policy is approximately equal to one year of all nondefense federal expenditures, or to a decade of total public and private spending on all other environmental quality objectives. Table 15.4 lists key factors which could affect those cost estimates.

The big policy question is whether we should spend all this money to clean up every toxic site. This larger policy question can be broken down into smaller policy questions, such as

TABLE 15.4

Key Factors Affecting Hazardous Waste Remediation Cost Estimates

Program	Factors which may lead to higher costs	Factors which may lead to lower costs
All	States may hinder transport of hazardous wastes. Contamination thresholds may be tightened.	Improvements in remediation technology may occur. Definitions of hazardous waste may narrow.
Superfund	Pumping and treatment may take longer than expected.[a] National priorities list sites could reach or exceed 6,000.	Incineration costs may be overestimated.[b]
Resource Conservation and Recovery Act	The number of solid waste management units may have been underreported.	Economies of scale were not considered.
Underground storage tanks	The Clean Air Act may restrict emissions of volatile organic compounds.	The percentage of registered tanks may be higher than is estimated.
Federal facilities	Contamination from mixed and radioactive wastes may be more widespread than was anticipated.	The public may allow more institutional controls than is expected.
State and private programs	The number and average costs of sites may be underestimated.	Sites may eventually fall under other programs.

SOURCE: *Environment* 34, 1992: 12–39. Reprinted with permission of the Helen Dwight Reid Educational Foundation. Published by Heldref Publications, 1319 Eighteenth Street, NW, Washington, DC 20036-1802. Copyright © 1992.

[a]C. B. Doty and C. C. Travis, *The Effectiveness of Groundwater Pumping as a Restoration Technology,* ORNL/TM-11849, Oak Ridge, TN: Oak Ridge National Laboratory, 1991.

[b]C. B. Doty, A. G. Crotwell, and C. C. Travis, *Cost Growth for Treatment Technologies at NPL Sites,* ORNL/TM-11849, Oak Ridge, TN: Oak Ridge National Laboratory, 1991.

1. What sites should be cleaned up first?
2. Should they be restored to pristine levels, or should we merely contain the wastes in place to limit their potential for damage?

The second question is particularly intriguing because many of the federal facilities are extremely large and geographically isolated from population centers. It might make sense to contain the wastes on the site and isolate the site from human and animal exposure.

Of course, the answers to these questions and other questions, such as whether we should spend our limited resources on completely cleaning all the sites, on mitigating other environmental problems (such as global warming and preserving habitat), on other social objectives (such as education, AIDS research, and homelessness), or on reducing the budget deficit, have to be answered. Unfortunately, it is extremely difficult to answer these questions at this point in time because the benefits of different levels of cleanup have not been estimated.

SUMMARY

The toxic waste problem is very different from many of the other environmental problems that we face. Toxic substances may persist for long periods, and can pervade the ecosystem as they are transported by physical means and through the food web. Another important difference between toxic waste and other types of pollution is that we have to deal with past actions as well as controlling present emissions.

Another important problem associated with toxic substances is that exposure to risk comes not just from release of wastes, but from the presence of toxic substances in economic processes and in goods. Both workers and consumers are exposed to toxic substances. Although this factor is not necessarily an externality, market failures may exist as imperfect information and the immobility of labor may imply that this type of exposure to toxic substances is greater than socially optimal.

Intentional or accidental release of toxic substances into the environment is a different type of problem, as it is a traditional environmental externality. Familiar command and control methods, as well as economic incentives, such as deposit–refund and liability systems, can be used to deal with this problem.

While it is conceptually possible to develop effective policies for dealing with present and future releases, the problem of dealing with our waste legacy may be much more complex. To what extent should we clean up existing toxic waste? Should we clean all sites? Should we restore sites to complete cleanliness, or should we merely try to isolate and contain the waste to keep it from spreading to other areas? These questions are made even more pressing by the huge price tag associated with remediating our legacy of toxic waste.

REVIEW QUESTIONS

1. What market failures are associated with toxic waste?
2. Why do past levels of toxic emissions matter for current policy?
3. What policy options exist for dealing with previously contaminated sites?
4. Why is the optimal level of toxic pollutants likely to be at or near zero?
5. Under what circumstances are deposit–refund systems most likely to work with toxic substances?

QUESTIONS FOR FURTHER STUDY

1. Why might people value the reduction of an externally opposed risk more than the reduction of a voluntarily accepted risk?
2. What is wrong with the "Delaney Clause"? How would you change the legislation to improve it?
3. How would you define the optimal time path of emissions of a toxic substance?
4. Do we need public regulation of workplace exposure to toxic substances? Why or why not?
5. How should we control public exposure to toxic substances through consumer goods, including food?

SUGGESTED PAPER TOPICS

1. The problem of where to locate facilities to store or process hazardous waste is an extremely difficult problem. We need the facilities, but everyone wants to place the facilities somewhere else. This dilemma leads to the "Not in my backyard" or NIMBY syndrome. Investigate the NIMBY syndrome, focusing on methods of turning potential Pareto improvements into actual Pareto improvements. Start with the AAAS book in the references and search bibliographic databases on waste and policy. See if you can devise your own creative solutions.
2. The oceans have been used as a dumping ground for hazardous wastes. The development of policy is even more complex due to the lack of sovereignty outside the "200-mile economic exclusion zone." Look at the global market failures associated with ocean dumping. Start with the Congressional report in the references and search bibliographic databases on ocean and dumping, law and oceans, and waste and international policy.
3. Investigate the question of whether toxic waste policy is discriminatory. Start with the works by Bullard, Bunyan and Mohai, and the Commission for Racial Justice. Search bibliographic databases and search on race and waste, environmental equity, equity and waste, and environmental justice. Use your knowledge of economics to develop potential policies that address both efficiency and equity.
4. Write a survey paper on methods to value the reduction in risk, paying particular attention to risks associated with toxic substances. Start with the book by Freeman for a general treatment and for further references.

WORKS CITED AND SELECTED READINGS

1. Alvares, Robert, and Arjum Makhijani. "Hidden Legacy of the Arms Race: Radioactive Waste." *Technology Review* (August/September 1988): 42–51.
2. American Association for the Advancement of Science (AAAS). *Hazardous Waste Management: In Whose Backyard?* Boulder: Westview Press, 1984.
3. Anonymous. "Status of Major Nuclear Weapons Production Facilities: 1990." *PSR Monitor* (September 1990).
4. Bullard, Robert D. *Dumping in Dixie: Race, Class, and Environmental Quality.* Boulder: Westview Press, 1990.
5. Bunyan, Bryant, and Paul Mohai, eds. *Race and the Incidence of Environmental Racism.* Boulder: Westview Press, 1992.
6. Carson, Rachel. *Silent Spring.* New York: Houghton Mifflin Company, 1962, p. 7.
7. Carter, Lester. *Nuclear Imperatives and Public Trust: Dealing with Radioactive Waste.* Washington, D.C.: Resources for the Future, 1987.
8. Commission for Racial Justice, United Church of Christ. *Toxic Wastes and Race in the United States,* 1987.
9. Davis, Charles E., and James P. Lester. *Dimensions of Hazardous Waste Politics and Policy.* New York: Greenwood Press, 1988.
10. Doty, C. B., A. G. Crotwell, and C. C. Travis. *Cost Growth for Treatment Technologies at NPL Sites.* ORNL/TM-11849. Oak Ridge, TN: Oak Ridge National Laboratory, 1991.
11. Doty, C. B., and C. C. Travis. *The Effectiveness of Groundwater Pumping as a Restoration Technology.* ORNL/TM-11849. Oak Ridge, TN: Oak Ridge National Laboratory, 1991.
12. Freeman, A. M. III. *The Measurement of Environmental and Resource Values.* Washington, D.C.: Resources for the Future, 1993.
13. Herzik, Eric B., and Alvin H. Mushkatel. *Problems and Prospects for Nuclear Waste Disposal Policy.* Westport, CT: Greenwood Press, 1993.
14. Jacob, Gerald. *Site Unseen: The Politics of Siting a Nuclear Waste Repository.* Pittsburgh: University of Pittsburgh Press, 1990.
15. Raven, Peter H., Linda R. Berg, and George B. Johnson. *Environment.* New York: Saunders College Publishing, 1993.

16. Renner, Michael. "Assessing the Military's War on the Environment." In *State of the World,* edited by Lester Brown, New York: Norton, 1991.
17. Russell, Milton, E. W. Colglazier, and Bruce E. Tonn. "The U.S. Hazardous Waste Legacy," *Environment* 34 (1992): 12–39.
18. U.S. Congress. House Committee on Energy and Commerce. *Radon Awareness and Disclosure Act of 1992: Report Together with Dissenting Views* (to accompany H.R. 3258). Washington, D.C.: Government Printing Office, 1992.
19. ———. House Committee on Merchant Marine and Fisheries. Subcommittee on Oceanography, Gulf of Mexico, and the Outer Continental Shelf. *Ocean Disposal of Contaminated Dredge Material: Hearing before the Subcommittee on Oceanography, Gulf of Mexico, and the Outer Continental Shelf of the Committee on Merchant Marine and Fisheries.* Washington, D.C.: Government Printing Office, 1993.
20. ———. House Committee on Science, Space, and Technology. Subcommittee on Energy. *Overview of the DOE's Environmental Restoration and Waste Management Program: Hearing before the Subcommittee on Energy of the Committee on Science, Space, and Technology.* Washington, D.C.: U.S. Government Printing Office, 1993.
21. U.S. Department of Energy. *Reassessment of the Civilian Radioactive Waste Management Program: Report to the Congress.* Washington, D.C.: U.S. Department of Energy, Office of Civilian Radioactive Waste Management, 1989.
22. *Webster's Ninth New Collegiate Dictionary,* 1989.

Agriculture and the Environment

Of all the occupations from which gain is secured, there is none better than agriculture, nothing more productive, nothing sweeter, nothing worthier of a free man.

Cicero, De Officiis

INTRODUCTION

Although our society tends to view agriculture as both a noble and a "green" lifestyle, the agricultural industry has profound environmental impacts. Agricultural practices often lead to environmental degradation that has a myriad of effects. Initially, society's focus was on the effect of this degradation on agriculture itself. Soil erosion has been of historical concern in the United States, and the loss of soil productivity has become a life-threatening condition in many developing countries. The importance of this loss of agricultural productivity, and its implications for feeding the world, cannot be overstated.

More recently, attention has also been devoted to other environmental consequences of agriculture. For example, agriculture is a major contributor to diminished water quality, where runoff from agriculture contains soil and agrochemicals such as fertilizers, pesticides, and herbicides. Irrigation also increases the salinity of surface waters. In addition to the effect on water quality, agriculture has other important environmental effects. The conversion of forest, prairie, and wetlands to farmland reduces habitat and threatens biodiversity. In addition, livestock production and rice paddy culture generate significant amounts of methane, an important greenhouse gas.

Although agriculture is an important source of environmental degradation, environmental degradation also has negative impacts on agriculture. In addition to the environmental degradation generated by agriculture that affects soil productivity, there are numerous other types of environmental change that hurt agricultural productivity. For example, tropospheric ozone reduces the growth of crops and trees. The increased ultraviolet radiation associated with the depletion of stratospheric ozone inhibits a plant's ability to photosynthesize. Potential global warming is likely to have both negative and positive effects on agriculture. The increase in carbon dioxide that is hypothesized to lead to global warming will also have mostly positive impacts on agriculture.

This chapter will proceed with a short discussion of the impacts of environmental degradation on agricultural productivity. Then, the impacts of agricultural activity on environmental quality will be examined, along with the nature of the market failures that lead to these impacts. Finally, current policies to deal with the environmental externalities of agriculture will be discussed, and new policies will be suggested.

THE EFFECT OF ENVIRONMENTAL QUALITY ON AGRICULTURE

Although the environmentally degrading activity that has the strongest effect on agriculture is probably agriculture itself, this effect will be examined in the next section. This section focuses on external activities that degrade the environment and have negative effects on agriculture.

An activity generating an effect that adversely affects agriculture will generally reduce the yield per acre of a crop or group of crops. This reduction will shift the supply curve upward, generating a direct loss of benefits equal to the shaded area in Figure 16.1. For example, if tropospheric ozone affects soybean production in Iowa, this model can be used to measure the direct loss in welfare.

However, in the case of agriculture, it is especially important to move past a partial equilibrium analysis (which holds conditions constant in related markets) and look at the indirect effects on other markets. The reason this viewpoint is important is that agricultural markets are very connected, as crops are good substitutes for each other, both in production and consumption. For example, the farmer in Iowa whose soybeans are affected by ozone could choose to plant another crop, such as corn, which might not be as badly affected. Additionally, soybean farmers in another area, such as Tennessee, might plant more soybeans as a result of the higher price. Similarly, consumers can choose options other than soybeans. Other grains can be used in animal feed and food products.

When economists measure the effects of impacts on agriculture, they model both the farmers' and the consumers' decisions, and they look at the demand and supply for all substitute crops throughout the United States. Then as a change in the supply function of one crop affects its price, the effect of this price change on the consumers' and producers' surplus in all other markets can be examined. For example, if the supply curve for soybeans shifts upward, it will have many effects on other markets. The higher market price for soybeans will increase the demand for other grains, increasing the

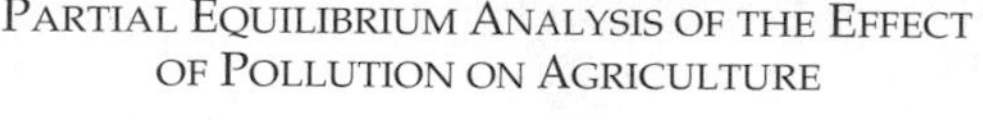

FIGURE 16.1

PARTIAL EQUILIBRIUM ANALYSIS OF THE EFFECT OF POLLUTION ON AGRICULTURE

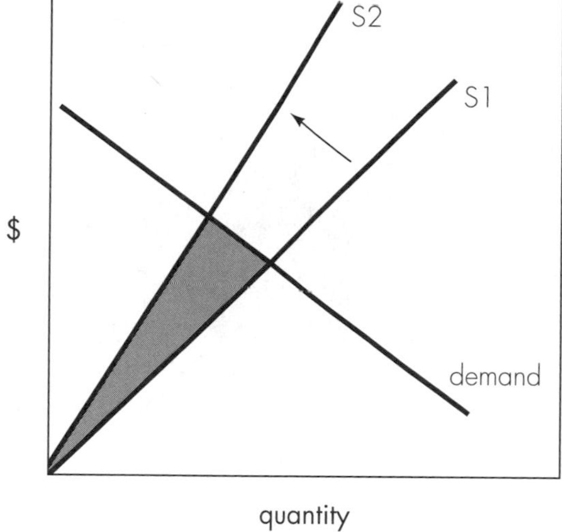

equilibrium price and quantity of these grains. For example, in Figure 16.2, the left-hand panel shows the reduction of consumers' and producers' surplus (dark shaded area) in the soybeans that are affected by pollution. The indirect effects through related markets are represented by the right-hand panel, which shows the increase in consumers' and producers' surplus in the market for a substitute crop. In addition, farmers in affected areas may switch to other crops, which will shift supply curves as well as demand curves. Also, the problem becomes even more complicated because price supports exist for many crops.

Estimates of this type of change in producers' and consumers' surplus are presented in Table 16.1, which contains estimates of the effects of tropospheric ozone on agricultural activity. Both losses in consumers' and producers' surplus are included in these estimates of net benefits. It is interesting to note that increases in pollution always hurt consumers but do not always hurt agricultural producers. It should be noted also that if price supports did not exist, the initial increases in pollution would hurt both producers and consumers.

Global warming and the increase in carbon dioxide emissions that will generate the global warming problem also have both positive and negative effects on agriculture. Increased temperature and the increased rainfall that

FIGURE 16.2

DIRECT AND INDIRECT EFFECTS OF A POLLUTANT THAT REDUCES SOYBEAN YIELDS

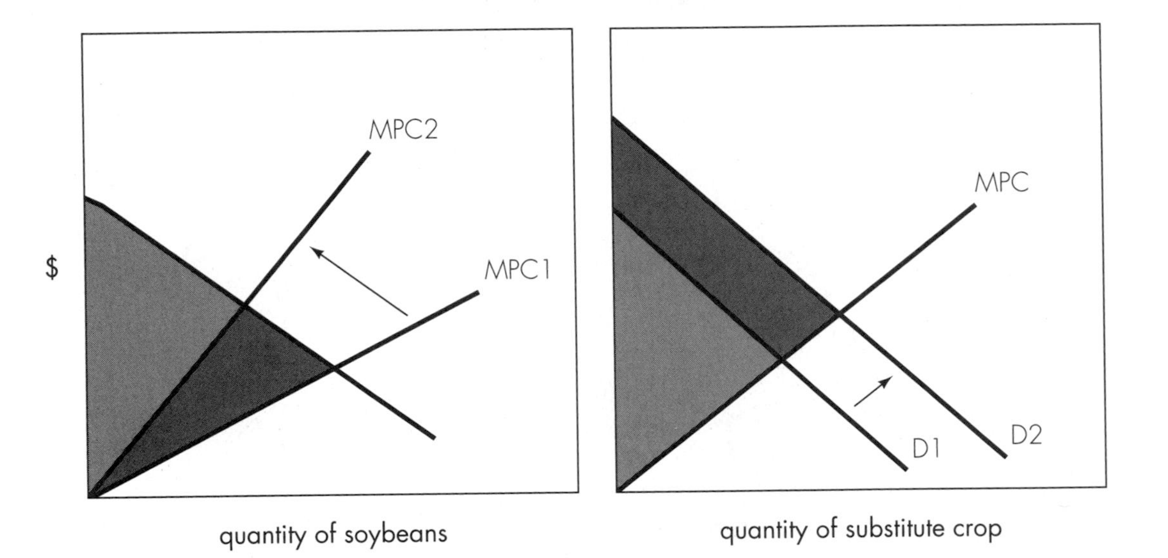

accompany global warming will generally have beneficial effects on agriculture, although specific effects may be negative, and the specific negative effects could outweigh the general positive effects. For example, although rainfall will increase in general, its distribution will change. Some areas will

TABLE 16.1

EFFECT ON AGRICULTURAL SECTOR OF CHANGES IN TROPOSPHERIC OZONE

% CHANGES IN OZONE	CHANGE IN SURPLUS (BILLIONS OF 1989 DOLLARS)		
	CONSUMERS	PRODUCERS	TOTAL
−10	0.785	−0.046	0.739
−25	1.637	0.095	1.732
+10	−1.044	0.215	−0.829
+25	−2.659	0.453	−2.206

SOURCE: NAPAP, *1990 Integrated Assessment Report*, p. 401–402.

receive less rainfall, and other areas will receive more rainfall. If the areas that lose precipitation are the fertile soil areas and the areas that gain rainfall are the infertile areas, the net effect on agriculture could be extremely negative. At this point in time, there is considerable scientific uncertainty over how the distribution of rainfall will change. Increased temperature can also generate costs, as areas become too hot for existing cropping patterns and new crops must be introduced. In addition, higher temperature may increase pest populations. Similarly, increased atmospheric concentrations of carbon dioxide will stimulate both crop and weed growth.

THE EFFECT OF AGRICULTURE ON THE ENVIRONMENT

Agriculture has several important effects on the environment. Soil erosion reduces the productivity of soil and leads to water quality problems. Runoff from agricultural fields leads to transportation of fertilizer, pesticides, and herbicides into ground and surface water. Irrigation reduces river flow, depletes aquifers, and increases the salinity of surface water. In addition, conversion of habitat to agricultural areas has led to reductions in biodiversity and losses of some of the world's ecological treasures.

SOIL EROSION

The environmental degradation associated with agriculture was blown into public consciousness with the dirt laden winds of the Dust Bowl, which drove many farmers from the land during the 1930s. Intensive plowing and cropping of the land led to the soil being exposed to water and (especially) wind, which led to massive loss of topsoil.[1] Much of this land is still barren.

Soil erosion is a problem for two major reasons. First, the top layers of the soil contain organic matter, which is the source of nutrients for plants. Lower layers of soil tend to be mineral soil or clay and have no organic content. As topsoil is lost, nutrients are lost, and the soil becomes less productive. Second, the roots of most crops extend about 1 meter into the soil. If the topsoil layer does not extend very far into the soil, the roots quickly penetrate into the infertile underlying soil, and crop productivity declines at a rapid rate (Heimlich, 1991). In addition, if the topsoil is completely removed, or if massive gullying occurs, it is impossible to profitably engage in agriculture, regardless of the input of synthetic fertilizers.

---

[1]There are two major types of water-related soil erosion. Sheet erosion is when the water moves across the land in a uniform fashion, in a thin sheet. This water movement removes a very thin layer of soil from the land. Rill erosion occurs when the path of the water is more concentrated and cuts gullies through the land. These gullies then capture more of the rainwater and the gullying is intensified.

The early public concern with soil erosion was with its internal (on-site) effect, not its external (off-site) effect. That is, public policy was concerned that if a farmer used inappropriate techniques, the farmer would lose the topsoil on his or her farm, and the farm would not be as productive. Public policy was not concerned with the external effects, such as the effect of soil erosion on water quality.

This focus on internal effects may seem strange to the reader of this book, who has devoted hundreds of pages to reading about the importance of public policy concerning external effects generated by market failures and lack of property rights. The question immediately arises, if the farmer has property rights to his or her farm, should not the market automatically generate the appropriate rate of soil erosion? The answer to this question is that even though there may not be an externality, there may be other market failures.

The first potential market failure is that of imperfect information. The farmer may not know the importance of conserving soil, or may not know of, or understand how to implement, agricultural techniques that are more conserving of the soil.

The second problem has to do with the constrained optimization problem that the farmer faces. The farmer may not be free to pursue all agricultural strategies, especially if a strategy involves current sacrifice for future gain. This lack of freedom is because the farmer has limited ability to support his or her family in the short run and may make decisions based on current cash flow rather than long-term profit maximization, particularly if the farmer does not have access to capital markets (because the farmer is already overextended, for example). The need to meet short-term needs will lead the farmer to have a much higher rate of time preference than society at large. The higher the rate of time preference, the less weight future costs and benefits have, and the less important the preservation of the soil for the future becomes.

These two factors exist both in developed countries, such as the United States, and in developing countries, such as in Latin America, Africa, and south Asia. In fact, the problem of soil erosion and loss of soil productivity is probably worse in developing countries for several reasons. Both the imperfect information and the short-run constraints may be more severe in developing countries than in developed countries.

There are two basic reasons why the imperfect information problem may be more severe in developing countries. First, it may be harder to get information into the hands of farmers. A greater percentage of the farming population is illiterate in developing countries; farmers are isolated by a lack of good roads and transportation; and mass communication media, such as television, radio, and newspapers, are less available. In addition, in countries such as the United States, government agencies (for example, Cooperative Extension, Land Grant Universities, and the National Resources Conservation Service [formerly the Soil Conservation Service]) provide information on less erosive agricultural techniques. Corresponding agencies do not exist in most developing countries, particularly the very low income developing countries. Second, the imperfect information problem is exacerbated by population pressures that lead to the cultivation of types of land that have not previously been intensively cultivated, such as steep hillsides and rain forest. The cultivation techniques that are acceptable in more conventional settings may be extremely degrading in these new settings, but the uneducated farmers are usually unaware of less erosive alternatives. For example, subsistence farmers in the Amazon clear-cut forests to plant conventional crops, rather than engaging in less destructive and more sustainable agroforestry (see Chapter 12).

The short-run constraints also may be more severe in developing countries as well. The banking system is more developed in countries such as the United States, and it is easier for farmers to obtain credit and store the surplus that is generated in good years. In addition, average consumption levels are much greater in the United States, and it is easier for farming families to meet basic needs, so they have a less pressing need to sacrifice the future in order to meet current consumption requirements. The problem is also mitigated by the existence of a "social safety net" in the United States (food stamps, Medicaid, welfare, unemployment insurance, etc.) that helps people meet minimum current consumption needs.

In addition to the internal (on the farm or on-site) effects of soil erosion, there are external (off-site) effects as well. Soil erosion has very detrimental effects on water quality. The soil is transported to rivers and lakes, where it becomes suspended in the water column. These suspended soil particles have several detrimental effects on aquatic ecosystems. The suspended sediment particles block light from reaching aquatic plants, they generate bottom sediments that change the nature of the stream or lake bottom and suffocate bottom life, and they interfere with the respiratory function of fish and other aquatic animals. In addition, the sediments lower the quality of drinking water and necessitate additional treatment before the water can be used in industrial processes. Another important problem is that the suspended sediments precipitate to the bottom of the stream or lake, where they have a series of undesirable effects, including clogging harbors and filling reservoirs with silt (reducing their ability to generate hydroelectric power and provide drinking water).

A major environmental problem associated with soil erosion is that nutrients, pesticides, and herbicides are carried into surface water bodies along with the soil particles. Pesticides and herbicides are toxic substances and interfere with aquatic life processes in a variety of ways (see Chapters 14 and 15). Nutrients from fertilizer and animal wastes have a number of important ecological effects, but the primary concern is their effect on dissolved oxygen levels in the water. The nutrients lead to growth of algae, which dies, decays, and removes dissolved oxygen from the water. The increased algae also blocks sunlight from reaching submerged plants.

AGRICULTURE, HABITAT, AND BIODIVERSITY

The establishment of agricultural land is generated through conversion of land from other uses. An important ecological problem arises when the land that is converted to agriculture has important ecological values. Although both agricultural areas and natural habitats (such as forests, wetlands, prairies, and other ecosystems) are green, they are not equivalent. They contribute to wildlife habitat, biodiversity, recreation, watershed protection, and existence values in very different ways. The farmer who is contemplating converting the land only compares the private costs and benefits of converting the land and does not consider the social costs. This problem is discussed in detail in Chapter 13 and is responsible for the loss of important habitat in developed countries and particularly in developing countries, where deforestation from agricultural activity (see Chapter 12) is an immense ecological problem.

AGRICULTURE AND GREENHOUSE GASES

Another important ecological effect of agriculture is its contribution to increases in the atmospheric concentrations of greenhouse gases. Agricultural activities lead to increases in carbon dioxide, methane, and nitrous oxide emissions.

Methane comes from a variety of agricultural sources, including manure piles, wet rice cultivation, and the digestive processes of ruminants, such as sheep, goats, and cattle. Carbon dioxide originates in the combustion of fossil fuels used in farm machinery and from the cutting and burning of forests to clear them for agricultural uses (particularly in tropical countries). Nitrous oxides are released by chemical and organic fertilizers.

AGRICULTURE AND PUBLIC POLICY

Although the types of ecological problems (water pollution, generation of greenhouse gases, habitat loss) generated by agriculture are analogous to those generated by other types of activities, there are some very important differences, particularly with respect to water pollution. The most important difference is that agricultural pollution of surface and groundwater originates from nonpoint sources rather than originating from point sources.

Point sources are sources of pollution where the pollution is released into the environment at a distinct location, such as the end of an effluent pipe. For example, paper and pulp manufacturing plants and municipal sewage treatment plants are point sources of water pollution, because the effluent flows into the rivers from a discharge pipe.

In contrast, agricultural pollution tends to be a nonpoint source of pollution. Rather than the pollution entering at a specific location, the pollution

Agriculture and the Florida Everglades

One of the most pronounced environmental effects of agriculture on U.S. ecosystems is the effect of agriculture on the Everglades, the largest wetland system in the United States and one of the nation's greatest ecological treasures.

Agriculture has affected these wetlands in two important ways. First, canals have been dredged to drain areas for cattle ranching and other agricultural activities. This process has interfered with the flow of water into the Everglades and caused them to become drier and smaller. The canals also provide pathways for nonendemic species (such as the cattail and melalucca tree) to invade these wetlands and displace native species. Second, excess nutrient loadings from sugar plantations and cattle ranching has degraded water quality in the Everglades and related surface water bodies (Kissimmee River and Lake Okeechobee). These problems were discussed as early as 1947 (Douglas). In addition to the problems that are occurring directly in the Everglades, the ecological changes in the Everglades are leading to increased water scarcity in southern Florida and environmental degradation of the saltwater areas surrounding the Florida Keys.

The federal government and the state of Florida have embarked on a large scale project to reengineer the landscape surrounding the Everglades to restore the integrity of the Everglades. Although the plans have not been completely finalized, this project would involve eliminating all sugar production in the area, creating buffer wetlands in former sugar and cattle growing areas, eliminating several of the existing canals, pumping water from remaining canals into the buffer wetlands, and limiting development in the areas that are upstream and adjacent to the Everglades. Although the cost of such a project is high, many people feel that protecting this area is important for both environmental and economic reasons.

For current information, search bibliographic databases and the Internet, using the following keywords in tandem with the keyword "Everglades": agriculture, sugar, restoration, cattle, environmental policy, nutrients, pollution, and hydrological cycle.

flows into the environment over a large area. For example, soil, nutrients, pesticides, and herbicides are carried by rainwater runoff into lakes and rivers. The runoff enters the lakes and rivers along the entire length of the interface between land and water.

The significance of the pollution being generated by nonpoint sources is that it is much more difficult to monitor and measure the release of pollution by a particular polluter, in this case, the farmer. For this reason, it is much more difficult to implement economic incentives, such as per unit pollution taxes or marketable pollution permits. Policy incentives could focus on command and control techniques (such as requiring "best available farming techniques") or on economic incentives that are oriented toward inputs (such as fertilizer or pesticide taxes) rather than waste outputs. Another possibility is for the government to encourage research and development of new agricultural techniques combined with programs to disseminate information about these techniques and encourage their use.

Another difference between agriculture and other polluting activities is that agriculture is often a price-supported industry. These price supports

FIGURE 16.3

WELFARE LOSSES FROM QUANTITY OR PURCHASE-BASED PRICE SUPPORTS (NO POLLUTION)

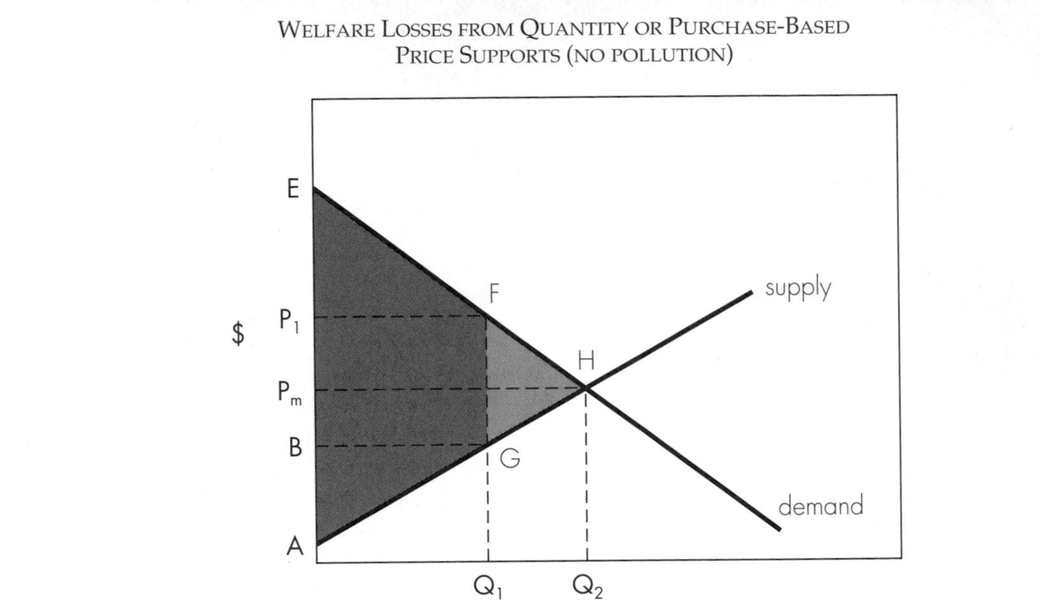

serve to raise price above marginal social cost. Price supports can take one of two major forms. In Figure 16.3, for example, if the government wanted to increase price from the market price (P_m) to P_1, it could either buy Q_2 minus Q_1 units of the crop, or it could pay farmers not to grow Q_2 minus Q_1 units of the crop. For the purposes of this discussion, the former type of price support will be called a purchase-based price support while the latter will be called a quantity-based price support. In the absence of any pollution or other market failures, the social losses associated with the price support that raises prices from P_m to P_1 are equal to the lightly shaded triangle in Figure 16.2. The heavily shaded trapezoid shows the net benefits of the Q_1 units of the crop that consumers purchase at the supported price of P_1. In this context, there are no differences between the types of price support in terms of the inefficiency (lightly shaded triangle) generated by the price supports.

There will, however, be a big difference between the two types of price supports in the presence of a pollutant. Pollution generates the disparity between the marginal private cost curve and the marginal social cost curve observed in Figures 16.4 and 16.5. In Figure 16.4, the price is supported at P_1 with a purchased-based price-support system. Although consumers only buy Q_1 units, Q_2 units are produced (with the difference purchased by the

FIGURE 16.4

Welfare Losses with Pollution and Purchase-Based Price Supports

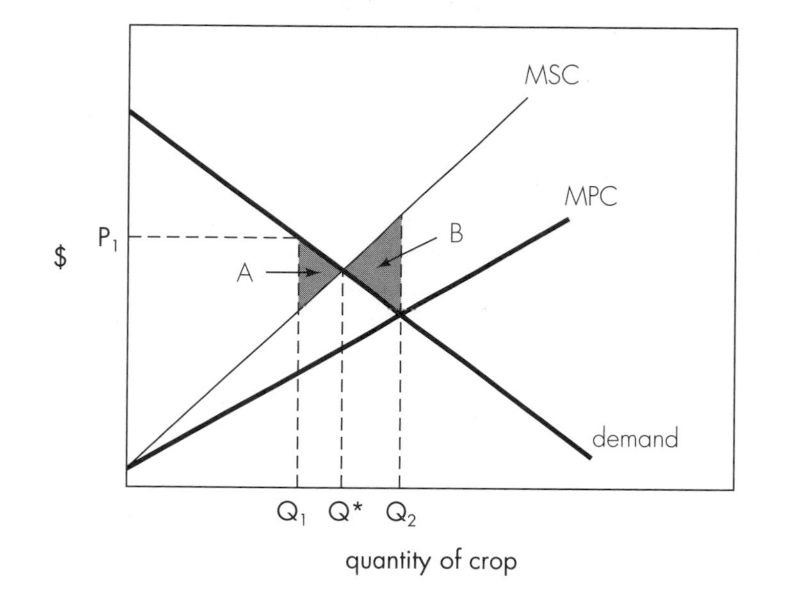

government). Neither Q_1 nor Q_2 are optimal in this case; the optimal level of output occurs at Q^*, where marginal social costs are equal to marginal social benefits. Since Q_2 units are produced, there are excess social costs generated by the amount of shaded triangle B, as marginal social costs are higher than marginal social benefits (given by the demand function) in this region. In addition, since consumers only consume Q_1 units, they lose consumers' surplus equal to the upper half of shaded triangle A, while producers lose producers' surplus equal to the lower half of triangle A. Thus, the losses from pollution in combination with a purchased-based price-support system are equal to the areas of triangles A and B.

In contrast, Figure 16.5 shows the social losses in the context of both pollution and a quantity-based price-support system. Since the area between Q_1 and Q_2 is not produced (farmers are paid to take this amount out of production), there is no pollution and no extra social costs associated with that nonexistent production. The social loss is equal only to the area of the shaded triangle, which is the lost consumers' surplus and producers' surplus associated with consuming below the optimal level.

The important point to this discussion is that the type of price-support system employed has an effect on pollution and other environmental effects

FIGURE 16.5

WELFARE LOSSES WITH POLLUTION AND QUANTITY-BASED PRICE SUPPORTS

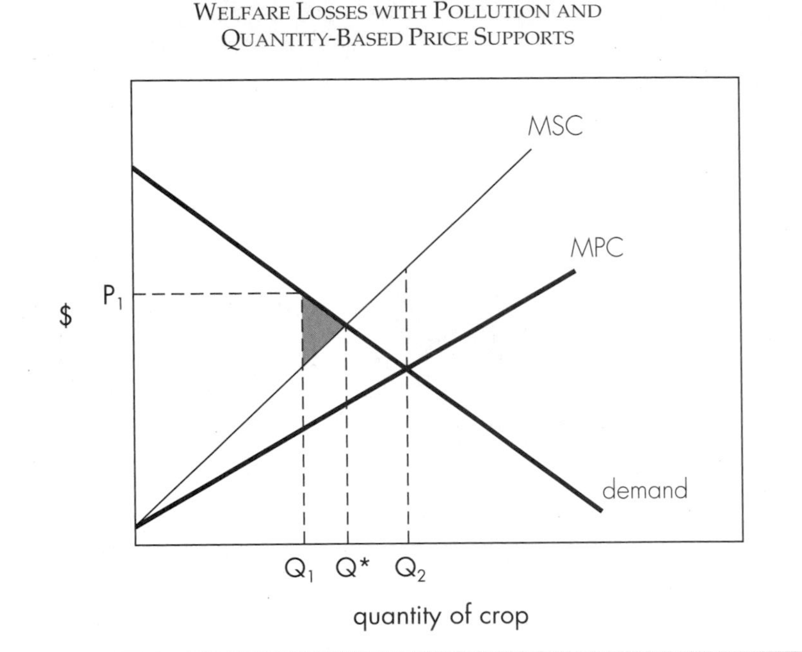

associated with agriculture. If the price supports are generated by taking land out of production, then the environmental externalities will be less severe than a price-support system that relies on government purchases of crops.

More recently, some agricultural subsidies have taken the form of income supports. These supports take the form of a payment based on the difference between a target price and the market clearing price. This per unit payment is issued over a quantity quota that is based on historical production levels for each participating farmer. Although an exact graphical representation of this program is quite complex, the most important effects are illustrated in Figure 16.6. If the primary effect of the income subsidy is to act as a lump-sum payment, the number of farmers will increase above the free market level (some farmers would have dropped out of farming without the income support). The increase in the number of farmers will increase supply and shift the MPC curve rightward, although MSC does not change. Since the shift of the supply function encourages more output in a range of output where MSC is above the demand curve (MSC>MSB), the policy increases welfare losses, which are measured as the shaded triangle in Figure 16.6.

FIGURE 16.6

WELFARE LOSSES WITH POLLUTION AND FARMER INCOME-SUPPORTS

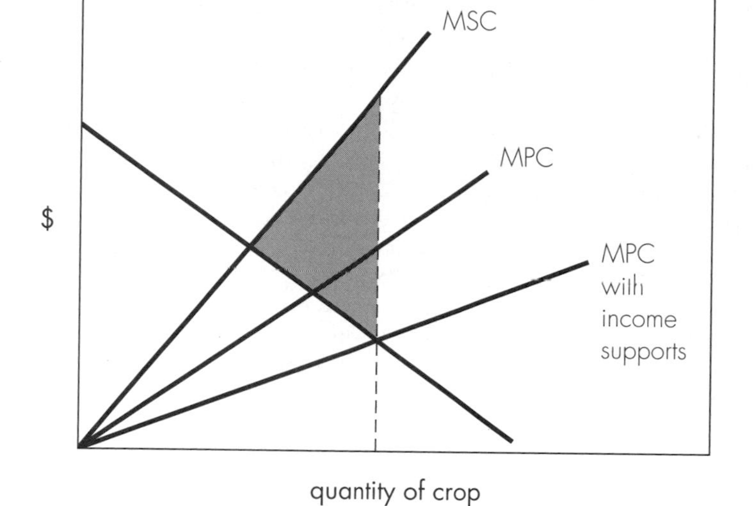

Of course, the price-support system itself is a source of inefficiency, but something that has been a part of modern public policy. The presence of price supports and associated welfare effects should be kept in mind when analyzing the efficiency of potential environmental policies for agriculture. Before analyzing future policy options, past and currently employed policies will be discussed.

PAST AND CURRENTLY EMPLOYED POLICIES

Past environmental policy in agriculture has been almost entirely directed at the internal effects of soil erosion. Policies have been developed to attempt to slow the rate at which soil is lost through wind and water erosion.

Modern awareness of the problems associated with soil erosion emerged in 1928, when Hammond Bennett published a U.S. Department of Agriculture report entitled "Soil Erosion: A National Menace." Shortly after that, Congress initiated a study of the causes of erosion and the means for its control. Of course, shortly after this study, the dust storms and massive erosion of the 1930s began, which led portions of the western United States to become known as the "Dust Bowl." In 1935, Congress established a permanent program of direct soil conservation aid under the auspices of the Soil Conservation Service of the U.S. Department of Agriculture (Heimlich, 1991).

AGRICULTURE AND FISHERIES

One of the most damaging effects of agricultural pollution is its effect on fisheries. This is a problem in both the developed and developing countries.

In the United States, many of our most important estuaries are degraded by upstream agricultural activities generating pollution that is carried downstream to the estuaries. Sediments, nutrients, pesticides, herbicides, and salinity are generated by agricultural activity. Many important fisheries in the United States have been affected by this type of pollution, including striped bass in the Chesapeake Bay, Atlantic salmon in many northeast rivers, and Pacific salmon in many northwest rivers.

This is also a problem in developing countries, where siltation of rivers, estuaries, lagoons, and coral reefs has substantially affected fish populations that are critically important in feeding people in these areas.

Ulf Silvander and Lars Drake (1991) have conducted a study of nitrogen from agricultural fertilizers affecting water quality and fisheries in Sweden (southern Baltic Sea). Not only has this pollution affected commercial and recreational fisheries that rely on wild fish stocks, but it has also affected the aquaculture of mussels and other shellfish. Silvander and Drake conduct empirical analysis of recreational and commercial fishing and of the willingness to pay to prevent nitrate contamination of groundwater. They find that the losses generated by agricultural sources of nitrogen are $22 million in recreational fishing, $93 million to $102 million in commercial fishing, and $140 million for impacts to drinking water.

To find more information about the impact of agricultural pollution on fisheries, search bibliographic databases on fisheries, fisheries management, fishery policy, and fish in combination with pollution, environmental policy, agriculture, nutrients, and pesticides. Also visit the National Marine Fisheries Service webpage at http://kingfish.ssp.nmfs.gov/, the American Fisheries Society homepage at http://www.esd.ornl.gov/societies/AFS/afshot.html, and the United States Department of Agriculture National Resources Conservation Service at http://www.ncg.nrcs.usda.gov/.

These initial conservation programs were designed to reduce erosion out of concern for the on-site impacts of soil erosion. The programs were designed to provide information on less erosive agricultural techniques (contour plowing, windbreaks, etc.) and to provide cost sharing for employment of anti-erosion measures. One of the barriers to developing appropriate public policy was that, until very recently, the extent of soil erosion was not known.

The focus of agricultural environmental policy on the on-site impacts of erosion has continued until very recently, where the major environmental programs were designed almost solely to eliminate soil loss. While this program also reduced impacts on water quality, other programs were needed to more specifically target water quality, habitat preservation, and biodiversity.

A good case in point is the Conservation Reserve Program (CRP), which until very recently was designed to take highly erodible land out of tillage and establish vegetative cover to control erosion. Under the CRP, the government pays farmers to take land out of crop production and plant protective vegetation on the land. Although improved water quality is an important goal (and result) of this program, water quality objectives may be better met with a program that is structured differently. For example, one of the best ways to improve water quality is to allow buffer strips of land near water bodies to become heavily vegetated with woody and nonwoody plants. These buffer strips trap much of the soil particles and absorb nutrients that would have run off the cultivated land.

The Conservation Reserve Program also did not promote low tillage or no-tillage agricultural techniques. These techniques do not plow the soil or remove the previous season's crop residues, so erosion is diminished. In addition, the Conservation Reserve Program does not attempt to preserve areas of important ecological significance.

Although the Conservation Reserve Program is not a comprehensive environmental program, it did generate net benefits for society. In addition to reducing erosion and improving surface water quality, the program provides support for participating farmers in a more efficient fashion than price-support payments. Table 16.2 lists the costs and benefits of the CRP.

A program that promotes habitat restoration is found in the Wetlands Reserve Program, which provides incentives for farmers to restore wetlands that were converted to agriculture before this type of conversion became restricted. Several other programs restrict the conversion of existing wetlands (see Chapter 13).

The 1996 Farm Bill generated a very important change in the orientation of the Conservation Reserve Program and the policy of the U.S. Department of Agriculture towards general environmental programs. The 1996 Farm Bill directed USDA programs in general (and the CRP specifically) to more directly target the off-site environmental benefits associated with farm programs. For example, farms are now chosen for participation in the CRP based on more broadly defined environmental impacts, including potential impacts on habitat, water quality and biodiversity. To find out more about how the 1996 Farm Bill and the USDA pursue environmental quality goals, visit the homepage of the National Resource Conservation Service (formerly the Soil Conservation Service) at http://www.ncg.nrcs.usda.gov/.

An interesting example of a comprehensive environmental policy for agriculture is the plan that has been developed to protect the Chesapeake Bay. The Chesapeake Bay, the largest estuary in the United States, is heavily polluted with nutrients from municipal treatment plants and agriculture. Although progress has been made in upgrading sewage treatment plants, water quality did not improve substantially due to nutrient loadings from agriculture. The problem is compounded by the fact that two states (Maryland and Virginia) have jurisdiction over the Chesapeake Bay, but six states (New York, Pennsylvania, West Virginia, Delaware, Virginia, and Maryland) and the

TABLE 16.2

COSTS AND BENEFITS OF 45 MILLION ACRE CRP ENROLLMENT, UNITED STATES 1986–1999

NATIONAL INCOME	PRESENT VALUE (BILLIONS OF DOLLARS) (NEGATIVE VALUES IN PARENTHESES)
Gains	
Landowners	
net farm income	9.2–20.3
timber production	4.1–5.4
Natural resources	
productivity	0.8–2.4
water quality	1.9–5.6
wind erosion	0.4–1.1
wildlife habitat	3.0–4.7
Losses	
Consumer costs	(12.7–25.2)
Cover crop establishment	
landowner share	(1.6)
government share	(1.6)
Technical assistance cost	(0.1)
Net Benefit	3.4–11.0
Federal Budget	
Government cost savings	
direct CCC savings	10.2–12.2
indirect (price effects)	6.0–7.3
Government expenses	
CRP rental payments	(19.5–20.8)
corn bonus payments	(0.3)
cover crop establishment	(1.6)
technical assistance	(0.1)
Net Government Expense	(2.0–6.6)

PRIMARY SOURCE: Young and Osborn, 1990; *The Conservation Reserve Program: An Economic Assessment*, AERO-626, EARS, USDA.
SECONDARY SOURCE: taken from Table 5.5 of Ralph E. Heimlich, 1991; "Soil Erosion and Conservation Policies in the United States." In *Farming and the Countryside: An Economic Analysis of External Costs and Benefits*, edited by N. Hanley, London, CAB International.

District of Columbia contribute to the pollution of the Chesapeake Bay through their pollution of rivers (Susquehanna, Potomac, and James, among others) that flow into the Chesapeake Bay.

Maryland, Virginia, Washington, D.C., Pennsylvania, the U.S. EPA, and the Chesapeake Bay Commission are signatories to the Chesapeake Bay

Agreement, which has a substantial nutrient reduction strategy. Approximately 15,000 projects covering 300,000 acres of agricultural land have been implemented to encourage farming practices to control nutrients and erosion. These farming practices include no-till farming, contour plowing, manure storage facilities, and other best management practices (BMPs) (Raven, Berg and Johnson, 1993). For more details about this program visit the webpages of the Alliance for the Chesapeake Bay (http://web.gmu.edu/bios/Bay/acb/papers/nutrient.htm) and the Environmental Protection Agency's Great Water Bodies and National Estuaries Programs (http://www.epa.gov/OWOW/BODIES/AppD.html).

A COMPREHENSIVE SET OF ENVIRONMENTAL POLICIES FOR AGRICULTURE

Environmental policy for agriculture should attempt to accomplish several goals. First, it should seek to discourage erosion and the on-site and off-site impacts of erosion. Second, it should discourage the excessive use of fertilizers and pesticides. Third, it should increase food safety. Fourth, it should restore marginal farmland to natural habitat and protect existing habitat. While accomplishing all these environmental goals, these policies should also help ensure the adequacy of food supplies, the well-being of consumers, and the profitability of farmers. Obviously, this task is extremely complex, but steps can be begun in these directions.

Fertilizer and Pesticide Taxes

The economic theory developed in Chapter 3 suggests that the efficient way of dealing with a pollution externality is to impose a per-unit pollution tax or to implement a system of marketable pollution permits. However, since the fertilizer and pesticide runoff associated with agriculture is nonpoint source pollution, this cannot be easily done. The amount of pollution per farm per unit time cannot be easily measured, so it is difficult to implement taxes or marketable permit systems.

Although this pollution cannot be taxed easily, economic theory suggests that fertilizer and pesticide applications will be excessive since the farmer chooses application levels by comparing private costs and private benefits rather than social costs and social benefits. At the margin, discouraging fertilizer and pesticide applications will increase social welfare.

Reductions in fertilizer and pesticide applications will be forthcoming if fertilizer and pesticides are taxed. Although some states have fertilizer taxes, they tend to be token taxes (from fractions of a percent of price to several percent of price) and are designed to raise revenues to fund farm programs rather than to internalize the full social cost of pesticide and fertilizer use (Carlin, 1992, pp. 3–5).

One advantage of pesticide taxes is that it will encourage a practice called integrated pest management to control insects and other agricultural pests. Integrated pest management is a full portfolio of techniques to control pests, including the use of biological controls, such as insect predators. If it becomes more expensive to rely totally on pesticides, more farmers will move to integrated pest management.

Pesticide and herbicide taxes also will have the benefit of increasing food safety. If pesticide and herbicide applications decrease as a result of taxes, less toxic residues will be found in food, *ceteris paribus*.

Although economic theory suggests that pesticide and fertilizer taxes could improve social welfare by mitigating the market failure, taxes that were too high would over control applications of these chemicals and lead to social losses from over control. Estimates of the marginal benefits to farmers and the marginal damages to the environment must be conducted in order to determine the appropriate tax levels for these agrochemicals.

Although pesticide and fertilizer taxes can reduce applications and therefore reduce the release of these substances into the ecosystem, they do not encourage the full range of pollution-mitigating actions. The reason is that even though they reduce applications, they do nothing to reduce the pollution associated with a given level of applications. For example, if the farmer leaves buffer strips of vegetation around the fields and between the creeks and the fields, the amount of fertilizer and pesticides that reach the creeks will be much less than if those buffer strips do not exist. In addition to input taxes, policies could be established to encourage "green" farming techniques.

PROMOTING "GREEN" FARMING METHODS

One of the most important ways of promoting green farming methods is by providing information on how to engage in these methods. Since the methods often reduce soil erosion and the need for fertilizer and pesticide inputs, they often reduce farmer's costs in the long run. Many agricultural extension agents are trained in conventional agricultural techniques and do not have the appropriate knowledge to pass on to farmers. The agricultural support services, such as the extension program, must become more oriented toward green methods if these methods are to be passed on to farmers. This reorientation of the focus of agricultural support services towards greener methods has accelerated during the past several years.

Some green methods are privately more efficient, but other green methods may reduce farmer benefits, even though they increase social benefits. Farmers have no incentive to adopt these methods, even if they have full information about them. Both command and control regulations and economic incentives can be used to induce farmers to adopt these methods.

Command and control regulations can be used to require or prohibit certain agricultural methods. For example, cultivating within a certain distance of a water body and filling of wetlands can be prohibited. In the same vein, a certain percentage of the farm area can be required to be covered by natural vegetation. Low tillage methods, holding ponds to contain runoff, and buffer strips surrounding fields can also be required.

Alternatively, farmers can be compensated for engaging in these techniques. Eligibility for price-support payments can be made contingent upon using best available farming methods on the land that is under cultivation. In addition, farmers can be paid for leaving critically important natural habitats undisturbed.

Adequately Priced Water

One of the biggest environmental impacts of agriculture is its effects on water scarcity, particularly in the west. The use of water in irrigation depletes aquifers at a much faster rate than they are being replenished, reduces stream flow, and increases the salinity of water. As extensively discussed in Chapter 14, this problem is caused by water being priced at a fraction of its true social cost. The problem can largely be eliminated by requiring farmers (and other water users) to pay the full social cost for the water they consume and by making water rights fully transferable.

Protecting Habitat and Biodiversity

There is an inherent conflict between farmers' profit objectives and the preservation of critical habitat and biodiversity. Since the cultivation of land requires a conversion from natural habitat, farmers' profit incentives lay toward clearing habitat. Again, policy to protect habitat and biodiversity can be undertaken using either direct controls or economic incentives. Conversion of certain types of habitat can be banned (this ban is already in existence for certain classes of wetlands and for habitat that supports endangered species). As a society, it makes sense for our food to be grown on ground that is already cleared, rather than clearing new natural habitat. However, generating this protection by banning agricultural activities in these areas places the financial burden of meeting this social goal entirely on the backs of farmers. Many people argue that in these cases, compensation must take place in order for these policies to be equitable. Government agencies (or nongovernmental organizations [NGOs]) could buy critical habitat from farmers, or farmers could retain ownership but could be paid to leave these areas undisturbed. In fact, many NGOs are already engaging in this practice, and new government programs, such as the Wetlands Reserves Program, are contributing to the preservation of important habitat.

In addition, many people suggest that abandoned or marginal agricultural land should be restored to natural habitat. Farmers could be paid to reintroduce native species on former farmland, or the government could purchase this land, restore it, and turn it into parkland or wildlife sanctuary.

For example, there is broad, grassroots support for the conversion of a large proportion of failed and marginal Great Plains farmland into natural prairie.

SUMMARY

Agriculture is impacted by environmental degradation, but agriculture is also a significant source of environmental degradation. Agriculture is adversely affected by air pollution, by global warming, and by agricultural activity itself. Agriculture generates many environmental problems, including soil erosion, degradation of water quality, increased water scarcity, contributions to global warming, toxic substances in the ecosystem and food supply, and loss of habitat and biodiversity.

Past agricultural policy toward the environment has focused on soil erosion, but this policy needs to be broadened to reduce other types of environmental degradation. The development of new policy is complicated by the fact that many agricultural pollutants are nonpoint source pollutants. An important component of policy could be the development of taxes on agricultural inputs, such as fertilizers and pesticides. However, since these taxes do not encourage the full portfolio of pollution-reducing responses, other techniques need to be implemented to reduce the release of pollutants into the ecosystem. A combination of direct controls and economic incentives could be used to promote the utilization of greener agricultural methods. In addition, policies must also be developed to promote the protection of habitat and biodiversity and to adequately price water.

REVIEW QUESTIONS

1. How do subsidy programs affect the net benefits from agriculture?
2. How do subsidy programs exacerbate social losses generated by the environmental externalities associated with agriculture?
3. Assume that an agricultural crop has an inverse demand function (marginal willingness to pay function) of

$$MWTP = 10 - 2Q.$$

Assume that $MPC = Q$, and $MSC = 1.5Q$.

Find the social losses from the externality, given the market clearing price and quantity. Now assume that price is supported at $7 by government purchases of the crop sufficient to drive P to $7. Find the losses in consumers' and producers' surplus associated with the pollution and the subsidy.
4. How does nonpoint source pollution differ in policy options from point source pollution?
5. What is the difference between on-site and off-site impacts of soil erosion? Identify the market failures associated with each type of impact.

QUESTIONS FOR FURTHER STUDY

1. Should agricultural activity be subsidized?
2. What type of subsidy program would best achieve environmental goals?
3. What are the strengths and weaknesses of the Conservation Reserve Program?
4. Is food safety adequately protected by current policies? Explain your answer.
5. Should pesticides be banned? Why or why not?

SUGGESTED PAPER TOPICS

1. Look at the economic rationale underlying integrated pest management. How does it affect private costs, social costs, and benefits from agriculture? Look in the *Journal of Environmental Economics and Management* and the *American Journal of Agricultural Economics* for current references.
2. Look at the costs and benefits of the Wetland Reserve Program. Start with the article by Heimlich for current references.
3. Should agricultural and environmental policy be more closely coordinated? Examine this question from an economic perspective. Start with the OECD publications for current references. Search bibliographic databases on agriculture, environment, and policy.
4. Many studies have been done that look at the effect of pollution on agriculture. Start with the NAPAP report for current references and critically examine studies that estimate the economic costs to agriculture of increased air pollution. Where would the agricultural sector most greatly benefit from reducing pollution?

WORKS CITED AND SELECTED READINGS

1. Bonnieux, F., and P. Rainelli. "Agricultural Policy and Environment in Developed Countries." *European Review of Agricultural Economics* 15 (1988): 263–280.
2. Bowers, J. K. *Agriculture, the Countryside and Land Use: An Economic Critique*. New York: Methune, 1983.
3. Carlin, Alan. *The United States Experience with Economic Incentives to Control Environmental Pollution*. Washington, D.C.: U.S. Environmental Protection Agency, EPA-230-R-92-001, 1992.
4. Douglas, Marjory Stoneman. *The Everglades: River of Grass*. New York: Rinehart, 1947.
5. Hanley, Nicholas, ed. *Farming and the Countryside: An Economic Analysis of External Costs and Benefits*. London: CAB International, 1991.
6. Heimlich, Ralph E. "Soil Erosion and Conservation Policies in the United States." In *Farming and the Countryside: An Economic Analysis of External Costs and Benefits*, edited by Nicholas Hanley. London: CAB International, 1991.
7. ———. "Costs of an Agricultural Wetland Reserve." *Land Economics* 70 (1994): 234–246.
8. OECD. *Agricultural and Environmental Policies: Opportunities for Integration*. Paris: Organization for Economic Cooperation and Development, 1989.
9. ———. *Agricultural and Environmental Policy Integration: Recent Progress and New Directions*. Paris: Organization for Economic Cooperation and Development, 1993.
10. Raven, Berg, and Johnson. *Regional Issues Sampler for Environment*. Philadelphia: Saunders College Publishing, 1993.
11. Shortle, James S., and J. W. Dunn. "The Economics of the Control of Non-Point Pollution from Agriculture." In *Farming and the Countryside: An Economic Analysis of External Costs and Benefits*, edited by Nicholas Hanley. London: CAB International, 1991.
12. Silvander, Ulf, and Lars Drake. "Nitrate Pollution and Fisheries Protection in Sweden." In *Farming and the Countryside: An Economic Analysis of External Costs and Benefits*, edited by Nicholas Hanley. London: CAB International, 1991.
13. U.S. National Acid Precipitation and Assessment Program (NAPAP). *1990 Integrated Assessment Report*. Washington, D.C.: Government Printing Office, 1991.
14. Young, C. E., and C. T. Osborn. *The Conservation Reserve Program: An Economic Assessment*. AERO-626, EARS, USDA, 1990.

The Environment and Economic Growth in Third World Countries

The achievement of sustained and equitable development is the greatest challenge facing the human race. . . .
Economic development and sound environmental management are complimentary aspects of the same agenda. Without adequate environmental protection, development will be undermined; without development, environmental protection will fail.

World Bank, World Development Report 1992: Development and the Environment

INTRODUCTION

The focus of environmental attention has been shifting in recent years from one that centers on developed countries to one that focuses on developing countries as well.[1] In fact, many who study environmental issues feel that addressing environmental degradation in developing countries should be the primary focus of environmental policy. A quick inventory of environmental problems associated with developing countries includes the massive environmental degradation in Eastern Europe, tropical deforestation, desertification, contamination of drinking water, soil erosion, loss of habitat of all

[1]Many terms are used to signify countries that have not yet achieved the standard of living of countries such as the United States, Western European countries, and Japan. In this chapter, we will use the term "developing countries" rather than "less developed countries (LDCs)" to emphasize the concept that a stage of development is not a permanent status, but a process of change. Occasionally, the term "Third World" countries will be used, where the term was originally conceived to differentiate developing countries from developed capitalistic/democratic countries (First World) and from socialist/communist countries (Second World). However, since the fall of the Iron Curtain and the end of the cold war, these distinctions have less meaning. When the term "Third World" countries is used in this chapter, it will be assumed to be synonymous with "developing countries" and includes those former Soviet republics and former Warsaw Pact countries whose economic status places them in the "low" (under $610 of annual per capita GDP) or "lower middle" ($611–$2,465 of annual per capita GDP) or "upper middle" ($2,466–$7,619 of annual per capita GDP) income designation of the World Bank.

types (including wetlands, grasslands, coral reefs, a variety of aquatic habitats, and forests), loss of biodiversity, and depletion of fisheries.

In the 1970s and early 1980s, many people argued that environmental quality was a luxury that only the rich nations could afford. However, beginning in the 1980s, it became apparent that deteriorated environmental quality was interfering with the economic development of many countries and that economic development cannot take place without environmental improvement.[2]

INCOME, GROWTH, AND THE ENVIRONMENT

In the 1960s, the field of development economics, which looks at how countries can increase their economic well-being, took on increasing importance as former colonies in Africa and Asia became independent. They began independence at a low level of development, and researchers and policy makers sought ways in which to increase the economic well-being of the people of these countries. The definition of well-being that people tended to focus on was per capita Gross Domestic Product (GDP), with capital accumulation as the primary tool to increase per capita GDP.

The focus was on capital accumulation because of how people viewed the economic process. Developing countries tended to have a low productivity of labor, which implied low per capita income. Since income was low, much or all of income was needed for consumption purposes, leaving little for saving and consequently little for investment. The lack of investment meant little capital accumulation, and since increasing capital is one way to increase the marginal product of labor, the lack of investment meant continued low productivity of labor. This process, which was viewed as keeping income low, was called the vicious cycle of poverty.

Note that the vicious cycle of poverty, as depicted in Figure 17.1, does not necessarily imply declining income; it just emphasizes the difficulty of increasing income due to the problems associated with a lack of capital formation.

The development projects initiated in the 1960s, 1970s, and early 1980s focused on capital formation. Large capital intensive projects involving dams, factories, energy facilities, and large-scale agriculture were initiated, but these projects were not really successful in improving the productivity of the economy. Many feel that a major reason for the continued—even worsening—standard of living in many developing countries is the continuing environmental degradation.

At first, this argument may seem to be strange, as North Americans tend to view environmental quality as an amenity or even as a luxury good. However, environmental quality is not only a consumption good, it is an input to

[2]The World Bank's *World Development Report 1992* focuses on the interaction between the environment and economic development. It is an excellent first reference for detailed examination of these issues.

FIGURE 17.1

TRADITIONAL VIEW OF VICIOUS CYCLE OF POVERTY

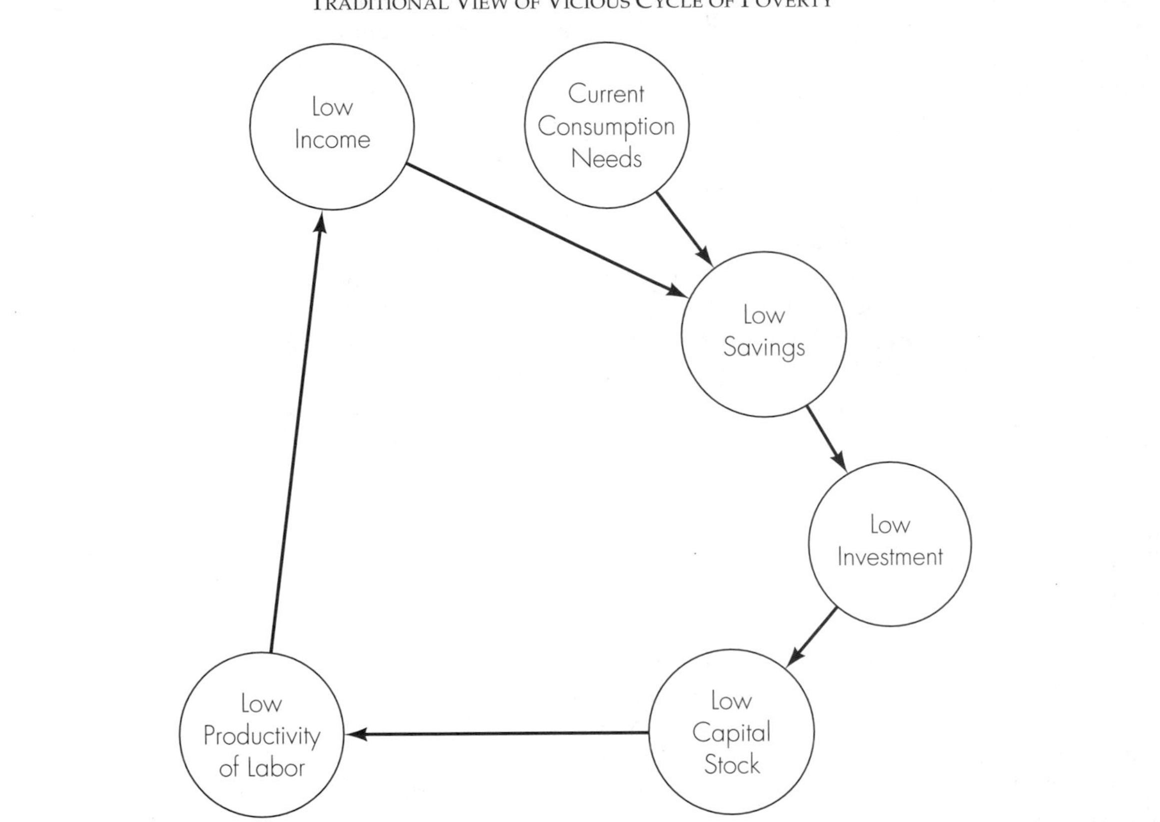

production processes (see Chapter 5 for further discussion of this point). As an input, the environment may be even more important in developing countries that are more dependent upon primary production activities, such as agriculture, forestry, and resource extraction.

THE EFFECT OF ENVIRONMENTAL QUALITY ON ECONOMIC PRODUCTIVITY

Environmental quality can affect economic productivity in two major ways. First, environmental degradation can drastically affect human health, which can affect the productivity of labor and consume resources in dealing with the adverse health effects. For example, intestinal disorders (cholera, dysentery, etc.) from contaminated drinking water are the leading causes of death of young children in many developing countries. Malaria is an important source of missed work days in agricultural areas, and its prevalence is increased by deforestation and other environmental changes (see boxed example on page 475). Second, the environmental resources are a direct input into many production processes, and environmental degradation interferes with these production activities. A good example of this type of degradation is deforestation, which leads to soil erosion, which inhibits agriculture and causes the siltation of rivers and embayments, which leads to diminished fishery output. The loss of the forests also precludes future economic activity that would directly use forest resources, such as timber and wood product industries.

THE EFFECT OF POVERTY ON THE ENVIRONMENT

Much of the environmental degradation is due to the low standard of living that currently exists in developing countries. Current low income makes it very difficult to meet current consumption needs, so environmental resources are unwisely exploited to produce current income. Forests, which could produce a steady stream of income, are decimated in order to provide a temporary burst of current income to meet current consumption needs. Similarly, agricultural areas are cultivated too intensely in order to produce food for the short run, which leads to soil erosion and declines in soil fertility that diminish the long-term ability to produce food. Current needs for food lead to large livestock herds, and the livestock overgraze rangeland, diminishing the productivity of the rangeland so that it can support fewer animals in the future. Both deforestation and overgrazing are significant factors in the process of desertification, where deserts expand into areas that were formerly vegetated.

This process of degradation is not simply limited to agricultural processes. In many countries, particularly Eastern Europe and populous developing countries (India, Brazil, Mexico, China, Indonesia, etc.), high levels of pollution are produced from industrial activity, because the government does not implement market failure–correcting environmental policies, as they hesitate in devoting resources to pollution control, since these resources are needed to meet current consumption needs. Unfortunately, high levels of air and water pollution degrade public health; reduce the productivity of agriculture, industry, fisheries, and forestry; and generate other social costs that reduce the future ability to produce income.

Figure 17.2 shows how the vicious cycle of poverty should be visualized with an environmental component. Low income leads to both low investment and high environmental degradation. Both the lack of investment and the environmental degradation lead to low productivity. Taken by itself, the decline in environmental quality can actually lead to a decline in the productivity of labor and a decline in income. In combination with the lack of investment, the decline in environmental quality will almost certainly lead to a decline in income.

CURRENT VIEW OF VICIOUS CYCLE OF POVERTY

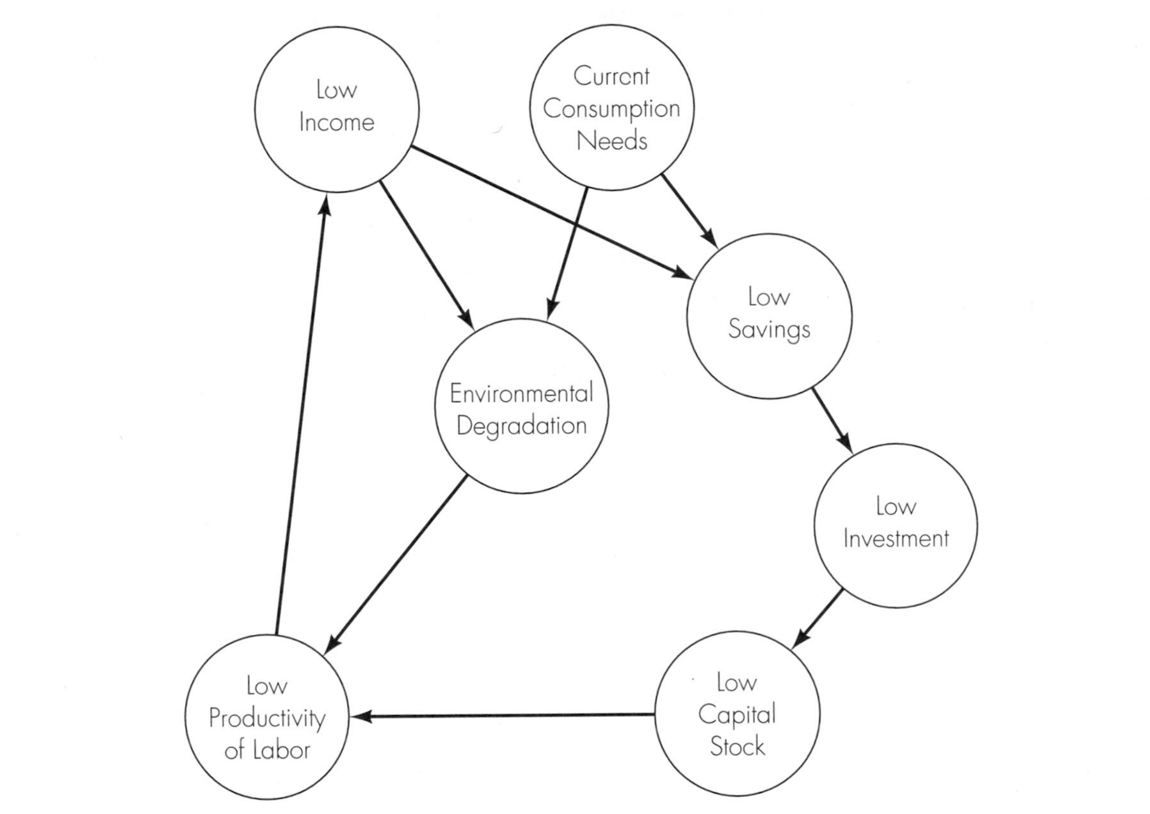

THE ROLE OF POPULATION GROWTH IN ENVIRONMENTAL DEGRADATION AND ECONOMIC DEVELOPMENT

Population growth is portrayed in the popular media as the root of all evil, in the sense that population growth is suggested as the primary cause of both the environmental degradation and the poverty that exists in many developing countries. This conclusion is allegedly supported by the high degree of negative correlation between the growth of income and the growth of population and the high degree of positive correlation between environmental degradation and the growth of population. However, correlation does not imply causation. Rather than rapid population growth leading to slow or negative growth in income, might the low income be leading to the high population growth? Or might the causation actually run in both directions (population growth affects income, and income affects population growth)? Since income and population growth are correlated, might the observed relationship between population growth and environmental degradation actually represent a relationship between income and environmental degradation, which runs in either or both directions? In order to fully understand the options that are available with which to raise the standard of living in developing countries, it is necessary to have a comprehensive understanding of the interrelationships among population, income, and environmental degradation.

The Effect of Population Growth on Income

In 1798, Thomas Robert Malthus (Appleman, 1975) authored *An Essay on the Principle of Population,* which still has a profound effect on the way society thinks about the relationship between population growth and income. In this essay, Malthus argued that the productivity of labor grows at a slower rate than population, because the increased population produces food in conjunction with a fixed amount of land, which implies a diminishing marginal product of labor.

Since population grows at a faster rate than labor productivity, population will grow faster than the amount of food that people can produce. As a result, Malthus postulated that population will only be constrained by the food supply. That is, population will grow until the lack of food keeps population from growing any further. Moreover, each individual will have just enough food to stay alive and, therefore, be always hungry and always at the brink of starvation. Incidently, this is why economics was given the nickname of "The Dismal Science," because of the dismal outcome it predicted for humankind.

However, even though there is much hunger and starvation in the world, Malthus' prediction has not come true (as of yet), as population has continually grown throughout history, and much of the world's population has a standard of living far in excess of the subsistence level. Why did Malthus' prediction not come true?

One possible explanation for the lack of fulfillment of this prediction is provided by Barnett and Morse (1963), who suggest that since Malthus only considered a two-input world (labor and land), his results are not truly generalizable to the real world. The reason is that the law of diminishing returns implies that if you hold all other inputs constant, and increase only one input, there must be diminishing returns to that input. If the only two inputs in the production process are land (which is fixed) and labor, then one would expect diminishing returns to increasing labor. However, if capital, energy, materials, renewable resources, and knowledge are important inputs in the production process, then holding land fixed (but allowing the capital and other inputs to increase) does not imply that there must be diminishing returns to labor. Also, when one considers capital, then there are prospects for technological innovation that can serve to increase both the marginal product of labor and the marginal product of capital. It is somewhat surprising that Malthus did not consider capital, because his work was written as the Industrial Revolution was speeding through England and other parts of Europe. Although there are many countries where the bulk of the population face Malthusian living conditions (existence on the brink of starvation), the simple version of the Malthusian law does not necessarily hold and does not necessarily imply a bleak existence.

However, even if one considers the other inputs, it is possible that population growth may reduce income in another fashion. This potential for reduction is because faster population growth implies greater current consumption needs, which inhibit the ability to invest and may lead to the destructive overexploitation of environmental resources. This possibility is especially true when population growth leads to a greater percentage of children and a lower percentage of workers in the population.[3] Population growth may also exacerbate the negative effects on income of inefficient government policies and market failures, such as poorly defined property rights. This phenomenon was discussed with respect to the decline in tropical forests in Chapter 12 and will be revisited in the discussion of current policies in this chapter.

Furthermore, population growth tends to exacerbate income inequality. There are several reasons for this factor. First, as population grows, the land available for the poorest elements of society remains constant. Since population growth tends to be highest among poor agrarian families, population growth tends to shrink the average size of an agrarian family's plot. This shrinkage not only reduces their productive capacity, but tends to lead to overly intense exploitation of the land. For example, with small landholdings, families do not have the luxury to leave parts of their land fallow so that it can renew its productivity. Over usage leads to soil erosion and declining soil fertility, which will obviously increase the poverty of these poorest families.

THE EFFECT OF INCOME ON THE GROWTH OF POPULATION

The effect of income on population growth has been characterized by a model known as the demographic transition. The basic premise behind this model is that population growth is related to the stages of development. In the traditional conception of the model, there are three stages of economic development. These are the traditional agrarian society, the beginning of industrialization, and the industrialized society. These three stages are la-

---

[3]In 1990, in low income countries, an average of 35.2 percent of the population was less than 14 years of age, while the corresponding figure for high income countries was 19.9 percent.

FIGURE 17.3

DEMOGRAPHIC TRANSITION

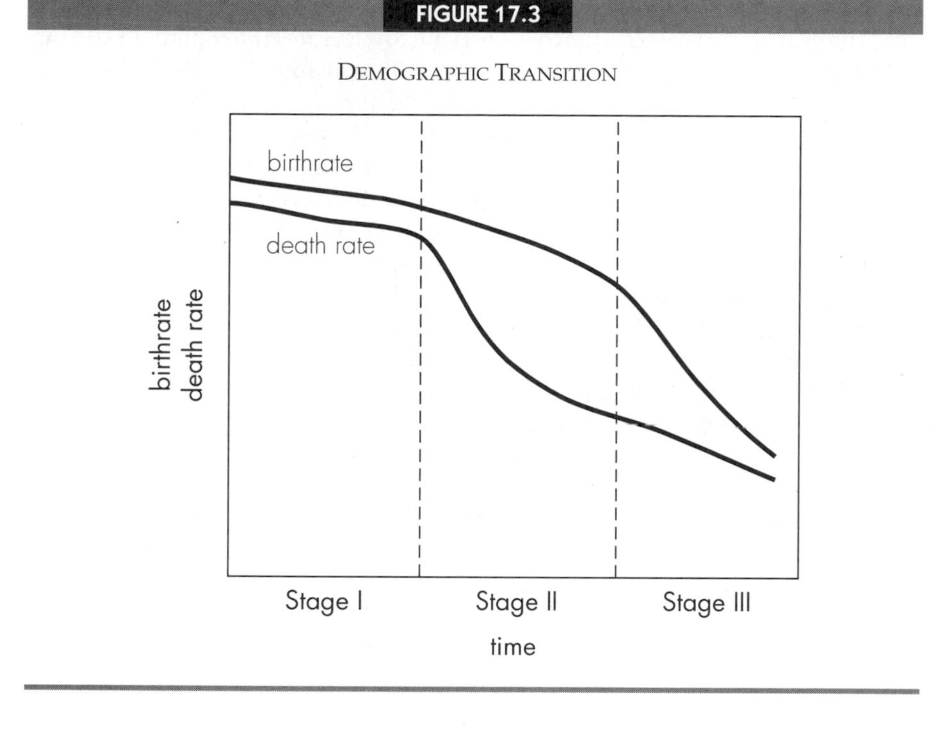

beled Stage I through Stage III on the horizontal axis of Figure 17.3, which represents time. The vertical axis measures the birthrate and the death rate.

In the first stage, both birthrates and death rates are relatively high, but birthrates are only slightly higher than death rates, so population grows slowly. As the society begins to become developed, death rates fall due to improvements in public health (vaccinations, pest control, availability of antibiotics, improved sanitation, etc.). However, birthrates do not fall commensurately, so population growth accelerates. In the third stage, as the countries become industrialized, the demographic transition occurs and birthrates fall to more closely parallel death rates. Once again, population growth slows. It should be pointed out that this model does not provide an explanation of the determinants of birth and death rates, but a way of organizing our past observations on population growth rates. This model fits the historical record of population growth in Europe and North America very well. It also seems to fit the rapidly industrializing countries of East Asia fairly well. For example, the average annual population growth rate in East Asia and the Pacific was 2.2 percent for the period 1965–1980, but fell to 1.6 percent for the period 1980–1990. The Republic of Korea (South Korea) had a population growth rate of 2.0 percent for the period 1965–1980, which fell to 1.1 percent in the period 1980–1990. Corresponding figures for Hong Kong were 2.0 percent and 1.4 percent. In contrast, countries that have been industrialized for some time (high income OECD countries) had a growth rate of 0.9 percent during the 1965–1980 period and 0.6 percent during the period 1980–1990.[4] In sharp contrast, the corresponding figures for sub-Saharan Africa were 2.7 percent and 3.1 percent, indicating both rapid and increasing population growth.[5]

There is evidence that there may be a fourth stage, and that as industrialized countries mature, population growth falls to equal or below the replacement rate, and zero population growth (ZPG) or negative population growth occurs. For example, in 1990 the total fertility rate (average number of children that women have during their lifetime) in high income OECD countries averaged 1.7 (*World Development Report 1992*). This rate means that once the effect of the post–World War II baby boomers (which includes the grandchildren of the baby boomers) works its way through the population, the population of this group of countries will stop growing and will probably even shrink. Based on current trends, this shrinkage will not happen until about 2030 (*World Development Report 1992*).

Since the model of the demographic transition is a model of what has been observed and not a model of underlying factors of causation, it cannot be conclusively stated that countries in Africa, Latin America, and South Asia will eventually undergo a demographic transition. In particular, environmental factors that were not present in Western Europe and North America may prevent population growth from slowing in developing countries in a manner analogous to the demographic transition that was observed in Western Europe and North America.

THE MICROECONOMICS OF REPRODUCTION

There are many factors that affect birthrates, including economic, cultural, social, and religious factors. The religious, social, and cultural factors are quite complex and vary significantly from one society to another. This section of the chapter will focus on economic factors that affect birthrates. Although the effect of economic factors on birthrates will generally hold, it is often difficult to modify birthrates through policies that are solely economic, as the social, cultural, and religious factors also affect outcomes.

Economists like to explain the decision to have children as a microeconomic utility maximizing decision, where the household (both parents) weigh

---

[4]OECD (Organization for Economic Cooperation and Development) high income member countries include Ireland, Spain, New Zealand, Belgium, United Kingdom, Italy, Luxembourg, Australia, Netherlands, Austria, France, Canada, United States, Denmark, Germany, Norway, Sweden, Japan, Iceland, Finland, and Switzerland. Middle income OECD countries include Portugal, Greece, and Turkey. OECD member countries include all high income countries except Israel, Hong Kong, Singapore, United Arab Emirates, and Kuwait.

[5]Population growth figures in this chapter are taken from *World Development Report, 1992,* by World Bank.

THE MARGINAL COSTS AND DEMAND FOR CHILDREN

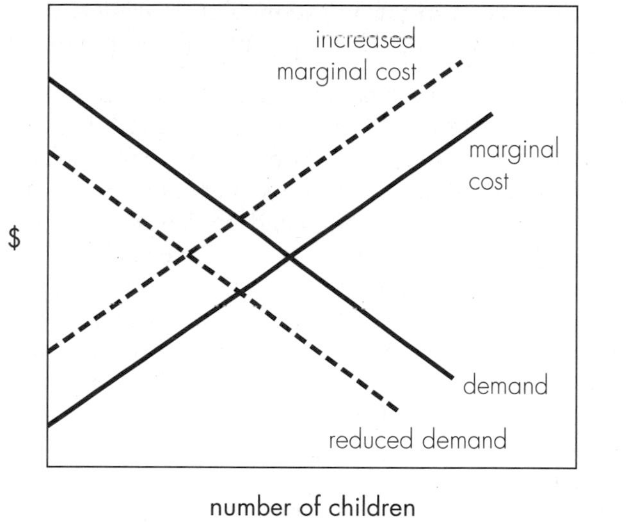

the costs and benefits of having an additional child.[6] Figure 17.4 contains a graphical representation of this process, with the marginal benefits of children declining with the number of children and the marginal costs of children increasing with the number of children.[7] The optimal number of children occurs where marginal costs are equal to marginal benefits. Any factors that serve to shift the marginal benefit function downward will reduce the number of children per family, and any factors that shift the marginal cost function upward will also reduce the number of children per family. It is enlightening to use this economic model to explain the economic behavior that may be underlying the demographic transition.

Factors that Affect the Marginal Benefits of Having Children. There are many benefits to having children, including the joy of having and raising children.

[6]See Becker and Lewis (1973) for a discussion of this decision in a more general context.

[7]The marginal cost function need not be upward sloping for this analysis to hold. The marginal cost function may be downward sloping, as long as it is less steeply sloped than the demand function. If the marginal cost function is more steeply sloped than the demand function, then the equilibrium number of children must be either zero or where the demand curve intersects the horizontal axis. Proof of this point is left as an exercise.

However, we will focus on the economic benefits and see how they change as a society undergoes economic development. Economic benefits of children include:

1. Children are labor inputs to production processes.
2. Children can provide for their parents in their parents' old age.

It is relatively easy to see how having more children can increase total agricultural production (although if environmental degradation occurs as a result of increasing intensity of cultivation, this may be a short-term effect). In a preindustrial agrarian society, children virtually always have a positive marginal product. Even very young children can help gather firewood and water, weed gardens, scare away pests, tend livestock, and watch the even younger children.

As a society becomes more industrialized and urbanized, the importance of children as an economic input diminishes. Modern manufacturing processes require an adult labor force with some degree of education. Child labor laws are established, and children are generally unproductive until they enter their middle to late teens. While some urban children can produce income through sweatshop labor[8] or through scavenging and other underground activities, these activities obviously have limited potential.

One interesting feature of the industrial/urban economy is that the importance of classroom learning increases. In a preindustrial agricultural society, education takes place by observing the parents' and siblings' activities in the field. However, in the industrial/urban society, classroom education becomes a much more important component of human capital. As the returns to education grow, an income maximizing strategy might be to increase the investment in each child (better education, nutrition, etc.) and to have fewer children. It is important to note that in many developing countries, all, or a significant part, of the cost of educating a child is the family's responsibility, particularly at the secondary level.

It is less easy to see the importance of children as a "social security system" for their parents; however, this is certainly the case in preindustrial agrarian societies. There are limited options available in agrarian societies to store wealth for the future, as most wealth is in the form of perishable foodstuffs, or livestock that have high maintenance costs. The ability to sell products and store the money for the future is limited by the lack of development of financial institutions. Similarly, because much of the economy is subsistence agriculture, it is virtually impossible to raise the tax money to implement a true social security system. Consequently, the best way for a husband

[8]In some developing countries, young children are paid very low wages to engage in repetitive tasks (such as sewing clothes or weaving rugs) under very poor conditions, providing some income but not contributing to the development of the child or society in general. This exploitive child labor has been of increasing concern in countries such as the U.S., which consume the results of this labor.

and wife to insure that their old age is not a completely bleak and miserable experience is to have a lot of children to provide for the parents when the parents are no longer capable of providing for themselves.

As an economy develops, it becomes easier to store wealth, and public social support systems improve (social security systems, public hospitals, etc.). Consequently, the need to have children to support one's old age diminishes.

Factors that Affect the Cost of Having Children. A variety of factors affect the marginal costs associated with increasing family size. These include:

1. the costs of education,
2. the costs of food and housing, and
3. the opportunity costs of the mother's time.

As an economy becomes more urbanized and industrialized, all of the above factors tend to increase the marginal costs associated with increasing family size. As classroom education becomes more important, the costs of making a child more productive increase. Similarly, as one moves from an agrarian to an urban society, the costs of food and housing increase. Both of these effects are rather straightforward and require little elaboration. However, the role of the opportunity costs of the mother's time is more complicated and requires detailed explanation.

The Economic Role of Women, Population Growth, and the Environment. In most rural economies in developing countries, women supply the bulk of the labor to the subsistence agricultural economy. Women (and their children) are responsible for preparing the fields, planting the seeds, weeding and protecting the crops from pests, harvesting the crops, gathering drinking water, gathering fuelwood, preparing meals, and sometimes herding the livestock (the livestock responsibilities are usually taken by adult and near-adult males). In such a society, women work at hard labor for the entire daylight period and into the night.

Under this type of economic system where women work so hard and long, it might seem strange to suggest that, as a society moves from a rural/agrarian society to an urban/industrial society, the opportunity costs of having children increase. However, this is exactly the case, as in the agrarian economic system, children are not a hindrance to the mother accomplishing her agricultural chores. Very small babies are strapped to the mother's breast while she works, and older children accompany the mother and are actually assigned some of the mother's responsibilities. In an urban economy where the woman might have a service, manufacturing, clerical, or professional job, the presence of children on the work site is not allowed. Consequently, having children interrupts a woman's employment (especially where child care is not readily available and in societies where men do not share these responsibilities), and reduces her productivity, development of her human capital, and chances for advancement. Thus, as equality of treatment for women increases, and as women's roles in the monetary economy increase, one would expect birthrates to fall.

However, this expectation is not a reason to be optimistic about further reductions in birthrates, because the integration of women into the market economy and the development of equal opportunity must first occur. Economic development does not guarantee that these conditions be met, but the lack of fulfillment of these conditions may limit economic development.

For example, many women do not have access to birth control measures. In some cases, this is because of their lack of availability (as many as 300 million women would like to prevent pregnancies, but have no access to birth control methods).[9] In other cases, birth control measures are available, but many women cannot employ birth control measures because of religious and social pressures.

TRADITIONAL MODELS OF DEVELOPMENT AND THEIR FLAWS

Traditional models of development tend to focus on increasing GDP or per capita GDP by increasing capital, transferring new technology to developing countries, and through increasing human capital. This process of increasing GDP is hypothesized to trigger other changes that will lead to the overall development of these societies. Although these methods seem very appropriate at first glance, important exclusions imply that relying solely on these methods will lead to less developmental progress than desired. In order to discuss whether or not these traditional methods work, it is necessary to state exactly what is meant by "development." We will adopt Michael Todaro's definition of development. Todaro (1989, pp. 90–91) suggests that development is

> *. . . both a physical reality and state of mind in which society has, through some combination of social, economic, and institutional process, secured the means for obtaining a better life. Whatever the specific components of this better life, development in all societies must have at least the following three objectives:*
>
> *(1) To increase the availability and widen the distribution of basic life-sustaining goods such as food, shelter, health and protection.*
> *(2) To raise levels of living including, in addition to higher incomes, the provision of more jobs, better education and greater attention to cultural and humanistic values, all of which will serve not only to enhance material well-being but also to generate greater individual and national self-esteem.*
> *(3) To expand the range of economic and social choices available to individuals by freeing them from servitude and dependence not only in relation to other*

[9]According to the U.N. Population Fund, as cited by Jacobsen (*Vital Signs, 1992*).

people and nation-states but also to the forces of ignorance and human misery.

Despite the articulation by Todaro and others of the more multidimensional aspects of development, development efforts tend to focus on GDP. In particular, both commercial banks and international loan agencies focus on GDP measures as an indicator of creditworthiness, which has several important ramifications. It is understandable that private banks should focus on this measure, as their primary responsibility is to their stockholders. However, it is less clear that international loan and development agencies should focus on this measure, for the following reasons. Remember that GDP represents total value of market goods produced in a country. In addition to the fact that GDP is not directly related to many of the development objectives categorized previously, the pursuit of GDP detracts from many development efforts. Obviously, more GDP is good, but policies that promote the development of GDP may do unintentional harm for two reasons. First, GDP does not include the subsistence agricultural sector, since subsistence crops are grown for the farmers' consumption and are not traded in markets. Since the subsistence agricultural sector provides for the basic needs of the majority of the population in many developing countries, its exclusion is significant. Second, although the measure of GDP is reduced to take into account the depreciation of human-made capital, it does not consider natural capital (see Chapter 5 for a discussion of this point). Consequently, when soils are exhausted and forests depleted, future GDP producing and subsistence producing assets are lost.

THE IMPACT OF THE EXCLUSION OF SUBSISTENCE AGRICULTURE FROM MEASURES OF GDP

The exclusion of subsistence agriculture from measures of GDP and the implementation of policies that are oriented to maximizing the growth of GDP has two major impacts on developing countries. First, it encourages a set of policies that favor cash crops (crops grown for sale, particularly in export markets, such as coffee, tea, sugar, bananas, cocoa, palm oil, and tobacco). Second, it may favor a set of policies that enhance the economic well-being of men at the expense of women.

Since the sale of coffee to the United States increases a country's GDP, but the production of corn for consumption by the farmer's family does not affect GDP, a variety of policies have developed that encourage cash crops over subsistence crops. Also, since export crops bring foreign exchange (desirable foreign currency such as U.S. dollars, British pounds, Japanese yen, and so on), that can be used to purchase imports or repay debt, there is even more incentive for governments to encourage the production of export crops. The policies that encourage these crops include fertilizer subsidies, agricultural extension activities, small loan programs, the development of export infrastructure, and price ceilings on food crops. Although the encouragement of export crops is not inherently bad (after all, holding everything else constant, higher GDP is good), it has several spillover effects that are very detrimental.

One of the most important spillovers is that more land is devoted to cash crops, at the expense of land devoted to subsistence crops, and at the expense of land devoted to natural habitats, such as forests. The loss of forests and other habitat has two primary impacts. First, it leads to environmental problems, such as soil erosion. The problems are exacerbated by the fact that the productivity of the land in cash crops is often very short (a matter of several years) due to losses in soil fertility and due to environmental problems associated with the development of monocultures. Because monocultures are very susceptible to pests and diseases in tropical climates, they require tremendous chemical applications to maintain their productivity. For example, the pesticide applications to banana plantations are very large and create toxic exposures for workers and downstream populations, as well as profound ecosystem effects as the pesticides work their way through the food chain. In addition to these environmental degradation effects, there is the loss of common-property resources, such as forests and savannas. These commons provide important resources (food, fuel, and fodder) and contribute to the subsistence agriculture that is managed primarily by women.

The loss of these commons, the environmental degradation of land in general, and the conversion of land from subsistence farming to cash crops may serve to increase the economic hardship and continued impoverishment of women in developing countries. Initially, when development agencies focused on market activities (which are the province of men), they felt that the economic benefits would trickle down to women. However, Jacobsen (1993) argues that this has not been the case. She feels that development programs need to be specifically oriented toward promoting the economic activity of women, and that given the social, cultural, and economic structures in many developing countries, one cannot expect a flow of benefits from the male dominated market economy to the female dominated subsistence economy.

Given that females are more closely tied to the land (although not often as owners), and that females are the exploiters/managers of common-property resources, it is not surprising to find that environmental movements in many Third World countries are dominated by women. For example, in Kenya, two of the most important grassroots environmental organizations were founded by women and have their activities implemented mainly by women. These are the Kenya Water for Health Organization (KWAHO) and the Green Belt Movement. KWAHO was founded as a women's cooperative to train women to build and maintain simple water systems. The Green Belt Movement was founded by women's rights and environmental advocate Wangari Maathiai. Its purpose is to encourage people to find public areas and plant trees on these areas. By 1989, over 20 million trees had been planted in 1,000 tree belts by over 50,000 Kenyan women (World Resources Institute, 1992, pp. 225–226).

Several decades ago, there was a feeling among those involved in international development that development efforts had to take place entirely within the cultural context of the developing country. It was considered to be interventionist to try to change aspects of the current cultures of these countries. While there is a lot of merit to these arguments in many contexts, many feel that it no longer applies in some contexts, particularly with respect to the role of women in developing countries. In the opinion of this author, when basic human rights for women conflict with current cultural norms, the development process should be constructed to help women obtain these rights. The same argument can be made about caste systems and other social structures that lead to discrimination against minority groups. Jacobsen (1993, pp. 76–77) argues that

> *Improving the status of women, and thereby the prospects for humanity, will require a complete reorientation of development efforts away from the current overemphasis on limiting women's reproduction. Instead, the focus needs to be on establishing an environment in which women and men together can prosper. This means creating mainstream development programs that seek to expand the women's control over income and household resources, improve their productivity, establish legal and social rights, and increase the social and economic choices they are able to make.*

SUSTAINABLE DEVELOPMENT

The focus on GDP also tends to promote activities that lead to the increase of current GDP at the expense of future GDP. For example, the cutting of forests to raise current income destroys a natural asset that could produce future income through a variety of activities (see Chapter 12) that do not destroy the forest. Monoculture agriculture is another example of an economic activity that can produce current income, but has a limited productive life and destroys a productive natural asset. The central point is that development policies that promote activities that are not environmentally sustainable do not really contribute to the economic development of the nation, as future possibilities are eroded by the current economic activities.

To emphasize the importance of the concept of sustainability in development, a considerable literature has developed that diagrams the objectives of and the conditions for sustainable development. David Pearce, Edward Barbier, and Anil Markandya (1990) list a set of development objectives that are similar to those defined by Todaro. These are

1. increases in real per capita income,
2. improvements in health and nutritional status,
3. educational achievement,

4. access to resources,
5. a "fairer" distribution of income, and
6. increases in basic freedoms.

Pearce and his coauthors suggest that sustainable development be defined as a process whereby the above objectives of development (or a modified version of the above list) be nondecreasing over time. The World Commission on Environment and Development (1987) defines sustainable development as a process by which the current generations can "meet their needs without compromising the ability of future generations to meet their needs." While either of these may constitute a suitable definition of sustainable development, Pearce, Barbier and Markandya point out that a definition of sustainability and the conditions for achieving sustainability are very different.

In Chapter 5, the role of capital in sustainable development was discussed. This discussion emphasized the role of environmental resources that generate environmental services in sustainable development. Maintaining these environmental resources has been a particular problem for many developing countries, because their lack of current income forces them to treat renewable resources as exhaustible resources, degrading the ability of the environmental resources to regenerate and to provide ecological services. Examples include intensive cultivation which leads to soil erosion, deforestation, intensive grazing which leads to desertification, destruction of coral reefs and overexploitation of fish stocks. How can sustainable growth of income be generated? The conceptual answer to this question can be developed by looking back to Figure 17.2. The marginal product of labor and current income must be increased while maintaining the integrity of environmental resources. In other words, a way must be found to meet current consumption needs, increase all types of capital (human-made, human and environmental), and protect the environment from degradation. Although each individual policy or development program might not be able to address all these problems simultaneously, the total package of development programs in a country must address all three of these problems. In addition, the other dimensions of development such as human health, education, basic freedoms, and so on, must be simultaneously addressed.

SUSTAINABLE DEVELOPMENT POLICY

Many past development projects emphasized the creation of human-made capital such as dams, roads, and factories. As originally conceived, these projects may have been thought to be sustainable, but they interact with the environment in a way that reduces the sustainability of other productive activities. For example, a hydroelectric dam seems like the epitome of a renewable, sustainable resource. However, even under ideal conditions, the dam will eventually fill with silt, which reduces its capacity to produce electricity. If the dam also encourages the agricultural development of the surrounding hillsides (because people move along the road that is built to the

WILDLIFE, ECOTOURISM, AND ECONOMIC DEVELOPMENT

The natural resources of many developing countries provide an opportunity to earn income and preserve the resources at the same time. For example, tourism is the leading earner of foreign exchange in Kenya, where tourists come to participate in photographic safaris and tropical beach vacations.

Ecotourism is growing rapidly in the Amazon region, the Galapagos Islands, parts of Africa, and many Caribbean and Central American countries. Many people with high incomes from Western Europe, North America, and Japan are willing to pay large amounts of money to see these "ecological wonders." This type of activity is often viewed as a way to "have your cake and eat it too" in the sense that it is possible to generate income and maintain resources at the same time.

However, one must be cautious in the development of these tourist activities for two reasons. First, it is possible the tourism takes a form where it generates environmental degradation that destroys the very resources the people are traveling to see. For example, in the fragile semi-arid grassland areas of Kenya, the vehicles that take tourists to see the wildlife spectacles can compress the soil and permanently render it unable to support vegetation. Sewage and solid wastes generated by hotels and resorts can pollute the water resources which the tourists come to see. Second, care must be taken to ensure the revenue that tourism generates raises the quality of life of the local inhabitants in the tourism areas. This factor is important for several reasons. If tourism does not raise the quality of life, it does little to enhance the development process. For example, much of a tourist's total tourism expenditure accrues to international airlines, as airfare is usually the largest component of expenditure. This obviously does little to aid in the development process. Care must be taken that other expenditures benefit the tourism region and lead to improvements in the quality of life in that region. Additionally, unless the local populations benefit from the tourism, they have little incentive to preserve the resource upon which the tourism is based. Since there is a scarcity of agricultural land in many of these areas, it is important that local residents see the benefits of preserving the resources. Giving the local residents a financial stake in the preservation of the environment is a very positive step toward environmental preservation. Instead of wild animals being seen as competitors to the residents' livestock, they are seen as a source of jobs and other benefits (schools, hospitals, etc.). Under these circumstances, local residents will often take positive steps to inhibit the operation of poachers (hunters who illegally take game to sell body parts, such as tusks, skins, horns, and internal organs, in illegal overseas markets).

A simple way of summarizing the policies which must be taken to preserve these ecological treasures is that a global potential Pareto improvement must be made into an actual Pareto improvement for those who have a stake in the environmental resources. The world as a whole benefits from the preservation of these environmental resources, but the country that contains the resource and the communities in proximity to the resource pay the opportunity costs (lost economic opportunities) of preserving the resource. If the world's willingness to pay for preservation can be translated into increased economic opportunities for those who are in proximity to the resources, preservation can be achieved.

In order to maximize the development benefits of ecotourism, more research needs to be conducted into what tourists value in an "ecotourism" vacation and how to structure ecotourism to preserve the ecological resources and improve the quality of life of local populations. If these questions can be answered and if ecotourism can be properly structured, it can play a substantial role in the sustainable development of tropical countries.

dam, or because factories are built near the dam to take advantage of cheap electricity), then the destruction of the hillside forests and exposure of the soil resulting from agriculture will lead to even faster siltation of the dam. Another example is the creation of factories that do not control their pollution emissions, and the pollution emissions adversely affect human health and agricultural productivity.

It is interesting to note that government development policies often create incentives for environmental degradation. In Chapter 12, inefficient forestry policies were discussed in great detail. These included poorly constructed concessionaire agreements, poorly defined property rights, and poorly defined incentives, such as subsidizing cattle ranching. These poorly defined policies are not limited to forests. The World Bank (1992, p. 65) lists several types of inefficient government development policies, including

1. subsidization of agricultural inputs, such as fertilizer,
2. subsidization of energy inputs,
3. subsidization of logging and cattle ranching,
4. nonaccountability of public sector polluters,
5. provision of services, such as water and electricity, at subsidized prices, and
6. inefficient management of public lands.

Therefore, a first step toward a well constructed development plan is the elimination of the above policy negatives. In addition, positive steps must also be initiated. The World Bank (1992) lists three classes of development policies that can increase the quality of life of developing countries, that will minimize environmental degradation, and that can be sustainable. These three classes are the elimination of inefficient policies, the provision of public and private investments that have net benefits independent of environmental benefits, and the correction of market failures that lead to reduced environmental quality.

Inefficient government policies that lead to the loss of GDP and the deterioration of the environment include subsidies for production inputs; subsidies for certain extractive activities, such as cattle ranching and logging; and poorly defined property rights to land for farmers. Each of these policies lead to a misallocation of the country's productive resources and to an inefficiently high level of environmental degradation.

Many governments subsidize the use of certain inputs in selected production activities in order to increase output in these activities. For example, energy, fertilizer, and irrigation water are often provided at subsidized prices. These subsidies encourage the excess use of these inputs and distort the economy in favor of those activities that use those inputs. For example, in the former Soviet Union, a massive irrigation program was developed using waters from rivers that flowed into the Aral Sea. The agricultural production never came close to meeting project goals and resulted in both environmental and economic damage. With so much water diverted to irrigation

Malaria is a tropical disease that has important impacts in the Amazonian rain forest and other tropical areas. The malarial season coincides with the beginning of the planting season, so when farmers contract debilitating malaria, it has important impacts on their ability to produce the entire year's crop.

Although malaria might be viewed as a natural phenomenon, it is profoundly affected by human activity. The construction of field-side and roadside ditches provides habitat for malarial mosquitos to breed. Deforestation reduces shading of surface waters, which also increases mosquito abundance. Additionally, greater population densities allow the disease to spread more easily.

Donald Jones and Robert O'Neill (1993) examine the economic decisions of farmers, who must allocate their labor between producing crops and engaging in preventative ecological activities (draining swamp areas, and so on).

They examine two alternative government policies, one which provides information on preventative ecological actions and one which provides medicine to suppress the effects of malaria. Jones and O'Neill find that both sets of policies are effective in improving agricultural productivity (because farmers are healthier), but the medication policy exacerbates deforestation while the information policy restrains it.

where evaporation takes place at a higher rate than in rivers, the Aral Sea shrank to a fraction of its former size.[10] This shrinkage caused immense ecosystem damage and destroyed a valuable commercial fishery that supplied both nutrition and income to the region. If the irrigation water was priced at its true social cost, the project would never have been viewed as having any economic benefits.

Similar arguments can be made with respect to both energy and fertilizer.[11] In fact, the artificially low price at which energy traded in Eastern Europe is one of the major factors that led to the abominably high levels of air pollution in that region.

It is relatively easy to see the importance of the elimination of policies that subsidize and encourage activities such as logging and cattle ranching. In many cases, these activities were only privately profitable because of the subsidies and resulted in net costs to the economy without even taking into consideration the environmental effects. The consideration of environmental costs implies that the subsidization of these activities leads to large social losses.

Finally, it is critically important that farmers' property rights to land be carefully defined. If property rights are appropriately defined, then farmers have an incentive to treat their land as a long-term asset and to protect their

---

[10]See Chapters 15 and 16 for a complete discussion of the environmental problems associated with irrigation.

[11]See Chapter 8 for a discussion of the environmental problems associated with energy use, and Chapter 15 for a discussion of the environmental problems associated with fertilizer use.

land from degradation. Notice that this process does not always imply the creation of private property rights. Many lands, particularly forestlands and rangelands may be most effectively used when maintained as a commons. However, property rights must still be defined to protect the land from open-access exploitation. If property rights to common land are given to local social organizations (such as villages, clans, tribes, or farmers' cooperatives), then the local social organizations can develop socially enforceable rules governing the use of these communal resources. This form of property rights is, in fact, the predominant form that existed among indigenous peoples of the world before they were incorporated into colonies or independent nation–states.

Policies that initiate public investment and that encourage private investment can also lead to important developmental benefits. The World Bank (1992) cites several investments that are representative of this type. These include investment in water supply and sanitation, soil conservation, and the education of women.

In addition, it is important to address market failures that cause a divergence between the marginal social costs and marginal private costs of market activities. These market failures were addressed in a general application in Chapter 3. Government policies to address these market failures could include taxes on emissions and wastes, regulation on the disposal of toxic and hazardous waste, and appropriately structured fees for timber and mineral extraction (World Bank, 1992).

SUMMARY

Many Third World countries find themselves in a poverty trap that is partially generated by environmental degradation. Poverty leads to overexploitation of environmental resources in order to generate current income, and the degradation of environmental resources lowers productivity and income. Population growth may exacerbate these pressures and lead to increased poverty and environmental degradation. In addition, population growth may make market failures more severe and make it more difficult to correct market failures.

Current development policies must be aimed at breaking this vicious cycle of poverty and environmental degradation by maintaining current income with environmentally sustainable development projects, such as agroforestry. In addition, direct government action aimed at eliminating inefficient government programs, stimulating appropriate investments, and addressing market failures are key components of a development program. Also, development policies must ensure that development benefits the entire population. Consequently, development policies must not ignore the role of women in the economy of developing nations and must seek to eliminate practices that rele-

ENVIRONMENTAL POLICIES FOR THE SUBSISTENCE SECTOR

This book has devoted much time to the development of policies such as economic incentives (pollution taxes, marketable pollution permits, deposit–refund systems, etc.). An interesting question is whether these systems can work in the non-cash subsistence sectors of developing countries. For example, although deposit–refund or liability systems may protect tropical forests from damage by commercial logging, can they be used to stop peasant farmers from cutting the forests to create cropland and to promote conservation of renewable resources among traditional forest dwellers who practice a combination of hunting, fishing and agriculture? Use of a system can be difficult because the subsistence inhabitants have no assets upon which to base a deposit–refund or liability system or to confiscate as the penalty associated with a command and control system.

This problem arose in the state of Amazonas, Brazil, where the range of a rare primate, the uakari, coincided with the area in which forest dwellers (of mixed indigenous and European ancestry) had established villages and engaged in hunting, fishing and agriculture for many generations. It was feared that habitat would be lost to the agricultural activities of the local population, which could cause the extinction of the uakari. It would have been difficult to enforce removing the people from the land (and it would have created a hardship for these people) to establish a sanctuary for the uakari, yet it would be difficult to protect the habitat of the uakari through either economic incentives or command and control.

The solution that was developed was to conduct scientific research on the fish and animals upon which the local people depended for a portion of their subsistence. The scientific findings were translated into practical guidelines on hunting and fishing practices, which increased both the abundance of the animals and the harvest of the people. This increase in hunting and fishing harvest lessened the need to create cropland, and the increase in the standard of living created credibility for the scientists in the eyes of the local people; additional conservation measures have been implemented. As with the example of goldmining and mercury (which was discussed in Chapter 9), and agroforestry and the rain forest (which was discussed in Chapter 12), solutions to environmental problems in the subsistence sector may be best developed by finding ways to make conservation and environmental protection economically beneficial, and then educating the local people in these methods.

gate women to second-class status. Although there is no "magic formula" or specific recipe for development, much progress has been made in recent years in more properly conceptualizing the development process, which can provide guidance for more enlightened policies.[12]

---

[12]For current information on the environment and development, investigate the webpages of the World Bank (http://www.worldbank.org/) and the United Nations Development Programme (http://www.undp.org/), and the 1997 Global State of the Environment Report of the United Nations Environment Programme (http://www.grid2.cr.usgs.gov/geo1/).

REVIEW QUESTIONS

1. What is the "vicious cycle of poverty," and how does environmental quality affect it?
2. What is the effect of income on population growth?
3. What is the effect of population growth on the environment and income?
4. What factors affect the costs of having children?
5. What factors affect the benefits of having children?
6. Prove footnote 7.
7. What is meant by "sustainable development"?

QUESTIONS FOR FURTHER STUDY

1. Can population growth be limited without redefining the role of women in society? Why or why not?
2. Should countries who provide development aid insist on increased rights for women and minorities, even if it conflicts with the developing countries' cultural norms?
3. How does a focus on GDP affect the development process?
4. Why should the focus of development be in sustainable development rather than in maximizing the present value of the time path of GDP? (Hint: How do the two alternatives treat the future differently?)
5. Why do property rights matter for the development process?
6. How does government subsidization of export crops affect the development process?
7. Is environmental quality a luxury that should be the concern of only wealthy nations?

SUGGESTED PAPER TOPICS

1. Investigate the economic paradigm of sustainable development. How does it differ from the neoclassical economic paradigm? Start with the book by Pearce and Warford. Also, check the webpages cited in the summary of this chapter.
2. Conduct an empirical analysis of the relationship between economic (or social variables) and environmental variables. *World Resources 1992–93* and *World Development Report 1992* are excellent sources of data on this subject.
3. Look at the problem of contaminated drinking water in Third World countries. What economic factors (market failures, etc.) lead to this problem? How do these economic factors affect the ability to develop solutions? Start with the *World Development Report 1992* for current references.

WORKS CITED AND SELECTED READINGS

1. Barnett, Harold J., and Chandler Morse. *Scarcity and Growth.* Baltimore: Johns Hopkins University Press, 1963.
2. Becker, Gary S., and H. Gregg Lewis. "On the Interaction between the Quantity and Quality of Children." *Journal of Political Economy* 81 (1973): 279–288.
3. Brown, Lester R., et al. *State of the World 1993.* New York: Norton, 1993.
4. Brown, Lester R., Christopher Flavin, and Hal Kane, eds. *Vital Signs, 1992.* New York: Norton, 1992.
5. Jacobsen, Jodi L. "Coerced Motherhood Increasing." In *Vital Signs, 1992,* edited by Brown, Flavin, and Kane. New York: Norton, 1992.
6. ———. "Closing the Gender Gap in Development." In *State of the World 1993,* edited by Brown et al. New York: Norton, 1993.
7. Jones, D. W., and R. V. O'Neill. "A Model of Neotropical Endogenous Malaria and Preventative Ecological Measures." *Environment and Planning* 25 (1993): 1677–1687.
8. Malthus, Thomas Robert. "An Essay on the Principle of Population," In *An Essay on the Principle of Population, Text, Sources and Background, Criticism,* edited by Phillip Appleman. New York: Norton, 1975.
9. Pearce, David, Edward Barbier, and Anil Markandya. *Sustainable Development.* London: Gower, 1990.
10. Pearce, David W., and Jeremy J. Warford. *World Without End.* Washington, D.C.: Oxford University Press for the World Bank, 1993.
11. Repetto, Robert. *World and Enough Time.* New Haven: Yale University Press, 1986.
12. Todaro, Michael P. *Economic Development in the Third World.* 4th ed. New York: Longman, 1989.
13. World Bank. *World Development Report 1992: Development and the Environment.* New York: Oxford University Press, 1992.

14. World Commission on Environment and Development. *Our Common Future: The Brundtland Report.* New York: Oxford University Press, 1987.
15. World Resource Institute. *World Resources 1992–93.* New York: Oxford University Press, 1992.
16. World Resource Institute. *World Resources 1994–95.* New York: Oxford University Press, 1994.
17. World Resource Institute. *World Resources 1996–97.* New York: Oxford University Press, 1996.

Prospects for the Future

Is the cup half empty or half full?

Anonymous

INTRODUCTION

The above quote is commonly used to emphasize that reality is a matter of perspective. One can look at an 8-ounce cup with 4 ounces of water in it and have either an optimistic or a pessimistic outlook. Both the pessimist and the optimist realize that there are 4 ounces of water. Of course, the surroundings in which one views the cup has an effect on the outlook. If one is dying of thirst in the middle of the desert, one would have a different outlook than if one was standing next to a faucet that provided clean drinking water. The surroundings in which one views the cup also has a great deal to do with the decisions that one makes about that 4 ounces of water. In one case, there is no need to worry about immediately drinking the water and no need to develop a plan to obtain more water. In the other case, decisions impact immediate prospects for survival.

One can look at the "environmental crises" facing the planet Earth in much the same fashion. We have made much progress in dealing with some environmental problems, particularly in developed countries. Conventional air and water pollutants have been regulated, birthrates have declined to the point where zero population growth will be realized in the near future, the ozone depletion problem has been addressed, solutions to the global warming problem are being negotiated, and governments are developing policies based upon principles of sustainability. There is a global societal recognition of the importance of protecting the environment. Although there are many problems remaining to be solved in developed countries, one can point to significant progress.

For all the reasons to be optimistic about the situation in developed countries, there are reasons to be pessimistic about the situation in many developing countries.[1] First and foremost, population growth is far from being under control. Second, environmental degradation is taking place at a massive and unprecedented rate, which is leading to the loss of much of the world's renewable resource base, including loss of tropical forest, desertification, fisheries depletion, and soil loss. Water pollution is a leading cause of death in many developing countries, leading to infant diarrhea and other intestinal problems. Third, the economic situation in many developing countries prevents the people from having a long-term perspective. This short-term perspective and a lack of understanding of the long-term economic importance of maintaining environmental quality implies that many developing nations have not recognized the need to take immediate steps to preserve the environment.

Will environmental degradation lead to a collapse of the world's ability to support its population, or will social responses mitigate environmental degradation and move us to a sustainable future? Obviously, this question cannot be answered in the short, concluding chapter to this book. Rather, this chapter will attempt to summarize some of the debate on this issue. Since it is not possible to test to see if the pessimistic or optimistic forecast (or some forecast in the middle) is correct, this chapter will present alternative outlooks on the issue, with a focus on a discussion of sustainability.

ABSOLUTE SCARCITY

The concept of absolute scarcity has shaped much of the debate about scarcity. This concept originated in the work of Malthus, who argued that since there was a fixed amount of land, scarcity was an inevitability. Malthus argued that the law of diminishing returns implies that if land is held constant, increased labor must ultimately be faced with diminishing marginal productivity. Thus, population will grow faster than the marginal product of labor, so population will grow faster than the food supply. Since Malthus believed that the only constraint on the growth of population is the availability of food, population will continually grow until it reaches the limits of the food supply. At this point, each individual will be at a subsistence level of consumption (just enough food to survive), and increased population growth will be limited by starvation.

Obviously, Malthus's prediction of nearly two centuries ago has not taken place. In some areas of the world, the population is at a subsistence level of consumption, but population growth has been continual and exponential. It is not likely that the food supply or other constraints on population

[1]While there are reasons to be pessimistic about the situation in most developing countries, some countries, particularly in Latin America, are making substantial progress in developing effective environmental policies.

will serve to limit population growth in the near future. In addition, when developed nations are also considered, there has been a large increase in the amount of consumption (both food and goods) per capita.

The idea of absolute scarcity has been revisited in modern times. In the 1960s and 1970s, as world population exploded, pollution intensified, hunger increased, and shortages of energy and other resources materialized, absolute scarcity was seriously considered in public debate. Many scholars citing the finite nature of the world and its resources, argued that scarcity was inescapable. Since these scholars viewed the finiteness of resources as a limit on the growth of population and economic activity, these scholars have come to be known as neo-Malthusians.

ENTROPY AND ECONOMICS

Nicholas Georgescu-Roegen (1971, 1975) has argued that entropy is the source of ultimate scarcity. Entropy refers to the amount of disorder in a system. The higher the entropy the higher the state of disorder, the lower the entropy the greater the state of order. The third law of thermodynamics (the entropy law) states that the universe (or a subsystem of the universe) is in a constant state of entropic degradation. A simpler way of phrasing this principal is that as time and the physical forces of the universe progress, energy and materials tend to dissipate and become more difficult to access.

The relevance of the entropy law for economic activity and prospects for the future is that useful goods and resources tend to be low entropy, and the economic process increases their entropy. Although we can switch from coal to oil to uranium, and we can substitute plastic for steel, we cannot substitute away from low entropy matter. According to Georgescu-Roegen, it is the ultimate source of scarcity.

For example, a fossil fuel, such as coal or oil, is characterized by low entropy. It is a relatively homogenous substance, with a much higher concentration of carbon than the environment as a whole. However, when the coal is burned in the economic process, the carbon is scattered throughout the atmosphere. The heat that is initially highly concentrated becomes highly dispersed. The process of using the coal increases its entropy and reduces its usefulness.

Since the amount of low entropy on earth is fixed, and since entropic degradation is irreversible, it will be the ultimate constraint on the human species.[2] Not only does the law of entropy limit the size of the social system, Georgescu-Roegen argues, but it constrains the life span of the human species as well. Entropic degradation implies that the time span of human exis-

---

[2]Entropic degradation is not irreversible if energy is available to reverse it. However, this energy must come from somewhere, so the entropic degradation of a system can be reversed, but only at the expense of the entropy of another system. For example, a person can increase his or her low entropy (grow), but only at the expense of the entropy of the rest of the system.

tence is limited because we will eventually run out of low entropy matter to support life.

There are two basic criticisms of the work by Georgescu-Roegen. First, since we do not really know the initial stock of low entropy matter, we do not know if our rate of consumption is high relative to the stock. Second, we receive constant infusions of energy from the sun, and this energy can be used to slow the process of entropic degradation.

Although the entropy constraint might not be binding in the foreseeable future, the relationship between entropy and the economic process provides some guidelines for policy. Activities should be encouraged that are less degrading of low entropy, as these impose less costs on the future. Of course, the current cost of encouraging these activities must be compared to the future costs of entropic degradation. Conserving low entropy is conceptually equivalent to saying that we should minimize use of inputs and minimize waste, which is what Daly calls for in his examination of the necessity and feasibility of moving our economic system to a steady state.

DOOMSDAY MODELS

As the fields of computer science, systems science, and operations research began to develop in the 1960s and early 1970s, these tools began to be applied to the questions of overall scarcity and environmental degradation. *The Limits to Growth* model of Meadows, Meadows and Behrens (1972) was the most widely cited of this class of models.

The basic premise of these models was to specify initial conditions (of population, wealth, arable land, capital, natural resources, etc.) and then allow these variables to grow. The implications of growth are computed (that is, more output implies more pollution; more output implies faster resource depletion; more people means more food cultivation and more soil erosion). What these models found was that in a rather short period of time (20–50 years), some sort of constraint interfered with the ability of the economic system to support population. Depending on the scenario that was examined, either resources were depleted, food shortages developed, or massive pollution developed. These constraints interfered with growth and implied collapse. Output per capita and food consumption per capita declined in these models and eventually was severe enough to constrain population growth. Because these models predicted such a dismal outcome for human society, they came to be known as doomsday models. Since they were similar to Malthus' work in having scarcity constraints limit population and impoverish the human condition, these models were also known as neo-Malthusian models.

These models caused considerable excitement for two reasons. First, society was just beginning to become aware of the problems of environmental degradation and population growth, as books such as Rachel Carson's *Silent Spring* (1962) and Paul Ehrlich's *Population Time Bomb* (1968) were published. Second, the public was enthralled by (and did not really understand the nature of) computers and computer models. Many people felt that if a computer predicted the collapse of humankind, this prediction must be taken seriously.

There are two criticisms of these doomsday models. First, they are tremendously sensitive to the assumed rates of exponential growth. Second, there is no endogeniety or positive feedback in the model, including the effect of scarcity on prices and prices on scarcity.

The sensitivity to growth rates can be illustrated by example. If population is increasing at 2 percent per year, it will double in 35 years, and if it is increasing at 3 percent a year, it will double every 23 years. If we start with a population of around 5 billion at the present, a population that grows at 2 percent for 100 years will be 30 billion, and a population that grows at 3 percent will be approximately 100 billion. The importance is apparent even for smaller growth rates and smaller differences between growth rates. For example, if the population is growing at 1.4 percent per year, it will double every 50 years, and if it is growing at 1.5 percent, it will double every 47 years. This difference might seem insignificant, but if we start at 5 billion people, there will be 20.3 billion in 100 years with a 1.4 percent growth rate, and 22.4 billion with a 1.5 percent growth rate. The 2 billion person differential does not seem as insignificant as the difference between 1.4 and 1.5 percent.

The other major criticism is that there is no endogeniety or feedback in the system. There is no mechanism for increasing scarcity, pollution, or population to affect behavior in these models. In other words, even as we are forced to wear gas masks because of the hideous pollution, we take no steps to reduce the level of pollution.

A major absence of feedback is that market prices are not included in these models. As scarcity increases, price will increase, which will trigger a full set of market responses that tend to lesson the scarcity problem. These effects are discussed more fully in the next section.

Before leaving this section, a comment on the importance of the neo-Malthusian models should be made. The people who developed these models did so to create a new intellectual paradigm that addresses the role of scarcity and the long-term prospects for humankind. It is easy to retrospectively criticize these models for omissions and shortcomings, because new ideas had not yet had a chance to be tested and perfected. These models served an important purpose in encouraging economists and other scientists to think about the issues of scarcity and the future of the planet, and they are the intellectual foundation for the models of sustainability that are discussed later in the chapter.

PRICE, SCARCITY, AND NEO-CLASSICAL ECONOMICS

The work of Hotelling (1931) (who identified user cost as the mechanism that makes price reflect scarcity) forms the basis of the neo-classical (modern microeconomic) economic approach to scarcity. This approach is epitomized by the work of Barnett and Morse, who investigated the scarcity question in the 1960s.

One of the most important contributions of Barnett and Morse (1963) is that they discussed mechanisms through which price mitigates scarcity. This aspect is extremely important because as price increases to reflect scarcity, it sets into motion mechanisms that reduce scarcity.[3] For example, if oil becomes more scarce, the price will rise. This higher price has several important effects. First, it generates added incentives for exploration of new sources of oil. Second, it makes it more profitable to remove oil from more difficult places. Third, it gives added incentives for research and development into new technologies for finding, extracting, and utilizing oil (more efficient engines, etc.). The higher price also discourages current consumption of oil. Finally, it encourages the development of substitutes for oil, such as solar energy.

There are many historical examples of what appears to be shortage that is mitigated by scarcity-induced forces, such as higher prices. The whole history of oil exploration and development is seeming scarcity mitigated by new discoveries. When it looked like the Pennsylvania oil fields were being depleted, more and larger fields where found in Texas and Oklahoma. Then as oil became more expensive, new discoveries where found in North Africa and the Persian Gulf. As oil became apparently scarce and very expensive in the 1970s, new fields came on line in Alaska, Mexico, and the North Sea. Currently, the price of oil is very low, at its lowest since World War II (inflation adjusted). When oil again becomes more scarce and more expensive, new adjustments will occur. These adjustments could include development of more oil producing areas or the development of alternatives to oil.

Although these mechanisms can tend to mitigate scarcity, they may not necessarily eliminate it or prevent it from impacting per capita output. Consequently, Barnett and Morse further investigated the question of scarcity. The first step was to develop conceptual models of scarcity, and the second step was to conduct empirical tests to determine if scarcity is increasing or decreasing.

The first step in Barnett and Morse's conceptual discussion of scarcity was to examine the Malthus hypothesis that the law of diminishing returns implies that as labor increases, the marginal product of labor must decline. As discussed in Chapter 17, Barnett and Morse point out that while Malthus correctly applied the law of diminishing returns to an economy where the only inputs are labor and land, Malthus's conclusion does not necessarily hold when one moves to an economy of more than two inputs. The law of diminishing returns states that if all but one input is held constant, the remaining input will exhibit diminishing returns. In a two-input economy, holding land constant while labor increases is equivalent to holding all

---

[3]See Chapters 2, 3, 8, and 9 for a discussion of price, user cost, and scarcity.

inputs equivalent while labor increases. However, if one considers an economy with more than two outputs (such as land, labor, capital, energy, etc.), increasing labor while only land is held constant does not imply that the marginal product of labor must be declining. Increases in capital (either quantitative or qualitative) can generate an increasing marginal product, even if land is held constant. The importance of this is that a finite amount of land (or natural resources) does not necessarily imply declining marginal product of labor. According to this discussion, the existence of scarcity is an empirical question that can only be answered by looking at real world phenomenon.

Barnett and Morse suggest two types of tests for the existence of scarcity and two methods for conducting each test. First, they stress that it is important to examine the scarcity of individual resources and scarcity of resources in general. It is important to examine resources in general, because as a particular resource becomes more scarce, other resources could be substituted for it. Both the scarcity of individual resources and the scarcity of substitution possibilities must be examined.

In examining these types of scarcity, Barnett and Morse suggest two hypotheses on which to build empirical tests. Both hypotheses are based on the supposition that if resources become more scarce, their opportunity cost should rise. Their strong scarcity hypothesis looks at whether the costs of extracting resources (labor and capital costs) have increased over time. Their weak scarcity hypothesis looks at whether the price of extractive materials has been rising relative to the price of other goods. Both of these hypotheses are tested using historical data, and their general conclusion is that scarcity is not increasing. This conclusion holds for most extractive resources when viewed separately and for extractive resources as a whole. The two major criticisms of this conclusion are discussed below.

First, it is difficult to make predictions into the future by looking at past trends, particularly when the forces underlying the trends are not included in the analysis. For example, the Barnett and Morse study looks at the change in extraction costs over time, but does not look at the factors that affect extraction cost, such as the characteristics of resource reserves, technological innovation, the availability of cheap energy, and so on. Second, Barnett and Morse choose a monotonic specification for their empirical tests. A monotonic function is one that is always increasing or always decreasing. If one fits a monotonic function through a longtime series of data, recent changes in the data will not be reflected in the sign of the slope of the function, as the preceding data outweighs the more recent change. This relationship is illustrated in Figure 18.1, where a stylized line-fitting will have a negative slope and not reflect the increases in costs in recent years. Slade (1982) has reestimated some of these functions for selected minerals with a quadratic (u-shaped) function and found that we are now on the upward sloping part of the "u."

Second, the measure employed by Barnett and Morse might not be the best indicator of scarcity. Different properties of scarcity indicators are dis-

FIGURE 18.1

FITTING A MONOTONIC FUNCTION TO NON-MONOTONIC DATA

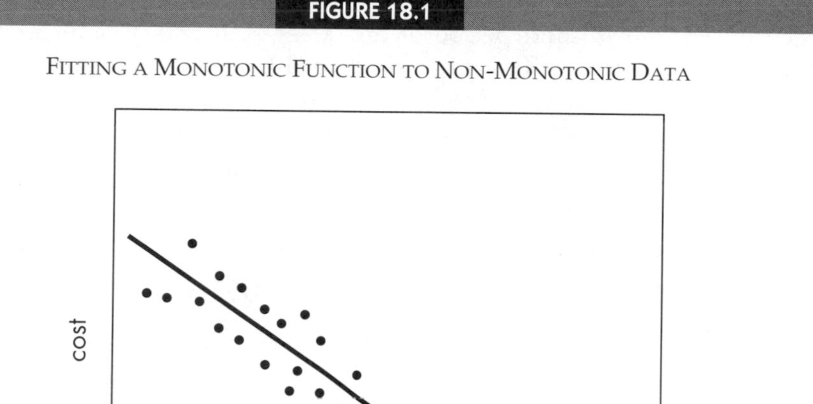

cussed in Smith (1979). For the purpose of this chapter, the most important missing dimension is that these measures do not include the full social cost of removing extractive resources, because they exclude environmental costs. The labor and capital cost associated with removing extractive resources could be falling at the same time that environmental costs are rising. This increase in environmental costs is likely to come about for three reasons, all of which have to do with the fact that the resources that are easiest to extract (closest to the surface, highest concentration, closest to market areas, etc.) are extracted first. First, as lower and lower grades of ore are extracted, the amount of waste material increases. Second, as lower grade ores and deeper deposits are extracted, the use of more energy is required. Third, as environmental degradation increases (from both extractive and nonextractive activities), the marginal social damages of degradation may be increasing. This increase is particularly true with the loss of unique natural areas, such as wilderness areas that are diminishing in number as more and more natural areas are lost to extractive activity and other development.

One might argue that if the price of the extractive good includes its environmental cost, then price would still serve as a good signal of resource scarcity. In fact, it is often argued that the current environmental regulations in the United States fully internalize the environmental cost of goods. However, even if one accepted this argument, one could not use it as a basis for measuring scarcity, because the strict environmental regulations have only been in place for a short period of time (less than two decades), which is insufficient time to conduct a time series analysis. However, since empirical tests of the change in total social costs are not available, it is difficult to say whether scarcity is increasing or decreasing.

SYSTEM-WIDE CHANGE AND SUSTAINABILITY

Recent concern with scarcity is not that we are running out of individual materials in general, but that economic activity is fundamentally changing the ecosystem and physical environment. This system-wide change, it is argued, interferes with the long-term ability of the planet to support human population. Global warming, acid rain, ozone depletion, deforestation, soil loss, water availability and quality problems, and the loss of biodiversity reflect this system-wide change. Current work by past neo-Malthusians (such as Meadows and Meadows and others, 1972, 1992) and by other authors, such as Gore (1992) and Brown (1991), cite this fundamental system-wide change as the potential source of limitations of future prospects.

These authors all argue that the current methods of economic production are not sustainable in that they generate this system-wide change. Since environmental resources and the state of the environment are fundamental to the production process, system-wide decline could lead to a great future loss of productive ability and severely limit the prospects of future generations.

Not everyone accepts the hypothesis that system-wide change is imminent nor significant even if it were to occur. Ray and Guzzo (1990), for example, challenge the scientific validity of many of the studies that predict global warming, attribute ozone depletion to economic activity, or measure the loss in biodiversity. This controversy over whether or not system-wide change exists leads to an extremely important policy question, and one that has not been adequately addressed. This question is: What level of proof is required before one acts?

The scientific method requires rigorous proof before claiming that a result holds. Typically, 95 percent or 99 percent confidence is required before rejecting a null hypothesis.[4] This high degree of burden of proof has been passed forward to the policy arena. For example, skeptics of global warming argue that there is not sufficient proof that global warming exists. However, do we wait for 99 percent confidence before taking action? For example, if

[4]A null hypothesis is the supposition that the researcher attempts to disprove, in order to accept the alternative hypothesis. In the global climate change area, the null hypothesis is that CO_2 emissions are not causing global warming.

we are 80 percent sure that there will be global warming as a result of carbon dioxide emissions, and 50 percent sure that there will be significant negative consequences, do we act now, or wait?

A risk-adverse posture would be to begin to take actions now to limit system-wide change in the future. People such as Pearce and Warford (1993), Gore (1992), Brown (1991), and Meadows, Meadows and Randers (1992) argue that the criterion of sustainability be used to evaluate present decisions. According to the criterion of sustainability, we do not need to prohibit economic growth, but we need to constrain growth and pursue growth paths that are sustainable, where the change that the economic activity engenders does not undermine the ability to engage in that activity or alternative economic activities in the future.

It should be noted that the concept of sustainable development is fundamentally different from the idea of Malthusian limits to growth. The concept of sustainable development implies that some growth paths are possible, while Malthusian concepts say that growth is fundamentally limited. It is rather interesting to note that some of the most noted neo-Malthusians have now become advocates of sustainability and have backed away from their predictions of societal collapse. In their newest book, *Beyond the Limits: Confronting Global Collapse, Envisioning a Sustainable Future,* Meadows, Meadows and Randers (1992) use computer models to contrast sustainable paths to non-sustainable paths. Sustainable paths would include such policies as controlling greenhouse gas emissions, limiting soil erosion, limiting population growth, and so on.

The current work on sustainability and the past work in traditional economics, which shows that price can limit scarcity, are not inconsistent. Their relationship can be found in John Krutilla's 1967 call to reconsider the concept of conservation. Krutilla argued that we need to refocus our old concern with the conservation of extractive resources and be more concerned with the conservation of unique natural environments. A current extension of Krutilla's reconsideration argument would be to extend the concern with conservation to the viability and sustainability of entire environmental systems. This need to focus on systems and not on individual resources or sites is the fundamental premise that underlies the concept of sustainable growth. *Growth that is conserving of systems is sustainable, even though individual resources may be depleted. Growth that degrades systems is fundamentally unsustainable.*

People who argue for movement toward a theory of sustainable economics do not argue that price is incapable of mitigating the types of scarcity that Hotelling (1931) and Barnett and Morse (1963) were concerned about. Rather, they would argue that system-wide environmental degradation will not be incorporated into market prices, so public policy is necessary.

The advocacy of sustainable economics is also not inconsistent with cost–benefit analysis. However, costs and benefits are treated much differently in the sustainability paradigm than in the traditional economics paradigm, as future costs and benefits are not necessarily discounted to present values in the sustainability paradigm. As discussed earlier (Appendix 2.A and Chapter 4), discounting reduces costs that are in the distant future to insignificant levels. The sustainability paradigm implies that these future costs and benefits should be more fully incorporated into current decision making.

Many scholars have developed plans to move the global economy onto a sustainable path. For example, Gore (1992), Brown (1991), and Meadows, Meadows, and Randers (1992), and others have developed point-by-point plans. These plans share many elements, including operational goals and attitudinal goals.

These attitudinal goals may be more important than the operational goals. Advocates of sustainable growth argue that sustainability will not occur unless values change. They state that we have to be more concerned with equity, both intratemporal and intertemporal, and be more willing to accept short-term sacrifice for long-term gain. Of course, since these attitudes are not synchronized with the way political systems have developed, political systems will have to change to be more future oriented. This would require leadership in this direction by both political leaders and citizens' groups. Meadows, Meadows and Randers (1992) speak of this change as a revolution. They discuss past revolutions in economic systems. The first of these was the agricultural revolution, where people moved from a hunting/gathering existence to domesticated agriculture. The second revolution that dramatically changed human existence was the Industrial Revolution. Meadows, Meadows and Randers argue that it is time for a new economic revolution and that is a sustainability revolution, where we fundamentally change our economic activity from attempting to maximize short-term output and focus on sustainable growth.

Although each author has suggested a different set of operational goals that must be met for sustainability to occur, there are some common threads. The reader is referred to the original books for complete descriptions of each author's plans. The common threads include:

1. Control population growth;
2. Eliminate reliance on fossil fuels;
3. Reduce pollution emissions and waste;
4. Eliminate the pressing Third World poverty that is the source of so much degradation;
5. Slow deforestation, desertification, soil erosion, and overexploitation of fisheries, other assaults on our renewable resource base;
6. Eliminate military conflict and the devotion of resources to military hardware; and
7. Develop new "appropriate" technology that is capable of meeting economic needs at minimal environmental costs, particularly in developing countries.

SUMMARY

It is relatively easy to read a book like this one and become depressed. Severe problems have been identified, and often little has been done to rectify these problems. Pollution, environmental degradation, global warming, loss of biodiversity, and other problems have been shown to have the potential to drastically affect the quality of life in the future. Many people speak of apocalyptic outcomes, and few are sanguine about the future.

However, the future is not already determined. Our actions today will determine the future. Sound policies and individual actions can make an important difference in the viability of our future. Through an understanding of scientific relationships, social institutions, and economic principles that determine people's behavior, we can understand the forces that govern environmental change and choose a set of policies that move us toward a sustainable future.

REVIEW QUESTIONS

1. What is meant by "absolute scarcity"?
2. What is the entropy law, and what is its relevance for environmental policy?
3. What are the primary criticisms of the "doomsday models"?
4. What does the empirical work of Barnett and Morse say about resource scarcity?
5. How can the invisible hand of the market mitigate scarcity? What are its shortcomings in this regard?

QUESTIONS FOR FURTHER STUDY

1. What are the shortcomings of Barnett and Morse's scarcity measures?
2. Why is the environmental cost of extracting resources increasing?
3. What are the necessary conditions for sustainability?
4. How do the sustainability paradigm and cost–benefit analysis differ in their treatments of costs?
5. How would you change political systems to make them more future oriented?

SUGGESTED PAPER TOPICS

1. Take a plan for sustainability from an author, such as Gore, Meadows, Meadows and Randers, or Brown. Evaluate each point in the plan, using the environmental economics concepts discussed in this book.
2. Look at the studies, such as Barnett and Morse and Smith, that develop empirical measures of resource scarcity. Compare and contrast the different measures and their methods of implementation. See current issues of the *Journal of Environmental Economics and Management* and other resource journals for the most recent studies.

WORKS CITED AND SELECTED READINGS

1. Barnett, Harold, and Chandler Morse. *Scarcity and Growth.* Washington, D.C.: Resources for the Future, 1963.
2. Brown, Lester. "Global Resource Scarcity Is a Serious Problem." In *Global Resources: Opposing Viewpoints,* edited by Matthew Poletsky. San Diego: Greenhaven Press, 1991.
3. Brown, Lester R., Christopher Flavin, and Sandra Postel. *Saving the Planet: How to Shape an Environmentally Sustainable Global Economy.* New York: Norton, 1991.
4. Carson, Rachel. *Silent Spring.* New York: Houghton-Mifflin, 1962.

5. Daly, Herman. *Steady State Economics: The Economics of Biophysical Equilibrium and Moral Growth.* San Francisco: Freeman, 1977.
6. Ehrlich, Paul R., and Anne H. Ehrlich. "Population Growth Threatens Global Resources." In *Global Resources: Opposing Viewpoints,* edited by Matthew Poletsky. San Diego: Greenhaven Press, 1991.
7. Ehrlich, Paul H. *The Population Time Bomb.* New York: Ballantine Books, 1968.
8. Georgescu-Roegen, Nicholas. "Energy and Economic Myths." *Southern Economic Journal* 41 (1975): 347–381.
9. ———. *The Entropy Law and the Economic Process.* Cambridge: Harvard University Press, 1971.
10. Gore, Albert. *Earth in the Balance.* New York: Houghton-Mifflin, 1992.
11. Hotelling, Harold. "The Economics of Exhaustible Resources." *Journal of Political Economy* 39 (1931): 137–175.
12. Krutilla, John. "Conservation Reconsidered." *American Economic Review* (1967): 777–787.
13. Malthus, Thomas Robert. "An Essay on the Principle of Population," In *An Essay on the Principles of Population, Text Sources and Background, Criticism,* edited by Phillip Appelman, New York: Norton, 1975.
14. Meadows, D. H., D. L. Meadows, and W. W. Behrens. *The Limits to Growth.* New York: Universe Books, 1972.
15. Meadows, D. H., D. L. Meadows, and Jorgen Randers. *Beyond the Limits: Confronting Global Collapse, Envisioning a Sustainable Future.* Post Mill, VT: Chelsea Green Publishing, 1992.
16. Pearce, David W. and Jeremy J. Warford. *World Without End.* New York: Oxford University Press, 1993.
17. Ray, Dixie Lee, and Lou Guzzo. *Trashing the Planet: How Science Can Help Us Deal with Acid Rain, Depletion of Ozone, and Nuclear Waste (Among Other Things).* Washington, D.C.: Regnery Gateway, 1990.
18. Simon, Julian L. "Global Resource Scarcity Is Not a Serious Problem." In *Global Resources: Opposing Viewpoints,* edited by Matthew Poletsky. San Diego: Greenhaven Press, 1991.
19. ———. *Population Matters: People, Resources, Environment and Immigration.* New Brunswick, NJ: Transaction Publishers, 1990.
20. Simon, Julian L., and Herman Kahn, eds. *The Resourceful Earth: A Response to Global 2000.* Oxford: Basil Blackwell, 1984.
21. Slade, Margaret E. "Trends in Natural Resource Commodity Prices: An Analysis of the Time Domain." *Journal of Environmental Economics and Management* 9 (1982): 122–137.
22. Smith, V. Kerry. "Measuring Natural Resource Scarcity." *Journal of Environmental Economics and Management* 5 (1978): 150–171.
23. ———. *Scarcity and Growth Reconsidered,* Washington, D.C.: Resources for the Future, 1979.
24. Zinsmeister, Karl. "Population Growth Does Not Threaten Global Resources." In *Global Resources: Opposing Viewpoints,* edited by Matthew Poletsky. San Diego: Greenhaven Press, 1991.

APPENDIX

The Internet—What Is It, How to Use It, What's on It

A GENERAL INTRODUCTION TO THE INTERNET

WHAT IS THE INTERNET?

The Internet is essentially a network of networks, that is, a system of interconnected computer resources that allows users to access literally millions of computer systems located throughout the world. This network is analogous to the modern highway system that consists of major backbones (interstates) carrying the heaviest traffic loads. These main arteries are connected, or interconnected, to smaller lines that may lead to colleges and universities or to commercial users. These backbone connectors can carry data at a rate of 45 million bits per second, and those that connect to regional networks are able to transmit at approximately 1.5 million bits per second. Modems, which allow users to connect to the Internet from their home or office, are able to transfer data at a rate of over 36,000 bits per second, and modems are becoming continually faster. Unfortunately, the rate of data transfer is directly related to the number of users that are simultaneously utilizing the same network or portion of the network as you are.

The actual size of the Internet, or "Net," is difficult to determine since it has been growing at an incredible rate. However, tens of millions of individual computers are connected to the Internet in virtually every country of the world. Practically any type of computer can be connected to the Internet, and machines ranging from personal microcomputers to powerful mainframe systems are interconnected and accessible.

The interconnectivity of the Internet is achieved by the use of a software protocol known as TCP/IP (Transmission Control Protocol/Internet Protocol). TCP/IP is comprised of a large number of lesser protocols that are utilized to efficiently organize computers and communication devices into a workable network. This protocol essentially breaks transmitted and received information into small pieces or "packets." These packets are given a designated order so the data can later be reassembled into the proper sequence. If some of the data packets are missing or omitted, the receiving computer asks for the data to be retransmitted. This entire system is similar to placing data packages within an envelope or, more precisely, a TCP envelope. This TCP envelope is in turn inserted into an IP envelope. The IP data container or envelope is transmitted onto the network and routed to the appropriate receiving computer. This ability to communicate and transmit data worldwide allows the user to obtain access to a myriad of shared information sources.

Computers that are connected to the Internet must be recognizable; that is, they must each have their own unique address so the packets of data will be transmitted to the proper location. What has made this possible is the development of a distributed, online system known as the Domain Name System. In a typical Internet address (for example: zzz@utkux.utk.edu) the domains are separated by the use of a period. The domains become increasingly larger as one reads the address left to right. The "top level" domains are organized into five distinct categories: .com (commercial), .edu (educational institutions), .gov (government institutions), .org (other organizations), and .net (network resources). In countries other than the U.S., an abbreviation of the country follows the .com or .edu. For example, the last part of an address of a Brazilian university would be .edu.br. You may have heard the term "client/server architecture" used with respect to the Internet. The primary use of a networked system is to allow for the sharing of resources. This sharing is actually accomplished by the division of the work between a "host" machine that serves out the desired information or resource and a "client" machine that receives and processes the output. This type of system is in direct contrast to the older terminal-host system where the mainframe computer completed all of the desired tasks and then sent the output directly to a display monitor. All of the services available on the Internet are based upon this client/server architecture.

OBTAINING ACCESS TO THE INTERNET

There are four main ways in which a student can obtain access to the Internet. The easiest way is to go to your university's library or microcomputer laboratory and sit down at a computer station that is directly connected to the Internet. If you have a computer account from your university, you can gain access to all Internet resources from these stations. The second way is if your dormitory room is wired for the Internet, and you can connect your personal computer directly to the university network. The third option is to obtain a "SLIP" account from your university and connect to the university network through a modem and the telephone lines. Most universities charge a monthly fee for this connection. The fourth option is to purchase Internet services from a private Internet provider and use your home computer, modem and telephone line to connect to the Internet. These services generally cost in the vicinity of $20 per month. Make sure you sign up with a service that provides a local telephone number to connect to so that you don't have to pay long distance telephone charges.

To become personally connected to the Internet you need to acquire four different items: a computer, a modem, telecommunications software, and a telephone line. The telephone line is typically your preexisting phone line and the modem will connect into the phone jack. The modem serves as a type of translation device between the computer and the telephone system. Computer systems operate by sending digital signals, while the telephone system involves the transmission of analog signals. The modem, or modulator/demodulator, converts digital information into analog signals when sending data, and converts analog data back into digital format when receiving. It is advisable that you purchase a modem that has a high rate of data transfer speed. Otherwise, you will be waiting forever to receive data or view images on your computer. A modem that operates at 28,800 bits per second transfer speed, or "baud," would be considered the bare minimum required for tolerable "Net" browsing. Digital phone lines (ISDN lines) and cable modems are now available which allow a much faster transmission of data. Currently, the cost of digital phone lines is much greater than analog phone lines, but as with all new technology, one would expect the cost to fall relatively quickly.

The modem will require telecommunications software that enables the computer and modem to operate together. Modem software is available for a reasonable price and is typically included when you purchase a new modem, when you sign up with an Internet provider, or when you obtain a "SLIP" account from your university. Additional software will be required to use many Internet applications, but this software can generally be downloaded free of charge on the Internet. Lastly, you need a computer system. You should be aware that Internet applications are generally designed to work well with the latest versions of computers and software. If you are using old equipment (such as a 386 processor or a slow 486 processor) you will experience some problems with many Internet applications.

"Net" Conduct and "Netiquette"

The Internet is essentially "owned and operated" by its users—that is, no one person or organization runs the network. The operation of the "Net" is due to the voluntary collective efforts of many thousands of users. These are individuals who lead discussions, maintain systems, and work to efficiently route network traffic. As a user you will be responsible for engaging in appropriate Internet behavior. You also will be required to uphold the rules and regulations set by your access provider. These may include restrictions upon which resources are accessible, certain time limitations, and the amount of disk storage space made available to you. These rules must be observed or you may forfeit your opportunity to access the Internet. Typically, these site policies will be presented to the user in the form of an AUP, or Acceptable Use Policy.

Perhaps the most important responsibility that a user must attend to is the issue of security. It is very important that you do not allow others to access your account. The best way to do this is to keep your password private. This will help prevent the unauthorized entry of individuals (i.e. hackers) onto the system. The concept of individualism is also valued highly on the Internet. Please do not attempt to restrict other individuals' freedom by sending incendiary electronic mail responses. Think about what you intend to say before sending it out onto the Net.

Also, e-mail, listservers and usenets allow you access to international experts in virtually every field. If you have a question that only these experts can answer, you should feel free to pose it. However, you should recognize that these people are extremely busy, and you should not ask them questions that you can answer through your own library research or investigation of Internet resources.

The concept of "netiquette" basically requires you to act as a responsible user. The Internet is a public place where tens of thousands of individuals may view your thoughts and comments. It is important to keep this in mind, and thus caution should be exercised before posting a comment. To summarize, the basic responsibilities to keep in mind are obeying the rules set down by your access provider, honoring the open discourse and non-commercial atmosphere of the Net, and realizing the truly public nature of the network.

INTERNET RESOURCES

ELECTRONIC MAIL

Electronic mail is perhaps the most heavily utilized type of information transfer on the Internet. If you have an account you have your own distinct electronic mail address. You can send messages to other individuals throughout the Net if you know their unique address. You can send requests to join discussion groups that focus upon nearly all imaginable topics. It is possible to send out requests for specific information and receive help almost immediately from many sources. Electronic mail can also be used to receive periodic postings and updated information regarding specific subject areas in which you would like to keep up-to-date. Your particular mail system will be determined by the services offered by the network provider and the type of computer system that you are accessing. It is extremely important that you know how to send and receive electronic mail. Please consult with your computer services personnel if you are unsure about how to use this Internet service.

WORLD WIDE WEB (WWW)

The World Wide Web is the newest information retrieval access application available on the Internet, and it is the most widely used method for making large amounts of information available. Virtually every university, government agency and large corporation now has an Internet presence through a

Web site on the World Wide Web. Most people access the World Wide Web, or Web, as it is more commonly called, through programs such as Netscape Navigator or Microsoft's Internet Explorer. Some of the larger Internet providers, such as America Online, have their own software for accessing the Web.

The Web sites are accessed through the specification of the address of the Web site, or its Uniform Resource Locator (URL). Most Web site URLs begin with http://, signifying that the Web site uses the hypertext link format. Most sites, but not all sites on the Web, begin with http://www. You must be very careful in typing in the URL, as addresses are case sensitive. (For example, there is a distinction between upper and lower case letters.) Hypertext will appear in the Web site as underlined text (also usually in a different color), and if you click on the hypertext with the computer mouse, you will move to a new site or resource specified in the hypertext. Most sites also have links to related sites, so you can easily go from site to site using the point and click method.

From the viewpoint of the student and scholar, the most useful feature of the Web is the availability of programs that allow you to search the Web by keyword. For example, in Netscape Navigator, if you click on the netsearch button, you will gain access to several different search programs which allow you to specify keywords and obtain a list of sites that contain the keyword. Since there are now so many sites on the Web, the more specific you can be the better. For example, using the Infoseek program on July 24, 1997, a search of the keyword "environment" yielded 1,185,068 sites. Obviously, the program will not list all the sites, but uses a variety of mechanisms to rank the sites and lists the ten highest ranked sites first. You can then ask to see the next ten, and so on. Adding additional keywords will generally decrease the number of sites that are found, but depending on the way the additional keywords are added, in some cases they will serve to increase the number of sites. However, even if the number of sites increases with the addition of more keywords, the additional keywords will allow the program to do a better job ranking the sites relative to your interests.

TELNET (REMOTE LOGIN)

Another of the most common Internet functions is the ability for you to access, or remotely login, to other computer systems throughout the network. Telnet is a program that allows you to access other information sources that are located outside your system. For example, by using Telnet you are able to connect to other on-line library systems, on-line databases, and other information sources. You can even login to your system if you happen to be travelling and would like to check your electronic mail. To use Telnet, you must know the Internet address of the target system and you also need to be connected to the Internet.

Just as you have a password that allows you to login to your home system, many systems restrict the amount of unauthorized access by asking you to supply an accepted password. Unless you have an account on that particular system, you will be denied access. However, many systems are open for full access and any necessary password will be supplied to you. Again, Telnet is the method of actually making the connection from your computer to another computer that is also connected to the Internet. You can Telnet using a specific Telnet program, or you can Telnet through Netscape Navigator, Internet Explorer or a similar program by typing in telnet:// and then the address of the computer to which you would like to link.

FILE TRANSFER PROTOCOL (FTP)

File transfer protocol allows the user to obtain data and transfer files between systems that are connected to the Internet. The FTP command can be used to copy files from your computer to a remote machine or from the remote machine to your own. Essentially all types of computer files can be copied including text files, software, and graphics files. The most common form of FTP is known as "anonymous FTP," which allows the user to access a specified area of disk space on a remote computer. With anonymous FTP, the user logs into the remote machine by typing "anonymous" and then uses his or her Internet address as the requested password. Once inside the remote machine, the user can move about within designated directories and acquire the desired files. This is an extremely efficient method for acquiring large amounts of desired information, such as large data sets collected by the government or working papers from a university. You can FTP using a specific FTP program, or you can FTP through Netscape Navigator, Internet Explorer or a similar program by typing in FTP:// and then the address of the computer to which you would like to link.

USENET NEWSGROUPS

Have you ever wanted to find out some information about the newest NASA space shuttle launches or the latest word on the status of your favorite athlete? Well, USENET newsgroups are good places for you to acquire that information. There are literally thousands of newsgroups with each particular group focusing upon a central theme or topic. The contributors to these newsgroups are people just like you, and the discussions can be quite informative and lively. USENET newsgroups are fed into your home computer system. You must then invoke a particular newsreader software program in order to access this information resource. Once you are able to access the newsgroups, you are able both to read and reply to the posted information. The postings appear in the form of collections of electronic mail messages.

The newsgroups are organized into a particular hierarchy with the broadest grouping appearing in the front, followed by any number of subheadings. The broad groupings are organized into eight major categories: comp (computer related), news (news network information), rec (recreational pursuits), sci (scientific information), soc (social issues discussion), talk (controversial debates), misc (other information), and alt (alternative topics).

For example, the newsgroup, "sci.bio.ecology" focuses upon a discussion of ecological topics and can be found under the "sci" hierarchy. These newsgroups often contain extremely useful information. Relevant questions related to the topic of the group are posted to the group and the user can view replies from other users located throughout the world. The entire USENET newsgroup system can best be compared to a bulletin board that is devoted to a single topic area. Electronic visitors are free to add to or merely read the postings that have been attached to the board. At a specific time as the board becomes full, messages are removed and the discussions move onward.

LISTSERVS

Listservs are yet another method for a user to acquire information about a given topic. These are similar to USENET newsgroups since each listserv is dedicated to a discussion about a certain subject. However, listservs are more analogous to a mailing list; thus, information is disseminated from one central routing source. Accessing the desired listserv is as easy as sending and reading mail. The message appears automatically once you have subscribed. Mailing lists, or listservs, have an administrator who may choose to moderate or not to moderate their lists. A moderated list means that postings are screened and irrelevant postings omitted. In an unmoderated list, all postings are passed along to you.

All mailing lists have an official Internet address, and to join or subscribe to the mailing list, you must send a message to that address expressing your desire to be included. To withdraw, you must send the opposite message. It is important to realize that many lists are very active, which means that you will receive many mail messages in a given time period. Other lists have a very light traffic flow and the number of messages will be few. It is your responsibility to read and delete incoming messages so that your mailbox does not overflow. Also, you should take the time to notify the listserv administrator whenever you have changed Internet addresses. This will prevent your electronic mail from bouncing back and accumulating in the listserv owner's account.

GOPHER

Gopher is an older program written at the University of Minnesota that allows you to browse Internet resources using a menu system. The Web has mostly displaced it, although you can still access Gopher sites through a specific Gopher program, or you can use Netscape Navigator, Internet Explorer or a similar program by typing in gopher:// and then the address of the Gopher site to which you would like to link. Gopher is a more user friendly program than FTP and is generally used like FTP to provide access to data files, text files and graphic files. When you select a menu item, you issue a command to receive that particular piece of information. The information resource may be located on a system physically located thousands of miles away. A Telnet or FTP command is issued in a way that is transparent to the user. After resolving the Internet address and making the connection, the information is retrieved and will be displayed on your monitor.

ENVIRONMENTAL INFORMATION AND THE TEXTBOOK WEB SITE

There is no way to list all the environmental resources available on the Internet in an appendix such as this one, as there are literally millions of environmentally related Web sites and other Internet resources. However, to provide a starting point for the student who is interested in environmental economics and environmental studies, a Web site has been developed to accompany the textbook. This Web site is entitled Enviro-Economics Gateway, and its URL is http://funnelweb.utcc.utk.edu~jrkahn/book.html. If you have trouble reaching this Web site, go to the University of Tennessee's primary Web site http://www.utk.edu, then go to the section on personal Web sites and click on Professor James R. Kahn's Web site. From there you will see a link to the textbook Web site.

The Web site is organized into two major sections. First, there is a listing of environmental resources of general interest, such as the U.S. Environmental Protection Agency's Web site. Second, Internet resources are organized according to the topics of the textbook, so you can click on Chapter 12—Tropical Forests, and receive a list of Internet resources related to tropical forests.

The Internet is a truly unique and wondrous means of obtaining information, but you should be aware of several important caveats. First, anyone can create an Internet site, so don't necessarily believe everything you see on the Internet. Second, some people of ill intention use the Internet and take delight in passing computer viruses to other people. You should make sure your computer is protected with anti-virus software, and avoid shady sites. Third, the Internet is constantly and rapidly changing, so almost everything in this appendix could be outdated before the next edition of this book is printed.

REFERENCES

1. Dern, Daniel D. *The Internet Guide for New Users.* New York, NY: McGraw-Hill, 1994.
2. Hahn, Harley & Rick Stout. *The Internet Complete Reference.* Berkeley, CA: Osborne McGraw-Hill, 1994.
3. Kehoe, Brenden P. *Zen and the Art of the Internet—A Beginner's Guide.* Englewood Cliffs, NJ: PTR Prentice-Hall, 1993.
4. Krol, Ed. *The Whole Internet.* Sebastpol, CA: O'Reilly and Associates, 1992.

CREDITS

Figure 5.4
From WORLD DEVELOPMENT REPORT, 1992 by WORLD BANK. Copyright © 1992 BY THE WORLD BANK. Used by permission of Oxford University Press, Inc.

Table 6.3
Nordhaus, William D., Global Warming: Economic Policy Responses, MIT Press. Copyright © 1991.

Table 6.5
Nordhaus, William D., Global Warming: Economic Policy Responses, MIT Press. Copyright © 1991.

Table 8.2
Energy Statistics Sourcebook, Sixth Edition, 1992. PennWell Publishing Co.

Table 8.3
Marcus, Controversial Issues in Energy Policy ©, p. 40, copyright © 1992. Reprinted by Permission of Sage Publications, Inc.

Page 242
Marcus, Controversial Issues in Energy Policy ©, p. xiii, copyright © 1992. Reprinted by Permission of Sage Publications, Inc.

Table 9.1
From STATE OF THE WORLD 1992: A Worldwatch Institute Report on Progress Toward a Sustainable Society by Lester R. Brown, et al. Copyright © 1992 by Worldwatch Institute. Reprinted by permission of W.W. Norton & Company, Inc.

Table 10.1
From WORLD RESOURCES, 1992–1993 by WORLD RESOURCE INSTITUTE. Copyright © 1993 BY WORLD RESOURCE INSTITUTE. Used by permission of Oxford University Press, Inc.

Table 10.3
Reprinted with permission of the *Southern Economic Journal.*

Table 12.1
Tropical Forestry Action Plan, FAO, June 1987, pp. 2–3. Reprinted with permission of the Food and Agriculture Organization of the United Nations.

Table 12.2
Tropical Forestry Action Plan, FAO, June 1987, pp. 2–3. Reprinted with permission of the Food and Agriculture Organization of the United Nations.

Page 383
Economic Value of Wetlands [box in Chapter 13]
Anderson, Robert and Mark Rockel, "Economic Value of Wetlands", Discussion Paper #065, American Petroleum Institute, Washington, DC, 1991. Reprinted courtesy of the American Petroleum Institute.

Table 13.1
From STATE OF THE WORLD 1985: A Worldwatch Institute Report on Progress Toward a Sustainable Society by Lester R. Brown, et al, eds. Copyright © 1985 by the Worldwatch Institute. Reprinted by permission of W.W. Norton & Company, Inc.

Table 13.2
From STATE OF THE WORLD 1992: A Worldwatch Institute Report on Progress Toward a Sustainable Society by Lester R. Brown, et al. Copyright © 1992 by Worldwatch Institute. Reprinted by permission of W.W. Norton & Company, Inc.

Table 14.1
From WORLD RESOURCES, 1992–1993 by WORLD RESOURCE INSTITUTE. Copyright © 1993 BY WORLD RESOURCE INSTITUTE. Used by permission of Oxford University Press, Inc.

Table 14.2
From STATE OF THE WORLD 1990: A Worldwatch Institute Report on Progress Toward a Sustainable Society by Lester R. Brown, et al. Copyright © 1990 by the Worldwatch Institute. Reprinted by permission of W.W. Norton & Company, Inc.

Table 14.3
From STATE OF THE WORLD 1985: A Worldwatch Institute Report on Progress Toward a Sustainable Society by Lester R. Brown, et al, eds. Copyright © 1985 by the Worldwatch Institute. Reprinted by permission of W.W. Norton & Company, Inc.

Table 15.1
From STATE OF THE WORLD 1991: A Worldwatch Institute Report on Progress Toward a Sustainable Society by Lester R. Brown, et al, eds. Copyright © 1991 by Worldwatch Institute. Reprinted by permission of W.W. Norton & Company, Inc.

Table 15.3
Environment, Volume 34, p. 12–39, 1992. Reprinted with permission of the Helen Dwight Reid Educational Foundation. Published by Heldref Publications, 1319 Eighteenth Street, NW, Washington, DC 20036-1802. Copyright © 1992.

Table 15.4
Environment, Volume 34, p. 12–39, 1992. Reprinted with permission of the Helen Dwight Reid Educational Foundation. Published by Heldref Publications, 1319 Eighteenth Street, NW, Washington, DC 20036-1802. Copyright © 1992.

Page 457
From WORLD DEVELOPMENT REPORT, 1992 by WORLD BANK. Copyright © 1992 BY THE WORLD BANK. Used by permission of Oxford University Press, Inc.

NAME INDEX

Adams, R. M., 283
Alvares, Robert, 427, 433
Anders, Gerhard, 283
Anderson, Frederick, 86
Anderson, Lee G., 300, 301, 304–305, 323
Anderson, Robert, 383, 391
Anderson, Terry Lee, 412
Appelman, Phillip, 492
Appleman, Phillip, 478
Ausabel, Jesse H., 178, 191

Balkan, Erol, 100, 128
Barbier, Edward, 471, 478
Barnett, Harold J., 461, 478, 485, 486, 489, 491
Bator, F. M., 33
Baumol, W. J., 22, 33, 86
Becker, Gary S., 478
Behrens, W. W., 483, 492
Bell, F. W., 100, 128, 129, 383, 391
Bennett, Hammond, 447
Berger, M. C., 129
Berg, Linda R., 412, 433
Bever, Michael, 266, 283
Bingham, Taylor, 224
Bird, G., 373
Bishop, R. C., 100, 391
Blomquist, G. C., 129
Bloomquist, 96
Bockstael, N. E., 100, 102, 103, 129
Bohi, Douglas R., 255
Bohm, Peter, 86
Bonnieux, F., 455
Boris, Constance, 412
Boulding, 133
Bowers, J. K., 455
Bowes, M. D., 335, 338, 339, 341, 350
Broadus, James, 392
Brookshire, D. S., 96, 129, 210, 224
Brown, G. M., 224, 391
Brown, Katrina, 373
Brown, Lester R., 10, 283, 368, 377, 392, 412, 426, 427, 434, 478, 488, 489, 490, 491
Brown, Peter G., 174–175, 191
Brown, Thomas C., 129
Buerger, R., 323
Bullard, Robert D., 433
Bunyan, Bryant, 433
Burrows, Paul, 86
Burton, Lloyd, 412
Byrd, Denise, 280, 283

Callicott, J. Baird, 129
Capistrano, A. D., 373
Carlin, Alan, 455
Carriker, Roy R., 412
Carson, Rachel, 415, 433, 483, 491
Carson, R. T., 130
Carter, Lester, 433
Chestnut, L. G., 130, 210, 211, 212, 213, 224
Clark, C. W., 323
Clark, D. E., 129, 323
Clawson, Marion, 350
Cline, William R., 176, 191
Coase, R. H., 33, 40, 42–48, 86
Colglazier, E. W., 428, 429, 430, 431, 434
Conroy, Patricia, 318, 324, 392
Constanza, R., 392
Coursey, D. L., 129, 130
Crotwell, A. G., 431, 433
Crowley, T. J., 191
Cummings, Ronald G., 129

Dales, J. H., 86
Dale, V. H., 373
Daly, Herman, 483, 492
d'Arge, R. L., 129
Darwin, Charles, 375
David, Martin Heidenhain, 86
Davis, Charles E., 433
Democritus, 87
De Officiis, Cicero, 435
Dern, Daniel D., 500
Desvouges, W. H., 320, 321, 324, 418
Devall, Bill, 129
Dewsnup, R. L., 412
Dorfman, N. S., 33, 86
Dorfman, R., 33, 86
Dornbusch, Rudiger, 191, 192
Doty, C. B., 431, 433
Douglas, Marjory Stoneman, 455
Dovido, J. F., 130
Dower, Roger C., 86
Drake, Lars, 448, 455
Driver, B. L., 128
Duffield, J. W., 129
Dumars, Charles T., 412
Dunn, J. W., 455

Eggert, R. R., 283
Ehrlich, Anne H., 492
Ehrlich, Paul R., 483, 492
Ehui, S. K., 373
Eisworth, M. C., 392
Ellefson, Paul V., 350
Ephraums, J. J., 191
Eppel, D., 129
Epstein, Joshua M., 174, 191

Farber, S., 383, 392
Faustmann, Martin, 334, 335, 350
Fisher, A. C., 33, 118, 130, 263, 283
Fisher, S. C., 34
Flader, Susan L., 129
Flavin, Christopher, 478, 491
Foster, Harold, 383, 412
Foster, J. H., 392
Freeman, A. M., III, 86, 129, 224, 323, 412, 433
Frielaender, Anne, 255

Gaffney, M., 335, 351
Gane, M., 350
Gardner, R., 129, 373
Gegax, D., 129
Georgescu-Roegen, Nicholas, 482, 483, 492
Gerking, S., 129
Gillis, Malcolm, 374
Gillis, T., 138, 139
Gocht, W. R., 283
Gold, Lou, 326
Goldstein, J. H., 391
Gordon, H., 323
Gore, Albert, 3, 10, 139, 179, 191, 488, 489, 490, 492
Gramlich, Edward M., 129, 389, 392
Green, David L., 238, 255
Greene, David L., 235, 236, 237
Green, Trellis, 322, 324
Gregory, Robin, 128
Griffen, James M., 30, 33, 255
Gupta, Raj, 174, 191, 383
Gupta, T. R., 392
Guzzo, Lou, 488, 492

Hahn, Harley, 500
Hahn, R. W., 86

Haney, J. C., 392
Hanley, Nicholas, 450, 455
Hannemann, W. M., 100, 129
Harrison, D., 96, 129
Harrison, Glenn W., 129
Head, Suzanne, 374
Heimlich, Ralph E., 450, 455
Heinzman, Robert, 374
Helfand, Gloria, 346, 351
Hemingway, Ernest, 287
Henning, J. A., 96, 130
Hertel, T. W., 373
Herzik, Eric B., 433
Hoehn, J. P., 129
Hong, S., 283
Hopper, J. R., 283
Hotelling, Harold, 30, 34, 129, 226, 255, 484, 489, 492
Houghton, J. T., 191
Hovis, J. L., 129
Howe, Charles, 338, 351
Huber, J., 131
Huppert, D. D., 102, 129

Icek, Ajzen, 128

Jacob, Gerald, 433
Jacobsen, Jodi L., 470, 471, 478
Jakus, Paul, 283
Jenkins, G. J., 191
Jensen, D. W., 412
Joeres, Erhard F., 86
Johnson, 283, 455
Johnson, George B., 412, 433
Johnson, Lawrence E., 130
Johnson, Lee Ann, 351
Jones, D. W., 478
Jordan, A. A., 283

Kahn, 100, 141, 146, 301, 321, 368
Kahn, Herman, 10, 492
Kahn, James R., 128, 129, 130, 323, 324, 374
Kaoru, Yoshiaki, 131
Kask, S. B., 130
Kealy, Mary Jo, 100, 130
Kehoe, Brenden P., 500
Kemp, W. M., 324
Kenkel, D., 129
Kiker, C. F., 373
Kilmarx, R. A., 283
Knudtson, Peter, 131
Krause, F., 191
Krol, Ed, 500
Krupnick, Alan, 80, 86
Krutilla, John V., 10, 118, 130, 261, 263, 283, 335, 339, 341, 350, 492

Lanky, Jean-Paul, 374
Leeworthy, V. R., 100, 128, 129
Leiby, Paul N., 191, 235, 238, 255
Leopold, Aldo, 123
Lester, James P., 433
Lewis, H. Gregg, 478
Loehman, E. D., 210, 224
Loomis, John, 324, 346, 351
Louviere, Jordan J., 130
Love, H. A., 283
Lovejoy, Thomas, 370
Lynne, Gary, 318
Lynne, G. D., 324, 392

McClelland, G. H., 130
McConnell, Kevin E., 91, 100, 102, 129, 130
McDonald, J. A., 141, 368, 374
McGartland, D., 138
McNeely, Jeffrey A., 392
Magat, W., 131
Makhijani, Arjum, 427, 433
Malthus, Thomas Robert, 226, 228, 255, 461, 462, 478, 481, 485, 492
Manne, Alan S., 191
Marcus, Alfred, 228, 229, 234, 250, 255
Markandya, Anil, 471, 478
Meade, James E., 34
Meadows, 228, 483, 489, 490
Meadows, D. H., 492
Meadows, D. L., 483, 490, 492
Meadows, Donella, 255
Mendelsohn, R., 359, 365, 374
Mieskowski, P., 130
Mills, 241
Mills, Edwin S., 255
Mills, Russell, 255
Milon, J. W., 103, 130
Mishan, E. J., 130
Mitchell, R. C., 130
Mohai, Paul, 433
Moi, Daniel Arap, 385
Montgomery, Mark, 130
Morey, E. R., 103, 130
Morgan, J. D., 266, 283
Morris, Glenn, 280, 283
Morse, Chandler, 461, 478, 485, 486, 489, 491
Mushkatel, Alvin H., 433
Myers, Norman, 10, 374

Nash, Roderick Frazier, 130
Nelson, Jon, 96, 130
Nestor, D., 138
Nielson, J. M., 283
Nordhaus, William D., 172, 176, 187, 188, 191

Oates, W. E., 22, 32, 80, 86
Odum, E. C., 130
Odum, H. T., 130
O'Neill, R. V., 478
Osborn, C. T., 450, 455
Ostro, B. D., 383, 392

Palmquist, R. B., 102, 131
Panaiotov, Todor, 374
Pasurka, C., 138
Patterson, D. A., 129
Pearce, David W., 10, 368, 373, 471, 478, 489, 492
Pearse, P. H., 335, 338, 351
Perry, Donald, 374
Peterson, F. M., 33, 34
Peterson, George L., 128
Pigou, A. C., 34, 40–41, 45, 48, 86
Plourde, Charles, 91, 130
Pochasta, F., 392
Polasky, Stephan, 392
Poletsky, Matthew, 491, 492
Porter, 139, 263, 277
Porter, Michael, 132
Porter, Richard C., 86, 130, 283
Portney, Paul R., 86
Posey, Dr., 366
Postel, Sandra, 393, 408, 410, 412, 491
Poterba, James M., 191, 192
Preckel, P. V., 373
Prochaska, Frederick J., 318, 324

Rae, D. A., 210, 224
Rainelli, P., 455

Randers, Jorgen, 489, 490, 492
Raven, Berg, 283, 455
Raven, Peter H., 412, 433
Ray, Dixie Lee, 488, 492
Reilly, William K., 39
Renner, Michael, 426, 427, 434
Repetto, Robert, 143, 374, 478
Ricardo, D., 34
Richels, Richard G., 191
Ricketts, Martin J., 256
Ridker, R. G., 96, 130
Rockel, Mark, 100, 130, 383, 391
Rosen, Sherwin, 130
Rowe, R. D., 130, 210, 211, 212, 213, 224
Rubinfeld, D. L., 129
Rubinfield, D. L., 96
Rubin, Jonathan, 346, 351
Rubin, Jonathon, 191
Rudawsky, Oded, 283
Russell, Clifford S., 324
Russell, Milton, 428, 429, 430, 431, 434
Ryan, John C., 377, 392

Schelling, Thomas C., 159
Schneider, S., 172, 191
Schultze, W. W., 129
Schulze, W. D., 129, 130
Scott, A., 324
Sedjo, Roger A., 351, 387, 392
Seip, Kalle, 130
Shaw, W. D., 103, 130
Shilling, J. D., 374
Shortle, James S., 455
Silvander, Ulf, 448, 455
Simon, Julian L., 10, 492
Slade, Margaret E., 486, 492
Smith, 102, 320, 321, 487
Smith, Adam, 11, 14, 34
Smith, V. Kerry, 91, 130, 131, 324, 418, 492
Solow, Andrew R., 192, 392
Sorenson, P. E., 100, 129
Sorg, Cindy, 324
Spiegel, Henry, 87
Steele, Henry B., 30, 33, 255
Stevens, William K., 366, 374
Stoll, John, 351
Stout, Rick, 500
Strand, Ivar E., 100, 102, 129
Strand, Jon, 130
Straszheim, M., 130
Suzuki, David, 131

Taylor, Kenneth Iain, 364, 374
Thayer, M. A., 129
Thibodeau, F. R., 383, 392
Tietenberg, Thomas H., 86
Tiller, Kelly, 283
Tilton, John E., 283
Todaro, Michael P., 468–469, 471, 478
Tolley, 210
Tolley, G. A., 224
Tolley, G. S., 129
Tonn, Bruce E., 224, 428, 429, 430, 434
Travis, C. C., 431, 433
Turvey, A., 301, 324

Van de Verg, Eric, 80, 86
Vaughn, William J., 324
Viscusi, W. K., 131
Von Moltke, Konrad, 374

Wahl, Richard W., 412
Waite, G. Grahame, 412
Warford, Jeremy J., 478, 489, 492
Webb, Michael, 256
Weiss, Edith Brown, 192
Weitzman, M. L., 86, 392
White, 241
White, J. C., 191
White, Lawrence J., 255
Wiggins, L., 138
Wilson, Edward O., 352
Witte, A. D., 130
Wolf, Edmund C., 375, 376, 392

Yohe, G. W., 192
Young, 262, 450
Young, C. E., 455
Young, John E., 283

Zantop, H., 283
Zinsmeister, Karl, 492

SUBJECT INDEX

Abatement costs
- command and control policies and, 62–67
- technology which lowers, 63

Absolute scarcity, 481–488
Access improvements policies, 319
Acid deposition
- aquatic impacts of, 202–206
- causes of, 194–198
- defining, 193–194
- health impacts, 212–213
- impact on health care costs, 200
- impacts of, 198–214
- materials impacts of, 213–214
- terrestrial impacts of, 206–210
- visibility impacts, 210–212

Acid deposition policies, 214–222
Acid Precipitation Act of 1980, 193
Acid rain, 194
Acid stress index (ASI), 203, 204
Adverse ecological impacts, 221
Agent Orange, 425
Agricultural conversion of land, 363–367
Agricultural sector
- acid deposition effect on, 207–210
- tropospheric ozone impact on, 438

Agriculture
- in developing countries, 409–410
- effect of environmental quality on, 436–439
- effect on environment by, 439–442
- environmental policies for, 451–454
- fisheries and, 448
- greenhouse gases and, 442
- habitat, biodiversity and, 442
- impact of exclusion of subsistence, 469–471
- income from sustainable/conventional, 367
- of indigenous Amazonians, 366
- public policy and, 442–451

Air pollutant emissions criteria, 227
Air quality
- decreasing rate of housing price function and, 96
- housing price and, 94
- increasing rate of housing price function and, 95

Alamo Bay (movie), 313
Alcohol taxes, 72
Alliance for the Chesapeake Bay webpages, 451
Alternative fuels, 246
Altruistic value, 91
Ambient based permit system, 78
American bison, 377–378
Anadromous fish, 308
Ancient growth forests, 345–349
Annual increment, 332–333
"Anonymous FTP," 498
Anthropogenic emissions of SO2, NOx, and VOCs (1985), 195
Aquaculture, 310
Aquatic impacts, 202–206
Arab oil embargo (1973), 234
Aral Sea, 474–475
Atmospheric Science and Climate (National Research Council), 171–172
Audio tape plastic containers, 278

Baby boomers, 464
Below-cost timber sales, 342–344
Benefits
- of beverage container deposit-refund systems, 277
- of climate change, 174–175
- of CRP enrollment (1986–1999), 450
- environmental taxation and macroeconomic, 146–148
- of having children, 465–467
- of hypothetical hydroelectric projects, 118
- improper disposal, 81
- jobs as, 116
- measuring cost-benefit analysis, 115–116
- potential, 110
- of temperate forests, 329
- time paths of timbering, 335
- of tropical forests, 356–358
- value of hypothetical project, 117
- when they cannot be measured, 117–120

Bequest value, 91
Best available technology (BAT), 406
Best practical technology (BPT), 406
Beverage container deposit-refund systems, 277
Beyond the Limits: Confronting Global Collapse, Envisioning a Sustainable Future (Meadows, Meadows, and Randers), 489
Bilateral Agreement on Air Quality (1991), 222
Biocentric model, 123
Biodiversity
- agriculture, habitat and, 442
- costs of losses in, 381
- policies for maintaining, 382–384
- protecting, 453–454
- species extinctions and, 375–381

Blue crabs, 318
Bohm, Peter, 166
Boise-Cascade, 330
Bonding systems, 81–82
Brazilian Amazon gold mining, 265
British Thermal Unit (BTU), 244
Brookshire, David, 129
Buffalo herds, 377–378

Cable modems, 495
CAFE (Corporate Automobile Fuel Efficiency), 240–241
Carbonated beverage containers, 268–270
Carbon cycle, 168–169, 327–328
Carbon dioxide (CO), 168–169, 173, 177, 181, 188–189, 226
Carbon monoxide (CO), 239
Carbon taxes, 189
Catch-based ITQs, 311–312
Catch per unit effort (CPUE), 203, 204
Catch and release policies, 319–321
CD plastic containers, 278
Centro Tecnologia Mineral, 265
Chemical weapons, 425

Chesapeake Bay, 303–304, 306, 314, 449–451
Child labor, 466
Children
  marginal benefit factors of having, 465–467
  marginal costs/demand for, 465, 467
Chlorofluorocarbons (CFCs), 69, 163–166, 166
Cigarette taxes, 72
Clean Air Act of 1972, 215, 226, 237
Clean Air Act Amendment of 1977, 215, 226
Clean Air Act Amendments of 1990, 77, 193, 202, 211, 212, 216–222, 226
Clean Air Act Amendments of 1990 (Title IV), 212, 217, 219, 220
Clean Water Act (CWA) of 1977, 406, 407
Clean Water Act (CWA) amendments of 1977 and 1987, 406
Client/server architecture, 494
Climatic models, 172
Climax community, 327
Closed seasons policies, 319
Coase Theorem, 47–48, 49
Command and control regulations
  catch and release programs based on, 320
  described, 51
  excess abatement cost and, 62–67
  pursuing environmental quality with, 62–70
  role of, 67–70
  to promote green farming methods, 452–453
  to reduce competition from nonnative species, 385
Commercial fishing, 203–207, 385
  See also Fisheries
Communication difficulties, 46
Competition from nonnative species, 380–381, 385
Comprehensive Environmental Response, Compensation and Liability Act of 1990 (CERCLA), 82, 104, 424–425, 428
Comprehensive materials policy, 276–281
Computer models, 483–484, 489
Conference of Parties (COP), 183
Conjoint analysis, 106–107
Consensus function, 211–211
Conservation Reserve Program (CRP), 448–449
Consumers' surplus estimates, 100, 101
Consumer surplus, 235–236
Contingent valuation method (CVM), 101–106
Contingent valuation studies, 210–211
"Convention of Climate Change," 182–184
Convention for International Trade in Endangered Species (CITES), 380
Cost-benefit analysis
  of climate change, 174–175
  environmental policy and, 109–110
  measuring benefits in, 115–116
  measuring costs in, 113–115
  sensitivity analysis and, 116–117
Cost-benefit caveats, 120
Costs
  abatement, 67–67
  acid deposition and health care, 200
  of beverage container deposit-refund systems, 277
  of biodiversity loss, 381
  for children, 465, 467
  of climate change, 174–175
  of CRP enrollment (1986–1999), 450
  environmental quality and health care, 137
  of hypothetical hydroelectric projects, 118
  improper disposal, 81
  invisible hand and marginal, 12–14
  of losses of habitat, 382
  of loss of habitat, 382
  marginal abatement, 59–61, 65
  measured in cost-benefit analysis, 113–115
  of pollution abatement in electric power generation, 201
  potential, 110
  of rising sea level, 176–177
  social, 62, 67, 238, 250, 268–273
  time paths of timbering, 335
  when they cannot be measured, 117–120
Cradle to grave manifest systems, 422
CTC (Conventional Travel Cost), 103
CV (Compensating Variation), 103
CVM (Contingent Valuation Method), 103
Cycle of poverty, 457, 458, 460

Deadweight loss, 235
Deep ecology, 123
Deforestation
  activities leading to, 358–366
  estimates of, 358
  macroeconomic reasons for, 367–368
Delaney Clause, 420–421
Demand
  for children, 465
  fishery model of, 302–303
  invisible hand and, 12
Demographic transition, 463–464
Deposit-refund incentive, 80–81
Deposit-refund systems, 166, 277, 423–424
Developed countries
  consumption per capita in, 482
  optimistic environment future in, 480
  soil erosion in, 440
Developing countries
  agriculture and water quality in, 409–410
  child labor in, 466
  described, 456n.1
  economic role of women in, 467–468
  pessimistic environment future in, 480–481
  soil erosion in, 440–441
    See also Third World countries
Digital phone lines (ISDN lines), 495

Direct production of
  environmental quality, 50
Direct use values, 92
Discounting, 34–36
Discount rates, 110–111
"The Dismal Science," 461
Dodo bird, 381
Dolphin safe tuna, 152
Domain Name System, 493
Dominant firm model, 230–231
Doomsday models, 483–484
Double dividend, 148
Dust Bowl, 447
Dynamic efficiency
  environmental/resource economics and, 31
  exhaustible resources and, 28–31, 37–38

Ecological constraints, 362
Economic development
  defining, 468–469
  population growth and, 460–468
  sustainable, 471–472, 474–476
  traditional models of, 468–469
  wildlife, ecotourism, and, 473
Economic incentives, 51, 362
Economic productivity, 135–139, 458–459
Economics
  entropy and, 482–483
  of mineral extraction, 259–267
  neo-classical, 484–488
  social system and, 7, 9
  sustainable, 488–490
The Economics of Natural Environments (Krutilla and Fisher), 263
Economy
  global warming and, 175–179
  impact on environment by, 141–142
  model of environment and, 133–146
  water as input to, 395
Electronic mail (e-mail), 496
Elephant population, 380
Elimination of all discharges (EOD), 407
Emergency Petroleum Allocation Act, 234
Emission-based permit system, 79
Endangered species, 378
Endangered Species Act of 1973, 387–389
Energy
  entropic degradation reversal and, 482n.2
  environment and, 237–241
  nuclear power issues, 248–251
  the Third World and, 253–254
  transition and future fuels, 251–253
Energy policy
  environment and, 241–249
  U.S., 226–237
Energy theory of value, 123
Entropy, 482–483
Entropy law, 260
Enviro-Economics Gateway website, 500
Environment
  economic productivity and, 135–139
  effect of agriculture on, 439–442
  energy policy and, 241–249
  energy and the, 237–241
  impact of economy on, 141–142
  impact of international trade policy on, 151–152
  international economic issues and, 149–152
  model of economy and, 133–146
  national energy strategy and the, 242–247
  Third World countries and the, 457–458
Environmental Assessment and Monitoring Program (EMAP), 145
Environmental degradation, 142–144
Environmental externalities
  government intervention to correct, 40–48
  mineral extraction and, 260–264
Environmental impact statement (EIS), 264
Environmental policies
  additional criteria for evaluating, 124–127
  for agriculture, 451–454
  cost-benefit analysis and, 109–110
  discount rates and, 110–111
  impact of international trade policy on, 151–152
  international trade and, 150–151
  malaria and, 475
  Porter hypothesis on, 139–141
  potential Pareto improvements and, 108–109
  for the subsistence sector, 477
  using value to determine, 107–122
  See also Policies
Environmental Protection Agency (EPA), 39, 420
Environmental quality
  choosing the correct level of, 51–61
  command and control policies for, 62–70
  direct production of, 50
  effect on agriculture on, 436–439
  GDP/sustainability and, 144–145
  health care costs and, 137
  impact on economic productivity by, 458–459
  measurement of, 145–146
  using economic incentives for, 70–84
Environmental resources, 5–10
Environmental taxation, 146–148
  See also Taxes
EPA Retrospective Study, 138
Equity
  global warming and, 186
  as policy criteria, 125
Equity across countries, 126
Equity across generations, 125–126
An Essay on the Principle of Population (Malthus), 461
European brown trout, 381
European House Sparrow, 381
EV (Equivalent Variation), 103
Excess abatement cost, 62–67
Excessive water withdrawals, 410
Exhaustible resources, 28–31, 37–38
Externalities

Externalities *(continued)*
  government intervention for environmental, 40–48
  market failure and, 21–25, 260–264
  pecuniary, 26
  pecuniary vs. technological, 26
  technological, 24
  transfrontier, 409
Externality taxes, 41, 72
Extinction. *See* Species extinctions
Exxon Valdez, 242

Farm Bill of 1996, 449
Federal materials policy, 276, 278–279
Fertilizer taxes, 451–452
File transfer protocol (FTP), 498
Fisheries
  agriculture and, 448
  biology of, 289–291
  environmental damage to, 287–289
  Gordon model and, 293–300
  limits to entry and, 313–314
  optimal harvest of, 291–293
  optimal and open-access, 300, 324–325
  pollution of, 317
  recreational resources of, 317, 319–321
  yield and estimated potential, 288
Fishery management, 314–321
Fishery models, 301–304
Fishery policy, 304–314
Fish farming (aquaculture), 310
Fishing industry, 203–207
Florida everglades, 443
Flow diagram of damage function, 122
Food and Agricultural Organization, 371
Food, Drug, and Cosmetics Act of 1958, 420–421
Forest ecology
  below-cost timber sales and, 342–344
  elements of, 327–329
  maximizing harvested wood, 330–335
  multiple use management in, 340–341
  optimal management of forests, 330
  optimal rotation in, 335–340
  temperate forests and, 329
Forests
  ancient growth, 345–349
  deforestation of, 358–368
  optimal management of, 330
  temperate, 329
  tropical, 352–358
Former Soviet Union, 474
Fox River system, 77
Fuel taxes, 244–249
Fuel transitions, 251–253
Fuelwood, 361, 363
Future fuels, 251–253

Gasoline prices, 248
Gasoline taxes, 248
GDP (Gross Domestic Product)
  defining, 133n.2
  discounting and, 113
  environmental analogues to, 145–146
  environmental degradation and, 142–144
  environmental quality and, 132–133
  global warming and, 176
  sustainability/environmental quality and, 144–145
General Agreement on Trade and Tariffs (GATT), 152
Georgia-Pacific, 330
Gill net, 315–316
Global climate
  benefit-cost analysis of, 174–175
  greenhouse gases and, 167–169
  increasing temperature of, 170–172
  threshold effects and, 178–179
Global public goods, 149–150, 369–372
Global temperature, 170–172
Global warming
  economy and, 175–179
  equity and, 186
  greenhouse gases and, 167–169
  ozone depletion and, 161
  significance of, 172–174
Global warming policy, 179–189
Global warming potential indexes (GWPI), 180, 182, 183
Gold mining (Brazilian Amazon), 265
Gopher, 499–500
Gordon model, 293–300
Government intervention
  mineral extraction and inappropriate, 264
  resolution of the issue of, 48–49
  to correct environmental externalities, 40–48
  types of, 49–51
Government regulations
  fishery limited-entry techniques, 310–313
  fishery open-access, 305–310
Great Water Bodies Program webpages, 451
Green Belt Movement, 470–471
"Green" farming methods, 452–453
Greenhouse effect, 159–161
Greenhouse gases, 167–169, 180, 442
Greenhouse gas reduction, 184–189
Gross Domestic Product (GDP), 457, 468–471
  See also Income
Gulf Coast redfish, 322
Gypsy Moth caterpillar, 381

Habitat
  biodiversity, agriculture and, 442
  preservation level of, 386
  protecting, 453–454
  uakari, 477
  See also Loss of habitat
Harvested wood
  maximizing physical quantities of, 330–335
  use of, 329
Harvest-replant-harvest cycle, 331
Health care costs, 137, 200
Health impacts, 212–213, 458–459
Hedonic pricing techniques, 93–95
Hedonic wage studies, 95, 97–98
Historical temperature record, 170
Housing price, 94

Housing price function, 95, 96
Hydrochlorflourocarbons (HCFCs), 165
Hydroflourocarbons (HFCs), 165
Hydrological cycles, 329, 394
Hypothetical hydroelectric projects, 118

Illegal waste disposal, 281
Imperfect competition, 15–17
Imperfect information, 17–18
Improper disposal costs/benefits, 81
Improper waste disposal, 425–428
Inappropriate government intervention, 20–21
Incidental catch, 315–317
Income
- effect on growth of population, 462–464
- effect of population growth on, 461–462
- growth, environment and Third World countries, 457–458
- of OECD countries, 464n.4
- *See also* Gross Domestic Product (GDP)

Increase per capita (GDP), 457, 468
Indirect use values, 91–92
Individual transferable quota (ITQ), 311–312
Indonesia, 233
Infoseek program, 497
Intergovernmental Panel on Climate Change (IPCC), 172, 180, 182
International economic issues, 149–152
International rain forest policy, 369–372
International trade, 150–151
International trade policy, 151–152
International water issues, 408–410
Internet
- Alliance for the Chesapeake Bay webpages on, 451
- described, 493–494
- "elephants and CITES" on the, 380
- Endangered Species Act information on, 388
- environmental information/textbook website on, 500
- fishery management using, 314
- Great Water Bodies Program webpages on the, 451
- information on environment and development on, 477n.12
- National Estuaries Program webpages, 451
- National Marine Fisheries Service website on, 448
- National Resource Conservation Service homepage on, 449
- "net" conduct and netiquette of the, 495–496
- obtaining access to the, 494–495
- resources available through the, 496–500
- USFS Web site address on, 342, 349

Intra-country equity
- global warming and, 186
- as policy criteria, 126–127

The invisible hand
- dynamic efficiency and, 28–31
- equity and, 27
- functions of, 11–14
- market failure and, 21–25

Iraq, 233
irrigated land damage, 408
Ivory trade ban, 380, 385

James River contamination, 415
Jobs (as benefits), 116

Kahn, James R.'s website, 500
Kayopo Indians, 364, 366
Kenya, 473
Kenya Water for Health Organization (KWAHO), 470–471

Labor market losses, 147–148
Land ethic, 123
Landfills, 271
Law of mass balance, 423–424
Lead (Pb), 226, 239
Liability systems, 82, 424–425
Libya, 233
Limited-entry techniques, 310–313
The Limits to Growth model, 483
Listservs, 499
Local solid waste policies, 279–281
Long-lining fishing method, 316
Losses
- association of pollution with agricultural, 209
- macroeconomic, 236
- of pollution abatement in electric power generation, 201
- price supports and welfare, 444–447

Loss of habitat
- costs of, 382
- policies to reduce, 386–390
- species extinctions due to, 379
- *See also* Habitat

Love Canal contamination, 415

Malaria, 459, 475
Manhattan Project, 250
Marginal abatement cost function, 57–59
Marginal abatement costs
- aggregating marginal, 65
- for CFC emissions from spray cans, 165
- environmental quality and, 59–61

Marginal cost (MC) curve of OPEC, 231
Marginal cost pricing of garbage disposal, 280
Marginal damage functions, 57–57, 120–122
Marginal damages, 59–61
Marginal private cost function (MPC), 268–269
Marginal private cost (MPC), 40–41
Marginal social cost function (MSC), 268–269, 344–345
Marginal social cost (MSC), 40–41
Marginal social harvesting cost function (MSC), 343–344
Marketable pollution permits (MPPs), 75–80, 240

Market equilibrium, 12–14
Market failure
  described, 14–15
  in exposure to toxic substances, 417–425
  externalities and, 21–25
  imperfect competition and, 15–17
  imperfect information and, 17–18
  Inappropriate government intervention and, 20–21
  of mineral extraction, 260–267
  monopoly and, 16
  property rights and, 25–27
  public goods and, 18–19
Market goods, 90
Market level of habitat preservation, 386
Mass balance law, 423–424
Materials impacts, 213–214
Materials policy, 276–281
Material use, 258
Mean annual increment, 332
Melting ice caps, 173
Mercury pollution, 265
MERCUSOL, 150
Methane, 181
Microeconomics of reproduction, 464–468
Miles per gallon (MPG), 241
Military hazardous waste sites, 426
Mineral extraction
  economics of, 259–267
  environmental impacts of, 262
  imperfect competition and, 264–265
  inappropriate government regulation and, 264
  market failures of, 260–267
  national security and, 265, 267
Mobile sources of pollution, 240–241
Models of development, 468–469
Modem, 495
Monopoly, 16
Monopoly profit, 235–236
Monotonic function, 486–487
Montreal Protocol (1987), 166
Moral suasion, 49–50, 320
Motor gasoline prices, 232
Multiple use management, 340–341
Multiple Use Sustained Yield (MUSY) Act of 1960, 340–341, 343

National Acid Precipitation Assessment Program (NAPAP), 193, 198, 209, 214
National energy strategy, 242–247
National Environmental Policy Act (NEPA), 264
National Estuaries Program webpages, 451
National Fish and Wildlife Foundation, 288
National Income and Product Accounts, 143
National Marine Fisheries Service website, 448
National Marine Fishery Service (NMFS), 322
National Pollution discharge Elimination System (NPDES), 406
National Resource Conservation Service homepage, 449
National Resources Defense Council, 342
National security, 265, 267
Native Americans, 123
Natural extinctions, 375–376
Natural Gas Act of 1938, 229
Natural resources
  CERCLA on, 82, 104
  damage assessment of, 104
  described, 4
  study of, 5–10
Nature Conservancy, 388
Nature swaps, 370
Neo-classical economics, 484–488
Neo-Malthusians, 228
Netiquette, 496
Netscape Navigator, 497
New York City water shortages, 399
Nigeria, 233
NIMBY (Not in My Back Yard) Syndrome, 272
1972 Clean Air Act Amendment, 215, 226, 237
1977 Clean Air Act Amendment, 215, 226
1990 Clean Air Act Amendments (CAAA), 77, 193, 202, 211, 212, 216–222, 226
1990 Integrated Assessment Report, 198, 203, 204–205
Nitrogen oxides (NOx), 194–197, 215, 218, 222, 226, 239
Nitrous oxide, 169
Nonexcludability, 18–19
Nongovernmental organizations (NGOs), 388–390, 453
Nonmarket goods
  defining, 90
  measuring value of, 92–107
  value and, 90–92
Nonnative species, 380–381
No regret policies, 188
North American Free Trade Agreement (NAFTA), 150, 151
NPD (net domestic product), 143
Nuclear power issues, 249–251
Null hypothesis, 488
Nutrient cycling, 327, 328, 394

Occupational Safety and Health Administration (OSHA), 420
OECD (Organization for Economic Cooperation and Development), 464
Offset system, 240
Oil spills, 242
Oligopolistic market structure, 264–265
On-the-job exposures, 419–420
OPEC (Organization of Petroleum Countries), 229–237
Open-access fisheries, 300, 324–325
Open-access harvesting, 376–379
Open-access policies, 384–390
Open-access regulations, 305–310
Opportunity cost of the land (OCL), 338
Optimal level of habitat preservation, 386
Optimal level of pollution, 59–62
Optimal rotation, 335–340
Option value, 91
Ozone, 195, 239
Ozone layer depletion
  CFCs and chemicals leading to, 163
  consequences of, 162–163

described, 161–162
policy toward, 163–166

Pamlico-Ablemarle Sound, 295
Pecuniary externality, 26
Persian Gulf War, 236–237, 425
Pesticide taxes, 451–452
PH levels, 205, 206
Photosynthesis, 327, 329
Pliny the Elder, 193
Policies
agriculture and, 442–451
comprehensive materials, 276–281
energy, 226–237, 241–249
equity as criteria for, 125
on fisheries, 304–314
on global warming, 179–189
international rain forest, 369–372
for maintaining biodiversity, 382–384
on materials, 276–281
sustainable development, 472, 474–476
to reduce loss of habitat, 386–390
to reduce open-access exploitation, 384–390
water pollution and U.S., 403, 404–408
See also Environmental policy
Pollutants
abatement costs/damages and, 197
impact on soybean yields by, 438
marginal damages of SO2 and other, 196
precursor, 193
regional, 194
Pollution
command and control regulations and, 62–70
economic incentives and, 70–84
of fishery habitat, 317
global warming policy for, 179–182
greenhouse effect and, 159–161
income urban concentrations of sulfur dioxide and, 140
mercury, 265
optimal level of, 121
prevention of, 50–51
regulations on mobile sources of, 240–241
regulations on stationary sources of, 237–240
sources/impacts of water, 404–405
transfrontier, 150
U.S. policy on water, 403, 406–408
welfare losses from price supports and, 445–447
Pollution bubbles, 239
Pollution permits, 75–80
Pollution subsidies, 82–84
Pollution tax, 71
Population growth
doomsday models applied to, 483–484
environmental degredation/economic development role of, 460–468
income and, 461–464
microeconomics of reproduction and, 464–468
Population Time Bomb (Ehrlich), 483
Porter hypothesis, 139–141
Potential costs/benefits, 110
Potential Pareto improvements, 108–109, 272, 316
Poverty
impact on the environment by, 459
vicious cycle of, 457, 458, 460
Precursor pollutants, 193
Present value, 34–36
Private goods spectrum, 19
Property rights
deforestation to establish, 365
market failure and, 25–27
water and, 400–402
Public goods
externalities as, 25
global, 149–150
market failure and, 18–19
optimal quantity of good as partial, 22
spectrum of, 19
Public policies. *See* Policies
Quantity
invisible hand and, 12–14
marginal/total willingness to pay and, 89

Radiative forcing
of hypothetical greenhouse gas, 180
of two gases with same time integral, 181
Radioactive and toxic contamination, 427
Radon exposure, 418
Recreational fishery resources, 317, 319–321
Recreational fishing quality value, 108
Recycling, 273–276, 280–281
Regional pollutants, 194, 199
Reproduction microeconomics, 464–468
Republic of Korea (South Korea), 463
Residential heating oil prices, 232
Resource Conservation and Recovery Act (RCRA), 428–429
Resource flows, 4
Resources
dynamic efficiency and exhaustible, 28–31, 37–38
the invisible hand and, 11–27
natural, 4, 5–10, 82, 104
taxonomy of, 4–10
See also Scarcity; Water resources
Revealed preference approaches, 92–93
Richter scale, 202
Right to know laws, 420
"The Rio Summit," 182
Risk-free market rate of interest, 111, 112–113
Rotation of the forest, 331
RUM (Random Utility Model), 103

Salt marshes, 318
Saudi Arabia, 233
Scarcity
absolute, 481–488
measuring, 486–488
See also Resources

Schultze, William D., 129
Sea level, 174, 176–176
Sensitivity analysis, 116–117
Sewer outflow, 401
SHR (Share Model), 103
Silent Spring (Carson), 415, 483
Size and creel limits policies, 321
"Slash and burn" agriculture, 363
"SLIP" accounts, 494, 495
Smokey the Bear program, 49–50
Snail darter, 389
Social costs
  of inefficient regulations, 67
  of nuclear power, 250
  of pollution, 62
  of U.S. monopoly of oil market, 238
  of waste disposal, 268–273
Social policy, 134
Social safety net, 441
Social system, 7, 9
Social welfare, 134
Soil Conservation Service of U.S. Department of Agriculture, 447
"Soil Erosion: A National Menace" (Bennett), 447
Soil erosion, 439–441
Solid waste, 267–273
Southwest pines, 210
Soybean yields, 438
Species declines (early 1990s), 377
Species extinctions
  anthropogenic causes of, 376–381
  due to competition of nonnative species, 380–381, 385
  due to loss of habitat, 379
  due to open-access harvesting, 376–379
  natural, 375–376
Sport fish catch rates, 102–103
Spotted owls, 346, 347
Squatters, 365–366
Stated preference valuation techniques, 101–107
State solid waste policies, 279–281
Stationary sources of pollution, 237–240
Stocking fish policies, 319
Stratosphere, 195
Stumpage value (V), 338
Sulfur dioxide permits, 219
Sulfur dioxide (SO2), 194–197, 215, 216–218, 221, 226, 239
Supernatural beliefs, 364
Supply curve
  for commercial fishing, 303
  fishery model of, 302
Sustainability
  activities leading to, 471–472, 474–476
  environmental quality/GDP and, 144–145
  as policy criteria, 126–127
  system-wide change and, 488–490
System-wide change, 488–490

Taxes
  alternative fuels and pollution, 246
  externality, 41, 72
  fertilizer, 451–452
  fuel, 244–247
  impact of different carbon, 189
  known aggregate MAC and, 73
  labor market losses and income, 147–148
  macroeconomic benefits and environmental, 146–148
  macroeconomic impact of, 247–249
  pesticide, 451–452
  unknown aggregate MAC and, 74
TCP/IP (Transmission Control Protocol/Internet Protocol), 493
Technological externality
  abatement cost and, 63
  described, 24
  pecuniary externality vs., 26
Tellico Project, 389
Telnet (remote login), 497–498
Temperate forests, 329
Tennessee Valley Authority (TVA), 217–218, 389
Terrestrial impacts, 206–210
Textbook website, 500
Third World countries
  described, 456n.1
  income, growth, and the environment in, 457–458
  sustainable development by, 471–472, 474–476
  See also Developing countries
Third World energy crisis, 253–254
Threatened species, 378
Three Mile Island incident, 250
Threshold effects, 56–57, 178–179
Timbering activities, 360–361
Times Beach contamination, 415
Tipping fee, 271
Title IV (1990 Clean Air Act Amendments), 212, 217, 219, 220
Tongass National Forest, 348
Toxic Substances Control Act of 1976 (TSCA), 419
Toxic waste
  deposit-refund systems and, 423–424
  described, 416–417
  government and improper disposal of, 425–428
  liability systems and, 424–425
  market failure in exposure to, 417–425
  optimal level of emissions of, 422
Toxic waste cleanup, 428–432
Transfrontier externalities, 409
Transfrontier pollution, 150
Transition fuels, 251–253
Travel cost demand curve, 98
Travel cost model, 99–101
Tropical dry forests, 356
Tropical Forestry Action Plan (TFAP), 371–372
Tropical forests
  benefits of, 356–358
  definition of, 352–353
  distribution of, 353
Tropical rain forests, 353–356, 369–372
Troposphere, 195
Tropospheric ozone, 438
Turtle Excluder Devices (TEDDIES), 316
200–mile economic exclusion zone, 312–313

Uakari habitat, 477
Uniform Resource Locator (URL), 497
United Nations Convention of the Law of the Sea, 312–313

United Nations Development Program, 371
United States
  endangered and threatened species in the, 378
  military hazardous waste sites of, 426–427
  soil erosion in, 441
  U.S. energy policy developed in post-World War II era, 226–237
  intellectual antecedents on, 226–229
  OPEC and, 229–237
  See also Energy
USENET newsgroups, 498–499
U.S. Environmental Protection Agency's website, 500
U.S. Fish and Wildlife Services (USFWS), 289
U.S. Forest Service (USFS), 340–341, 342, 368
U.S. Forest Service Website, 342, 349
U.S. net mineral imports, 266
U.S. water pollution policy, 403, 406–408

Value
  additional approaches to, 122–127
  conjoint analysis for recreational fishing quality, 108
  defining, 88–92
  for determining environmental policy, 107–122
  energy theory of, 123
  estimating wetlands, 383
  loss from cutting ancient forest, 348
  measuring nonmarket goods, 92–107
  measuring spotted owls, 346
  of net benefits to hypothetical project, 117
  nonmarket goods and, 90–92
Venezuela, 233
Vicious cycle of poverty, 457, 458, 460
Visibility impacts, 210–212
Volatile organic compounds (VOCs), 195, 226, 239

Waste
  proper/improper disposal of, 270
  recycling and, 273–276
Waste disposal
  government and improper, 425–428
  illegal, 281
  marginal cost pricing of, 280
  private/social costs of, 270–273
  of solid waste, 267–273
Waste policies, 279–281
Water consumption, 395–400
Water pollution
  sources and impacts of, 404–405
  U.S. policy toward, 403, 406–408
Water Pollution Control Act of 1972, 406
Water prices, 396, 398, 453
Water quality
  comparison of improvements for, 320
  consumers' surplus associated with, 101
  in developing countries, 409–410
  factors degrading, 402–403
Water resources
  adequately priced, 453
  availability of, 393
  excessive withdrawals of, 410
  exhaustible nature of, 399–400
  flow of, 396–399
  hydrological cycles of, 394
  nutrient cycling of, 394
  property rights and, 400–402
  sewer outflow and, 401
  See also Resources
Water shortages
  caused by low price, 398
  of New York City, 399
  in Southern California, 400
The Wealth of Nations (Smith), 11
Wetlands, 318, 382, 383
Wetlands criteria, 388
Wetlands Reserve Program, 449
Wet nitrate ion deposition (1991), 203
Wet sulfate ion deposition (1991), 204
Weyerhauser, 330
White Cloud Peaks Mountains, 263
Women
  economic role of, 467–468
  improving the status of, 471, 477
World Bank, 141, 371, 456, 474, 476
World Commission on Environment and Development (1987), 472
World Development Report 1992 (World Bank), 456–457
World Health Organization (WHO), 393
World Resource Institute, 371
World Resources 1992–93, 139, 225
World Trade Organization (WTO), 152
World Wide Web (WWW), 496–497

Yanomami Indians, 364
Yard waste, 281
Yom Kippur War (1973), 233

Zero population growth (ZPG), 464